POINT DEFECTS IN SOLIDS

Volume 2
Semiconductors and Molecular Crystals

POINT DEFECTS IN SOLIDS

Volume 1: General and Ionic Crystals
Volume 2: Semiconductors and Molecular Crystals
Volume 3: Defects in Metals

POINT DEFECTS
IN SOLIDS

Edited by

James H. Crawford, Jr.

Chairman, Department of Physics
University of North Carolina
Chapel Hill, North Carolina

and

Lawrence M. Slifkin

Department of Physics
University of North Carolina
Chapel Hill, North Carolina

Volume 2
Semiconductors and
Molecular Crystals

PLENUM PRESS • NEW YORK AND LONDON

Library of Congress Cataloging in Publication Data

Crawford, James Homer, 1922-
  Point defects in solids.

  Includes bibliographical references.
  CONTENTS: v. 1. General and ionic crystals.—v. 2. Semiconductors and
Molecular Crystals.
  1. Crystals — Defects. 2. Metals — Defects. 3. Semiconductors — Defects. I.
Slifkin, Lawrence M., joint author. II. Title.
QD931.C72   548'.81   72-183562

ISBN-13: 978-1-4684-0906-2   e-ISBN-13: 978-1-4684-0904-8
DOI: 10.1007/978-1-4684-0904-8

CONTRIBUTORS TO THIS VOLUME

Jacques C. Bourgoin
Laboratoire de Physique
Ecole Normale Superieure
Paris, France

H. C. Casey, Jr.
Bell Telephone Laboratories, Inc.
Murray Hill, New Jersey

A. V. Chadwick
University Chemical Laboratory
University of Kent, Canterbury
Kent, England

James W. Corbett
Physics Department
State University of New York at Albany
Albany, New York

O. L. Curtis, Jr.
Northrop Corporate Laboratories
Hawthorne, California

G. L. Pearson
Stanford University
Stanford, California

J. N. Sherwood
Department of Pure and Applied Chemistry
University of Strathclyde
Glasgow, Scotland

D. Walton
Physics Department
McMaster University
Hamilton, Ontario, Canada

George D. Watkins
Solid State and Electronics Laboratory
General Electric Corporate Research and Development
Schenectady, New York

PREFACE

Volume 1 of *Point Defects in Solids* has as its major emphasis defects in ionic solids. Volume 2 now extends this emphasis to semiconductors. The first four chapters treat in some detail the creation, kinetic behavior, interactions, and physical properties of both simple and composite defects in a variety of semiconducting systems. Also included, as in Vol. 1, are chapters on special topics, namely phonon–defect interactions and defects in organic crystals.

Defect behavior in semiconductors has been a subject of considerable interest since the discovery some twenty-five years ago that fast neutron irradiation profoundly affected the electrical characteristics of germanium and silicon. Present-day interest has been stimulated by such semiconductor applications as solar cell power plants for space stations and satellites and semiconductor particle and γ-ray detectors, since in both radiation damage can cause serious deterioration. Of even greater practical concern is the need to understand particle damage in order to capitalize upon the developing technique of ion implantation as a means of device fabrication. Although the periodic international conferences on radiation effects in semiconductors have served the valuable function of summarizing the extensive work being done in this field, these proceedings are much too detailed and lack the background discussion needed to make them useful to the novice. Therefore, in keeping with the original aim of this series, Chapters 1–3 represent a reasonably ordered development of the topic covering elemental and compound semiconductors, with special attention to the effect of lattice defects upon electrical and optical behavior. Chapter 4 is an in-depth treatment of one of the very valuable tools for defect identification, namely ESR. Taken together, we feel that these chapters are a valuable and reasonably comprehensive treatment of defects in semiconductors which will have more than a short-term value.

The last two chapters are devoted to two areas of defect research which have recently come into prominence. Phonon–defect interactions, a subject of long-standing interest, has undergone rapid progress in the last several years because of the development of appropriate mathematical techniques. For this reason any comprehensive treatment of recent work must unavoidably be mathematically complex in approach. However, we believe that this subject has become of such importance to solid state physics that it should be included, even though there is insufficient space for a thorough mathematical development. The confrontation of experiment with theory is particularly impressive and in our opinion gives additional justification for including phonon–defect interactions in what is otherwise a predominantly introductory series.

Although scarcely at home in a treatise devoted to semiconductors, the last chapter, which treats imperfections in molecular solids, demonstrates the scope and power of the defect concept when applied to diffusion and transport. The significant recent progress in this field makes it a rewarding exception to the main theme of the volume. We trust the reader will agree.

J. H. CRAWFORD, JR.

January, 1975   L. M. SLIFKIN

CONTENTS

Chapter 1. Defect Creation in Semiconductors

James W. Corbett and Jacques C. Bourgoin

1. Introduction 1
2. Defect Properties 4
   2.1. Introduction 4
   2.2. Silicon 26
   2.3. Germanium 28
   2.4. Diamond 31
   2.5. III–V Compounds 32
   2.6. II–VI Compounds 35
   2.7. Other Materials 36
3. Damage Phenomenology 38
   3.1. Displacement Damage 38
   3.2. Thermal Spikes 45
   3.3. Ionization Phenomenology 47
   3.4. Plastic Deformation 54
   3.5. Heat Treatment Effects 56
4. Interactions of Particles with Semiconductors 60
   4.1. Introduction 60
   4.2. Photon Interactions 61
   4.3. Heavy Ions and Atoms 66
   4.4. Electrons 80
   4.5. Neutrons 91
5. Survey of Displacement Damage Results 95
   5.1. Introduction 95

5.2. Germanium 97

5.3. Silicon 102

5.4. Diamond and Graphite 111

5.5. III–V Compounds 115

    5.5.1. InSb, InAs, InP 115

    5.5.2. GaAs, GaSb 118

5.6. II–VI Compounds 120

    5.6.1. ZnSe, ZnTe, ZnS 121

    5.6.2. CdS, CdSe, CdTe 123

    5.6.3. MgO, ZnO, BeO 126

6. Subthreshold and Ionization Damage 128

7. Theories of Displacement Damage Production 134

8. Summary 148

References 149

Chapter 2. Diffusion in Semiconductors

H. C. Casey, Jr. and G. L. Pearson

1. Introduction 163

2. Atomic Theory of Diffusion 164

  2.1. Diffusion Mechanisms 164

  2.2. The Flux Equation and Diffusivity 167

    2.2.1. The Diffusive Flux 167

    2.2.2. Jump Frequency 168

    2.2.3. Diffusivity and Activation Energy 170

3. Continuum Theory of Diffusion 170

  3.1. Fick's Laws 170

  3.2. The Diffusion Profile 171

  3.3. Impurity Diffusion for a Moving Boundary 174

    3.3.1. Experimental Conditions for a Moving Boundary . . 174

    3.3.2. Diffusion into a Growing Layer 174

    3.3.3. Diffusion into an Evaporating Surface 176

  3.4. The Boltzmann–Matano Analysis 177

4. Diffusion in Ge and Si 179

  4.1. Introduction 179

  4.2. Self-Diffusion in Ge and Si 181

4.3. Diffusion of the Group III and V Impurities in Ge and Si   187
   4.3.1. Low-Concentration Case   187
   4.3.2. High-Concentration Case   191
4.4. Interstitial Diffusion of the Alkali Elements and Inert Gases   197
4.5. Interstitial–Substitutional Diffusion of the Transition Elements   198
   4.5.1. Preliminary Considerations   198
   4.5.2. Low Solubility and Slow Interstitial Diffusion   199
   4.5.3. Intermediate Solubility and Rapid Interstitial Diffusion   200
   4.5.4. High Solubility and Slow Interstitial Diffusion . . .   200
4.6 Graphical Summary of the Diffusion Coefficients in Si . . .   201

5. Diffusion in the III–V Compounds   201
  5.1. Introduction   201
  5.2. Ternary Considerations   204
  5.3. Concentration Gradient Diffusion of Zn in GaAs   207
  5.4. Interstitial–Substitutional Diffusion   208
   5.4.1. The Flux Equation   208
   5.4.2. The Built-in Field   209
   5.4.3. The Effective Diffusion Coefficient   211
   5.4.4. Interstitial–Substitutional Equilibrium   212
   5.4.5. Analysis of Experimental Data   213
   5.4.6. Isoconcentration Diffusion   217
  5.5. Effects of Arsenic Pressure   219
  5.6. Departure from Equilibrium   221
  5.7. Compilation of Diffusion Coefficients in the III–V Compounds   222

6. Diffusion in the II–VI Compounds   227
  6.1. Introduction   227
  6.2. Solid–Liquid–Vapor Equilibrium   230
   6.2.1. Component Partial Pressures   230
   6.2.2. Congruent Evaporation and Minimum Pressure . .   232
   6.2.3. The Solidus Curve   233
  6.3. Self-Diffusion   235
   6.3.1. General Comments   235
   6.3.2. Effective Diffusion Coefficient   236
   6.3.3. Self-Diffusion in CdS   237
   6.3.4. Compilation of Self-Diffusion Coefficients in the II–VI
       Compounds   245

7. Summary and Conclusions   247

   References   248

Chapter 3. Effects of Point Defects on Electrical and Optical Properties of Semiconductors

O. L. Curtis, Jr.

1. Introduction 257
2. Carrier Concentration 259
   2.1. Germanium 261
      2.1.1. Energy Levels 261
      2.1.2. Thermal Annealing 262
      2.1.3. Radiation-Induced Annealing 264
   2.2. Silicon 269
      2.2.1. Energy Levels 269
      2.2.2. Temperature Dependence of Defect Introduction Rates 270
   2.3. Compound Semiconductors 273
3. Carrier Mobility 274
   3.1. Germanium 276
   3.2. Silicon 278
   3.3. Other Materials 279
4. Minority Carrier Lifetime 280
   4.1. Two-Level Model for Recombination 283
   4.2. Recombination in Irradiated Germanium 284
      4.2.1. Moderate-Resistivity n-Type Material 284
      4.2.2. Low-Resistivity n-Type Material 287
      4.2.3. p-Type Material 289
   4.3. Silicon 291
      4.3.1. Analysis for n-Type Material with Two Levels Effective at Low Excess Density 291
      4.3.2. p-Type Material 294
      4.3.3. Annealing Studies 296
5. Optical Absorption 300
   5.1. Germanium 301
   5.2. Silicon 303
   5.3. III–V Compound Semiconductors 311
6. Photoconductivity 311
   6.1. Germanium 312
   6.2. Silicon 313
   6.3. Other Materials 316
7. Luminescence 316

7.1. Recombination Luminescence in Irradiated Silicon 317
7.2. Effects of Irradiation on Luminescence in Gallium Arsenide 320
8. Conclusion 323
References 328

Chapter 4. Electron Paramagnetic Resonance of Point Defects in Solids, with Emphasis on Semiconductors

George D. Watkins

1. Introduction 333
2. Basic Concepts 335
    2.1. Origin of Paramagnetism 335
    2.2. Diamagnetic Solids 336
    2.3. Concept of Resonance 336
    2.4. Experimental Apparatus 339
3. Theory of EPR for Defects in Solids 340
    3.1. Quenching of Orbital Angular Momentum 340
    3.2. Hyperfine Interactions 342
        3.2.1. Changes in the Spectrum 342
        3.2.2. Examples 344
        3.2.3. Quantitative Aspects of the hf Interaction 348
        3.2.4. ENDOR 350
    3.3. The g-Tensor 351
        3.3.1. Changes in the Spectrum 351
        3.3.2. Quantitative Treatment of the g-Tensor 353
        3.3.3. The g-Shift of the V_k Center 354
    3.4. Fine Structure Terms for $S > 1/2$ 355
        3.4.1. Changes in Spectrum 355
        3.4.2. Origin of D 356
        3.4.3. Higher Order Terms for $S > 3/2$ 358
    3.5. Orbital Angular Momentum Not Quenched 358
    3.6. Summary 359
4. Additional Examples 360
    4.1. Defects in Irradiated Silicon 360
        4.1.1. Annealing of Interstitial Aluminum 360
        4.1.2. The Oxygen–Vacancy Pair 360
        4.1.3. The Phosphorus–Vacancy Pair 364
        4.1.4. Multiple-Vacancy Defects 367
    4.2. Transition Elements in Silicon 369

5. Auxiliary Techniques 373
   5.1. Applied Uniaxial Stress 373
      5.1.1. Jahn–Teller Alignment 373
      5.1.2. Defect Alignment 376
      5.1.3. Correlation with Optical Dichroism 378
      5.1.4. Electrical Level Determination 379
      5.1.5. Removing Orbital Degeneracy 380
      5.1.6. Distortion of the Wave Function 380
   5.2. Applied Electric Fields 381
   5.3. Optical Illumination *in situ* 382
      5.3.1. Excited States 382
      5.3.2. Metastable Charge States 382
      5.3.3. Optical Alignment 383
   5.4. Studies of the Effect of Temperature 384
      5.4.1. Annealing 384
      5.4.2. Linewidth 385
      5.4.3. Relaxation Times 387
   5.5. Defect Production 389
   5.6. Optical Detection of EPR 390
   References 391

Chapter 5. Phonon–Defect Interaction

D. Walton

1. Introduction 393
2. Theoretical Background 397
   2.1. General 397
   2.2. Multiple Scattering 401
      2.2.1. The Coherent Potential Approximation (CPA) . . . 401
      2.2.2. Range of Validity of CPA 403
      2.2.3. Summary and Conclusion 404
3. External Interactions 404
   3.1. Theory 404
      3.1.1. Mass-Defect and Force Constant Changes 404
      3.1.2. Strain-Field Scattering 407
   3.2. Experiment 408
      3.2.1. Mass-Defect Scattering—Isotopes 408
      3.2.2. Scattering Due to Mass and Force Constant Changes 410
      3.2.3. Strain Fields 413

4. Internal Interactions 415
    4.1. Introduction 415
    4.2. Theoretical Background 416
        4.2.1. Partial Wave Analysis 417
        4.2.2. Formal Scattering Theory 419
        4.2.3. Form of the Interaction 421
        4.2.4. Dispersion Relations 423
        4.2.5. Scattering Cross Section 423
    4.3. Paramagnetic Ions 424
        4.3.1. Theory 426
        4.3.2. Experiment 428
    4.4. "Molecular" Defects 430
        4.4.1. Experimental 430
        4.4.2. Comparison with Experiment 432
        4.4.3. Some Other Probable Consequences 437
5. Summary and Conclusion 438
    References 439

Chapter 6. Point Defects in Molecular Solids

A. V. Chadwick and J. N. Sherwood

1. Introduction 441
2. Self-Diffusion 442
    2.1. Radiotracer Measurements 443
        2.1.1. Tracers 443
        2.1.2. Specimen Preparation 443
        2.1.3. Sectioning Experiments 444
        2.1.4. Penetration Profiles 445
        2.1.5. Integrating Techniques 448
        2.1.6. Results 450
    2.2. Nuclear Magnetic Resonance Measurements 451
    2.3. Radical Recombination Studies 453
    2.4. Summary 454
3. Experimental Determination of the Formation and Migration
    Parameters for Point Defects 455
    3.1. Excess Specific Heat Studies 457
    3.2. Thermal Expansivity Measurements 458
    3.3. Compressibility Measurements 462
    3.4. Summary 462

4. Theoretical Calculations of Point Defect Parameters 463
   4.1. Results 463
   4.2. Summary 466
5. Jump Correlation Experiments 466
   5.1. Comparison of NMR and Tracer Diffusion Measurements . 467
   5.2. Isotope-Mass Effect Measurements 468
      5.2.1. Experimental Measurement of the Mass Factor . . . 469
      5.2.2. Results 470
   5.3. Summary 472
6. Conclusions 472
   References 473

Index 477

Chapter 1

DEFECT CREATION IN SEMICONDUCTORS*

James W. Corbett

Physics Department
State University of New York at Albany
Albany, New York

and

Jacques C. Bourgoin

Laboratoire de Physique
École Normale Superieure
Paris, France

1. INTRODUCTION

In this chapter we will focus on the physics of the defect creation process in semiconductors. In its broadest interpretation this topic can encompass all of the physics of radiation effects in semiconductors; other chapters in this treatise cover many of these other aspects, and consequently we will adopt a narrower view of the topic, considering the question, "How are defects created?," reserving for ourselves the right to draw on other areas where necessary.

There have, of course, been previous reviews of damage in semi-conductors[1-24] and of the subject of the creation of defects in semiconductors.[3,15,21,25,26] In this survey, however, we will do more than simply include the body of experimental data which has appeared in the inter-

* Supported in part by the Office of Naval Research under Contract No. 0014-70-C-0296.

vening period of time; we will include a number of new developments
which have promise of explaining a number of the baffling points in the
field. Here we have in mind the new concepts of the interstitial configuration, and the phenomenon of ionization-enhanced diffusion in particular.
These developments are so recent that their impact has only begun to be
felt. There is still much to be learned in the field; the new developments,
however, as we will indicate, have the character of greatly sharpening the
questions which need to be answered.

There are a variety of means in which defects can be introduced into
lattices. In the case of compound semiconductors defects can appear in the
growth process quite readily due to a lack of stoichiometry; we will not
treat defects that arise in this way in detail since they are discussed separately in a chapter planned for a later volume in this treatise.[27] Defects
can enter a material along with the diffusion of wanted, or unwanted,
impurity atoms; here again such defects will not be considered directly in
this chapter since they are treated elsewhere in this volume by Casey and
Pearson.[28] As was treated in general in Vol. 1 of this treatise by Franklin,[29]
at high temperatures there is an equilibrium distribution of defects in a
material; we shall treat defects that can arise in this way, but only briefly
because an apparent characteristic of semiconductors is that the concentrations of defects that arise in this way is very small. Defects can also be
introduced into materials by the processes of plastic deformation; here
again our treatment will be brief, because plastic deformation apparently
does not introduce large concentrations of defects in semiconductors.

The bulk of the material in this chapter deals with defects introduced
by the interaction of energetic particles with the lattice, and the major
portion deals with what is called *displacement damage*. Energetic particles
interact with the lattice in two ways which will be of interest to us: (1) ionization of the lattice (in most cases this is the dominant process) and (2) direct
collisions with the nucleus which result in a recoil energy T being imparted
to that nucleus (atom). These recoil energies range from very small fractions of an electron volt to, in some cases, millions of electron volts. The
consequences of this recoil energy depend strongly on the magnitude of
the energy. If the recoil energy exceeds a value called the displacement
threshold energy T_d, then the atom can be displaced from its lattice site.
If the recoil energy is just above T_d, then the type of damage that can be
expected is relatively simple, i.e., single vacancy–interstitial pairs, which
may or may not be in close enough proximity that they are effectively
bound to one another. If the vacancy and interstitial are bound, then they
are termed "close pairs" and their behavior is substantially different from

that of unbound pairs. As the recoil energy becomes larger than the threshold energy, the recoiling atom retains enough energy not only to be displaced itself but to displace other nearby atoms, resulting in multiple vacancy defects and several interstitials. As the recoil energy increases further these multiple defects can become rather large clusters of defects, and, as we will see, in some cases can result in a substantial region of the lattice being disordered, or even rendered amorphous.

In principle, then, the investigation of the physics of the defect creation process should be quite straightforward: Perform the experiment at very low temperatures where the defects that are created will be frozen in the lattice, vary the recoil energy, and study the defects that are created. But things are not so simple! Developments over the past several years have resulted in a number of complexities which we will try to synthesize into a coherent picture; the reader should recognize, however, that there is no assurance that all the pieces in this jigsaw puzzle have been discovered. One major complication is that in many instances *defect mobility occurs at the lowest temperatures* obtainable, and may be occurring via an *athermal* process. As we will discuss, low-recoil-energy collisions themselves can contribute to such mobility, but a more likely mechanism is the large amount of ionization that can be present. As one might expect, the results of the displacement process differ depending upon the direction of the recoil atom with respect to its neighboring atoms, i.e., the lattice structure itself. In addition, experiments now indicate that the physics of the displacement process also depends, in some instances, upon the temperature and the charge state of the defects.

Once the defects are mobile, a number of reactions can occur, and hence what is observed experimentally can change. The mobile defects can experience annihilation, i.e., an interstitial, say, can annihilate with its own or another vacancy. The interstitials or vacancies can encounter one another and cluster or agglomerate. Interstitials and/or vacancies can encounter impurity atoms and be trapped.

Again the purpose of this chapter is to focus on the creation of defects. We will discuss these other aspects only to the extent that they cast light on our understanding of the physics of the creation of defects.

In Section 2 we survey, in turn for each semiconductor system, what is known about the defects in those systems; here it would be inappropriate for us to be comprehensive, since such matters are discussed in other chapters in this volume; we introduce such matters primarily to permit us to assess the experiments on defect creation which have been performed. In essence, we must lift ourselves by our own bootstraps, utilizing what is known of the properties of the defects to try to understand the physics of damage production. In Section 3 we discuss the phenomenology of the displacement damage description, considering single displacements, multiple displacements, displacement spikes, and thermal and ionization spikes. This background will permit us to assess, in Section 4, the importance of various aspects of the interactions of particles with semiconductors. We will consider photons, protons, electrons, heavy ions, and neutrons as energetic particles capable of damaging semiconductors; of course, in this section we of necessity consider the role of ionization, channeling, replacement collisions, and assisted focusing. In Section 5 we present a survey of the experimental displacement damage results. In Section 6 we consider subthreshold and ionization damage, including experiment and theory. In Section 7 we consider the theories of displacement damage production. Section 8 presents a summary of the status of the field.

2. DEFECT PROPERTIES

2.1. Introduction

The focus of this review is on damage production, but we cannot assess the creation of defects without a knowledge of the nature of those defects. In this section we provide a brief survey of the properties of defects in semiconductors, treating primarily elemental semiconductors, which have the diamond lattice, and compound semiconductors, which have both the zinc-blende and wurtzite lattices.

There are many things about the mechanisms of defect creation and the properties of defects which are not known; progress in both areas involves substantial boot-strapping—an advance in one area facilitating an advance in the other, and vice versa. The properties of relatively few defects are well established experimentally. Using a melange of these experimental results and recent theoretical insights, we will attempt in this section to synthesize a broad picture of the nature of defects in semiconductors. In subsequent portions of this section we will survey the results for defects in specific semiconductors.

Defects in semiconductors have presented a formidable problem for solid-state theory. The effective mass theory and corrections[30-33] to it have been highly successful for the shallow impurity levels in silicon and germanium, but for the ubiquitous and important defects with energy levels more remote from the carrier band edges, no general theoretical description has yet been devised. A number of calculations have been carried out. These include studies of the electronic properties of vacancies V, divacancies V_2, and interstitials I as well as of the formation energy E_F, migration energy E_m, and divacancy binding energy $E_B(V_2)$. For diamond the calculations have been for all the above, V,[34-63] I,[45,64-67] $E_F(V)$,[45,68-74] $E_m(V)$,[51-53,68,71,73,75] $E_F(I)$,[71,73,75] $E_m(I)$,[64-66,71,73,75] V_2,[76] $E_F(V_2)$,[69,70,72,77,78] and $E_B(V_2)$[69,70,72,77,78]; for graphite, V,[79] I,[80] $E_F(V)$,[81,82] $E_m(V)$,[81,83] $E_F(I)$,[84] and $E_m(I)$[84,85]; for silicon, V,[36,41,42,45,56,57,86-92] I,[45,65,86,87,93] $E_F(V)$,[45,68-75,77,94-97] $E_m(V)$,[68,71,73,75] $E_F(I)$,[71,73,75] $E_m(I)$,[71,73,75,98-100] V_2,[94,95] $E_F(V_2)$,[69,70,72,77,78] and $E_B(V_2)$[69,70,72,77,78]; and for germanium, V,[42,45,86,87,101] I,[42,45,86,87,101] $E_F(V)$,[45,68-75,77,96,98-100] $E_m(V)$,[45,68,71,73,75,102,103] $E_F(I)$,[71,73,75,103] $E_m(I)$,[71,73,75,98-100,102] $E_F(V_2)$,[69,70,72,77,78] and $E_B(V_2)$.[69,70,72,77,78] These calculations range from model calculations and Morse function calculations, to band structure and molecular orbital treatments. The early calculations had, in many instances, only a limited validity, but increasing sophistication is resulting in a closer and closer coordination between experiment and theory.

We will begin our discussion with the vacancy and vacancy-related defects, then go on to the interstitial and interstitial-related defects. With that knowledge we will be able to discuss interstitial–vacancy close pairs. We then complete our discussion with spike-type damage, the damage associated with dislocations, and finally the damage states of amorphous material. We discuss these defects in elemental semiconductors first, then go on to the consideration of compound semiconductors.

More is probably known about the vacancy in the diamond lattice than any other defect we will discuss. With the aid of Fig. 1 we can describe what is known about the vacancy in the diamond lattice. Watkins has made experimental observations on the V^+ and V^- states in silicon and indirect measurements on the properties of V^0. He argued that there are strong

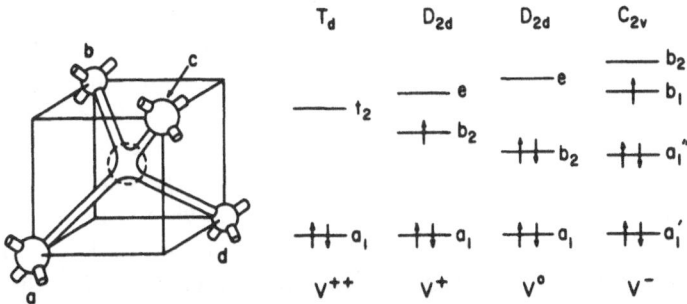

Fig. 1. Configuration of the vacancy in diamond with the symmetry of the various charge states indicated.

Jahn–Teller and bonding effects in these and other defect states.[60] (Although this point is subject to some controversy,[61] we believe it is well established experimentally and theoretically and will employ it throughout this review.) Watkins has shown that one-electron LCAO–MO states are useful in understanding the behavior of the defects. In Fig. 1 we show the symmetry of the electronic states of the vacancy and their filling for various charge states. He argued that for V^{2+} the symmetry of the defect is T_d with two electrons in the a_1 orbital [for the atoms labeled in Fig. 1 this orbital would be $(a + b + c + d)$], the t_2 orbital being unoccupied. In the V^+ charge state an electron goes into this t_2 orbital, giving rise to a tetragonal Jahn–Teller distortion—a bonding between the atoms in pairs; the symmetry now becomes D_{2d} with the additional electron in a b_2 orbital which, for example, may be $((a + d) - (b + c))$. In the V^0 charge state an additional electron goes into this b_2 orbital, augmenting the bonding energy, but leaving the same D_{2d} symmetry. In the V^- charge state an additional electron goes into the e orbital, giving rise to an orthorhombic Jahn–Teller distortion reducing the symmetry to C_{2v}; now the a_1' orbital is $(a + b + c + d)$, the a_1'' is $(a + d) - (b + c)$, and the b_1 state is an antibonding orbital $(a - d)$. We note that an additional electron may give rise to such a large electron–electron interaction is the V^{2-} charge state that the whole symmetry of the defect is altered. This is shown in Figs. 2a and 2b, in which we again show the V^- configuration, this time with a set of three next-nearest-neighbor atoms. We also show how the bonding can change to accommodate the additional electron in the V^{2-} charge state by moving a lattice atom adjacent to the vacancy to a position halfway between lattice sites, thereby permitting extended bonds to be formed with the six nearest neighbors and the excess charge to be distributed more widely. The gist of this suggestion is that the electron–electron interaction term may favor this configuration over a V^{2-} configuration centered on a substitutional lattice site, but there is no evidence at present which indicates that this is the configuration of the V^{2-} charge state. The configuration with a lattice atom halfway between two vacant lattice sites is of course the canonical saddle-point configuration for vacancy migration in the diamond lattice.

As we discuss in more detail in Section 3.3, if the equilibrium configurations of the V^- charge state and the V^{2-} charge state are as shown in Figs. 2a and 2b, then the vacancy will migrate via the Bourgoin mechanism[65,104,105]; that is, a charge-state change from V^- to V^{2-} to V^- can result in the net transport of the vacancy through the lattice, i.e., the migration proceeds through the charge-state changes rather than via thermal activation over a barrier. Masters[106] has suggested that this saddle-point

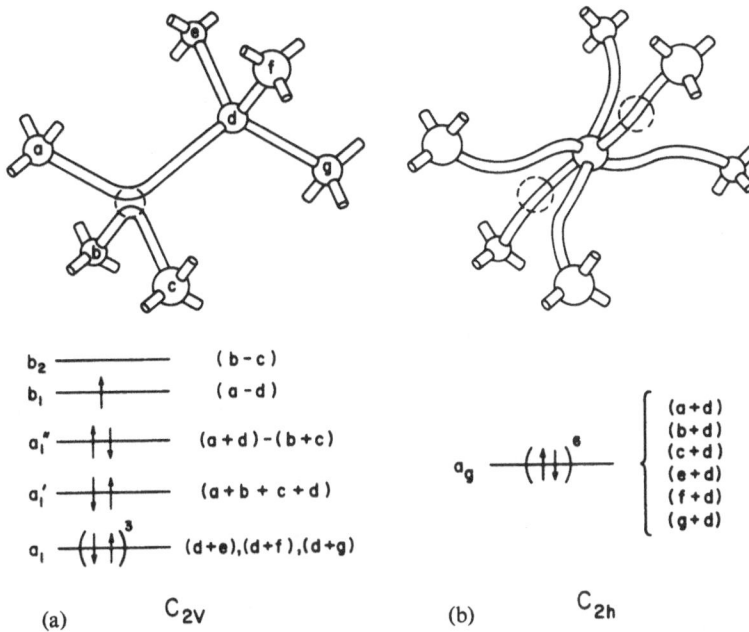

C_{2v}

C_{2h}

(a)   (b)

Fig. 2. (a) Configuration of the V^- in the diamond lattice with some of the molecular orbital states shown. (b) A possible configuration for V^{2-} with some of the molecular orbitals shown. This configuration is the saddle-point configuration for normal vacancy migration, but it is argued in the text that it may be the stable configuration for this charge state.

configuration (Fig. 2b), which he calls a pair of semivacancies, is an alternate configuration which a vacancy can assume in the diamond lattice. (Our argument about the electron–electron interaction argues that his proposal has particular force for the V^{2-} charge state.)

The divacancy in the diamond lattice has the configuration shown in Fig. 3. As we will see in Section 2.2, there is substantial experimental information on the divacancy in silicon. Migration of the divacancy presumably involves a saddle-point configuration such as is shown in Fig. 4, with at least a partial dissociation of the divacancy. The remarks made above about the electron–electron interaction have even more force for the divacancy. As indicated in Fig. 3, the V^{2-} state involves two electrons in an antibonding orbital between atoms a and f; the same number of electrons can be accommodated in the bonds of the split divacancy in Fig. 4, but now the charge state is the neutral charge state, the reason being that the atom which splits the divacancy can accommodate two electrons so the

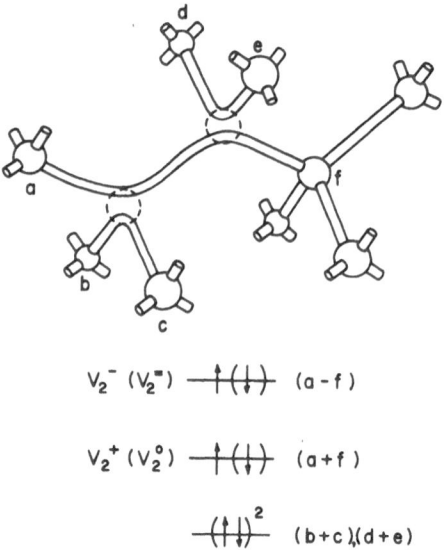

$$V_2^- \ (V_2^=) \ \text{---}\!\!\!\frac{\ }{\ }\!\!\!(\!\!\frac{\ }{\ }\!\!\!) \ \ (a-f)$$

$$V_2^+ \ (V_2^0) \ \text{---}\!\!\!\frac{\ }{\ }\!\!\!(\!\!\frac{\ }{\ }\!\!\!) \ \ (a+f)$$

$$\text{---}\!\!\!(\!\!\frac{\ }{\ }\!\!\!)^2 \ \ (b+c)(d+e)$$

Fig. 3. Configuration of the divacancy in the diamond lattice with some of the molecular orbitals shown. The filling of the molecular orbitals for the various charge states is indicated, the second electron in a given level being denoted by parentheses, as is the corresponding charge state.

eight electrons are now spread over eight bonds. If this is the more stable configuration for V^{2-}, then, as discussed above for the vacancy, the divacancy would migrate via the Bourgoin mechanism, i.e., a charge change from V_2^- to V_2^{2-} to V_2^- could result in the net transport of the divacancy through the lattice.

The dissociated divacancy is the simplest prototype of a type of defect proposed by Hornstra.[107] He argued that if the divacancy was dissociated one step beyond the configuration shown in Fig. 4, a new possibility for bonding occurs as is shown in Fig. 5, where the two intervening atoms can now experience double bonding. He argued that a whole family of such defects could be generated by incorporating more and more vacancies each remotely coupled to the other. He termed such defects *sponge* defects. Such defects have not been observed experimentally, but we should continue to bear in mind the possibility of their existence.

Aggregates of vacancies in the diamond lattice have also been discussed. Our understanding of bonding in other vacancy defects suggests a bonding

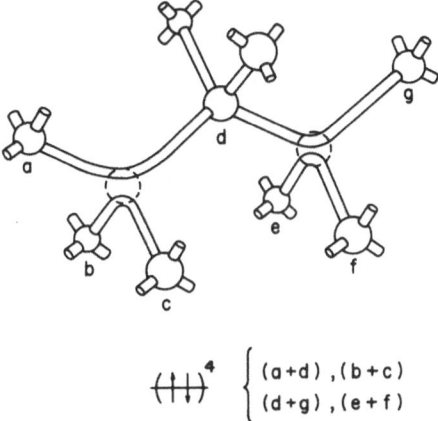

$$(\uparrow\downarrow\uparrow\downarrow)^4 \quad \begin{cases} (a+d),(b+c) \\ (d+g),(e+f) \end{cases}$$

Fig. 4. Configuration of the split-divacancy in the diamond lattice, with some of the molecular orbitals indicated. This configuration is the saddle-point configuration for normal divacancy migration, but it is argued in the text that for the filling of the electronic states as shown, the V_2^{2-} charge state, this configuration may be the stable configuration.

in multivacancy defects. Figure 6 shows a V_3 defect with pair bonding between the atoms above and below the plane of vacancies, leaving non-bonded electrons at the ends of the vacancy chain. We note that in V_3 the angle between the axes of these nonbonded electrons is 109° (and similarly for defects with an odd number of vacancies), whereas in the V_4 defect shown in Fig. 7 these axes are parallel to one another (as would be the case for an even number of vacancies in a chain). Presumably there is no

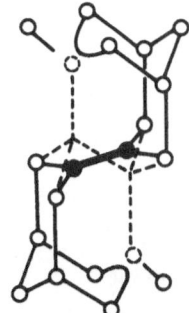

Fig. 5. The sponge divacancy.

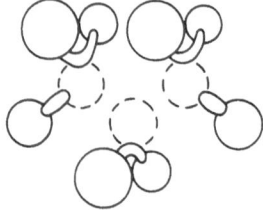

3-VACANCY

Fig. 6. Configuration of the trivacancy in the
diamond lattice.

impediment to forming a chain of vacancies of this sort of an indefinite length. We will note later that experimental evidence exists in support of these configurations for V_3 and V_4; however, the evidence that exists for V_5 suggests that it adopts a nonplanar configuration such as is shown in Fig. 8. We note as well that V_6 could have what appears to be a natural configuration, as shown in Fig. 9, with the vacancies occupying the six-membered (chair) puckered ring which is the natural building block of the diamond lattice; as yet no experimental information exists on such a configuration.

We go on to discuss interstitial and interstitial-related defects in the diamond lattice, returning later to discuss vacancy- and interstitial-related defects in compound semiconductors.

There is much less experimental information concerning the interstitial than the vacancy in semiconductors. It was long held that the interstitial would occupy the rather large tetrahedral (T) "hole" in the diamond lattice shown in Fig. 10. Experiments support the view that some impurity atoms do occupy this site. Weiser,[108] in treating the migration of the interstitial impurities, argued that the saddle-point configuration for the migration of T interstitials would be the "hexagonal" (H) site shown in Fig. 11, i.e.,

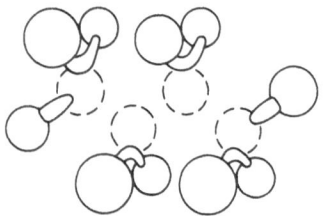

4 - VACANCY

Fig. 7. Configuration of the quadrivacancy in
the diamond lattice.

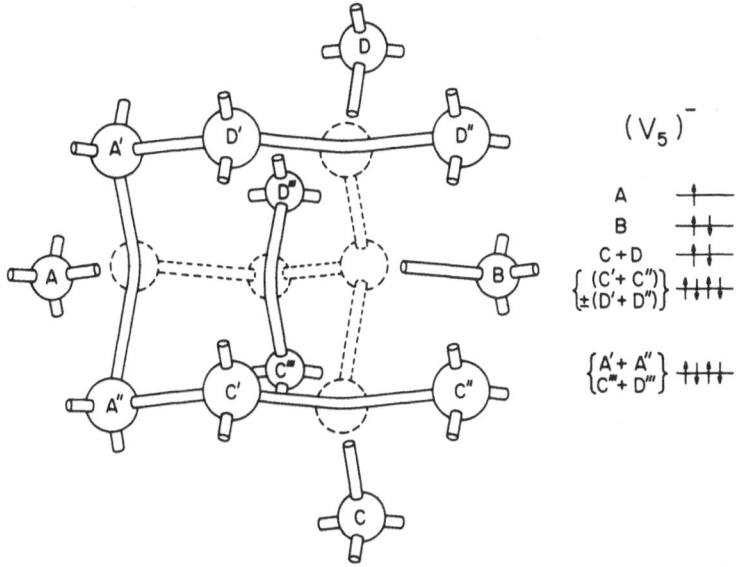

Fig. 8. Configuration of a nonplanar pentavacancy in the diamond lattice. Some molecular orbital states corresponding to the negative charge state are shown.

the site at the center of the puckered, six-membered ring; he also argued that some impurities might find their stable position at the H site, migrating through the T site as a saddle point (and we have argued[65,104,105] that a given impurity may be stable at a T site for one charge state and at an H site for

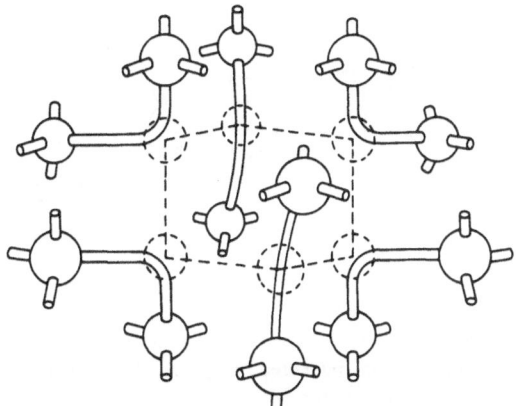

Fig. 9. Possible configuration of a hexavacancy in the diamond lattice.

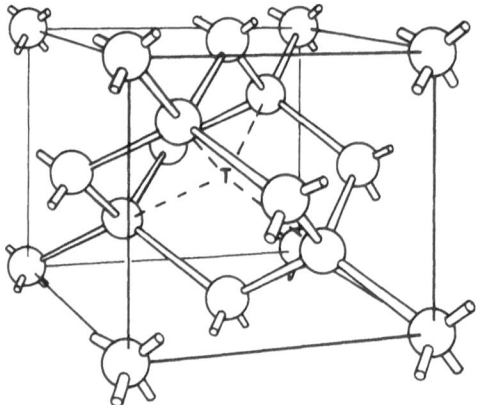

Fig. 10. The tetrahedral T interstitial site in the
diamond lattice.

another charge state). Watkins and Friedel have separately argued that
interstitials may prefer other sites to these essentially nonbonding sites,
T and H. Watkins* argued that the double-plus charge state of the self-
interstitial in the diamond lattice could readily form bonds with its two
remaining valence electrons in the bond-centered configuration shown in
Fig. 12. Friedel[45] argued that another bonded configuration of the self-

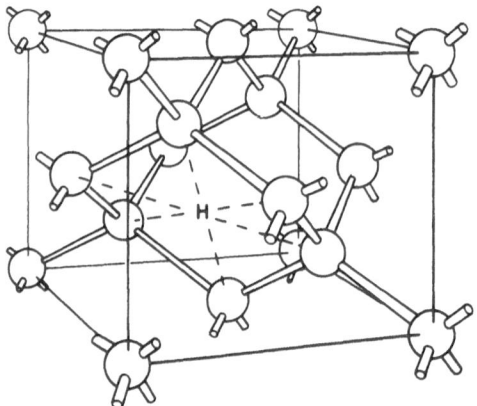

Fig. 11. The hexagonal H interstitial site in the
diamond lattice. This site is in the center of a
six-membered puckered ring in the so-called
"chair" configuration.

* See discussion in Ref. 15, p. 99.

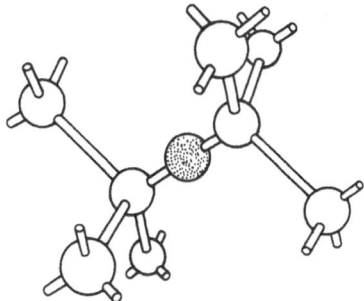

Fig. 12. The bond-centered B interstitial configuration in the diamond lattice, with the
interstitial atom shown stippled.

interstitial could be as shown in Fig. 13. This configuration is referred to
as a split-interstitial, i.e., two atoms occupy a substitutional site, neither
being identifiable as *the* substitutional or *the* interstitial atom. Recent theoretical calculations[53,60,64,65,67] support the physical arguments advanced by
Watkins and Friedel. Specifically, the calculations indicate that in both
diamond and silicon, the split-$\langle 100 \rangle$ interstitial shown in Fig. 13 is the
lowest energy configuration for the neutral, single-plus, and single-minus
charge states of the self-interstitial, the bond-centered interstitial (B) being
next higher in energy, followed by the hexagonal configuration, with the
tetrahedral configuration the highest of all; in fact for these charge states B
and H are saddle-point configurations and T is at a potential energy
maximum. The migration path for the split-$\langle 100 \rangle$ interstitial in these charge
states is from the minimum at the split-$\langle 100 \rangle$ configuration through a B
configuration to another split-$\langle 100 \rangle$ configuration. For the double-plus
charge state of the self-interstitial the calculations indicated that the B site
is the energy minimum with migration proceeding through a split-$\langle 100 \rangle$
saddle point; the H site remains a saddle point and T a potential maximum.

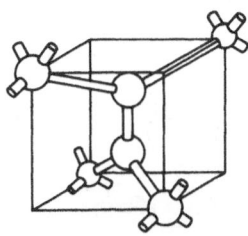

Fig. 13. Configuration of the split-$\langle 100 \rangle$ interstitial
in the diamond lattice.

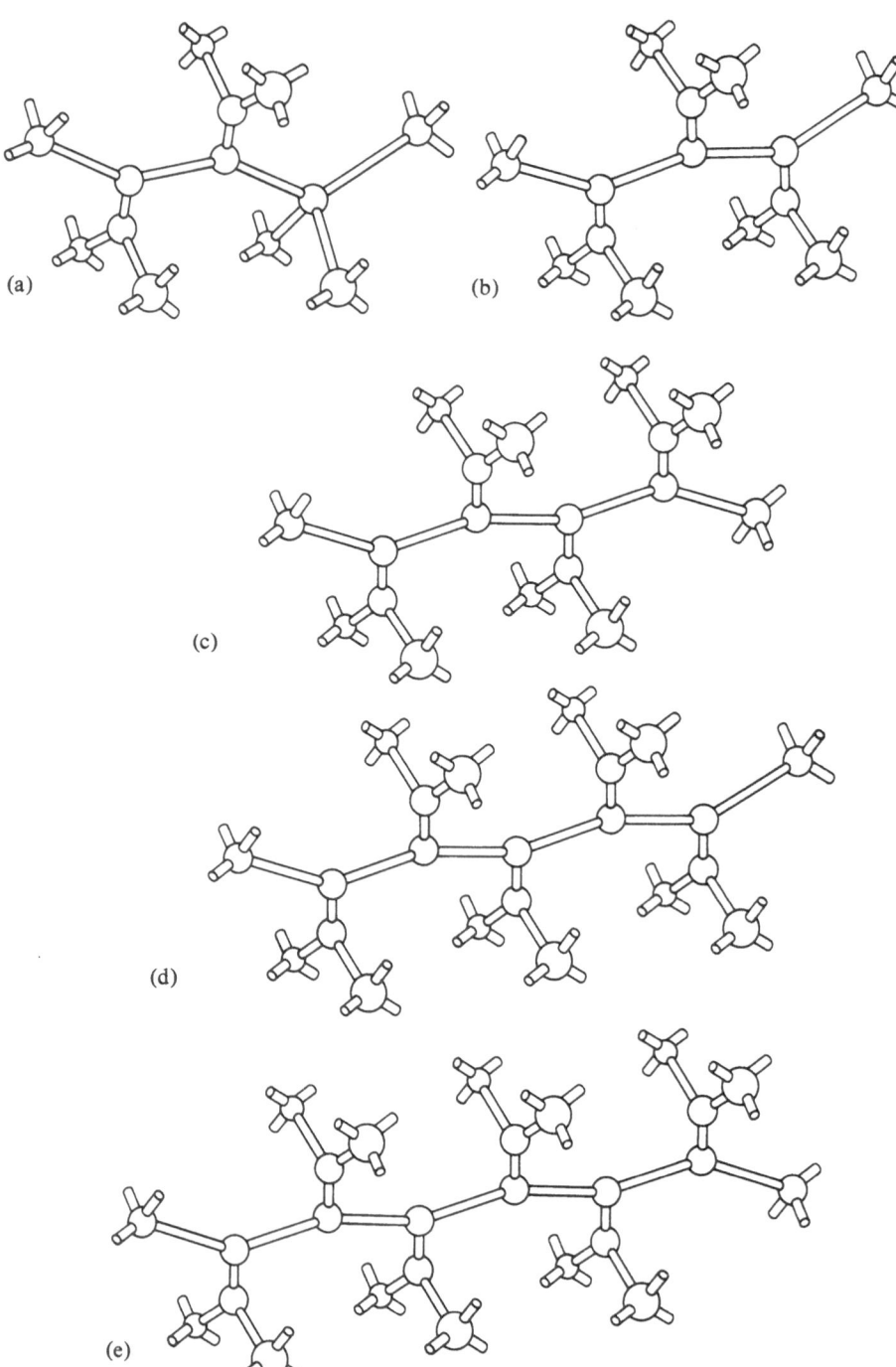

Fig. 14. Possible configurations of "close-packed" interstitial agglomerates in the {110} plane. Atoms in split-interstitials are identifiable by having only three chemical bonds.

We note that the calculation indicates a change of stable position with charge state; consequently a charge-change sequence from single minus to double minus to single minus would result in a net transport of the defect through the lattice, i.e., migration via the Bourgoin mechanism.[65,104,105]

The minimum-energy interstitial configurations in metals and alkali halides also seem to be split configurations, so that the split-interstitial increasingly appears to be universal.

Interstitial aggregates apparently occur. Calculations have not yet been done, but the bonding nature which the calculations suggest argues for aggregates of bonded interstitials either closely packed as shown in Fig. 14 or at next-nearest-neighbor sites as shown in Fig. 15. We note in particular that the unbonded p orbitals on two of the atoms in split-$\langle 100 \rangle$ interstitials at next-nearest-neighbor sites point directly at one another and can form an extended bond; the five-membered ring thereby formed occurs in a number of instances of high-pressure and impurity phases of carbon, silicon, and germanium, in some cases forming an almost pentagonal ring.

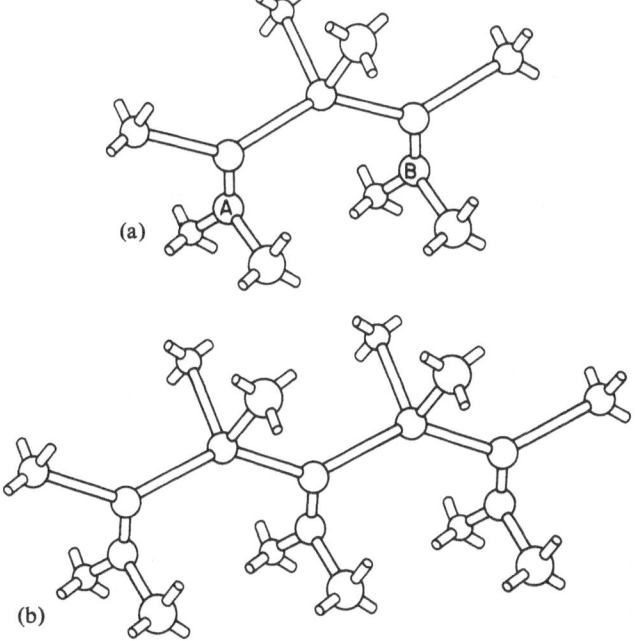

Fig. 15. Possible configurations of interstitial agglomerates in the {110} plane on next-nearest neighbor sites. Bonding of the p orbitals of atoms such as A and B can occur in either this configuration or the configurations shown in Fig. 14, and will enhance the stability of these configurations.

   Almost no information exists on interstitial–vacancy pairs. Our knowledge of the configuration of interstitials and of vacancies suggests that
interstitial vacancy pairs such as those shown in Fig. 16 could form, but
gives us no inkling if any of these are stable. The calculations on the elec-

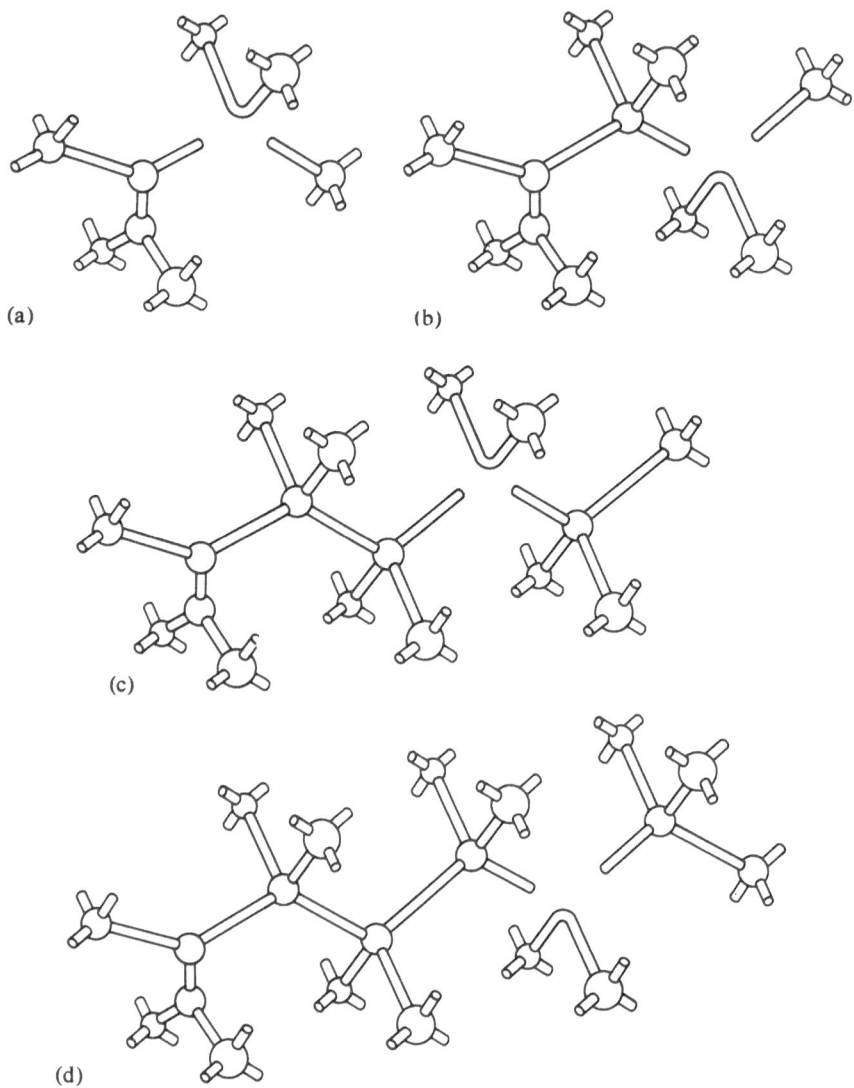

(a)   (b)

(c)

(d)

Fig. 16. Possible close-pair configurations in the {110} plane of diamond. Presumably
the configuration in (a) in unstable, but neither theory nor experiment comments on the
stability of the other configurations. Again, interstitial atoms are those that participate
only in three bonds.

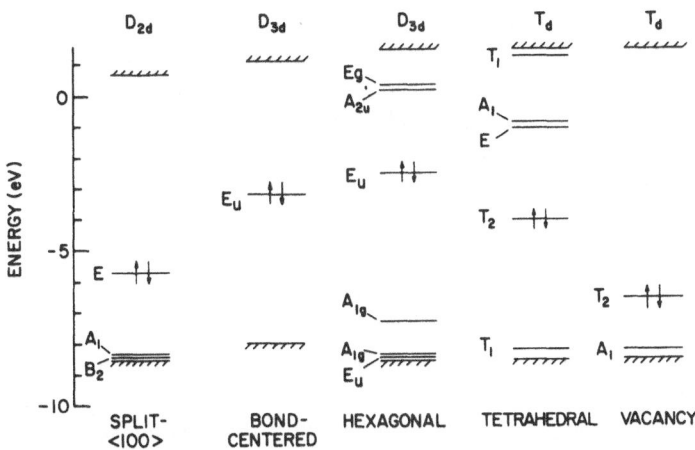

Fig. 17. The ground-state electronic levels for the neutral split-⟨100⟩, bond-centered, hexagonal, and tetrahedral interstitials and the vacancy in diamond as given by recent LCAO calculations. The arrows indicate the filling of the levels for the neutral charge state.

tronic levels of interstitials and vacancies (see, for example, the results on diamond[65,67] in Fig. 17) argue that in general the electrical levels of the interstitial and of the vacancy will not occur at the same place in the forbidden gap. Consequently, an interstitial and vacancy in close proximity, i.e., interacting strongly enough to be considered a close pair, will in all likelihood experience a charge exchange. If this results in I^+ and V^-, for example, then the stability of the close pair will have an additional Coulomb interaction term.

Finally in our consideration of intrinsic defects in the diamond lattice we must note that high-energy collisions can result in extensive damaged regions, called *spikes* (see Section 3). The structure of such spikes is not well established but we must recognize that there is a wide variety of possibilities in semiconductors. The same possibilities can occur as a result of the annealing of defects and hence can be encountered in irradiation carried out at a temperature where one or more defects are mobile or in which annealing is carried out after the irradiation. The spike structure may consist solely of distinct point defects and their aggregates as already discussed, but may be a variant of three other structures: an amorphous structure, a dislocation structure, or that of another phase of the material. An amorphous nature has often been ascribed to spike regions; most experimental information, particularly in close-packed materials, argues that a damaged region in contact with a perfect crystal will revert toward a crystalline structure, but we must remember that the distinction between an amorphous region and a highly relaxed damaged region is a slight one. The occurrence of new phases as the result of the recovery of damaged semiconductors is well established.* Most semiconductor systems have a number of phases which they assume in different temperature and pressure regimes. Consequently, the possibility of there being vestiges of another phase in the damaged region cannot be ignored, particularly when the phase being investigated is unstable, e.g., diamond at standard temperature and pressure. Also the occurrence of loops of interstitial and vacancy character is well documented in electron microscope observations and hence it must be recognized that dislocation structures are a natural consequence of damage and its recovery.

Impurities can play a dominant role in irradiation damage experiments. The literature is full of instances where the unexpected, pivotal role of impurities was only belatedly and painfully uncovered. Impurities can interact with vacancy-type defects. The occurrence of vacancy–impurity pairs is well established; a substitutional impurity can occupy a nearest-neighbor vacancy site or an interstitial impurity can, at least partially, occupy the vacancy site. For example, in silicon the (vacancy + group V impurity) pair has the configuration shown in Fig. 18a, while the (vacancy + group III impurity) pair has the configuration shown in Fig. 18b; the normally "interstitial" oxygen in silicon forms a pair with a vacancy as shown in Fig. 18c. There are also suggestions in the literature of impurities associated with high-order vacancy defects, e.g., impurity–divacancy pairs.

We indicated that experimental evidence suggests that some interstitial impurities occupy T and H sites. Calculations[93] and experimental evidence[109,110] also suggest that other states of impurity interstitials can occupy bonded configurations. It may be that the impurity participates in a bond-centered-like configuration. In other cases the impurity may participate in a split-interstitial-like configuration.

We will not treat the different phases which can occur in these systems except to note that a number of natural building blocks occur. Coherent twins can occur naturally in the diamond lattice (and, as we will see in SiC, give rise to extensive polytypism). Successive twins in the cubic diamond lattice can leave a portion of the lattice in the hexagonal diamond structure. Figure 19 shows the distinction between these structures; the zinc-blende structure with all the same atoms is the cubic diamond structure, whereas the wurtzite structure with all the same atoms is the hexagonal

* See, e.g., the discussion of phase change in Ref. 15.

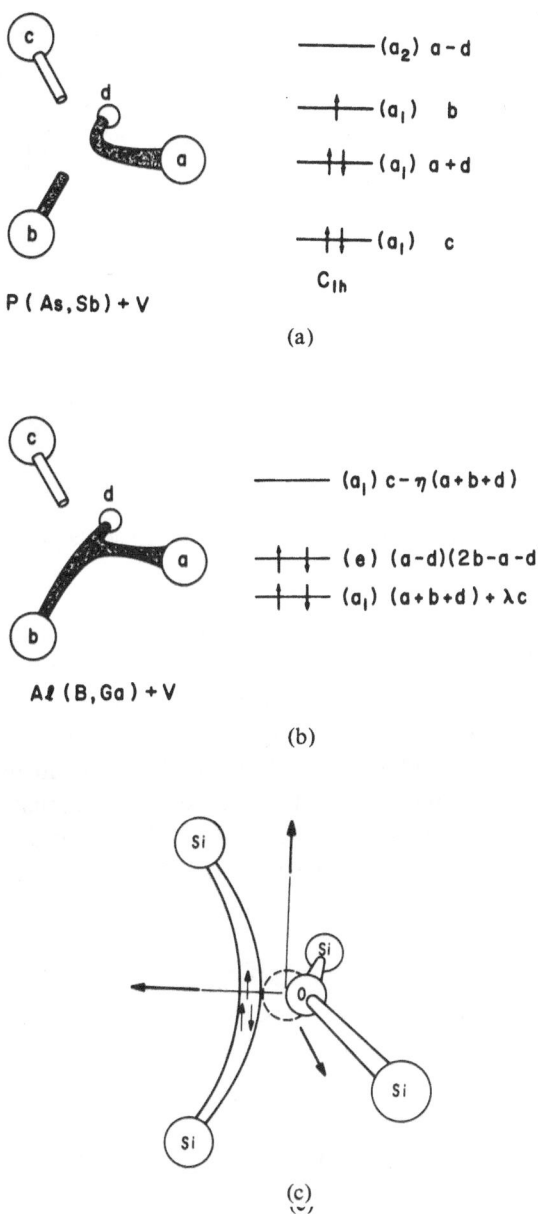

Fig. 18. (a) Configuration of the (vacancy + group V impurity) pair; atom C is the impurity; (b) Configuration of the (vacancy + group III impurity); atom C is the impurity; (c) Configuration of the (vacancy + oxygen) pair.

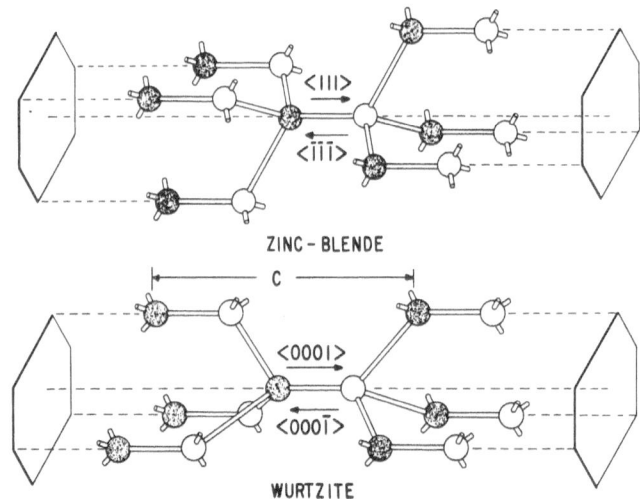

Fig. 19. The zinc-blende, or sphalerite, structure and the wurtzite
structure.

diamond structure. More distorted, and even broken-bond, structures occur
for dislocations.[111-113] It is well recognized that the "chair" and "boat"
puckered, six-membered rings are basic building blocks in tetrahedrally
coordinated structures. We show in Figs. 20 and 21 how these elements
naturally conjoin with a planar and "rowboat" five-membered ring. Of
course, five-membered rings can conjoin to form pentagonal icosahedra
which are found in some of the dense phases of these materials.[114] The
planar hexagonal ring is the building block of the graphite structure and
vestiges of this structure can occur in damage regions in other materials.

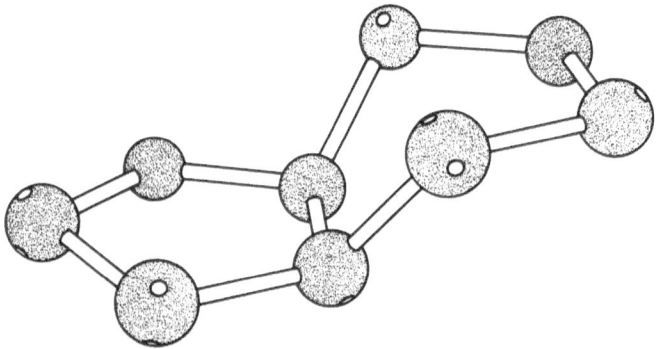

Fig. 20. Conjunction of the "boat" six-membered ring and a planar
five-membered ring.

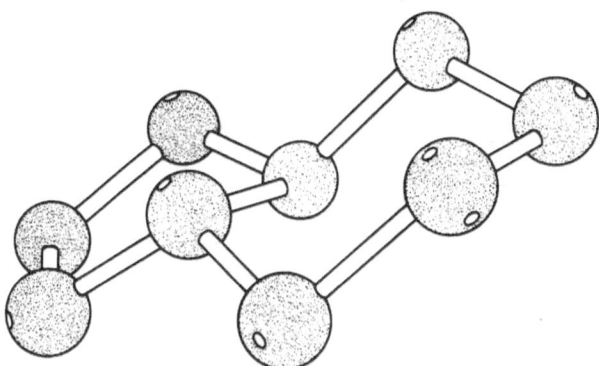

Fig. 21. Conjunction of the "chair" six-membered ring and a
"row-boat" five-membered ring.

(We defer further discussion of graphite and of the helical semiconductors, tellurium and selenium, until Section 2.7.)

The compound semiconductors occur primarily in either of two structures, the zinc-blende (or sphalerite) structure or the wurtzite structure, both of which are shown in Fig. 19. There has been relatively little theoretical work done on the defect states of the compound semiconducting materials. We will treat the problem of defect configurations from the viewpoint of our understanding gained in the diamond lattice, i.e., emphasizing the covalent bonding aspects, and ignoring size difference effects; we must note, however, that there may be places where the size effects and ionic effects dominate and would substantially alter our conclusions. We recognize that compound semiconductors involve both covalent and ionic bonding, to various degrees. As Watkins has emphasized,[60] there are many similarities between the principles governing defect configurations in the ionic alkali halide materials and in covalent materials. We presume these principles apply to compound semiconductors as well. In the compound semiconductors we denote the generic composition by RX, where R represents the group II or group III atom and X denotes the group VI or group V atom. We see in Fig. 19 that the zinc-blende and wurtzite structures are both characterized by R atoms being tetrahedrally surrounded by X atoms and vice versa. The distinction between the two systems comes from a rotation of the bond directions as shown. We note that, as indicated in Fig. 19, the compound nature of these materials implies that they present a different aspect depending on how they are viewed; for example, in a zinc-blende structure of an RX compound, looking into the crystal on a {111} face, the R atoms have empty spaces behind them, whereas looking into the crystal along the $\{\bar{1}\bar{1}\bar{1}\}$ direction, the X atoms have empty spaces behind them.

The occurrence of two different types of atom in the lattice substantially increases the number of possible defect configurations. We will consider the zinc-blende lattice first. For the zinc-blende lattices we know more about the vacancy defects in II–IV materials than in III–V materials. In the II–VI materials the observations[60,115] indicate that the group VI vacancy does not undergo a Jahn–Teller distortion and retains the tetrahedral symmetry. The group II vacancy, on the other hand, undergoes a static trigonal Jahn–Teller distortion. There are no direct observations on group III or group V vacancies in III–V compounds but Watkins[60,115] has speculated that the group III vacancies will also undergo a substantial Jahn–Teller distortion. Thus the types of vacancy configuration are not substantially different from what we have discussed for the diamond lattice. There are substantially increased complexities when we consider the interstitial configurations. There are two distinct types of T sites in the zinc-blende RX structure, one (T_R) surrounded by four R atoms and one (T_X) surrounded by four X atoms, as shown in Figs. 22 and 23. The midpoint between a T_R and a T_X site is an H site, a "hexagonal" site in the center of a chair, six-membered ring with three R atoms and three X atoms as shown in Fig. 24; in fact, in Fig. 24 the sequence of sites along the $\langle 111 \rangle$ axis is X–T_X–H–T_R–R–X. Migration of interstitial atoms which do not bond with the lattice presumably must go through the T_X–H–T_R–H–T_X–$\cdots$ sequence. We note that in these materials the electronic configuration of a

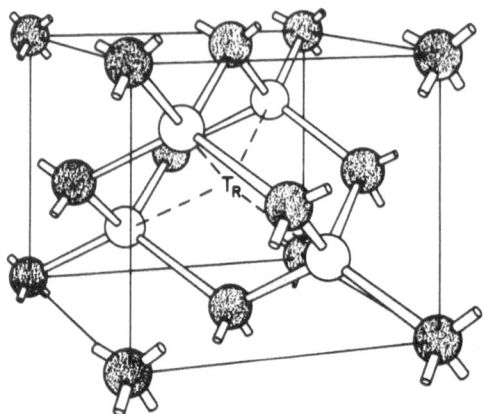

Fig. 22. The T_R tetrahedral interstitial position in the RX zinc-blende lattice.

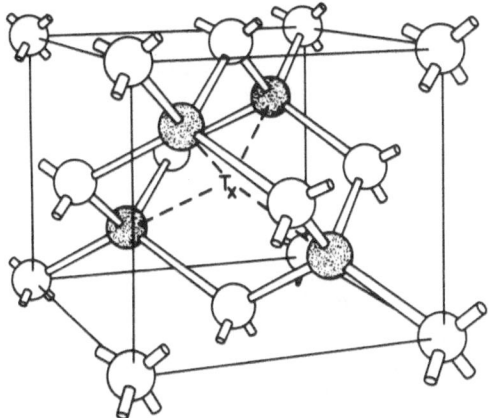

Fig. 23. The T_X tetrahedral interstitial position in
the RX zinc-blende lattice.

group VI atom is ns^2, np^4; consequently such an atom in the double-minus
charge state has an ns^2, np^6 configuration, i.e., the electron levels are filled
so that it looks like a noble gas atom; we presume that such an interstitial
would occupy an unbonded interstitial site and would migrate rather readily
through the lattice.

In the event that the interstitial undergoes bonding with the lattice,
we must expect bonded interstitials as in the diamond lattice. The bond-
centered interstitial would not be symmetric, i.e., if the bond-center atom
is an R atom, its neighbors would be an X and an R atom and equivalently

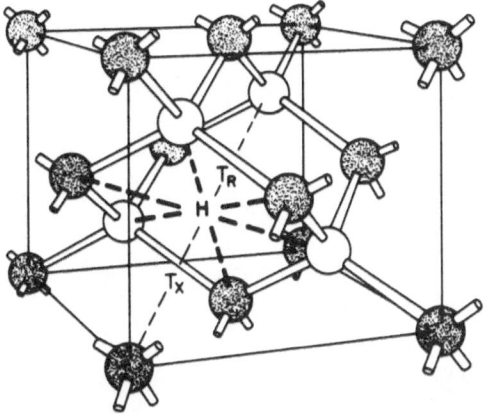

Fig. 24. The H hexagonal interstitial site in the RX
zinc-blende lattice, shown joining a T_R and a T_X site.

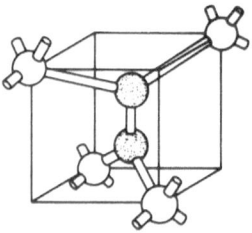

Fig. 25. A possible configuration of an interstitial in the RX zinc-blende lattice, namely
a split-$\langle 100 \rangle$ interstitial made up of X atoms
(or R atoms).

for X bond-centered atoms. Consequently we speculate, since there are no
calculations as yet, that a split-$\langle 100 \rangle$ interstitial configuration could be
assumed by either an X interstitial as is shown in Fig. 25, or by an R interstitial, which would have the same structure shown in Fig. 25 with the
atom type exchanged. Motion of a split-$\langle 100 \rangle$ X interstitial presents a new

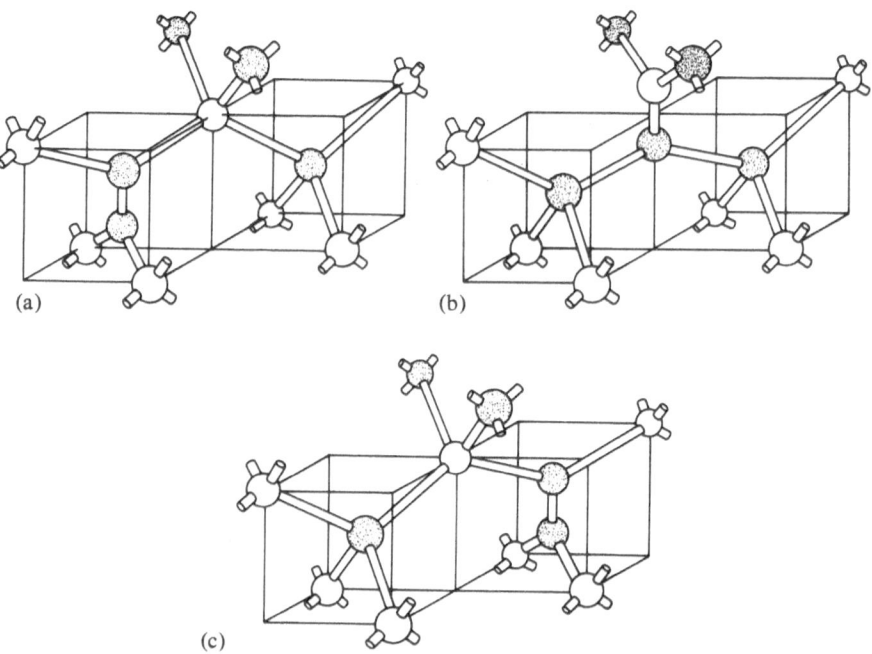

(a)

(b)

(c)

Fig. 26. A possible migration mechanism of an interstitial in the RX zinc-blende lattice.
Migration from a split-$\langle 100 \rangle$ interstitial made up of, say, X atoms proceeds from one
site shown in (a), to another site shown in (c), via a mixed split-interstitial shown in (b).

problem. We presume that the interstitial centered at an X site will proceed to an equivalent position via a *mixed* split-$\langle 100 \rangle$ interstitial as is shown in the sequence in Fig. 26. We presume interstitial aggregates can form in the same way we have argued they can form in the diamond lattice.

We can now return to consider the migration of a vacancy in the zinc-blende lattice. The migration of, say, an R vacancy requires a jump to a next-nearest-neighbor site since the nearest-neighbor sites are X sites; this jump can occur by an R atom moving to the vacant R site via their common T_R site, but we note that it can also occur via a mixed split-interstitial as shown in Fig. 26b. A mixed divacancy can in fact be more mobile than either a single vacancy or higher-order vacancy agglomerates. More complicated defects such as mixed interstitial–vacancy pairs can occur but we will not consider them further here.

The wurtzite lattice presents even more complexities. The vacancy properties we presume will be approximately the same as in the zinc-blende lattice since they are to first order determined by the nearest neighbors. The analogous site to the T interstitial site now becomes the center of a triangular prism; in fact there are two such sites as shown in Fig. 27, a "capped" site T_C (where T now stands for trigonal) and an "uncapped" site T_U. The T_C sites are surrounded on three sides by T_U sites, being connected to them via the analog of the H site, now in the center of the base of the "boat" puckered ring. There are open channels along the c axis consisting of T_U sites connected to one another via genuine H sites (i.e., sites which are the center of a "chair" puckered ring) with no intervening

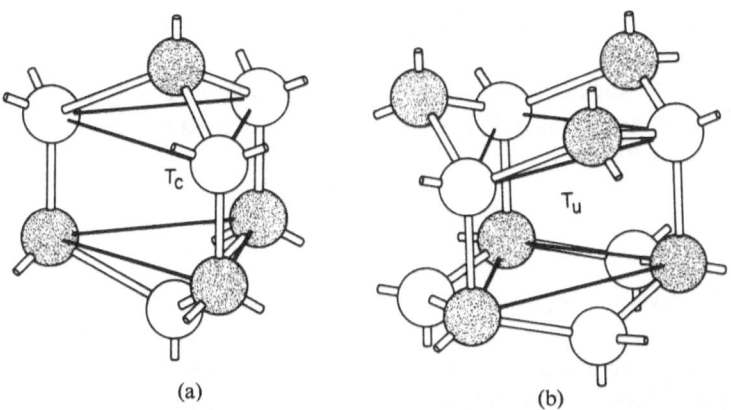

(a)   (b)

Fig. 27. Trigonal T interstitial sites in the wurtzite lattice. (a) The capped site T_C; (b) the uncapped site T_U.

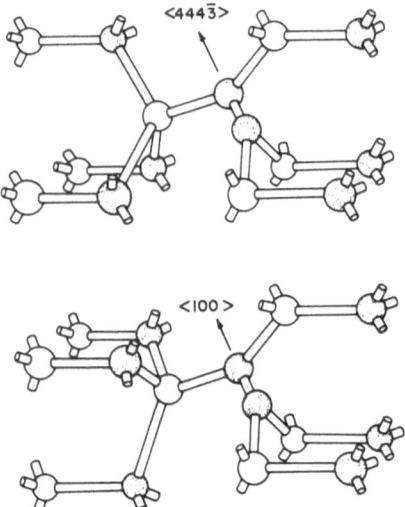

Fig. 28. Below: the configuration of the split-$\langle 100 \rangle$ interstitial in the zinc-blende lattice. Above: the corresponding configuration in the wurtzite lattice, the split-$\langle 444\bar{3} \rangle$ interstitial.

atoms in the sequence. We also presume that bonded interstitials can occur in the wurtzite lattice, where the analog of the split-$\langle 100 \rangle$ interstitial is now the split-$\langle 444\bar{3} \rangle$ interstitial as shown in Fig. 28. We presume that pairs of split-interstitials equivalent to those shown in Fig. 15 for next-nearest-neighbor sites can occur, and in fact there can be chains of such interstitials connecting sites in the basal plane.

Impurities certainly can play an important role in compound semi-conductors. Impurities can have size effects, ionic effects, and covalency effects. Consequently we cannot give a general treatment of impurity-related defects in these materials beyond saying that we expect all of the complexity associated with intrinsic defects and more!

2.2. Silicon

We review here what is known about defect properties in silicon; information specifically related to damage production mechanisms is deferred until Section 5. The configuration of a number of vacancy-related defects has been well established by EPR measurements in silicon: [V^- and V^+][109,115-119] (see Fig. 1), [V_2^- and V_2^+][120-123] (see Fig. 3), V_4[124] (see Fig. 7), V_5[125] (Fig. 8), $(V + P)$,[109,117,126] $(V + As)$,[109,127,128] $(V + Sb)$,[109,127,128]

$(V + \mathrm{Al})$,[109,118,127,129] $(V + \mathrm{B})$,[109,127] $(V + \mathrm{Ga})$,[127] and $(V + \mathrm{O})$[129,130] (see Fig. 18). There is also evidence of $(V + \mathrm{Ge})$,[119,131,132] $(V + \mathrm{Sn})$,[132] $(V + \mathrm{C})$, [133–135] and vacancy aggregates with oxygen impurities.[124] Figure 29 shows schematically the temperature dependence of the recovery of these defects. Figure 30 shows what is known of the recovery of interstitial-related defects in silicon. Watkins[116] found that when he irradiated aluminum-doped silicon at 4°K he observed, not silicon interstitials, but aluminum interstitials. He recognized that this required a long-distance migration of the silicon interstitial, in what is now thought to be an athermal, ionization-enhanced diffusion process (see Section 3.3), with the silicon interstitial finding an aluminum substitutional atom and undergoing a conversion with the silicon assuming the substitutional site and the aluminum the interstitial site. As shown in Fig. 30 at high temperatures this aluminum interstitial becomes mobile and aluminum interstitial plus aluminum substitutional pairs are

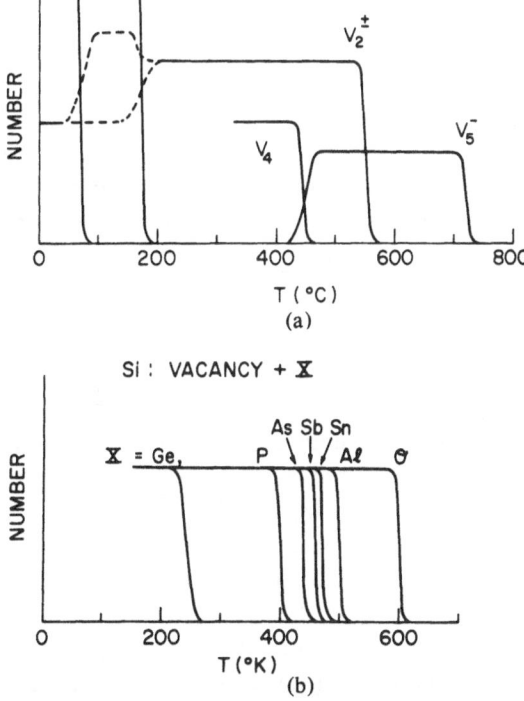

Fig. 29. Schematic representation of the recovery stages of various vacancy-related defects in silicon. (a) Vacancy and vacancy clusters; (b) vacancy + impurity defects.

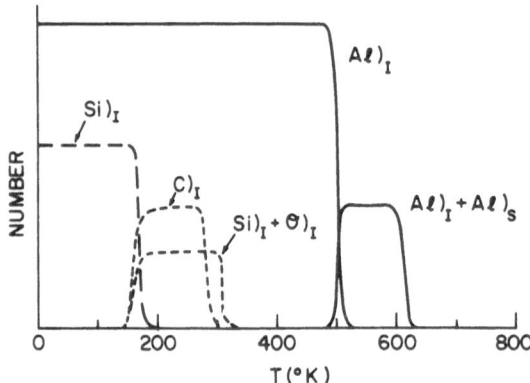

Fig. 30. Schematic representation of the various recovery stages of interstitial-related defects in silicon. The dashed lines indicate that a tentative, or indirect, identification of the defect was made.

formed and in turn decay at higher temperatures; similar pairs are observed for boron–boron and gallium–gallium in suitably doped material. In n-type material Watkins[109] observed a spectrum which he tentatively attributed to the silicon interstitial; the spectrum disappears at approximately 170°K. Other workers[132–135] have found the formation of infrared bands at this temperature which they attribute to an interstitial carbon atom and to an interstitial silicon associated with an oxygen atom, as shown in Fig. 30. No interstitial vacancy pairs have been observed in silicon.

2.3. Germanium

Not as much detailed information is available about defects in germanium as in silicon. Figure 31 summarizes schematically the various types of recovery that have been observed following a low-temperature radiation in germanium. In high-purity and lightly doped n-type and p-type germanium (depending on how the experiment is carried out) one can observe a prominent recovery stage at 65°K. The character of the recovery that is observed depends very much upon the degree of ionization and the initial irradiation; Ge is the system in which the Purdue workers discovered radiation annealing.[136] If one takes a sample which has defects that could undergo a 65°K recovery and irradiates the samples with light which has an energy greater than the band gap energy,[137] then the recovery occurs at 27°K. If one irradiates the sample instead with broad-band light which has less than

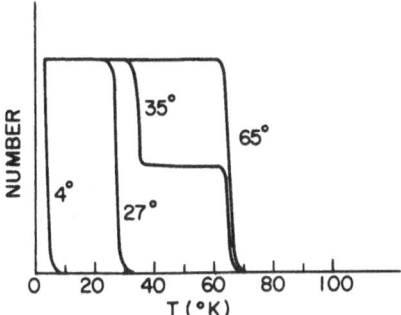

Fig. 31. Schematic representation of the low-temperature recovery stages in germanium.

the band gap energy[137],* or with high-energy (but subthreshold) electrons, [136,140] then the recovery occurs at 4°K. Curiously monochromatic light with energy less than the band gap energy produces almost *no* recovery[136] below 65°K; presumably both carriers are required. The 65°K recovery process exhibits a modest dependence of the recovery rate on the nature of the doping impurity, particularly at higher temperatures.

In more heavily doped *n*-type material, an irradiation carried out at low temperatures will result in a 35°K process[141–146] as well as the 65°K process. The amount of the 35°K recovery is proportional to the impurity concentration and clearly the 35°K process is impurity dependent.[145,146] The 65, 27, and 4°K processes are thought to be intrinsic processes. In the early days, the 65 and 27°K processes were thought to be close pairs, with the conversion between them reflecting a charge-state-dependent stability of the close pairs.[136,147] Zizine[139,148] has shown that the 65°K process exhibits a square root of time dependence which is characteristic of a diffusion process; other workers have shown that the radiation annealing at 4°K has a square root of fluence dependence[140] and the optical conversion to 4°K exhibits a $\sqrt{t}$ dependence.[149] The 27°K process also is thought to reflect a $\sqrt{t}$ dependence.[149] These intrinsic processes are attributed[150] to the long-range migration of an interstitial and its annihilation at a vacancy or its reaction at another site. We have argued[65] that the 65 and 27°K processes are the thermally activated processes for migration of two different charge states of the interstitial and the 4°K process is the athermal Bourgoin-

---

* See, however, Refs. 138 and 139. Ishino and Mitchell[138] found, with a light filtered through 1 mm of Ge, rather than the 1 cm used by Arimura and MacKay, that there was no effect of illumination below ~25°K.

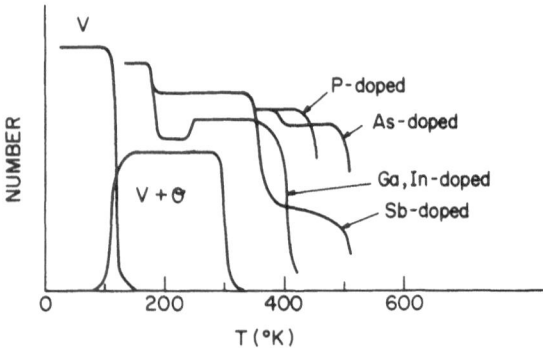

Fig. 32. Schematic representation of various high-temperature recovery stages in germanium. The only firm defect identifications are the vacancy V and the vacancy + oxygen.

mechanism migration which occurs between these two different charge states. At low bombarding energies a small portion of the damage remains after annealing to 80°K; at higher bombarding energies a larger fraction of the damage remains after annealing to 80°K; it is argued that this increase is due in part to the increased probability of production of divacancies.

The higher temperature processes are shown schematically in Fig. 32. Probably the most firmly identified defect in germanium results from infrared work by Whan.[151,152] She observed the growth at 65°K of a band at 719 cm^{-1} which subsequently decays at 120°K. She argued that this band is not an electronic transition, but the defect responsible for this band has not been identified. At 120°K she observed the growth of a band at 620 cm^{-1} which she argued very persuasively is the analog of the 12-μm band in silicon, namely that it is due to a vacancy–oxygen pair; Baldwin[153] did EPR work in support of this assignment. This work clearly indicates that vacancy motion (at least for one charge state) occurs at 120°K. The vacancy–oxygen center disappears at about room temperature. The remaining curves in Fig. 32 indicate that for various dopings various sorts of recovery are observed. Now, following from the understanding of the processes in silicon, many workers on germanium have attributed these recovery processes to vacancy-, divacancy-, and interstitial-impurity-related defects. In a number of instances on the basis of a limited number of experiments a reasonable interpretation in these terms has been achieved; however, we must note that no broad consensus has yet emerged on these defect properties.

2.4. Diamond

Diamond differs from silicon and germanium in that sample purity is very much less under control, since most experiments have of necessity been performed on natural diamonds. The work has involved optical studies, electrical studies, and EPR studies. There have now been EPR and electrical measurements following low-temperature irradiation. Lomer and Wild[154] have carried out irradiation at 17°K, while Brosious, Bourgoin, and Corbett[155] have carried out some at 80°K. Figure 33 shows the Brosious *et al.* results; the Lomer and Wild results are consistent with these data but they did not show the detailed annealing curves. Brosious *et al.* also carried out electrical resistivity measurements which correlate with the EPR results and indicate that the observed changes are not due to ionization alone but require irradiation by energetic particles and presumably therefore are due to displacement collisions. They had no EPR spectrum prior to irradiation nor was one created by ionization. They observed the growth of a line at $g = 2.0$, with the growth rate apparently following a square root of flux dependence, which they argue may indicate defect motion at liquid nitrogen temperature. They also observed the creation of a pair of lines of 60 G separation (as observed by Lomer and Wild) which grew with a linear production rate. Upon annealing they observed a new set of lines of approximately 33 G separation which grew at 130°K and disappeared at approximately 200°K. The $g = 2$ line disappears by 300°K. The Lomer–Wild and Brosious *et al.* results suggest that substantial defect motion occurs

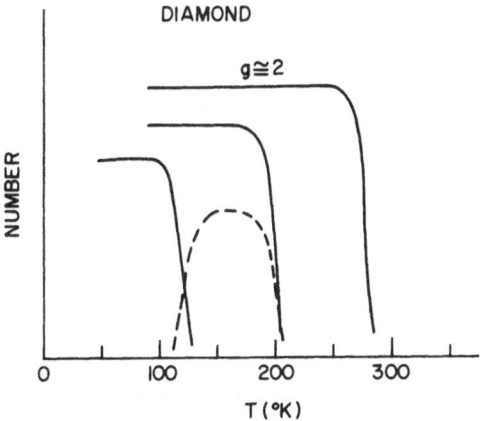

Fig. 33. Schematic representation of various low-temperature recovery stages observed in EPR studies of diamond.

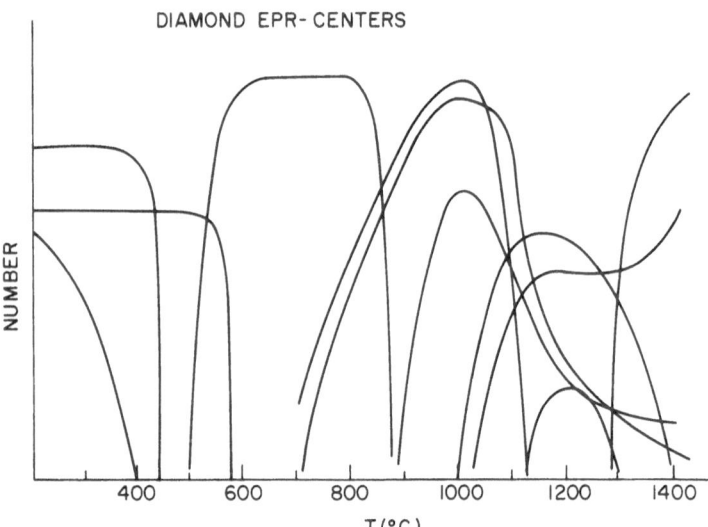

Fig. 34. Schematic representation of various high-temperature recovery stages observed in EPR studies of diamond. (After Lomer and Wild.[158])

below room temperature, but more work needs to be done to determine its nature.

The annealing behavior of a number of other EPR centers[156] is indicated in Fig. 34. None of these spectra has been firmly established as belonging to a specific defect. There have been a number of suggested configurations for some of these defects: close pairs,[157–159] the split-$\langle 100 \rangle$ interstitial,[65] multivacancy, and multivacancy–impurity complexes.[65,156] From optical studies (in particular on the GR-1 band) and theoretical work it is thought that the vacancy becomes mobile (see, e.g., Ref. 160) in the vicinity of 800°C.

The panorama of successive recovery stages is reminiscent of a similar panorama in silicon, but much more work will be needed to establish the details of specific defect processes.

2.5. III–V Compounds

The III–V compounds occur in both zinc-blende and wurtzite structures as shown in Table I. Most of the defect studies have been carried out on four compounds: GaAs, GaSb, InAs, and InSb; relatively little work has been done on other compounds.[15,161] Much has been established about the nature of the recovery, but the detailed nature of recovery processes remains

Table I. Structures of III–V Compounds[a]

	N	P	As	Sb
Al	W	Z	Z	Z
Ga	W	Z	Z	Z
In	W	Z	Z	Z

[a] Z, zinc-blende; W, wurtzite.

to be fully established. No recovery has been observed below 50°K in these materials. Figure 35 shows a typical recovery curve for these materials, in this case for InSb obtained by Eisen.[162] This pattern of recovery stages seems to be canonical in these materials, as shown[163] in Table II for other compounds; as is shown in these other compounds various authors argue that Stages III and IV overlap. As indicated in Table II the lower temperature stages have been generally attributed to close-pair processes. Often the recovery kinetics is not simple, i.e., is neither first order nor second order. Specifically the following stages do not have simple recovery kinetics: GaAs, III and V; GaSb, II, III, and V; InSb, I, II, III, IV, and V. There is strong evidence that Stage V involves long-range migration in InAs,[164,165] InSb,[163] and, in particular, GaAs,[163,166] where Jeong, Shirafuji, and Inuishi[166] argued that migration in this stage resulted in the formation of a defect complex

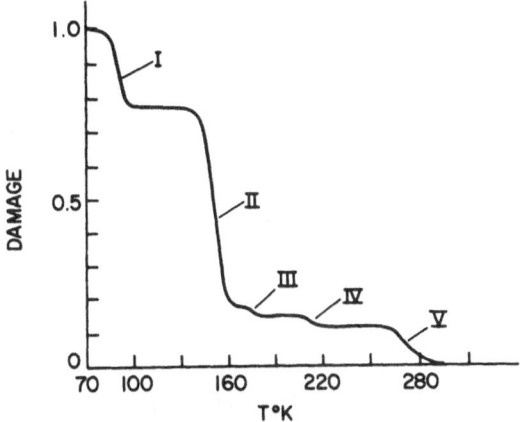

Fig. 35. Isochronal annealing in InSb showing Stages I–V. (After Eisen.[162])

Table II. Recovery Stages in III–V Compounds[a]

Stage	GaAs	GaSb	InAs	InSb
I	90[b]	120[b,c]	(100)	90[c]
II	220[b]	160[c]	(120)	150[b,c]
III	{280}	{200}	(190)	175
IV				210
V	500	360	300	270

[a] Stages III and IV are shown as overlapping in all but InSb. The stages in parentheses in InAs are reverse annealing stages.
[b] Attributed to close pairs.
[c] Charge-state-dependent recovery.

involving an As vacancy with a Si on an As site. Long-range motion in at least some of the lower temperature stages represents a real possibility, since even first-order kinetics need not imply close-pair recovery, i.e., it can arise from the long-range migration of a defect in the presence of a dominant trap, as is well recognized by the workers in this field. We note that in Table II a number of the recovery processes are charge-state dependent, i.e., the recovery process varies with doping impurity. If the charge-state dependence really reflects a dependence upon the trapping by the doping impurity, then it is reminiscent[146] of the impurity dependence observed in the 65°K process in germanium, which is due to long-range migration. If the charge-state dependence reflects a change of the apparent recovery energy with Fermi level, which is favored in the literature, it is reminiscent of the long-range migration in silicon and germanium via the Bourgoin mechanism. In this regard we note that InSb exhibits an anomalous damage creation[163] by X-rays; we have suggested[65] that this damage is due to ionization-enhanced diffusion of impurities; independently Mashovets, Vikhlii, and Vitovskii[167] observed ion pairing in p-type InSb induced by X-rays, which they argued is due to the radiation-induced migration of germanium atoms and donor–acceptor ion pairing.

These compounds can also have stoichiometric defects, the concentration of which can vary with heat treatment. Blanc, Bube, and Weisberg[168] observed the introduction of lattice defects by heat treatment. Further, Goldstein and Almeleh[169] found that heat treatment introduced a strong EPR spectrum which they speculated might be a Ga vacancy, an As vacancy, or an As atom on a Ga site with a neighboring Ga interstitial. Combined thermal treatment and irradiation studies therefore loom as a very powerful tool in studying these materials.

In summary, then, consideration of the results in III–V compounds in the light of our understanding in silicon and germanium shows that in the compound materials not only can we observe all the complexities observed in the elemental materials but that the problems may be further compounded.

2.6. II–VI Compounds

The II–VI compounds usually occur in either the zinc-blende or wurtzite structure as shown in Table III. Most of the work in these materials has been in CdS, CdTe, ZnSe, and ZnTe; relatively little work has been done in the other materials.[15,170] Watkins[115] has recently reviewed work in these compounds and has emphasized the importance of stoichiometry defects in these compounds. There have been a number of irradiations of these compounds at low temperatures but a common annealing pattern has not emerged. Recovery stages as low as 60°K have been observed with complex annealing behavior.

In the II–VI compounds a number of defects have been identified using EPR. The chalcogen vacancy has been reported in ZnS,[171] ZnO,[172,173] and BeO[174]; it is argued that these defects are in the single positive charge state. It is found that they have the full symmetry of the lattice, i.e., have no distortion, and that they are stable well above room temperature. Using EPR, the metal vacancy has been observed in ZnSe,[115] ZnS,[60] ZnO,[175,176] BeO,[177,178] and CdS[179]; some of these defects are in the single negative charge state, while others are in a neutral charge state and are observed as a spin-1 center. Indications[65] are that this vacancy is also stable at room

Table III. Structure of II–VI Compounds[a]

	O	S	Se	Te
Be	W	Z	Z	Z
Zn	W	Z,W	Z,W	Z
Cd	NaCl	Z,W	Z,W	Z

[a] Z, zinc-blende; W, wurtzite; NaCl, sodium chloride. Some compounds occur in either of two forms.

temperature. No direct measurements on interstitials have been made but it has been argued[115] that interstitial migration occurs at low temperatures, e.g., 60°K. These defect assignments seem consistent with those proposed on the basis of other measurements.

We should also note that in these materials there are indications that ionization-enhanced diffusion occurs. Kaneev[180] studied radiation-stimulated diffusion in CdS, arguing that he observed the diffusion of Cd interstitials migrating under X- or gamma-irradiation. Kulp and Kelley[181] found that subthreshold electron irradiation caused the long-distance diffusion of sulfur atoms which they argued occurred in the interstitial form. We note that the neutral charge state of a group VI atom has the electronic configuration ns^2, np^4; in these compounds the large electron affinity of the group VI atoms can cause them to become negatively charged. A double negatively charged group VI atom would have the electron configurations ns^2, np^6, i.e., the closed shell configuration of a noble gas atom. We argue that should such a charge state exist in the semiconductor such an atom would be a likely candidate for a nonbonded interstitial in the lattice and, we would argue, may have a very low migration energy.

2.7. Other Materials

While a number of other semiconducting materials could be considered, in most cases there have been very few radiation damage studies carried out on them. In this section we briefly discuss results on three materials of interest to the topic of this review, SiC, Te, and graphite.

Silicon carbide crystallizes in various structures called polytypes. Two of the more common polytypes are the cubic, or zinc-blende, structure and the $6H$ polytype which has a hexagonal structure. There have not been threshold measurements made on these materials; consequently, we will not consider them in the subsequent sections. We note, however, that the state of theory is approaching the point where it should be able to treat these compounds, that a variety of measurements have been made which provide a background for further study,[15] and that a number of EPR measurements have been made on silicon carbide with a variety of identified centers. For example, Balona and Loubser[182] studied the EPR centers in room-temperature, electron- or neutron-irradiated material and found seven different radiation-induced spectra. Three of the spectra they tentatively ascribed to (1) the uncharged carbon vacancy, (2) the negatively charged carbon divacancy, and (3) the uncharged silicon vacancy trapped at a boron site. We recognize that silicon carbide can exhibit very complex annealing, including low-temperature annealing, but we conclude that the prospect for studies in silicon carbide are quite promising.

Tellurium has a structure which can be visualized as made up of parallel chains of helices of atoms, there being three atoms per repeat unit in the chain. The material is difficult to handle but lends itself to a variety of sophisticated measurements. Threshold measurements have not been made on this material either; consequently, it will not be considered in subsequent sections. We note, however, that workers (e.g., Refs. 183 and 184) have carried out irradiations at 10°K, with subsequent annealing revealing a recovery stage at 50°K, that Kronmüller, Jaumann, and Seiler[185] made measurements on quenched tellurium, and that Weigel[186] is currently carrying out extended Hückel theory calculations on defects in tellurium. We conclude that prospects for progress in tellurium look quite good.

Graphite is not a semiconductor. It is, however, the stable form of carbon at room temperature, and consequently in subsequent sections we make comparisons between results on graphite and those on diamond. There has been a substantial amount of work on the irradiation of graphite. (For a review of earlier work see Corbett.[15]) The first low-temperature (below 10°K) irradiation experiment on graphite was carried out by Austerman and Hove,[187] using 1.25-MeV electrons. They showed that there was practically no recovery in the damage below 80°K. Other workers[188,189] also concluded that there was not a major amount of recovery between 10 and 80°K. On the other hand, Iwata, Fujita, and Suzuki[84] calculated the migration energy of a single interstitial atom and obtained 0.016 eV. Iwata and Suzuki[190] proposed that the single interstitial atom would be able to easily migrate below liquid nitrogen temperatures. Thrower and Loader,[85] following the method of calculation of Coulson et al.,[82] obtained a migration energy of a single interstitial of 0.02 eV, i.e., in agreement with Iwata and co-workers.

The theoretical treatment makes a natural assumption concerning the form of the interstitial, but we feel we must enter a word of warning. Graphite has a more open lattice than diamond. The graphite lattice has a layered structure, the layer planes being made up of six-membered rings of carbon atoms, the bonds in the rings involving sp^2 orbitals. The bonding between layers is weak, being due to the p orbitals perpendicular to the layers. There have been a number of calculations of defect properties in graphite, particularly of the vacancy,[79,81-83] with a few concerning the interstitial.[80,84,85] The form of the interstitial assumed is a nonbonding interstitial between the layer of planes. We argue that that assumption is very similar to the assumption of the tetrahedral interstitial in the diamond lattice.

A warning then is that a bonded interstitial, i.e., an interstitialcy type configuration, may well occur in the graphite lattice and may have a very high mobility. Calculations on this point are now being carried out by Bailey.[191]

Bochirol and Bonjour[192] found a stored energy release peak at approximately 70°K. Iwata, Nihira, and Matsuo,[193] in a preliminary report, found two annealing stages below 90°K. In their more complete study[194] they reported four substages: I_A (5–15°K); I_B (15–45°K); I_C (45–65°K); I_D (65–85°K). Their electron irradiations were carried out with the sample temperature below 5°K. They studied the recovery for 0.2- and 0.8-MeV irradiations, both being carried out with the initial beam direction along the c axis. They also studied the effect of radiation doping and of pre-heat-treatment of their pyrolytic graphite. Substages I_A and I_B are the larger substages. They argued that in stage I the interstitial could not recombine with the vacancy, and that interstitials could not form diinterstitials due to the electronic and elastic repulsion between the defects. They concluded that substages I_A and I_B were caused by the correlated rearrangement of close interstitial–vacancy pairs, where the migrating interstitial atom forms a loose coupling with its own vacancy, but does not recombine. They also argued that long-range migration of the interstitial occurs in I_C, with I_C and I_D being due to the formation of loosely coupled clusters of interstitial–interstitials and interstitial–vacancies.

Austerman and Hove[187] observed the I_D annealing and another stage which extends from approximately 170 to 300°K. There is substantial recovery in this stage, but no consensus as to what defect is mobile. The earlier workers had attributed it to interstitial migration. Montet[195] and Montet and Myers[196] used the electron microscope to study defect production in room-temperature irradiated natural graphite crystals and observed vacancy clusters. This argues strongly that vacancy migration occurs below room temperature and suggests that the small recovery peak at approximately 300°C observed by Tsuzuku and Arai[197] is due to the disappearance of interstitial clusters, while the large recovery peak at approximately 1300°C is due to the dissociation of vacancy clusters.

3. DAMAGE PHENOMENOLOGY

3.1. Displacement Damage

We will discuss here the phenomenology used in describing displacement damage. In a sense this whole chapter relates to the discussion of displacement damage and so we will return a number of times to the general question. Here we will discuss the simple phenomenology, which will be useful in assessing the importance of the interaction of particles with semiconductors in Section 4. We will return in detail to the specific examination of displacement damage mechanisms when we consider the experimental data on given systems and the corresponding theory which has arisen in discussing those data in Section 7.

Seitz[198] used the concept of an isotropic square-well potential to describe the binding of an atom to its substitutional site. He argued that the displacement of an atom from its substitutional site as a result of recoil energy received from a nuclear collision with an energetic particle would be a highly nonadiabatic process. He estimated that the binding energy would be approximately four times that for the sublimation of the material. This binding energy is then referred to as the threshold energy since a recoil energy greater than that binding energy is necessary to create damage. This threshold displacement energy, according to his estimate, is approximately 25 eV for most materials.

The isotropic square-well approximation then translates itself into a probability for damage creation as shown in Fig. 36a; the probability rises from zero to unity abruptly at the threshold energy T_d. More complicated probability functions have been considered, such as that shown in Fig. 36b, where the probability rises linearly from zero to unity over an energy range of 2Δ. Corbett et al.[200] showed that for a displacement probability such as that shown in Fig. 36b, including those where the deviations from linearity are not too marked, and for energies significantly above threshold the displacement process can be described by an effective square-well threshold energy which characterizes the energy at which the probability function is 1/2.

More complicated displacement probabilities have been considered. [199–207] The more complicated displacement probabilities were introduced in an effort to fit experimental data better. Of course, as we shall discuss in more detail later, the threshold energy is not an isotropic square-well potential, but can vary with orientation in the crystal, as is characteristic of the surroundings of a substitutional atom. Recent years have seen a decline in the use of these arbitrary phenomenological displacement probabilities and the increasing use of direct calculations attempting to describe the displacement process in terms of interatomic collisions.

The displacement probabilities shown in Fig. 36 are appropriate for single displacement processes; that is, they show that there will be only one atom displaced regardless of how large the recoil energy is above the threshold energy. Physically we know that the recoiling energetic atom,

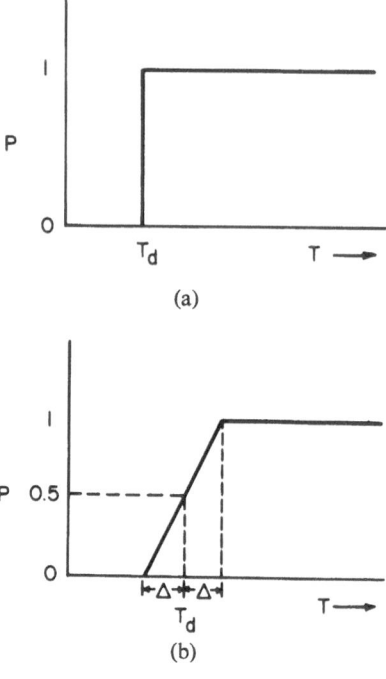

Fig. 36. Displacement damage probability P versus atom recoil energy T. (a) Isotropic well. (b) Linear probability rising from zero at $T_d - \Delta$ to unity at $T_d + \Delta$.

often referred to as the primary knock-on atom (PKA), can itself, if it has sufficient energy, displace other atoms by interatomic collisions. The early theories[1,199,201,208–213] of the phenomenology of multiple-atom displacement were statistical in nature, but more recent work has included the direct modeling of the displacement damage process on computers, as we discuss in more detail in Section 7. The early work has been reviewed recently by Robinson.[214] The mathematical approaches of these early theories differ somewhat, but the physical assumptions fall into two distinct types. The models of Snyder and Neufeld[209,211,212] and Harrison and Seitz[210] assume that the binding energy (the threshold displacement energy) is lost when the atom is displaced; the primary knock-on atom consequently recoils with a kinetic energy $T - T_d$ for T greater than T_d; the PKA and its progeny continue to produce displacements until their energies fall below T_d. In the model of Kinchin and Pease[1] the displacement threshold energy is not lost to the lattice when the atom recoils; the PKA recoils with the full kinetic

energy T if this energy exceeds T_d. Kinchin and Pease introduced and included replacement collisions; that is, if the PKA imparts an energy greater
than T_d to another lattice atom and retains less kinetic energy than T_d,
then the secondary atom leaves its lattice site and the primary knock-on
atom is considered to have replaced that atom on its lattice site. The result
of the replacement collision is that no new net displacements are created
when the recoil energy falls below $2T_d$. Subsequently Neufeld and Snyder
included replacement collisions in their model.

In all of these models it is assumed that the atomic scattering is
described by hard-sphere scattering, i.e., there will be an equal probability
of all partitions of the energy between the atoms after the collision. This
permitted Robinson to give the following approximate value of $g(T)$, the
mean number of displacements produced by a recoil energy T. Let the
energy which is lost when an atom is ejected from its lattice site be B and
let C be the energy at which the multiplication of displacements begins.
He then obtained the equation

$$g(T) = (T + B)/(C + B), \qquad T > C \tag{1}$$

and he obtained the following results for the various cases:

Snyder–Neufeld and Harrison–Seitz assume $B = C = T_d$, giving

$$g(T) = (T + T_d)/2T_d, \qquad T > T_d \tag{2}$$

Neufeld and Snyder assume $B = T_d$ and $C = 2T_d$, giving

$$g(T) = (T + T_d)/3T_d, \qquad T > 2T_d \tag{3}$$

Kinchin and Pease assume $B = 0$ and $C = 2T_d$, giving

$$g(T) = T/2T_d, \qquad T > 2T_d \tag{4}$$

Equations (2) and (3) are in fact approximate equations which are valid
for high recoil energies. Figure 37 shows the dependence of $g(T)$ on T
for these various cases and including the more accurate treatment obtained
by Snyder and Neufeld and by Harrison and Seitz for their case. As was
observed by these early workers, the Snyder–Neufeld and Harrison–Seitz
treatment differs little from the Kinchin–Pease treatment; that is, the inclusion of a binding energy and the omission of the replacement collisions
tend to compensate one another.

Figure 37 also shows results obtained by Fein[201] in which he assumes
that an already displaced recoiled atom can displace other atoms which

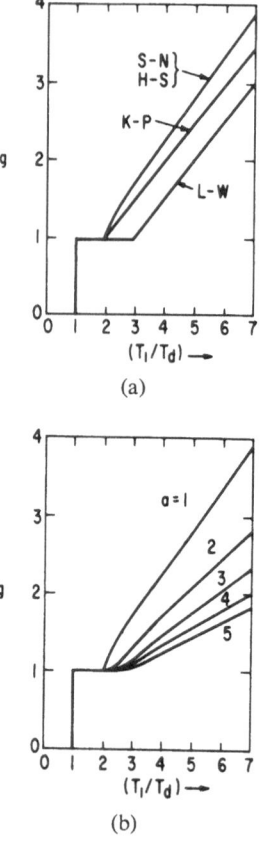

Fig. 37. Total displacement probability g, i.e., the probability of primary and secondary displacement, versus initial primary atom recoil energy T_1. (a) g for a step-function displacement probability for the Snyder–Neufeld[209,211,212] and Harrison–Seitz[210] (S-N, H-S), Kinchin–Pease[1] (K-P), and Lucasson and Walker[203] (L-W) models. (b) g for the displacement probability shown in Fig. 36(b) as obtained by Fein.[201]

are bound with a displacement probability such as that shown in Fig. 36b; we note here that his assumption that the primary recoil atom is already displaced means that $g(T)$ for $T < 2T_d$ does *not* reflect the fact that the displacement of the primary atom follows the displacement probability shown in Fig. 36.

These considerations indicate that the primary knock-on atom can have a number of energetic progeny which in turn can cause displacement damage, as was originally discussed by Brinkman,[215,216] who termed the whole collection of damage a displacement spike. As Seeger[2] has discussed, the nature of the damage process is such that the high-energy recoil atoms move off into the lattice, ending up as interstitials and leaving vacancies behind. The vacancies are created primarily along the trail of the higher energy recoil atoms, particularly the primary knock-on atom, with the consequence that there is a zone which is depleted in atoms, surrounded by a shell enriched in interstitial atoms. The actual physical extent of these depleted zones and the displacement spikes are subjects which have been examined in recent computer studies.*

We have discussed how the displacement damage phenomenology depends upon the incident energy. It also depends upon the sample temperature and upon the presence of other defects in the sample. Both effects are in general small. Of course, temperature, when it activates migration of defects, can play a major role in characterizing the nature of the defects observed. We refer here instead to the contribution of the thermal vibrations to the displacement process.

Brown and Augustyniak[202] were apparently the first to appreciate that the thermal vibration of the atoms in the crystal could contribute to the recoil energy of an atom and hence to its displacement probability. In fact, what needs to be considered is the momentum of the lattice atom, not the energy. The total recoil momentum of an atom after a collision is the sum of its vibrational momentum and the momentum received in the collision. If these momenta are parallel, the corresponding energies are then related by

$$T = T_c + 2(T_c T_v)^{1/2} + T_v \tag{5}$$

where T is the total recoil energy, T_C is the collision recoil energy, and T_V is the vibrational energy. The last term in Eq. (5) can usually be ignored, since, for example, at room temperature it is approximately $1/40$ eV, which is small in comparison to the many-electron-volt value of T_C. But the cross term cannot be ignored, in particular, near threshold.

Iwata and Nihira[217] explicitly employed the Debye model of lattice vibrations in treating the anisotropy of the threshold in graphite. In their experiments (see Section 5.4) they observed a temperature-dependent threshold energy, in addition to the contribution due to thermal lattice vibrations. They argued that the vibrations of the surrounding atoms could

---

* See Ref. 214 for a discussion of this work.

impede the displacement process, effectively *raising* the threshold energy. In view of their experimental results that argument seems irrefutable, but it strikes the present authors that their conclusion may not be completely general, i.e., there may be lattices in which thermal vibrations reduce the threshold energy. Their conclusions are summarized in Fig. 38, which shows the displacement versus energy with temperature dependence.

The presence of other defects can also alter the displacement damage process. A notable example of this occurred in the early studies of radiation damage in semiconductors. At that time it was assumed that the damage observed at room temperature was intrinsic, i.e., the *immobile*, isolated vacancy and interstitial. The damage production rate was studied as a function of impurity concentration, with the view to studying the electronic energy level structure of the intrinsic defects. It was found that the carrier removal rate rose dramatically when the Fermi level approached either the conduction band or the valence band. It was assumed that this dramatic change in carrier removal rate reflected energy levels of the defects near the band edges, but we now know that the defects were in fact *mobile* and that the effect of the impurities was to enhance the trapping, and retention, of the damage. This does not mean, however, that a Fermi level dependence cannot exist; as we will see in subsequent sections, the damage mechanism is Fermi level dependent, i.e., it does depend upon the presence of other defects in the crystal which both establish the Fermi level and alter it as new defects are introduced.

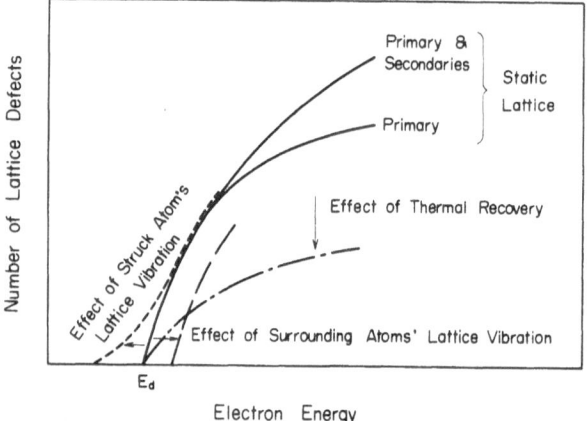

Fig. 38. Schematic representation by Iwata and Nihara[217] of the total number of defects versus electron energy with the inclusion of various effects.

The presence of defects can also alter the nature of the damage production process by interrupting channeling, and, in the case of some lattices, by interrupting collision chains. (See Sosin and Bauer[21] for discussion of these effects.) Dramatic effects occur in semiconductors with the production of damage at "soft spots," i.e., subthreshold events which we will discuss in more detail in Section 6.

Finally the presence of defects in a sample can alter the damage production rate by radiation-induced recombination. This is a process in which either an unstable defect configuration is caused to undergo recovery by a nearby collision with the bombarding particles and the lattice atoms, or the same collision causes an enhanced migration of a defect with its consequent recovery. These processes have not been observed to occur in semiconductors, but are well documented in metals and may eventually be observed in semiconductors. They are discussed fully by Wollenberger[218] and Sosin and Bauer.[21]

3.2. Thermal Spikes

A high-energy particle interacting with a condensed medium will deposit a great deal of energy in a localized volume. This energy can result in a much increased lattice temperature which then decays by thermal conduction. Such a pulse of localized temperature is called a *thermal spike*. The release of energy can arise from either ionization loss of the energetic particle[219] or from direct displacement collisions. The amount of energy which can be released locally is quite substantial. For example, bombardment of silicon with a 1-MeV proton can impart over 0.1 MeV to a lattice atom, which in its recoil will be stopped in a very short distance, liberating all of this energy in the lattice. For silicon with a melting temperature T_m of approximately $1700°K$ (with $3kT_m = 0.4$ eV) the release energy is more than sufficient to melt the lattice locally. The cascade of displacement collisions due to a high-energy recoil can be treated as a *displacement spike*. The thermal energy release should be treated separately since it makes a somewhat distinct contribution, and can occur when a displacement spike does not occur. The thermal energy can contribute to damage (or a change) of the material by causing a phase change (from a less stable phase to a more stable phase, or vice versa), by causing the material to order or disorder, by causing the dissociation or association of complexes, and by facilitating the diffusion of entities in the material.

Seitz and Brooks did substantial work on the theory of thermal spikes (see Ref. 3). The treatment considered two distinct geometries, a spherical spike and a cylindrical spike. They assumed that an amount of thermal energy Q is released at a point in an isotropic medium at time $t = 0$ and that the diffusion of heat can be described by a constant diffusivity K ($K = \varkappa/C\varrho$, where $\varkappa$ is the thermal conductivity, C is the specific heat, and ϱ is the mass density). For times t not too close to $t = 0$ and for not too small distances r from the origin, the temperature T for the spherical spike is given by

$$T(r, t) = \frac{Q}{8\pi^{3/2}C\varrho(Kt)^{3/2}} \exp\left(-\frac{r^2}{4Kt}\right) \tag{6}$$

(The limitations on time and distance simply recognize the discreteness of the medium and the process.) We know now from dynamic collision calculations on model lattices that, particularly in close-packed lattices,[220] direct collision sequences can funnel a great deal of energy away from the origin, but Eq. (5) will satisfactorily describe the diffusion of the remaining energy. As Seitz and Koehler[3] have discussed, the thermal spike persists only $\sim 10^{-11}$ sec, i.e., of the order of 100 lattice vibration periods. They argued that for a rate process in which the jump frequency ν is given by

$$\nu = \nu_0 \exp(-E_m/kT) \tag{7}$$

where E_m is the activation energy and ν_0 is the frequency coefficient ($\nu_0 \sim kT/h$, i.e., $\sim 10^{13}$ sec^{-1}), the number of jumps n_j which this process makes in the central zone of the spike is given by

$$n_j = \int_0^\infty \nu_0 \exp[-E_m/kT(t)]\, dt \tag{8}$$

They evaluated the integral, and making some approximations, arrived at

$$n_j = 0.093\,\xi(Q/E_m)^{2/3} \tag{9}$$

where ξ is a dimensionless quantity which gives the ratio of the atomic frequency factor ν_0 and the frequency with which translational energy is transferred from moving atom to moving atom; presumably this ratio will lie near unity for thermal processes. Seitz and Koehler discussed in detail the number of transitions per atom typically made in a thermal spike and the relationship of this to disordering processes.

Spike-type damaged regions are observed in semiconductors, but the evidence at present seems to indicate that they are due to displacement spikes. Thermal spikes can contribute to diffusion processes and consequently are of great interest in the consideration of damage creation processes.

3.3. Ionization Phenomenology

As we will discuss in Section 4, there are a number of ways that energetic particles can produce ionization in semiconductors and thus alter the properties of the semiconductors by creating (sometimes energetic) electrons, holes, and ions. We note that the act of ionization can create energetic electrons which in turn slow down in the medium primarily by ionization, but with an occasional displacement collision from a scattering by a nucleus. This damage creation by high-energy secondary electrons occurs with a probability reduced by the cross section for creation of the secondary electron. The essential physics is described in Section 2.1 and consequently will not occupy us further here.

Energetic charged particles lose energy in the medium primarily by ionization and in some cases can leave a more or less continuous trail of ionization, sometimes called an *ionization spike*. We note that Fleischer, Price, and Walker[221] argued that in ionic materials this trail of ionization causes the knocking out of electrons, leaving behind a column of positive ions which experience a very strong mutual repulsion; this mutual repulsion, they argued, like the Varley mechanism would cause the ejection of a number of the positive ions, thereby resulting in a damaged region which they call the *ion-explosion spike*. The same sort of thing can occur in semiconductors, but, we argue, is unlikely to create damage for the following reason. Consider a rare event in which two nearest-neighbor, say, silicon atoms are ionized by a single ionizing particle, yielding two Si^+ ions at nearest-neighbor sites. True, they would experience a substantial Coulombic repulsion, but in the semiconductor there is a high probability of the *holes being repelled* from one another, leaving the neutralized silicon atoms undisplaced, i.e., no damage production.

As Sonder and Sibley[222] have discussed in Vol. 1 of this Treatise there is a variety of ways that the act of ionization itself is thought to create Frenkel defects in ionic materials. Briefly, there are two prominent mechanisms: the Varley mechanism[223–225] and what we will call the X_2^- mechanism.[226–230] Varley argued that the ionization might convert a normally negative halogen ion into a positive halogen ion by a double-ionization process (see Fig. 39), at which point the positive halogen finds itself surrounded by positive metal ions; i.e., it is at a potential maximum rather than a minimum, and can be ejected. As discussed by Sonder and Sibley,[222]

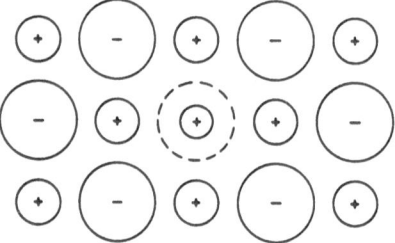

Fig. 39. Configuration just prior to atom displacement in the Varley[223–225] model. The central atom, which was negatively charged, has become positively charged by a double-ionization event.

the present thinking is that the Varley mechanism does not account for Frenkel pair production in alkali halides, and consequently, we expect it will be of even less importance in the semiconductors, including those which are quite ionic. The X_2^- mechanism occurs because in a number of ionic crystals a hole is self-trapped, creating an X_2^- molecule (instead of the normal state, X_2^{2-}) with a very substantial relaxation along the X–X distance, as is shown schematically in Fig. 40. Viewing the X_2^- state as an excited state, we can show in Fig. 41 the potential energy for the two states X_2^- and X_2^{2-} versus distortion along their internuclear distance. The mechanism assumes that a nonradiative electron capture can take place causing a transition from X_2^- to X_2^{2-} and that the system relaxes nonradiatively as shown in Fig. 41; this relaxation results in a linear impulse of momentum along the $\langle 110 \rangle$ axis, and it is argued that this impulse can have sufficient energy to create a Frenkel pair. As Sonder and Sibley discussed, there is

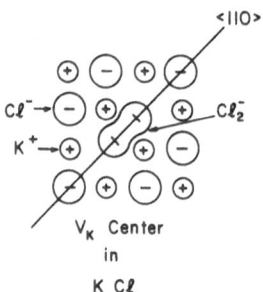

Fig. 40. Configuration prior to the displacement event in the X_2^- mechanism, in this case shown for KCl.

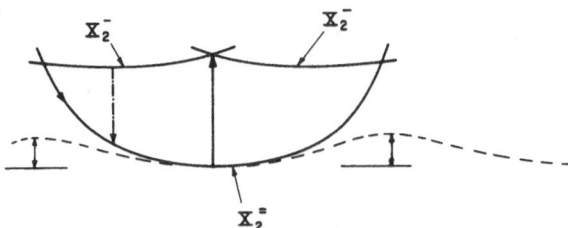

Fig. 41. Potential energy diagram versus configuration co-ordinate for the X_2^{2-} and X_2^- states. The potential energy barrier for displacement is schematically indicated by the dashed curve.

substantial evidence in favor of this mechanism for the creation of Frenkel pairs in alkali halides. We see, however, that this mechanism depends on the detailed properties of the alkali halide lattice and defects contained therein. A comparable mechanism has not been proposed for any semiconductor system.

The preponderance of the evidence is that ionization alone does not create Frenkel pairs in semiconductors. As we will see, however, there are a few instances of experimental results which suggest that ionization can, in some semiconductors, create damage. We will return to this question in Section 6.

Ionization can, in addition to altering the electrical conductivity, change the defect structure in the material. It can, for example, change the charge state of preexisting complexes of defects, possibly leading to the dissociation of the complex and the diffusing away of the trapped defect. Alternatively, the presence of ionization can cause the redistribution of defects, including the diffusion of impurities into the bulk of the sample from the surface, by means of ionization-enhanced diffusion. Bourgoin, Corbett, and Frisch[231] have recently analyzed several enhanced diffusion mechanisms. These mechanisms are quite important in defect studies.

Bourgoin and Corbett[232] have reviewed a number of enhanced-diffusion mechanisms: the normal ionization-enhanced diffusion mechanism, and the Bourgoin, energy release, and recoil mechanisms. (They pointed out that conduction processes can give rise to electromigration* and impact ionization† processes as well but these processes will not concern us here.) The normal ionization-enhanced diffusion process occurs when the potential energy for the relevant charge state is as is shown in Fig. 42, i.e., when the

* For a recent review see Ref. 233.
† For recent reviews see Refs. 234.

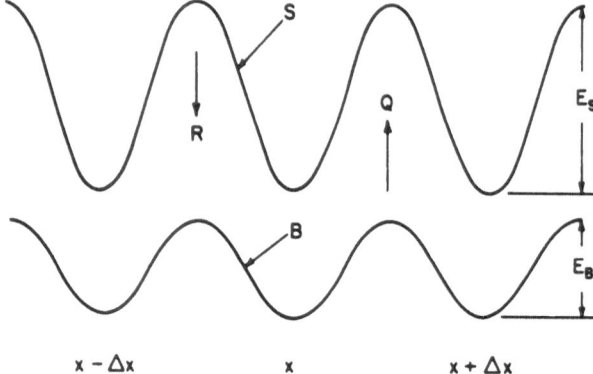

Fig. 42. Potential energy curve versus distance through the lattice for two channels S and B for the normal ionization-enhanced diffusion mechanism. Q and R are the transition rates between the two channels and the corresponding migration energies E_S and E_B are shown.

potential minimum position in the lattice is the same for both charge states (S and B). In this case ionization has the result of changing a defect from a relatively immobile charge state to a more mobile charge state, thereby facilitating the diffusion. In both charge states for normal ionization-enhanced diffusion the diffusion occurs by a thermally activated process. The remaining enhanced diffusion mechanisms are *athermal*, i.e., do not require an athermal activation event for migration.

We can assume without loss of generality that the state B has one more electron than state S. As shown in Fig. 43, if the diffusing defect has an energy level in the forbidden gap at the position E_T, then the charge state can change between S and B by capturing an electron e^- with a capture rate given by k_e, by emitting an electron with a generation rate constant g_e, by capturing (k_h) a hole e^+ or emitting (g_h) a hole, as is summarized in the equations

$$S + e^- \underset{g_e}{\overset{k_e}{\rightleftarrows}} B \tag{10}$$

$$B + e^+ \underset{g_h}{\overset{k_h}{\rightleftarrows}} S \tag{11}$$

The probability per unit time Q for going from state of ionization B to S and the probability per unit time R for going from S to B are given by[231,232]

$$Q = k_h + g_e \tag{12a}$$

$$R = k_e + g_h \tag{12b}$$

In terms of the usual semiconductor notation for capture and generation rates these quantities can be expressed as

$$k_e = \sigma^n v_n n \tag{13a}$$

$$k_h = \sigma^p v_p p \tag{13b}$$

$$g_e = \sigma^n v_n N_C \exp[-(E_C - E_T)/kT] \tag{14a}$$

$$g_h = \sigma^p v_p N_V \exp[-(E_T - E_V)/kT] \tag{14b}$$

with σ^n and σ^p the cross sections for electron and hole capture; v_n and v_p the electron and hole velocities; and N_C and N_V the density states in the conduction and valence bands, respectively. The concentration s in the S channel and the concentration b in the B channel are given by the equations

$$\frac{\partial s}{\partial t} = D_S \frac{\partial^2 s}{\partial x^2} - Rs + Qb \tag{15}$$

$$\frac{\partial b}{\partial t} = D_B \frac{\partial^2 b}{\partial x^2} + Rs - Qb \tag{16}$$

which are the equations for dissociative diffusion.[235,236] Unless $Rs = Qb$, these equations lead to non-Fickian diffusion, due to the reaction terms. If the two channels are in thermal equilibrium ($Rs = Qb$), then the same

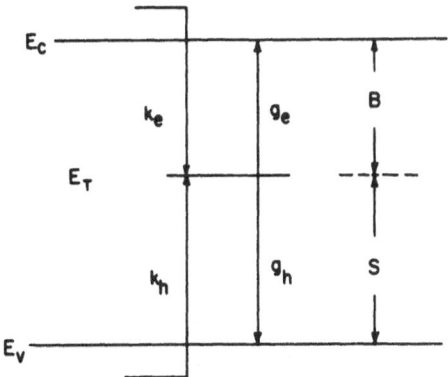

Fig. 43. Energy level E_T for the transition between state B and state S. The transition for the liberation g_e of an electron to the conduction band E_C and of a hole g_h to the valence band E_V, and the capture of an electron k_e and the capture of a hole k_h, are as shown.

symmetrized effective diffusion coefficient D_e is used in Fick's diffusion equation whether we monitor just one channel (either channel) or both channels simultaneously:

$$\partial s/\partial t = D_e \, \partial^2 s/\partial x^2 \tag{17}$$

with

$$D_e = (D_S Q + D_B R)/(Q + R) \tag{18}$$

The Bourgoin mechanism depends upon the potential energy for the two channels being different in character, as shown in Fig. 44 for a one-dimensional case. For the Bourgoin mechanism to occur, the potential energy minimum for the B system must occur at the saddle-point position for the S system and vice versa. Mass transport by the Bourgoin mechanism then occurs by successive charge-state changes between the S and B systems, without any thermally activated process occurring. For example, if a defect finds itself in the charge state S at position x, a charge-state change carries it to charge state B, at which point it is at a saddle point in energy and can relax, say, to a potential energy minimum $(x + \Delta x/2)$. A subsequent charge change to state S again results in the defect finding itself again at a potential energy saddle point; it can relax, say, to position $x + \Delta x$. Thus by cycling from charge state S to B and back to S, the defect has moved from the potential energy minimum at x to the potential energy minimum at $x + \Delta x$. As we have noted,[231,232] Weiser's theory[108] of the interstitial in the diamond

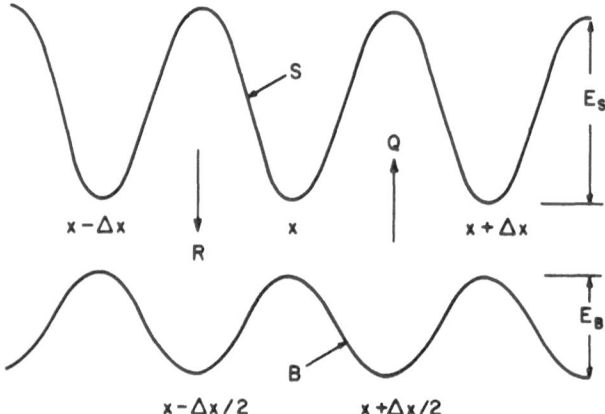

Fig. 44. Potential energy curve versus distance in the lattice for two channels S and B for the Bourgoin mechanism. Q and R are the transition rates between the channels. The thermally activated migration energies E_S and E_B are shown.

lattice implies that impurity interstitial diffusion via the tetrahedral and hexagonal interstitial sites will occur via the Bourgoin mechanism; further, the extended Hückel theory of the intrinsic interstitial in the diamond lattice[64,67] suggests that diffusion of the intrinsic interstitial can occur via the Bourgoin mechanism. We note as well that vacancy diffusion might take place via the Bourgoin mechanism in the diamond lattice. The vacancy in the diamond lattice is centered at the substitutional site (see Fig. 1) for the V^- and the V^+ charge states, but we argued in Section 2 that the V^{2-} charge state may have a split-vacancy equilibrium configuration (see Fig. 2), i.e., the equilibrium configuration for the V^{2-} occurs at the saddle point for the V^- charge states and vice versa, which are the conditions for the Bourgoin mechanism. The diffusion equations for the Bourgoin mechanism are more complicated than the normal ionization enhanced diffusion mechanism, but we have shown[231,232] that in the case of strict local equilibrium, a symmetrized diffusion coefficient D_e can be defined for either or both channels

$$\partial s/\partial t = D_e \, \partial^2 s/\partial x^2 \tag{19}$$

with

$$D_e = \frac{(D_S + D_R)Q + (D_B + D_Q)R}{Q + R} \tag{20}$$

where

$$D_R = Ra^2/8, \qquad D_Q = Qa^2/8 \tag{21}$$

and a is the normal diffusive jump distance and R and Q are as defined earlier.

We have also noted[232] that a localized release of energy can enhance diffusion. The localized release of energy can occur, like the X_2^- mechanism in the alkali halides as we discussed earlier, or may result from the nonradiative release of energy in an electronic transition. Lax[237] has argued that the large capture cross sections which occur in silicon and germanium result from the capture of electrons or holes into giant-orbit, excited states which then decay to the ground state via a cascade of low-energy phonons. Thus this capture of a charge carrier results in the localized release of phonons, i.e., heat. We have pointed out that this localized release of heat is like a thermal spike, and that it can be treated by Eqs. (6)–(9) with the recognition that these equations provide only a rough estimate of the effectiveness of the localized energy release in enhancing diffusion. If k_E is the rate per unit time at which the energy release occurs, then the probability per unit time that this energy results in a jump is given by $P_E = k_E n_j$

and we can use the ordinary diffusion equations with a modified diffusion coefficient, say, for the S channel given by

$$D_S' = (P_S + P_E)a^2/2 \qquad (22)$$

where, as before, P_S is the probability per unit time that a thermally activated jump occurs in the S channel.

Another enhanced diffusion mechanism arises from the direct collision of high-energy particles with the diffusing defect. The enhanced migration which results from the direct Rutherford collision between a high-energy electron and a nucleus is now well documented in the subthreshold recovery of close pairs in metals,[218,238–240] e.g., copper, aluminum, platinum, tantalum, and gold. The evaluation of the probability of this process for a given current density of high-energy electrons of a given energy is carried out exactly as is the calculation for the damage production process; it will be treated in Section 4, where we will also evaluate the damage production process probability. We simply note here that the direction of recoil of the diffusing atom is related to the incident beam direction of the high-energy electrons. Consequently there is a preferential direction for the net momentum interchange which can give rise to a drift force K (and an appropriate frictional coefficient β) which must be added to the diffusion equation,

$$\partial s/\partial t = \text{div}[D \text{ grad } s - (Ks/\beta)] \qquad (23)$$

Finally we have discussed[232] how the photoionization of an electron will require that an atom recoil, which in turn can aid diffusion. The probability of this process is smaller than that of ionization-enhanced diffusion processes, and consequently this mechanism is more important in metals than in semiconductors.

3.4. Plastic Deformation

Semiconductors can be produced with no (or few) dislocations per unit area. Plastic deformation creates dislocations[112,113,241,242] and consequently creates damage in the lattice. The dislocations themselves introduce electrical levels[242–245] into the lattice and have associated EPR spectra.[246–250] In semiconductors with the diamond structure the dislocation velocity depends exponentially on the temperature,[251–257] for example, with an activation energy of 2.3 eV in silicon. Generally, then, dislocation generation and motion processes require that plastic deformation in semiconductors take place at very elevated temperatures.

It has long been recognized that various dislocation processes can generate point defects in the lattice. In fact Seitz[241] has argued that since there are two lattice atoms per unit cell in the diamond lattice, the dislocation processes are very likely to produce divacancies and diinterstitials. The plastic deformation experiments are carried out at such high temperatures that isolated point defects are not observed. Experiments which directly confirm the generation of point defects have not proven possible in semiconductors but indirect confirmation has been observed. For example, Wöhler, Alexander, and Sander[248] observed that the recovery in their EPR line occurred with an activation energy of 1.35 eV, which they identify with the migration energy of the divacancy in silicon (which is observed[120] to be 1.2 eV). Wöhler[251] found that the recovery of dislocations at deformation temperature was governed by an activation energy of $\sim$1.2 eV, again apparently controlled by divacancies which had been formed during the deformation and which dissolved slip barriers on the dislocations during recovery, permitting the dislocations to slip into an arrangement of reduced strain energy.

The problem is, of course, that the point defects, including the divacancies, are very mobile at the deformation temperature and are not retained in the lattice. Garber et al.[258,259] have shown that plastic deformation can also be observed at much lower temperatures by carrying out a great number of oscillatory applications of stress. Such experiments raise again the possibility of the observation of point defects created by plastic deformation, but such experiments have not been carried out.

In Section 2.3 we discussed how ionization can enhance diffusion processes in some instances. Frisch and Patel[255] have argued that dislocation motion occurs at charged sites on dislocations and have used this model

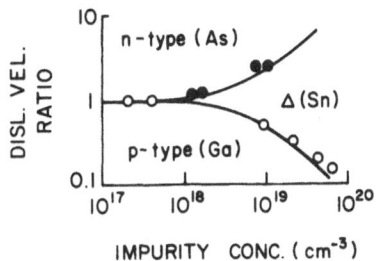

Fig. 45. Ratio of the dislocation velocity in intrinsic germanium to that in doped germanium for various dopings. (After Frisch and Patel.[255])

to treat the strong dependence of the dislocation velocity on the concentration of the doping impurity, as is shown in Fig. 45. This figure shows that there is a marked dependence for both n-type dopant and p-type dopant, while there is none for a neutral impurity (tin). Bourgoin and Corbett[232] have argued that this Fermi level dependence of the dislocation velocity may in fact indicate that an ionization-enhanced process is facilitating the dislocation motion. Experiments of the Garber type in which the dependence of ionization is specifically probed should answer this question and may be very pertinent in establishing the nature of point defect generation by plastic deformation in semiconductors. Much work in this area clearly remains to be carried out.

3.5. Heat Treatment Effects

Heat treatment can change or damage semiconductor materials in two ways: It can alter preexisting defects or it can introduce new ones. A classic example of the former is the "heat treatment effect" which occurred in the early silicon technology. At the time the only way of growing silicon crystals was the Czochralski technique, in which the molten silicon was held in a quartz vessel. Crystals were obtained with a very low concentration of electrical levels (of the order of 10^{13} cm^{-3}) and it was presumed that this concentration of levels reflected the impurity concentration in the material. In the course of fabricating devices, various heat treatment steps resulted in the "spontaneous generation" of electrical levels characteristic of a defect concentration $\sim 10^{16}$ cm^{-3}. Since this occurred in spite of extraordinary precautions to prevent the introduction of impurities in the heat treatment steps, it was the cause of considerable consternation. Of course, the resolution of the dilemma was found in the recognition that such pulled crystals contained of the order of 10^{18} cm^{-3} oxygen atoms and that upon suitable heat treatment four oxygens could combine to create an electrical level. (The precise structure of this center is still not worked out.) One could equally easily imagine that the growth of the semiconductor material would result in "inert" defect complexes being present in the sample material, only to be "activated" by a heat treatment and/or irradiation. Since the history of defect studies in semiconductor materials continues to produce examples of the discovery of the importance of unsuspected impurities, the moral of these lessons is something we ignore at our peril.

Heat treatment can also introduce new defects into the material, either by placing the material in a temperature regime where there is a natural concentration of intrinsic defects, or by causing plastic deformation of the material which in turn can introduce defects. Since plastic deformation is discussed in Section 2.4, we will not discuss that aspect any further here. The question of the concentration and properties of intrinsic defects in silicon and germanium at high temperatures has been extensively reviewed[96,260-263] and is the subject of considerable controversy. Information about the equilibrium concentration of defects at high temperatures can be obtained by carrying out measurements at high temperatures on the diffusion of these defects (usually through self-diffusion measurements), by studying the interaction of these defects with impurity atoms, or by making measurements at low temperatures on samples which have been quenched from high temperatures in such a way as to retain some of the equilibrium concentration of defects.

The results of the high-temperature self-diffusion measurements in silicon[263] are summarized in Fig. 46. The dashed curve is obtained primarily from tracer measurements of self-diffusion, the solid curve is obtained primarily from the analysis of impurity diffusion and impurity precipitation. For a monovacancy diffusion mechanism the self-diffusion equation is

$$D = D_V f_V C_V = D_0 \exp(-E_D/kT) \qquad (24)$$

where D is the self-diffusion coefficient, D_V is the diffusion coefficient for single vacancies, f_V is the correlation factor for single vacancies (in the

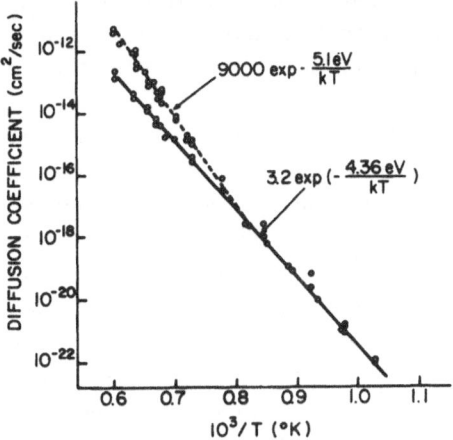

Fig. 46. Self-diffusion coefficient in silicon versus temperature. The dashed curve is derived from tracer measurements; the solid curve from impurity diffusion and precipitation. (After Seeger.[263])

diamond lattice $f_V = 1/2$), and C_V is the equilibrium concentration of vacancies. In turn

$$D_V = \tfrac{1}{8}a^2\nu_V \exp(S_m/k) \exp(-E_m/kT) \tag{25}$$

where a is the lattice constant, ν_V is the frequency factor for vacancy migration, S_m is the entropy factor for vacancy migration, and E_m is the migration energy for vacancy migration. For the concentration of vacancies we have

$$C_V = \exp(S_f/k) \exp(-E_f/kT) \tag{26}$$

where S_f and E_f are the entropy and energy of formation of a single vacancy. Seeger and co-workers have noted that there are several anomalies associated with the data on self-diffusion in Si and Ge: (1) The self-diffusion coefficient at the melting temperature is several orders of magnitude smaller than in metals (where it is typically of the order of 10^{-9} cm/sec); (2) the preexponential factor is about a factor of 10^3 larger in silicon than in germanium, which in turn is of the order of 50 times a typical value in metals; and (3) the activation energies of self-diffusion in silicon and germanium are substantially higher than in metals with comparable melting temperatures, which are approximately 2 eV. Further, they have argued that inserting appropriate numbers into Eqs. (24)–(26) results in the values of $S_f + S_m \sim 15k$ for silicon and $S_f + S_m \sim 10k$ for germanium. They argued that these values are substantially larger than those obtained in metals and they seriously questioned whether the monovacancy mechanism is appropriate for silicon and germanium. Seeger reviewed the three proposals that had been made to explain the entropy problem. Peart[264] and Goshtagore[265] suggested that in silicon the divacancy caused diffusion and that this accounts for the large entropy factor. Masters[106] proposed that the split-vacancy (which we discussed in Section 2 as the saddle-point configuration for vacancy migration) was the high-temperature form of the vacancy. Seeger argued persuasively that neither of these explanations would account for the large entropy factor. Seeger and co-workers have argued that the large entropies could only be understood if "extended defects" were involved. They noted that an ordinary vacancy with a small relaxation of neighboring atoms is not expected to have a formation entropy much larger than k nor an entropy of migration which could exceed a few k. They concluded, therefore, that a large formation entropy indicates the participation of many atoms, ergo an extended defect. Specifically they proposed that this new defect in germanium will be an extended vacancy, while in silicon they argued that the much larger preexponential factor requires an alternative possibility, namely an extended self-interstitial as the source of self-diffusion in silicon at high temperatures. Recently Corbett, Bourgoin, and Frisch[266] have suggested another explanation for the entropy problem. They proposed that the self-diffusion process described by the solid line in Fig. 46 is due to divacancy migration. The dashed line in Fig. 46 they attributed to an additional contribution to the diffusion which arises from the diffusion of nonequilibrium vacancies which arise from the dissociation of the divacancies. Now the preexponential factor does not reflect directly only the entropy, but also factors related to the divacancy dissociation and subsequent vacancy motion. Consequently, this explanation does not have a entropy problem. They noted that self-diffusion in germanium could proceed in the same manner, thereby accounting for the entropy problem in silicon and germanium. They showed that reasonable parameters will describe self-diffusion in their model, but that further work is required to verify the model.

There have been a number of quenching experiments in silicon and germanium. In a quenching experiment the sample is left for a prolonged period at high temperatures so that the equilibrium concentration of defects can be established, and then the sample temperature is quickly lowered, say to room temperature or lower, with the hope that some of the equilibrium defects will be frozen in the lattice. In silicon, several authors[267-269] have observed that quenching defects in p-type Si created defects with an energy level at $E_v + 0.4$ eV, which has an activation energy for migration of 0.81 eV. Seeger[263] noted that several workers had observed the same sort of defects in irradiation experiments.[270-272] Seeger proposed that this defect is an interstitial atom, in keeping with the proposal that self-diffusion in silicon occurs at high temperatures via an interstitial atom. Swenson[273] noted that the earlier experiments on quenching p-type silicon were carried out with such a low acceptor concentration ($<10^{15}$ cm^{-3}) that oxygen probably was the predominant impurity even in those cases where floating zone material was used. He carried out quenching experiments (to liquid nitrogen temperatures) on vacuum-floating-zone material doped with 10^{17} boron atoms cm^{-3} and found energy levels close to the edge of the valence band, i.e., below $E_v + 0.05$ eV; namely, he found energy levels precisely characteristic of those suggested for the vacancy by Watkins[116] in the EPR experiments. These results clearly indicate that further work needs to be done on understanding *all* of the ramifications of quenching work in silicon.

It seems generally agreed that quenching experiments in germanium result in vacancy defects, although experiments require special precautions to avoid diffusion of copper into the germanium. All heat treatment and quenching experiments must be carried out with the recognition that a number of impurities are "fast diffusers" in semiconductors; consequently trace amounts of these impurities on the surface of sample can be uniformly distributed to the sample as the result of a modest heat treatment. The history[96,274-281] of quenching in Ge is replete with controversies arising from this problem. Whan[152] has shown that, *for at least one charge state*, the vacancy is mobile at low temperatures. This means that it is unlikely that the quenching experiments will result in isolated vacancies. The more recent work[280,281] has interpreted the quenching experiments as resulting in vacancy related defects, namely (vacancy plus donor) pairs and divacancies.

Compound semiconductors represent a special problem. Stoichiometric defects are very common, particularly in the II–VI compounds where rapid changes in electrical and optical properties can be created by heating alternately in an excess chalcogen or metal vapor. As discussed in Section 2, very little is actually known about the defects with certainty, and often experimental results appear contradictory and defy simple interpretation. Radiation effect studies and diffusion studies have been employed to provide the understanding which has been discussed in Section 2; systematic, careful quenching studies have not been carried out in compound semiconductors.

4. INTERACTIONS OF PARTICLES WITH SEMICONDUCTORS

4.1. Introduction

In this section we discuss the interaction of particles with semiconductors. We are particularly concerned with those interactions that can create damage, but, as we have discussed, there is a variety of ways that ionization can influence damage production processes, so for the sake of completeness we will consider the particle interactions that also create ionization. We consider first the various interactions due to photons (Section 4.2). We next discuss charged particle interactions, beginning with the relatively simpler case of proton interactions in Section 4.3. Then we proceed to the interactions of heavy ions and atoms (also Section 4.3) and then consider the case of high-energy electrons (Section 4.4), which necessitates a relativistic treatment. Finally, we consider the interactions of neutrons (Section 4.5). Throughout, we will be concerned primarily with those interactions that are useful probes for, or are inextricably involved in, studying the damage production process.

4.2. Photon Interactions

Photon interactions can be both intrinsic and extrinsic (i.e., depend upon impurities in the material). Photon interactions can be ionizing or nonionizing. Intrinsic, ionizing interactions require photon energies larger than the forbidden gap for the material. Table IV shows approximate room-temperature values for the forbidden gap of various elemental and compound semiconductors; in detail the value of the band gap depends upon the type of measurement made, so we here indicate optical gaps.

We will indicate the magnitudes of the various photon interactions by considering a specific case, namely germanium. Germanium has a band structure[282] with the conduction band minimum at the edge of the Brillouin zone in the [111] direction, an indirect band gap of 0.771 eV, and a direct band gap of 0.806 eV. Figure 47 shows that the absorption coefficient rises abruptly with energies greater than the band gap. We can relate absorption coefficient to a cross section as follows. We recognize that in an absorption experiment the transmitted intensity I is related to the incident intensity I_0 by $I = I_0 \exp(-\alpha x)$, where α is the absorption coefficient and x is the sample thickness; but we can also express the absorption coefficient in terms of a cross section per atom σ by the equation $\alpha = N\sigma$, where N is

Table IV. Approximate Room-Temperature Values for the Forbidden Gap (eV) for Various Elemental and Compound Semiconductors[a]

Diamond	Si	Ge	α-Sn	Te	Se
5.5	1.12	0.77	0.08	0.33	1.85

	P	As	Sb
Al	3.0	2.3	1.5
Ga	2.25	1.4	0.7
In	1.3	0.33	0.17

	S	Se	Te
Zn	3.6 (Z)	2.6 (Z)	2.2
Cd	2.4 (W)	1.7 (W)	1.4

[a] Z, zinc-blende; W, wurtzite.

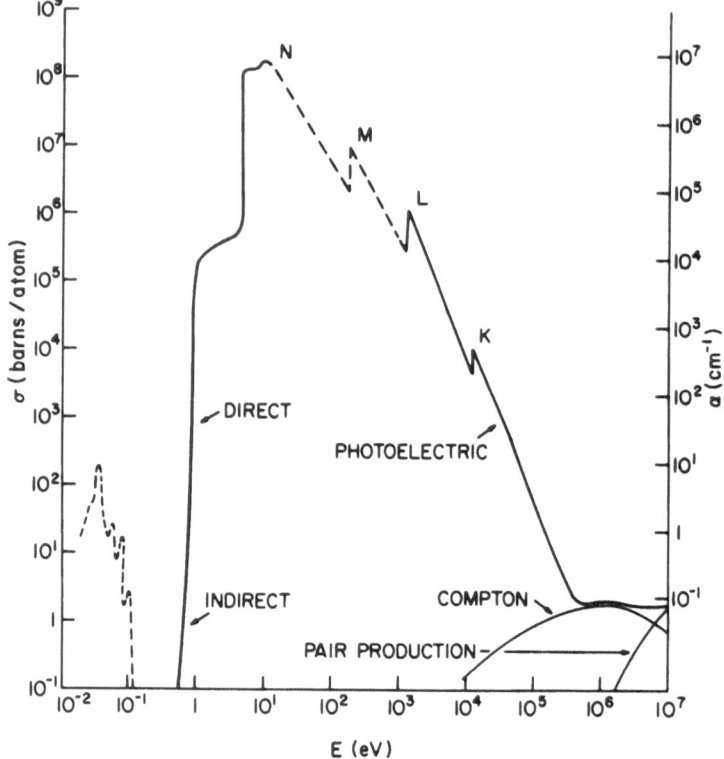

Fig. 47. Photon absorption coefficient α and cross section σ in germanium versus photon energy for various processes. The low-energy dashed line indicates lattice absorption. Absorption at the indirect and direct band gap is indicated, as is the absorption at the various X-ray shells K–N. The dashed line between the L shell and the N shell indicates where the cross section is interpolated.

the number of atoms cm^{-3} and σ is in units of cm^2. The corresponding values for Ge are shown in Fig. 47.

Below the band gap there are no intrinsic ionizing interactions. As shown in Fig. 47, the lattice can absorb photons of energy less than the band gap; in the elemental semiconductor the lattice has no permanent dipole moment and consequently the lattice absorption, as shown, is due to multiple phonon processes; in compound semiconductors the absorption coefficient can be larger because of the permanent dipole moment in the lattice, but the interaction is still nonionizing. Impurity and defect states can be ionized by photons of less than band gap energy, but the total ionization thereby produced tends to be small since there is a limited number of such impurity or defect states available in the material. In case the interaction to be considered is the ionization of that specific impurity or defect, then it is not total ionization that is important but the probability of ionization of that defect. The cross section for impurity ionization will vary strongly with energy, but at resonance will approximate the geometric cross section of the orbits involved in the state.

In the vicinity of the band gap, of course, absorption due to excitons can also occur, depending upon the sample temperature and exciton binding energy; whether these processes contribute to ionization of the material also depends upon the sample temperature and exciton binding energy.

Absorption of photons just above the band gap energy can occur with a rather high cross section, as indicated in Fig. 47, but can still be limited by the details of the band structure (i.e., density of states effects). Consequently in Ge the absorption coefficient rises abruptly above the band gap to a cross section of $\sim 10^5$ b, but not until an energy of ~ 4.5 eV does it rise to a value approximately corresponding to the geometric cross section of the germanium atom, i.e., 10^8 b. (The rise at 4.5 eV is called the E_2 absorption process.) Germanium has an electronic configuration of $1s^2\,2s^2\,2p^6\,3s^2\,3p^6\,3d^{10}$ $4s^2\,4p^2$. The absorption just above 4.5 eV represents the ionization of the $4s^2\,4p^2$ states, i.e., in X-ray terminology the N shell. The results up to approximately 10 eV are accessible experimentally[281] and are shown in Fig. 47 by a heavy line. The region beyond that accessible by vacuum-ultraviolet techniques has not been measured directly and consequently is shown by a dashed line up to the energy where the results again become available experimentally in the X-ray region. The cross section declines from that of the N-shell ionization with the cross section characteristic of the photoelectric ionization of the corresponding M $(3s^2\,3p^6\,3d^{10})$, L $(2s^2\,2p^6)$, and K $(1s^2)$ shells. The data shown in the region beyond 10^3 eV are taken from the compilation of theoretical and experimental results given by Storm and Israel.[284] In particular, since we are not interested in cross sections for scattering, etc., we have considered only the total energy-absorption cross section, which they denote $\sigma_{tot.\,en.}$. Figures 48 and 49 present the comparable X-ray cross sections for carbon, silicon, and gold, the latter being a common impurity in semiconductors. At high energies the photoelectric cross section no longer dominates and the Compton and pair production processes successively contribute. The Compton process varies proportionately to the Z of the scattering material, while the pair production process varies as Z^2.

The cross section shown in Fig. 47 describes the probability of ionization of the material versus photon energy. These are the pertinent cross

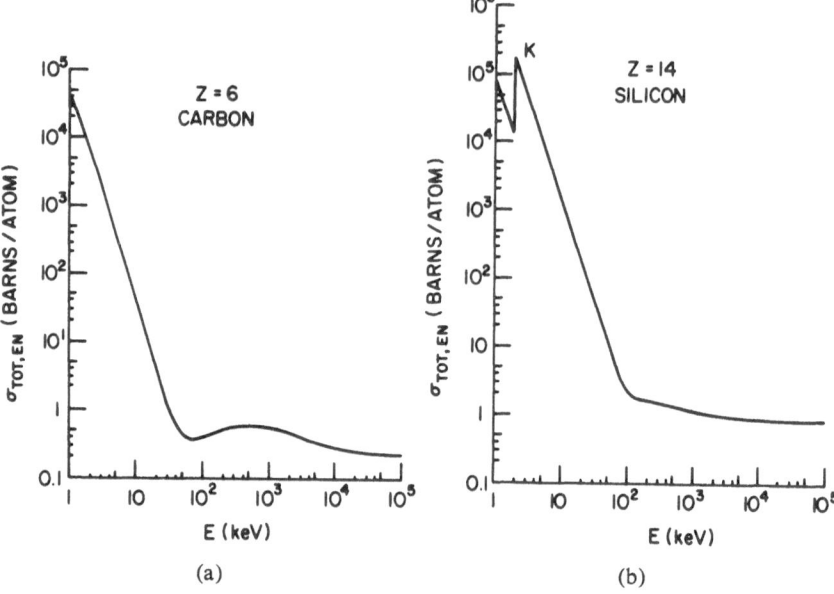

Fig. 48. Total energy absorption cross section for photons versus photon energy in (a) carbon and (b) silicon.

sections for processes in which the ionization itself is of direct interest, i.e., in ionization-enhanced diffusion and ionization damage production processes. As we discussed in Section 3, there are some damage production mechanisms which are only ionization dependent. These mechanisms are not thought to apply to germanium, but, in the context of the data shown in Fig. 47, were these mechanisms to apply to germanium, then the cross sections shown would be the appropriate ones.

We can also discuss the Varley[223-225] mechanism in terms of the cross sections presented in Fig. 47. It is often asserted that a double-ionization process is very unlikely and consequently the Varley mechanism itself is very unlikely. The phrase "very unlikely" is somewhat elastic in its meaning. In any event X-ray data have long shown that double ionization of a low-Z element can occur with a probability of approximately 10% of the single-ionization probability. Double ionization gives rise to the so-called satellite, or nondiagram X-ray lines.[285] These are lines which occur in *doubly ionized atoms* and hence the energy is not characteristic of the singly ionized energy levels. The probability of these satellite lines rises sharply from the threshold of their occurrence, i.e., the energy for double ionization, and saturates to an approximately constant value for increasing photon energy; thus above the double-ionization threshold, the double-ionization cross section will be a constant fraction of that cross section shown in Fig. 47. As far as we know the *total* double-ionization cross-section probability is not available, but Parratt[285] gives the percentage that the double-ionization probability is of the single-ionization probability for the K shell; for germanium this percentage is only 0.4% but for lower Z elements, e.g., silicon, the probability rises to 17%. Parratt points out that Richtmeyer[286] calculated these K-shell ionization probabilities and showed that the ratio of the intensity of the double ionization I_D to the intensity of the single ionization I_S can be fit by

$$I_D/I_S = C/(Z - \sigma)^3 \tag{27}$$

where σ is the internal screening constant, taken for all elements to be equal to 4.5, and the empirical value of C is 91. Parratt showed that the experimental data fit this functional dependence quite well. Sachenko and co-workers[287,288] have treated the relative probability of ionization of the L shell given that the K shell is ionized, but experimental data are not available. They indicated that $(K + L)$ multiple ionization is more probable than the

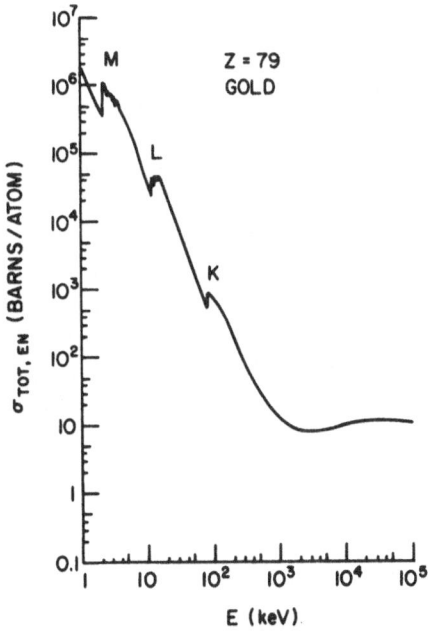

Fig. 49. The total energy absorption cross section for photons on gold atoms, as a common impurity in semiconductors, versus photon energy E.

$(K + K)$ multiple ionization, as the single-ionization cross sections shown in Fig. 47 would indicate; we would therefore argue that the $(K + L, M$ or $N)$ probability is even larger. This kind of argument would also suggest that in Eq. (27) the Z^{-3} dependence would be reduced by a factor of Z in the numerator to account for the total sum of electrons on the atom, yielding a Z^{-2} dependence for the double-ionization probability.

Turning to a general discussion of the cross sections in Figs. 47–49, we note that the Compton and pair production processes not only contribute to ionization, they also result in high-energy electrons (and positrons in the pair production case) which can create damage by a displacement process, as treated in subsequent sections.

In the Compton scattering of the gamma ray, the electron which participates in the scattering process receives a fraction of the energy of the gamma ray. The relationship between the scattered and primary photon energies is

$$E_\mu'/E_\gamma = 1/[1 + \alpha(1 - \cos\theta)] \tag{28}$$

where E_γ and E_γ' are the primary and scattered gamma-ray energies, θ is the photon scattering angle, and $\alpha = E_\gamma/(m_0 c^2)$, with m_0 the electron mass. The kinetic energy T of the recoil electron is given by[289]

$$\frac{T}{E_\gamma} = \frac{\alpha(1 - \cos\theta)}{1 + \alpha(1 - \cos\theta)} \tag{29}$$

Equation (29) implies that for $\theta = 180°$ (i.e., the maximum recoil energy) and for a 1-MeV gamma ray, the recoil energy is 780 keV; it implies that the fractional energy rises rapidly for higher gamma-ray energies.

Pair production can occur in the field of the nucleus or in the field of the atomic electrons; both processes are included in the cross sections shown in Figs. 47–49. The partitioning of energy between the electron and positron depends upon the details of the production process; i.e., the attraction of the nucleus for the electron yields an asymmetric distribution favoring higher energy positrons in the case of production at the nucleus. We will not go into the details here since they are treated fully in the article by Bethe and Ashkin.[290]

4.3. Heavy Ions and Atoms

High-energy ions lose energy in two ways of importance to us: (1) ionization, which is the dominant energy loss mechanism; (2) energetic Rutherford scattering events, which give rise to displacement damage.

Ionization, or electron excitation, is the major source of energy loss of moving ions, particularly at higher energies. The rate of energy loss is given[290],* by

$$-\frac{dE}{dx} = 2\pi e^4 Z_1^2 Z_i N \frac{M_2}{m} \frac{1}{E} \ln \frac{4E}{I} \tag{30}$$

where Z_1 and Z_2 are the atomic charges of the incident and lattice atoms, respectively, N is the atomic density of the lattice, M_2 is the mass of the lattice atoms, m is the mass of the electron, E is the energy of the incident ion, and I is the mean ionization potential, which is approximately equal to $8.8Z_2$ for all except the lightest elements. The value of I should be energy dependent in that it is an average over the electrons of each atom in the solid. We will ignore this dependence for now. By dividing the energy loss by the approximate mean ionization potential and by the atomic density we can express these results in terms of a cross section per atom, which will facilitate comparison to other cross sections,

$$\sigma_I = -\frac{1}{NI} \frac{dE}{dx} = 8\pi Z_1^2 Z_2 \left[\frac{R_H a_H}{I}\right]^2 \frac{M_2}{m} \left[\frac{1}{x} \ln 4x\right] \tag{31}$$

with

$$x \equiv E/I \tag{32}$$

with R_H the Rydberg energy (13.54 eV), and a_H the Bohr radius of hydrogen. Figure 50 shows how this cross section varies versus energy for the case of a high-energy proton in diamond, silicon, and germanium; we also indicate the proton ionization probability for gold, a common impurity in these materials. In the case that the proton energy is high enough to require a relativistic treatment, Eq. (30) becomes

$$-\frac{dE}{dx} = 4\pi N Z_1^2 Z_2 \frac{e^4}{m\beta^2 c^2} \left\{\ln\left[\frac{mc^2\beta^2}{(1-\beta^2)I}\right] - \beta^2 - F(Z,\beta)\right\} \tag{33}$$

The last term is a correction for nonparticipating electron shells. We will not require the relativistic expression and hence will not employ it further.

The ionization cross sections given by Eqs. (30)–(33) are not correct at low energies because the moving ion begins to capture electrons from the solid, and its effective charge changes. Bohr[292] made estimates of the energy dependence of Z_1 by assuming that the moving atom will lose all

---

* Experimentally corrected values are given by Williamson and Boujot.[291]

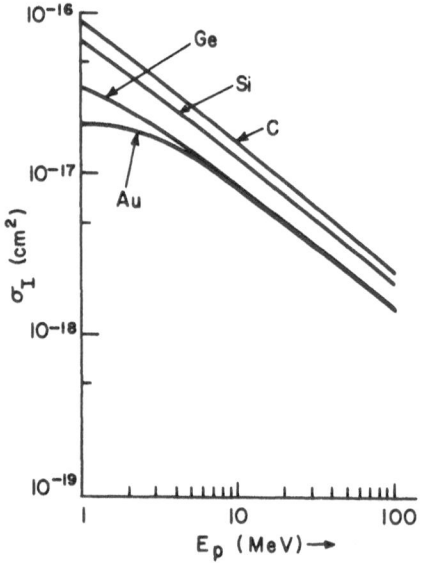

Fig. 50. Cross section for ionization σ_I by a proton impinging on carbon, silicon, germanium, and gold atoms versus energy of the proton E_p. At low energies the cross section is adjusted to conform to experiment.

electrons whose orbital velocity is less than the velocity of the moving atom. He concluded that

$$Z_1(\text{eff}) = Z_1^{1/3} \frac{\hbar}{e^2} \left(\frac{2E}{M}\right)^{1/2} = Z_1^{1/3} \frac{137\sqrt{2}}{\sqrt{A}} \left(\frac{E}{M_p C^2}\right)^{1/2} \qquad (34)$$

Figure 51 shows how the effective charge $Z_1(\text{eff})$ varies for various ions versus the energy of the moving ion. We recognize, of course, that Eq. (34) cannot be precisely correct because the effective charge cannot rise above Z_1. In more detail, in fact, the energy dependence of the effective charged should reflect the shell structure of the electrons in the moving ion, i.e., the energy dependence of the mean ionization energy of the moving ion. Various estimates have been made to supplement the energy dependence of $Z_1(\text{eff})$ shown in Fig. 51 by including better estimates of the ionization potential of the ion. In particular Leibfried[293] has argued that the energy of the moving atom necessary to singly the ionize the atom $E_{\min}$ is given by

$$E_{\min} \simeq A \text{ keV} \qquad (35)$$

and the energy to fully ionize the atom E_{max} is given by

$$E_{max} \simeq 500 A Z_2^2 I_H \tag{36}$$

with I_H the ionization potential of hydrogen. For the moving atom, then, Eqs. (35) and (36) delimit the region in which the atom is neutral, partially ionized, and fully ionized as shown in Fig. 52. There has been extensive work calculating the energy loss in this low-energy region. Robinson[214] has critically reviewed the main theories in this area: the Lindhard theory[294–297] and the Firsov[298–300] theory. In both theories the low-energy stopping cross section is proportional to the velocity of the moving particle. The Lindhard cross section σ_L is given by

$$\sigma_L = 8\pi e^2 a_H \xi_e (Z_1 Z_2 / Z)(v/v_0)(1/I) \tag{37}$$

where ξ_e is $\sim Z_1^{1/6}$, v is the velocity of the particle, $v_0 = \alpha c = c/137$, with c the velocity of light, I is the mean ionization energy (taken equal to $8.8 Z_2$ here), and

$$Z^{2/3} = Z_1^{2/3} + Z_2^{2/3} \tag{38}$$

Expressing these values in simpler units, we obtain for σ_L

$$\sigma_L(\text{in Å}^2) = 1.216 \left(\frac{Z_1^{7/6} Z_2}{ZI} \right) \left[\frac{E(\text{in eV})}{A(\text{in amu})} \right]^{1/2} \tag{39}$$

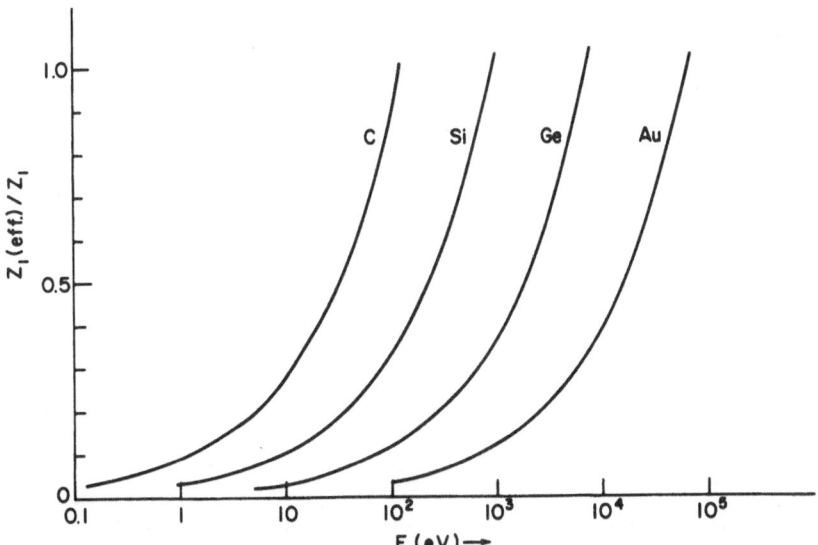

Fig. 51. Effective charge Z_1(eff) of various ions as a function of their energy.

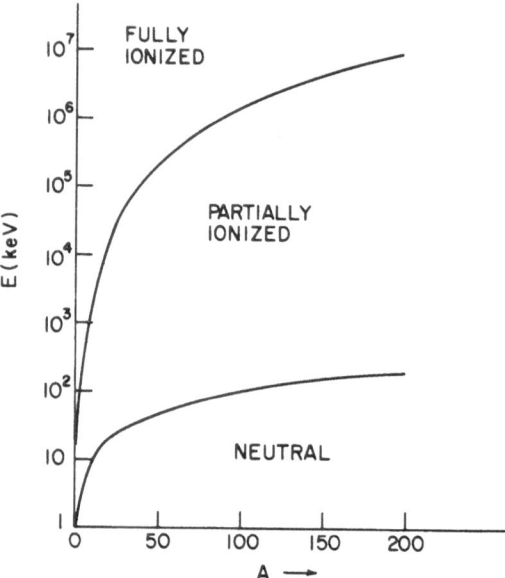

Fig. 52. Energy regimes in which a particle of atomic number A is either fully ionized, partially ionized, or neutral.

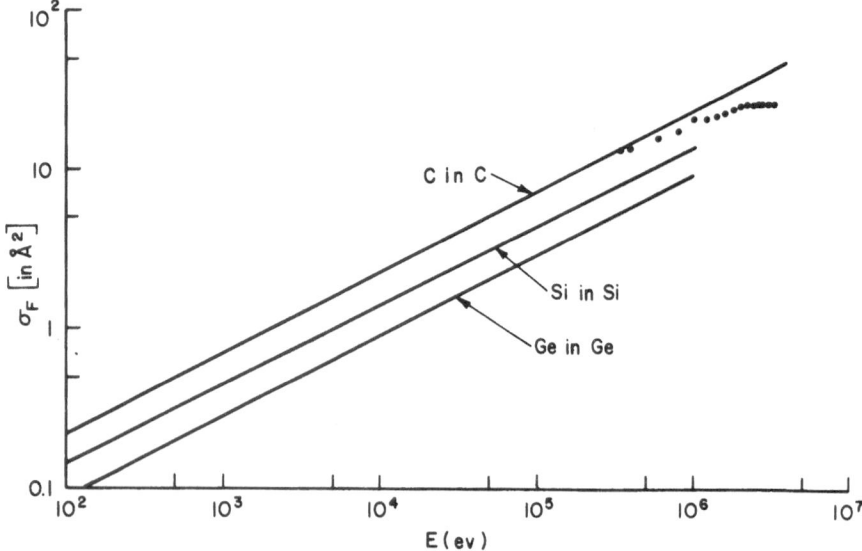

Fig. 53. The cross section for ionization given by Firsov (σ_F) versus particle energy E. Curves are given for carbon impinging on carbon, silicon on silicon, and germanium on germanium. Experimental data points for carbon on carbon[301] are given.

The corresponding expression for the Firsov theory is given by

$$\sigma_F(\text{in Å}^2) = 0.3362 \frac{(Z_1 + Z_2)}{I} \left[\frac{E(\text{in eV})}{A(\text{in amu})} \right]^{1/2} \qquad (40)$$

Figure 53 shows how the stopping cross section varies versus energy for carbon in carbon, silicon in silicon, and germanium in germanium. Also shown are the experimental data for carbon in carbon, the data being those of Porat and Ramavataram.[301] It can be seen that both theories give a good approximation to the low-energy electronic stopping power but, in detail there remain some difficulties:

(1) At low energies the electronic stopping power deviates from being proportional to the velocity.

(2) There is a Z_1 and Z_2 dependence[302,303] of the electronic stopping power which is not given by either theory, but which is due (to first order) to the electronic shell structure of both atoms, as is shown in Fig. 54 for various singly ionized atoms in silicon.[302]

(3) Finally, there is channeling!

Channeling studies have proven to be a major research technique which, for example, produces data such as shown in Fig. 54. Channeling occurs because the regular lattice periodicity in many crystals results in relatively open columns (and planes) through the crystal. An energetic atom (or ion) moving directly down the axis of one of these channels undergoes collisions with the atoms neighboring the channel at the maximum impact parameter possible, i.e., the minimum energy interchange possible. An atom proceeding down the channel in a direction which makes a small angle with respect to the channel axis will undergo a succession of very gentle collisions with

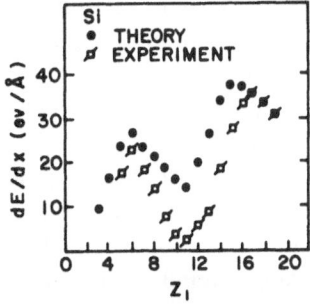

Fig. 54. Dependence of the energy loss dE/dx in silicon upon the charge Z_1 of the impinging particle. (After Cheshire and Poate.[302])

the atoms surrounding the channel, which collisions successively steer the channeling atom along the channel direction. If the trajectory of the channeling atom deviates from the channel axis by too large an angle (the so-called critical angle), then the atom makes a collision with the lattice which is at too large an angle to remain in the channel and the collision is termed a dechanneling collision. Robinson and Oen,[304] and now a number of other authors, have studied the channeling trajectories by computer, following trajectories corresponding to several hundred collisions and a corresponding penetration into the crystal of several hundred angstroms. A channeling experiment is carried out with a well-collimated beam of particles carefully oriented with respect to the channeling axis. Even so, relatively few ions enter the channels; the remaining ions impinge on the crystal surface so far displaced from the channel axis that they undergo energetic collisions with the surface atoms and are dechanneled. In the ordinary irradiation experiment the incident beam is not well collimated and an even smaller fraction of ions enters into channels. Lattice atoms that have undergone energetic collisions can also be scattered into channels, but the probability of such a scattering event is in general small. It should be recognized, however, that channeling is one mechanism by which a displaced atom can be created a long distance from the initial damage site, i.e., the residual vacancy. Electrons can also channel, and in electron microscope studies where the incident beam direction is very well defined, channeling results in a substantial change in the displacement probability near a channeling direction.

The ionization loss of a heavy ion can result in the excitation of tightly bound electrons with the resultant creation of X-rays. The production of X-rays by energetic heavy ions is currently a very active research area. Some calculations exist for the energy dependence of the excitation of X-rays, particularly for protons (see, for example, Hansteen and Mosebekk,[305] with calculations for tin, neodymium, ytterbium, gold, and thorium) and some work on the probability of double ionization such as would be required for the Varley mechanism.

As Fig. 51 indicates, at low energies a high-energy ion will pick up enough electrons that it becomes neutral, i.e., an atom; at this point the collisions must be described by an atom–atom potential. The proper atom–atom potential will include a relatively long-range, and relatively shallow, attractive interaction characteristic of the binding between the atoms, as well as a strong, short-range, repulsive potential. In many instances in radiation damage studies the attractive interaction is essentially ignored and the atom–atom potential is simply approximated as a repulsive potential. In metals this is a good approximation since the binding energy

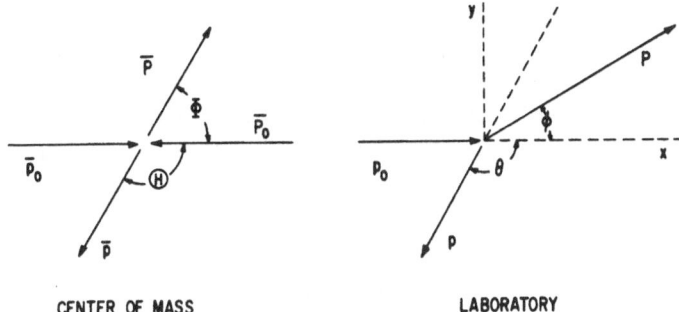

CENTER OF MASS   LABORATORY

Fig. 55. Definition of angles in the laboratory and center-of-mass (designated by the bars) frames of reference. The momentum of the incident particle is denoted by p, the momentum of the recoil atom by P. The recoil atom is assumed to be at rest prior to the collision.

is primarily due to the conduction electrons which are not localized. Even in semiconductors *for energetic interactions* the approximation of the potential by the repulsive interaction alone is quite justified. The repulsive interaction of atoms is sometimes approximately treated by a hard-sphere interaction, the radius of the hard sphere being energy dependent in some cases. A variety of more sophisticated repulsive potentials has been used, ranging from Fermi–Thomas and relativistic Hartree, screened Coulomb interactions to more phenomenological potentials such as the Born–Mayer potential. The subject of these interatomic potentials has recently been reviewed[306] from a variety of points of view.

While the dominant interaction of energetic ions with the lattice is electron excitation, our main concern is with the more rare, direct Rutherford scattering between the ion and the nucleus of the lattice atom which results in the displacement of a lattice atom, i.e., the displacement damage event. We can describe* the Rutherford scattering process with the aid of the coordinates shown in Fig. 55 for the laboratory system and the center-of-mass system. The kinetic energy T of the recoil atom is given by

$$T = T_m \cos^2 \phi = T_m \sin^2(\Theta/2) \tag{41}$$

where

$$T_m = [4M_1M_2/(M_1 + M_2)^2]E \tag{42}$$

$$E = \tfrac{1}{2}M_1V^2 \tag{43}$$

* For a more extensive discussion see Refs. 3, 15, and 21.

with M_1 the mass of the incident particle and M_2 the mass of the recoil atom, and V the velocity of the incident (nonrelativistic) particle. The maximum recoil energy T_m is shown in Fig. 56 versus E, the energy of the incident particle, and A, the atomic number of the lattice, or recoil, atom.

As is well known, the differential cross section for Rutherford scattering[307] can be written

$$d\sigma_R = \tfrac{1}{4}\pi b^2 \cos(\tfrac{1}{2}\Theta)\, \csc^3(\tfrac{1}{2}\Theta)\, d\Theta \tag{44}$$

$$\equiv R(\Theta)(2\pi \sin\Theta\, d\Theta) \tag{45}$$

or expressed in terms of the recoil energy

$$d\sigma_R = \tfrac{1}{4}\pi b^2 T_m\, dT/T^2 \tag{46}$$

The quantity b is the impact parameter, which is given by

$$|Z_1 Z_2|\, e^2/b = \tfrac{1}{4}(M_1 + M_2)V^2 \tag{47}$$

which means we can express the Rutherford cross section as

$$d\sigma_R = \frac{4Z_1^2 Z_2^2}{E}\, \frac{M_1}{M_i}\, \frac{dT}{T^2} \tag{48}$$

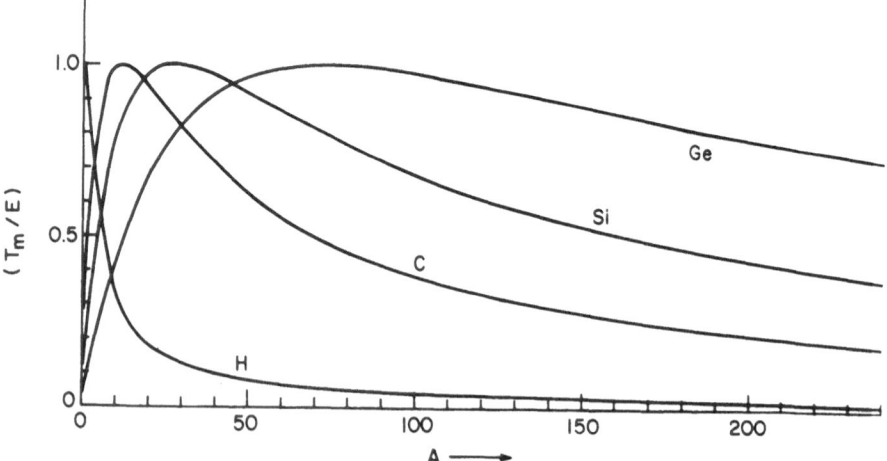

Fig. 56. The ratio of the maximum recoil energy T_m of a lattice atom of atomic number A to the energy E of an incident particle for the cases of incident hydrogen, carbon, silicon, and germanium.

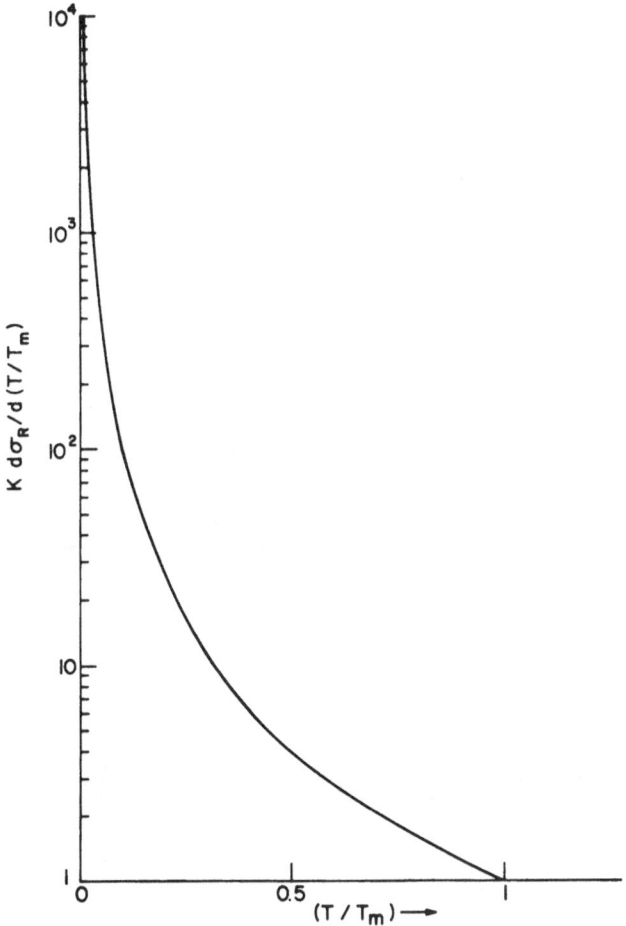

Fig. 57. Value of the differential Rutherford scattering cross section versus the ratio of the recoil energy T of the particle to the maximum recoil energy T_m. The value of K is the coefficient of $d\sigma_R$ in Eq. (50).

or in terms of a_H, the Bohr radius, and the Rydberg $R_H = 13.54$ eV,

$$\left(\frac{E}{R_H}\right)^2 \frac{1}{16Z_1{}^2Z_2{}^2} \frac{M_2}{M_1} \frac{d\sigma_R}{a_H{}^2} = \frac{dT}{T^2} \tag{49}$$

or

$$\left(\frac{M_1 + M_2}{M_1}\right)^2 \frac{1}{64Z_1{}^2Z_2{}^2} \left(\frac{T_m}{R_H}\right)^2 \frac{d\sigma_R}{a_H{}^2} = \frac{d(T/T_m)}{(T/T_m)^2} \tag{50}$$

As can be seen in Fig. 57, this cross section diverges for small recoil energies;

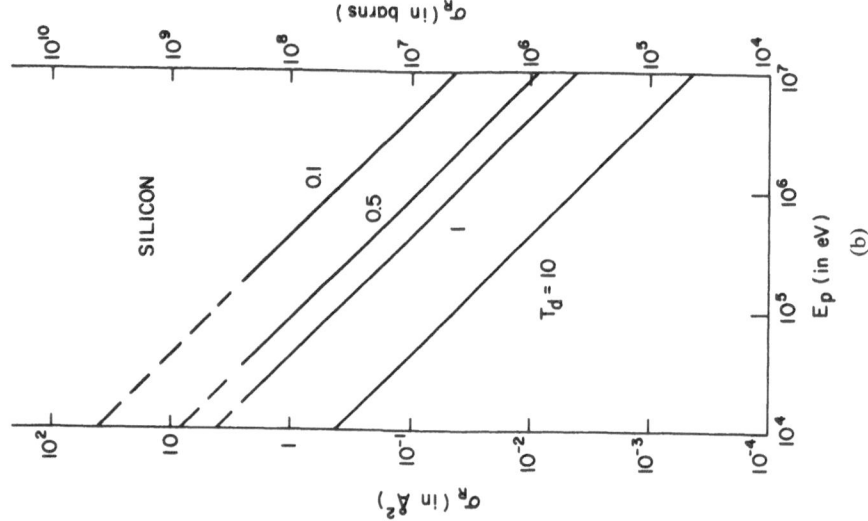

(b)

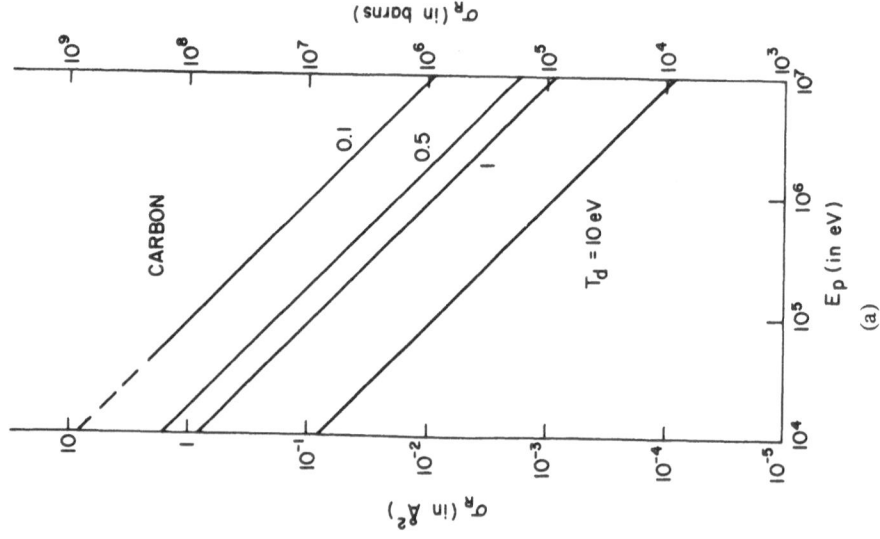

(a)

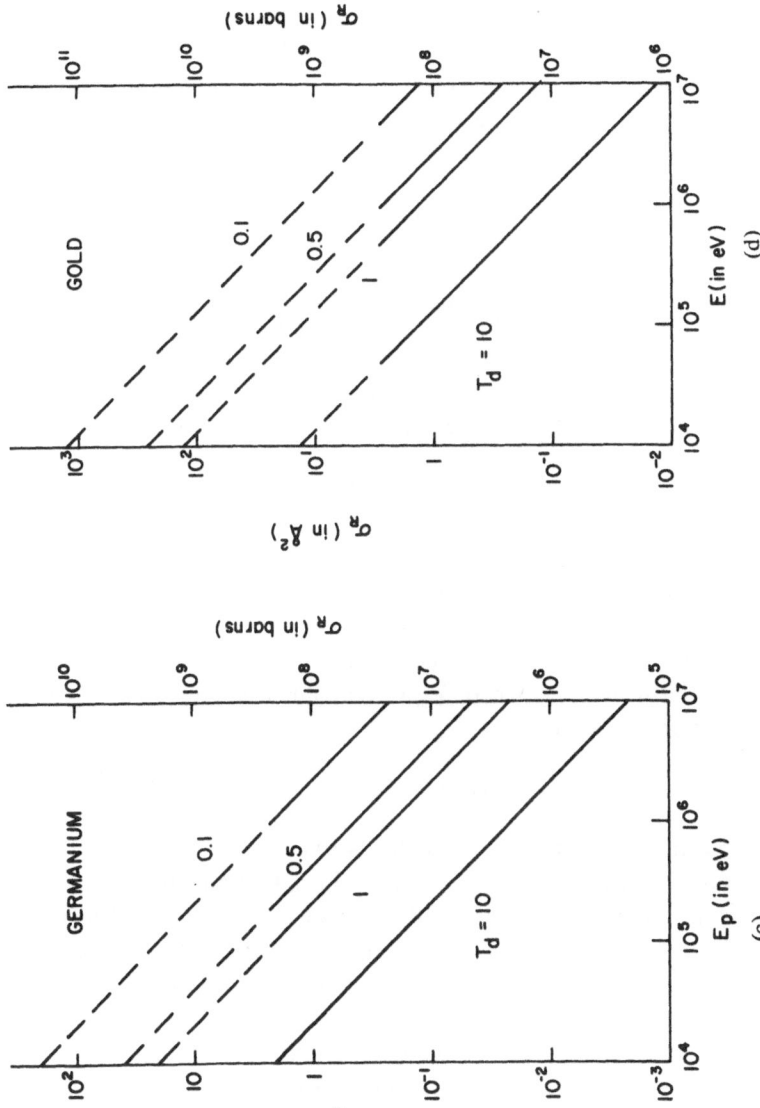

Fig. 58. Value of the total Rutherford displacement cross section, in angstroms squared and in barns, versus the energy E_p of the incident proton. Different curves are given for different values of the displacement threshold T_d. Where the curves are dashed, the cross section has exceeded the geometric cross section of the atom, and the formalism breaks down, i.e., screening must be included. (a) carbon; (b) silicon; (c) germanium; (d) gold.

as recognized by Bohr[292] and discussed fully by Seitz and Koehler,[3] this divergence is cut off by the screening due to the atomic electrons. The screening cutoff depends upon the detailed nature of the collision, but in any event occurs typically for values of the recoil energy of the order of 10^{-2} eV, which is substantially smaller than we will be concerned with.

By integrating the differential cross section from a minimum energy T_0 to the maximum energy T_m we can obtain the total cross section for scattering with a recoil energy larger than T_0. We obtain

$$\sigma_R(T_m, T_0) = \tfrac{1}{4}\pi b^2 [(T_m/T_0) - 1] \tag{51}$$

or

$$\frac{E^2}{R_H^2} \frac{1}{16Z_1^2 Z_2^2} \frac{M_2}{M_1} \frac{\sigma_R}{a_H^2} = \frac{E}{E_0} - 1 \tag{52}$$

or

$$\left(\frac{M_1 + M_2}{M_1}\right)^2 \frac{1}{64Z_1^2 Z_i^2} \left(\frac{T_m}{R_H}\right)^2 \frac{\sigma_R}{a_H^2} = \frac{T_m}{T_0} - 1 \tag{53}$$

As we have indicated previously, Rutherford collisions can result in enhanced diffusion as well as displacement damage. Figure 58 shows how the cross section varies with energy for carbon, silicon, germanium, and gold. Of course the threshold energy for a displacement process is typically 10 eV or larger; we will consider more specific cases later. The smaller values of T_0 were chosen as typical of those values of interest in enhanced diffusion processes. For the usual case where T_m is very much greater than T_0 the mean recoil energy is given by

$$\bar{T} = T_0 \ln(T_m/T_0) \tag{54}$$

that is, the mean recoil energy is typically $\sim$100 eV.

In general, recoil atom trajectories start in a random direction and subsequent collisions occur randomly. Because of the regularity of atoms in lattices, however, we must consider certain special types of collision sequences as well: (a) focusing, (b) replacement collisions, and (c) assisted focusing.

Silsbee[308] first showed that under certain conditions there was a strong tendency for energy to be focused along close-packed directions, such as occur in metallic lattices. He considered a line of hard spheres of diameter D uniformly spaced at the distance R. He showed that the successive scattering angles were related by

$$\theta_i = \sin^{-1}[(R/D) \sin \theta_{i-1}] - \theta_{i-1} \tag{55}$$

where θ_i is the angle made by the motion of the ith particle. For a sufficiently small initial angle we have $\theta_1 > \theta_2 > \theta_3 \cdots$, i.e., energy is focused down the chain. For a critical angle θ_c we have $\theta_c = \theta_1 = \theta_2 = \theta_3 = \cdots$; for angles smaller than θ_c focusing will occur, for angles greater than θ_c defocusing occurs.

Replacement collisions were first discussed by Kinchin and Pease.[208] The sequence of events in a replacement sequence is readily envisioned. At $t = 0$ a recoil energy is imparted to one atom in a chain, say, for sake of illustration, with the momentum going directly down the chain. The recoil atom collides with its nearest neighbor, displacing the neighboring atom and occupying its site, while the neighboring atom goes on to collide with the next neighboring atom; this sequence of collisions and replacements can continue for some distance, resulting in a vacancy quite remote from the interstitial atom.

As can be seen, both focusing collisions and replacement collisions result in the transport of energy down a close-packed direction. The distinction between a focusing collision and a replacement collision sequence is that the latter results in a metastable mass transport down the line of atoms; computer calculations have shown that replacement collision sequences can progress quite far down a line of atoms only to relax back to the initial configuration, i.e., return to the lattice without a vacancy or an interstitial; such a collision sequence is sometimes referred to as a dynamic crowdion. It should be recognized that the fate of a dynamic crowdion may depend upon the presence of scattering impurities and phonons, i.e., the scattering of a dynamic crowdion may result in a successful damage creation.

Focusing and replacement collision sequences require essentially close-packed lines of atoms in the lattice. These close-packed axes do not occur in semiconductor lattices. A different type of focusing can occur in the more open semiconductor lattice, namely assisted focusing. Figure 59 illus-

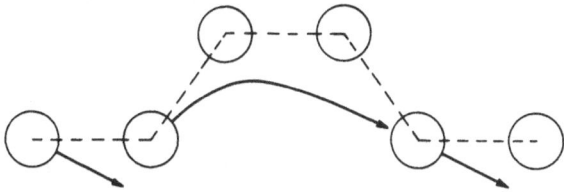

Fig. 59. A schematic example of assisted focusing, showing how the presence of intermediate atoms in the wurtzite lattice can assist the propagation of the angle of recoil of an atom through the lattice.

trates schematically how assisted focusing can occur. Basically all that is required is that the atoms which surround an open space in the sequence assist the focusing of energy along the chain. As might readily be appreciated, assisted focusing occurs more rarely than normal focusing, and relaxation back along the focusing axis does not occur in assisted focusing because of the open spaces in the sequence.

In summary, then, we can describe the type of damage left by a high-energy proton in its wake. Most of the energy-loss events will be due to electron excitation; since such energy loss is small, the proton path will be relatively straight. Occasionally a high-energy proton–nuclear collision will occur, resulting in a recoil atom and a substantial deviation in the proton path. Although the maximum recoil energy can be hundreds of kilovolts, depending upon the proton energy and the mass of the lattice atom (see Fig. 56), the mean recoil energy is approximately several hundred electron volts [see Eq. (54)]. Consequently most of the recoil atoms will have a rather low energy and result in relatively simple damage, i.e., single isolated Frenkel pairs, or small vacancy clusters with associated interstitials. The more energetic recoil atoms may be ionized, in which case the main energy-loss mechanism is ionization or electron excitation. For recoil atoms the energy-loss mechanism is atom–atom collisions. In the case of a very energetic recoil atom the number of atomic collisions which occur in stopping the atom can become quite large and the resultant damage properly described as a displacement spike.

The damage in the wake of a high-energy ion closely parallels that due to a proton. The primary energy-loss mechanism of the ion is electron excitation, or ionization loss, which results in a relatively straight damage track. As with the proton, an occasional energetic displacement collision can occur. As the ion energy decreases, the state of ionization of the ion also decreases, ultimately becoming that of a neutral atom. At this point the interaction of the atom with the lattice becomes relatively strong, resulting very often in a displacement spike at the end of the ion range.

4.4. Electrons

Like heavy charged particles, high-energy electrons undergo both ionization and displacement-collision energy losses, with the ionization energy losses dominating. As we will see, the smallness of the electron mass implies that the maximum recoil energy in a displacement also tends to be small, dictating that the damage produced will tend to be simple, i.e., usually isolated Frenkel pairs. Moreover, the electron bombarding energy can be reduced to the point that the maximum recoil energy does not exceed the minimum energy for damage production, and therefore damage production by electrons can be readily used to probe the damage threshold and the various dependences of the damage production probability. The energy loss rate experienced by high-energy electrons is small enough that thin samples can be fabricated and, at least a little above threshold, essentially mono-energetic experiments done; near threshold, of course, any energy loss will contribute to a gradient of damage across the sample, as we will discuss in more detail later. The small mass of the electron does imply, however, that each energy-loss collision can substantially scatter the direction of the high-energy electron; consequently corrections for the multiple scattering of the electron beam and the attendant change in electron path length must be made. The corrections for multiple scattering and the electron path length change have been discussed in detail elsewhere[15,21]; since they primarily concern experimental details we will not repeat that discussion here.

There are detailed theories which treat the energy distribution function for the energy lost by high-energy electrons, including the most probable energy loss as well as the average energy loss. In the ionization loss domain, the energy is lost primarily through collisions involving small incremental energy losses. Williams,[309] Landau,[310] and Blunck and Leisengang[311] have all given theories of the distribution of the energy losses. Because of the difficulty of the measurements, the theory has not yet had an extensive experimental test, but what measurements have been made support the theory with some minor corrections.[312] At high energies, e.g., above 1 MeV typically, the ionization loss no longer is the dominant energy-loss mechanism for the high-energy electrons, energy loss by Bremsstrahlung radiation production becoming important at $\sim$1 MeV and dominating at the higher energies. Heitler[313] treats the energy dependence of the average energy loss $-dE/dx$ (expressed here in MeV/cm of the target material). Figure 60 shows his results for the reduced average energy loss, i.e., the average energy loss divided by the product of $ZN\phi_0$, where ϕ_0 ($= 8\pi R_0{}^2/3 = 6.653 \times 10^{-25}$ cm^2) is the cross section of the classical electron, Z is the nuclear charge of the lattice atoms, and N is the number of atoms cm^{-3}. The Bremsstrahlung radiation energy will have a complicated dependence upon the electron energy and scattering angle; in any event the Bremsstrahlung radiation will tend to have a small absorption coefficient and will consequently escape the vicinity of the electron beam, i.e., not contribute to local ionization.

The theory of the displacement collisions produced by high-energy electrons closely parallels that for high-energy heavy ions, except that the

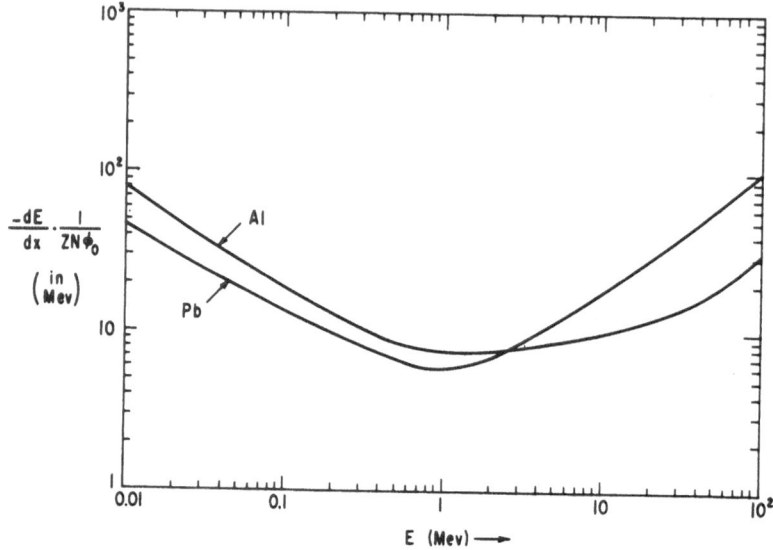

Fig. 60. The reduced average energy loss for electrons versus electron energy. Z is the nuclear charge, N is the number of atoms per cm^3, and ϕ_0 is the cross section of the classical electron.

electrons require a relativistic treatment. A more extensive discussion of the relativistic collision has been given earlier.[15] We can still use the same coordinates (see Fig. 55 in Section 43). Now, if the incident electron momentum p is small, then the recoil energy is given by

$$T = (2p^2/M) \cos^2 \phi = (2p^2/M) \sin^2(\Theta/2) \tag{56}$$

where M is the mass of the lattice atom, or, as before [Eq. (41) in Section 4.3],

$$T = T_m \sin^2(\Theta/2) \tag{57}$$

with

$$T_m = 2p^2c^2/Mc^2 \tag{58}$$

Expressing the maximum recoil energy in terms of the energy of the incident electron, we obtain

$$T_m = (2m/M)[(E + 2mc^2)E/mc^2] \tag{59}$$

or expressing the maximum recoil energy in eV and the incident electron energy in MeV, we obtain

$$AT_m = 2147.8(E + 1.0220)E \tag{60}$$

where A is the atomic number of the lattice atom. Figure 61 shows how the maximum recoil energy varies versus A and the incident electron energy E.

The Rutherford differential cross section for scattering of a non-relativistic classical electron by a nucleus into the solid angle $d\Omega$ is given by

$$d\sigma_{\mathrm{R}} = (Ze^2/2mv^2)^2 \csc^4(\Theta/2)\, d\Omega \qquad (61)$$

where we have explicitly included the incident electron velocity v. Since the solid angle for an annular cone of width $d\Theta$ about Θ is

$$d\Omega = 2\pi \sin\Theta\, d\Theta = 4\pi \sin(\Theta/2)\cos(\Theta/2)\, d\Theta \qquad (62)$$

we obtain, corresponding to Eq. (44)

$$d\sigma_{\mathrm{R}} = \pi(Ze^2/mv^2)^2 \cos(\Theta/2)\csc^3(\Theta/2)\, d\Theta \qquad (63)$$

The first attempt to make this expression relativistic was carried out by

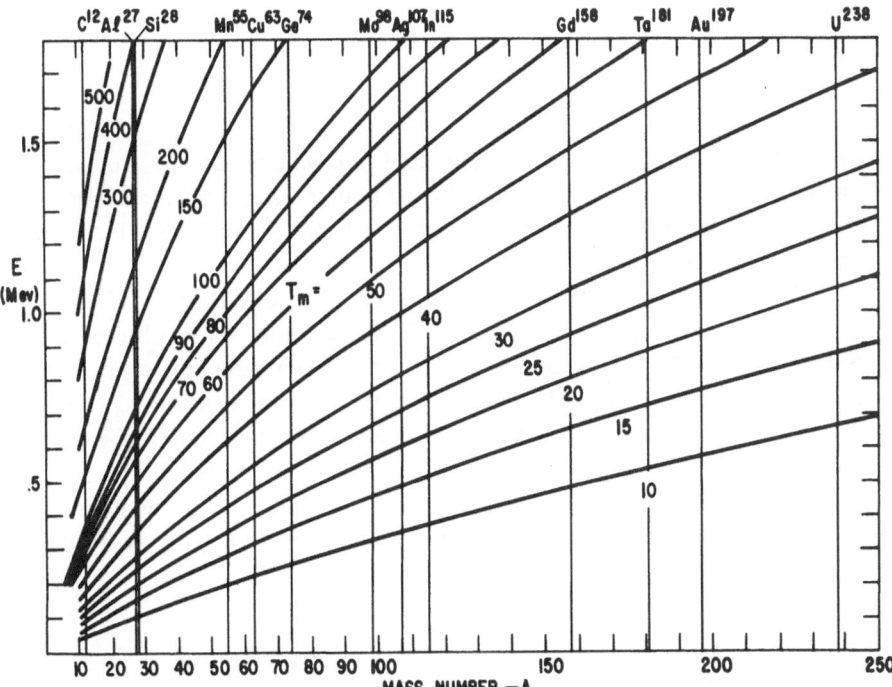

Fig. 61. Maximum recoil energy T_m (in eV) versus electron bombarding energy E and mass number A.

Darwin.[314] He replaced the velocity by the corresponding relativistic expression $v = c\beta\gamma$ [$\beta = v/c$; $\gamma = (1 - \beta^2)^{1/2}$]. He obtained

$$d\sigma_{\mathrm{DR}} = \pi\left(\frac{Ze^2}{mc^2}\right)^2 \frac{1}{\beta^4\gamma^2} \cos\frac{\Theta}{2} \csc^3\frac{\Theta}{2} \, d\Theta \tag{64}$$

Evaluating the numerical factors, we can express the Darwin–Rutherford cross section in barns, obtaining

$$d\sigma_{\mathrm{DR}} = 0.2495(Z^2/\beta^4\gamma^2) \cos(\Theta/2) \csc^3(\Theta/2) \, d\Theta \tag{65}$$

Mott[315,316] used the Dirac equation for electrons to derive the formula for the Coulomb scattering of electrons by a point nucleus. The formula is quite complex, being the sum of two conditionally convergent infinite series. McKinley and Feshbach[317] give an approximate expression for the Mott formula which is valid for low-Z elements. Their expression is

$$d\sigma_{\mathrm{McF}} = R_{\mathrm{McF}} \, d\sigma_{\mathrm{DR}} \tag{66}$$

where

$$R_{\mathrm{McF}} = [1 - \beta^2 \sin \tfrac{1}{2}\Theta + \alpha\beta Z\pi(\sin \tfrac{1}{2}\Theta)(1 - \sin \tfrac{1}{2}\Theta)] \tag{67}$$

with α is the fine structure constant ($\alpha \sim 1/137$). The McKinley–Feshbach approximate formula has been widely used to calculate damage production cross sections. Oen[318] has developed a computer program which uses the full Mott formula. We shall subsequently use both the McKinley–Feshbach and Oen results.

To obtain a cross section for displacement damage, we must make some assumptions about the threshold probability function. As we discussed in Section 3.1, the simplest choice is to assume a step function probability, i.e., an isotropic square-well potential of depth T_d, the threshold energy. Thus for a recoil energy below T_d the displacement probability is zero, while for energies greater than T_d the displacement probability is unity, as shown in Fig. 36a. With this assumption we can readily integrate the expression for the Darwin–Rutherford cross section [Eq. (64)] and obtain

$$\sigma_{\mathrm{DR}} = \pi\left(\frac{Ze^2}{mc^2}\right)^2 \frac{1}{\beta^4\gamma^2} \left[\frac{T_m}{T_d} - 1\right] \tag{68}$$

or from Eq. (65)

$$\sigma_{\mathrm{DR}}(\mathrm{barns}) = 0.2495\left(\frac{Z^2}{\beta^4\gamma^2}\right) \left[\frac{T_m}{T_d} - 1\right] \tag{69}$$

The corresponding integration for the McKinley–Feshbach expression differs from those given in Eqs. (68) and (69) by only a change in the expression in the square brackets, i.e., corresponding to Eq. (68) we obtain

$$\sigma_{\text{McF}} = \pi\left(\frac{Ze^2}{mc^2}\right)^2 \left(\frac{1}{\beta^4\gamma^2}\right)\left[\left(\frac{T_m}{T_d} - 1\right) - \beta^2 \ln \frac{T_m}{T_d}\right.$$

$$\left. + \alpha Z\beta\pi\left\{2\left[\left(\frac{T_m}{T_d}\right)^{1/2} - 1\right] - \ln \frac{T_m}{T_d}\right\}\right] \tag{70}$$

As we have discussed previously (Section 3.3), the recoil energies which arise from the collisions between a high-energy electron and a nucleus can give rise to both enhanced diffusion and to displacement damage, depending upon the threshold energy chosen. Figures 62–66 show the corresponding

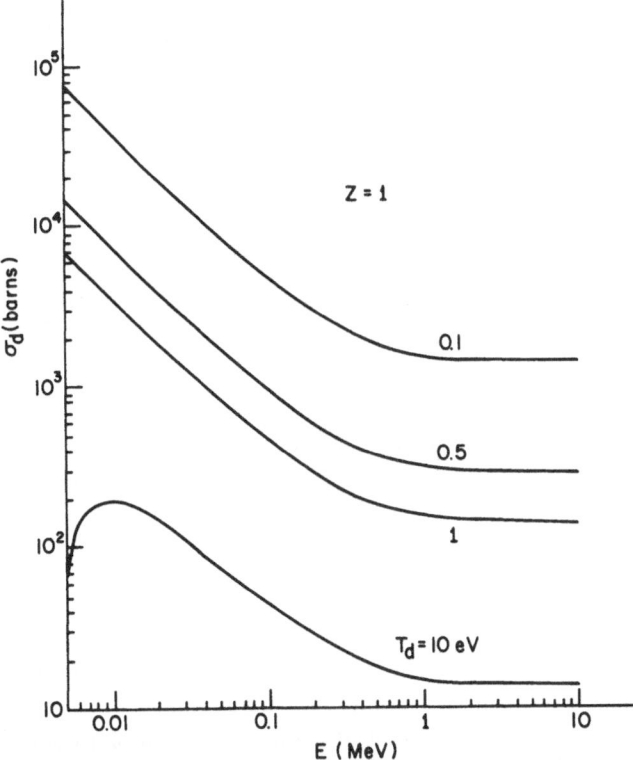

Fig. 62. The displacement cross section σ_d for electron bombardment of hydrogen versus electron energy. The results are given for different values of the threshold energy T_d.

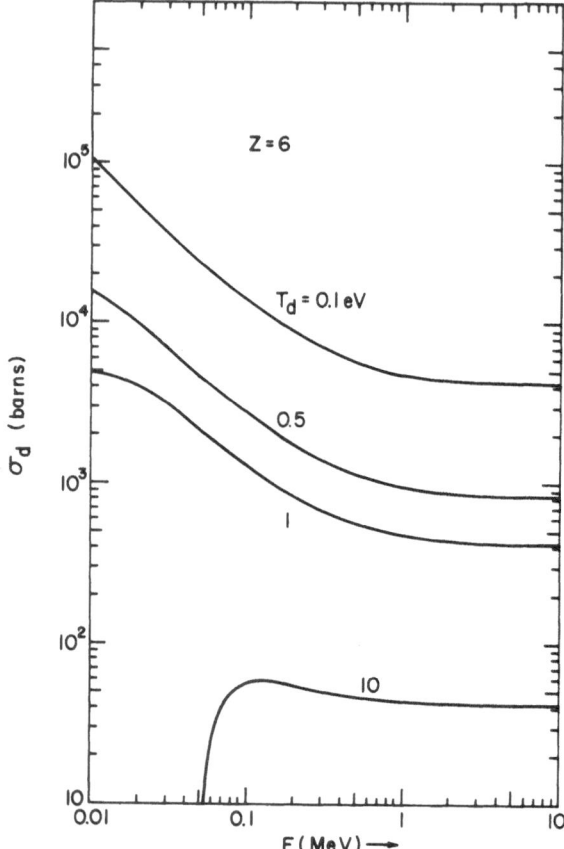

Fig. 63. The displacement cross section σ_d for electron bombardment of carbon versus electron energy. The results are given for different values of the threshold energy T_d.

displacement cross section calculated using the exact Mott expression (i.e., using the Oen program) for hydrogen, carbon, silicon, germanium, and gold; values are shown for $T_d = 10$ eV, as is typical for displacement damage processes, and for $T_d = 0.1$, 0.5, and 1.0 eV, as may be typical for enhanced diffusion processes. The logarithmic presentation in Figs. 62–66 makes it difficult to examine the energy dependence near threshold. By normalizing the cross sections at 1.5 MeV, we can plot the cross section on a linear scale and examine its functional dependence for a variety of threshold energies. Figure 67 shows the cross section for beryllium normalized at 1.5 MeV. The peaked character of the cross-section curve is a purely classical result, i.e., it occurs in the Darwin–Rutherford cross section. For low threshold energies, the cross section is linear in $T_m - T_d$ near the threshold energy. For higher values of the threshold energy (in Figs. 67 the 500-eV value typifies the case we are considering) the near-threshold dependence becomes quadratic in $T_m - T_d$. While Fig. 67 illustrates this quadratic dependence for beryllium with $T_d = 500$ eV, similar effects are found for lower (and more reasonable threshold energies) in higher Z materials. The linear and quadratic near-threshold dependences that we have just discussed will be observed only in an infinitesimally thin sample in which no loss of energy in the bombarding electron beam occurs. In a

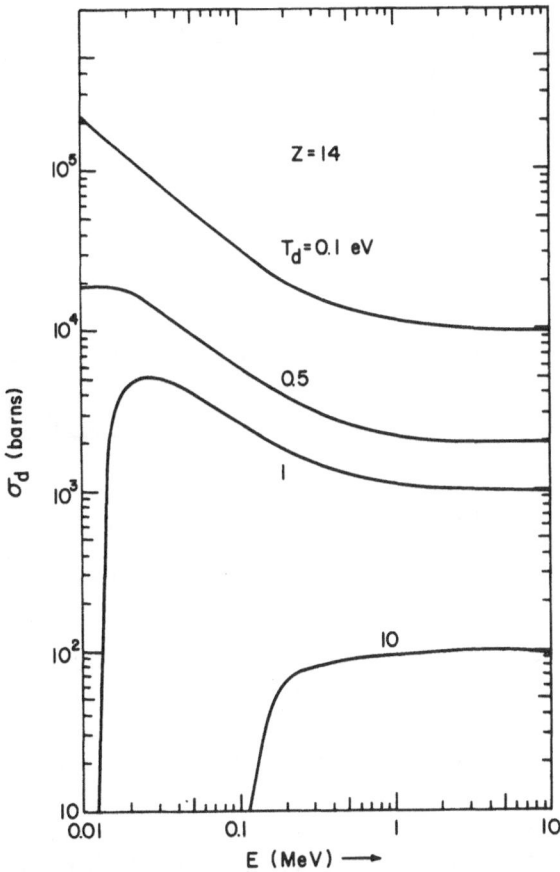

Fig. 64. The displacement cross section σ_d for electron bombardment of silicon versus electron energy. The results are given for different values of the threshold energy T_d.

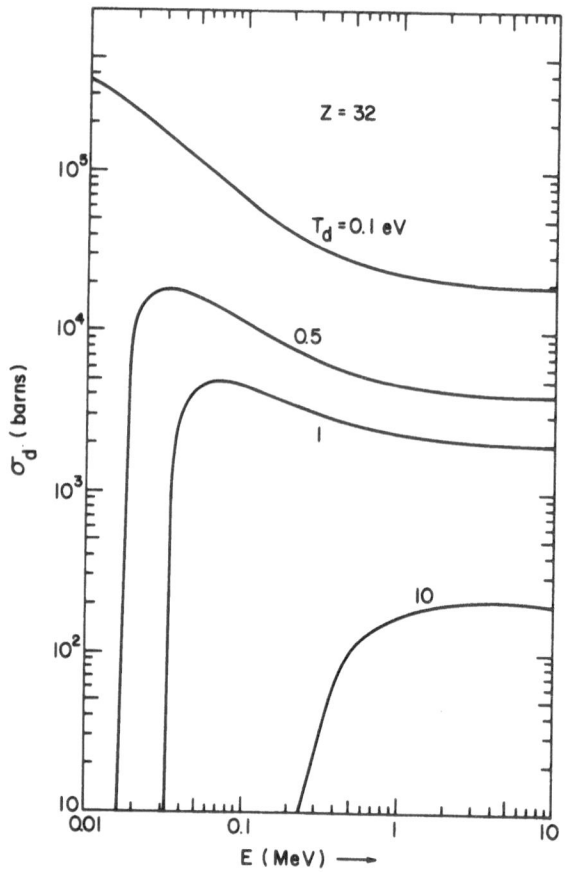

Fig. 65. The displacement cross section σ_d for electron bombardment of germanium versus electron energy. The results are given for different values of the threshold energy T_d.

sample of finite thickness the energy loss of the bombarding particles means that there will be a variation in the production of damage as the beam penetrates the sample. For thin samples the differential energy loss can be considered constant over the samples thickness. Then for low threshold energies, since the cross section for displacement is linear in $T_m - T_d$, integration over the sample thickness will give a number of displaced atoms which is quadratic[25] in $T_m - T_d$ near threshold. Similarly for high values of a threshold energy, where the cross section is quadratic in $T_m - T_d$, integration causes the measured number of defects to depend cubically[25] on $T_m - T_d$ near threshold.

Returning to Figs. 62–66, it will be noticed that far above threshold σ_d approaches a constant value. This is because at very high energies the cross section is largely determined by small-angle scattering. For very small angles the differential scattering cross section approaches the classical value, i.e., in Eqs. (69) and (70), T_m/T_d is very much greater than unity. But for $E \gg mc^2$, in Eq. (59) we obtain $T_m \sim E^2/Mc^2$ and we obtain for the displacement cross section

$$\sigma_d \xrightarrow[E \to \infty]{} \frac{2\pi}{T_d} \frac{Z^2 e^4}{Mc^2} = \frac{140 Z^2}{T_d(\text{eV}) A} \quad \text{(barns)} \qquad (71)$$

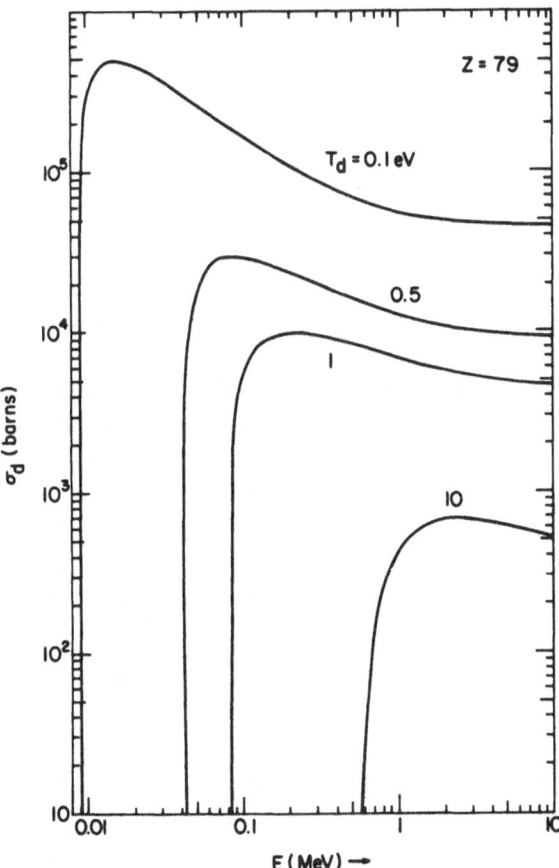

Fig. 66. The displacement cross-section σ_d for electron bombardment of gold versus electron energy. The results are given for different values of the threshold energy T_d.

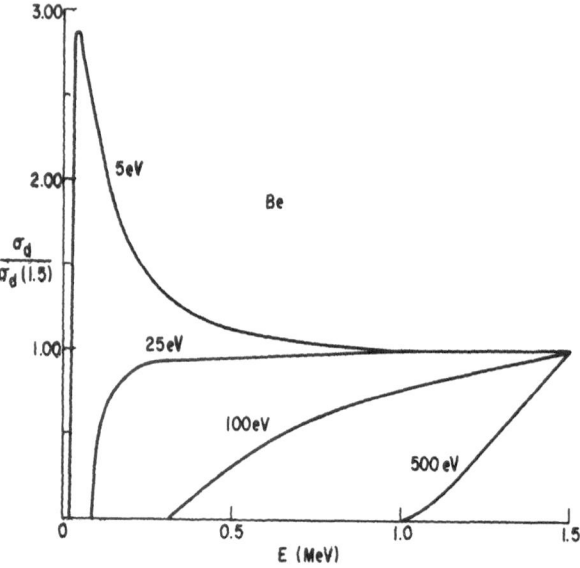

Fig. 67. The relative displacement cross section for electron bombardment of beryllium versus electron energy, where the displacement cross sections have been normalized to the value for an electron energy of 1.5 MeV. The different curves are parameterized by the value of the displacement threshold T_d.

Figure 68 presents a comparison of the energy dependence of the displacement cross section for a constant value of threshold energy (10 eV) for several cases, hydrogen, carbon, silicon, germanium, tin, and gold.

Seitz and Koehler[3] give an expression for the mean recoil energy $\bar{T}$ for the McKinley–Feshbach approximate cross section. They obtain

$$\frac{\bar{T}}{T_m} = \left(\ln\left(\frac{T_m}{T_d}\right) - \beta^2\left(1 - \frac{T_d}{T_m}\right) + \pi\alpha Z\beta\left\{2\left[1 - \left(\frac{T_d}{T_m}\right)^{1/2}\right] - \left(1 - \frac{T_d}{T_m}\right)\right\}\right)$$

$$\times \left(\left(\frac{T_m}{T_d} - 1\right) - \beta^2 \ln \frac{T_m}{T_d} + \pi\alpha Z\beta\left\{2\left[\left(\frac{T_m}{T_d}\right)^{1/2} - 1\right] - \ln \frac{T_m}{T_d}\right\}\right)^{-1}$$

(72)

Because the cross section emphasizes the low-energy collision, for all but very high maximum recoil energies the mean recoil energy will be very close to the threshold value. An equivalent expression for mean recoil energy does not exist for the Mott cross section.

The calculations that we have presented thus far are for the step-function displacement probability. Calculations for more complicated displacement phenomenology have been carried out (see, for example, Ref. 15) as well as for multiple displacement damage phenomenology (as discussed in Section 3.1). We will not discuss them in detail here.

4.5. Neutrons

Neutrons themselves produce no ionization in passing through a material and the damage they produce is due to the direct interaction of the neutrons with the nuclei of the lattice. The damage produced by the direct nuclear interaction can occur from the recoil of the nucleus either as the result of a direct scattering event or as the result of a nuclear reaction, e.g., (n, γ), (n, α), etc. The recoil energy due to a reaction is fixed by the energetics of the reaction; while damage production by neutron reactions can be important, it is not a useful probe for exploring

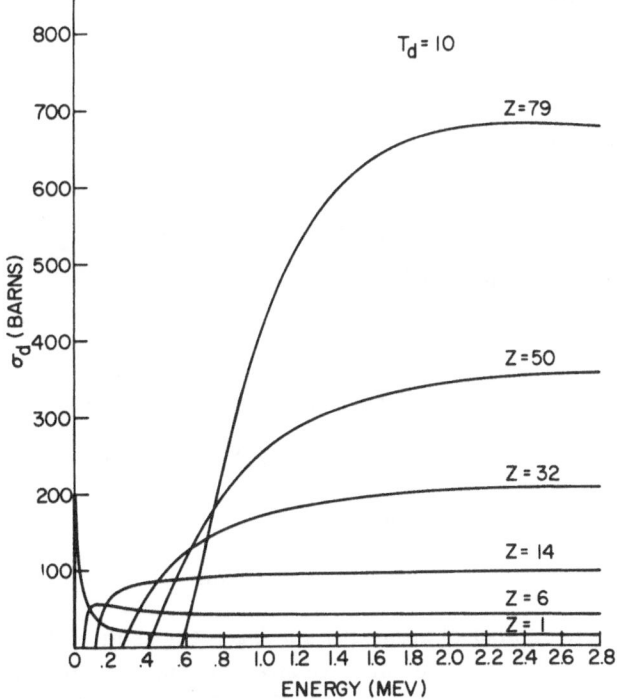

Fig. 68. The displacement cross section σ_d for hydrogen, carbon, silicon, germanium, tin, and gold versus incident electron energy. All of the calculations are for a value of the displacement threshold T_d of 10 eV.

the details of the damage production process and will not concern us further.

Neutron scattering by nuclei is predominately elastic scattering for neutron energies below several MeV. (See Kelly[319] for a discussion of the modifications required for inelastic scattering contributions.) To first order, neutron scattering by nuclei is hard-sphere scattering, i.e., isotropic in the center-of-mass system. To a very good approximation the total elastic scattering cross section σ_E is given by

$$\sigma_E = 4\pi R^2 = 4\pi A^{2/3}(1.5 \times 10^{-13} \text{ cm})^2 \tag{73}$$

where R is the nuclear radius and we have inserted the dependence of the nuclear radius on the atomic number of the nucleus. The recoil energy T of the nucleus is again given by Eq. (41) and the maximum recoil energy T_m is given by Eq. (42). The isotropic scattering cross section means that the probability of scattering into any increment dT of recoil energy is the same, i.e., constant over the range of $0 \leq T \leq T_m$, or the differential scattering cross section is given by

$$d\sigma = \sigma_E \, dT/T_m \tag{74}$$

The damage cross section is then given by

$$\sigma_d = \sigma_E \int_{T_d}^{T_m} \frac{dT}{T_m} = \sigma_E\left(1 - \frac{T_d}{T_m}\right) \tag{75}$$

The mean recoil energy $\bar{T}$ is given by

$$\bar{T} = T_m/2 \tag{76}$$

and the mean energy of the displaced atoms is given by

$$\bar{T}_d = (T_m - T_d)/2 \tag{77}$$

The damage rate R is then given by

$$R = \sigma_d N_0 \phi = \sigma_d N_0 n v \tag{78}$$

where N_0 is the number of atoms per cm^{-3}, ϕ is the neutron flux, and in the second expression in Eq. (78) we have given the more common form of the neutron flux $\phi = nv$, where n is the number of neutrons of a given velocity and v is their velocity.

The above equations give quite a good estimate of the damage production by neutrons, but for more quantitative estimates the actual energy

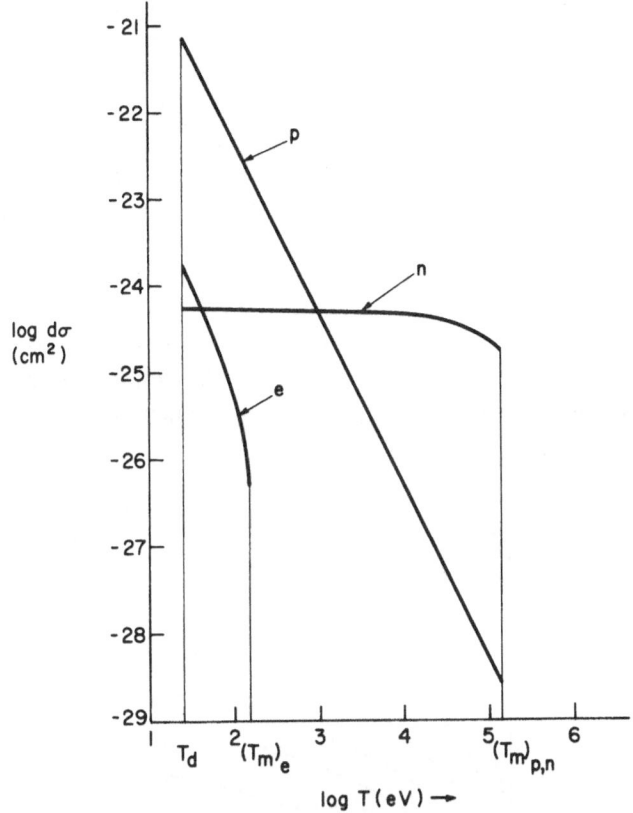

Fig. 69. Comparison of the differential scattering cross section $d\sigma$ for 1-MeV electrons, protons, and neutrons versus the recoil-atom energy T. The total displacement cross section σ_d involves an integration from the threshold energy T_d to the maximum recoil energy T_m.

dependence of the neutron scattering cross section must be included, where it is available. In detail the differential scattering cross section is often not isotropic, and is energy dependent. Moreover, for a given energy, the character of the differential scattering cross section will vary with the Z of the lattice nuclei.

In a nuclear reactor we must modify Eq. (78) to account for the energy dependence of the flux of neutrons as given by

$$R = N_0 \int \phi\sigma \, dE_n \qquad (79)$$

One of the common, almost intractable problems in reactor experiments is determining the energy dependence of the flux of neutrons.

Monoenergetic neutron sources are available as the result of certain reaction processes and are now becoming available in sufficient fluxes so that they may soon be useful in damage production studies. Neutrons have a number of advantages as damaging particles. First and foremost is the fact that they do not directly create ionization in the lattice; consequently they do not have an attendant ionization-enhanced diffusion, with the result that the damage is unmodified by diffusion. (Here we assume that the neutrons are available without an attendant gamma-ray or X-ray flux.) Figure 69 compares the differential cross section for 1-MeV electrons, protons, and neutrons. As can be seen, the electron and proton damage is created predominately by low-energy recoils. When one is trying to study

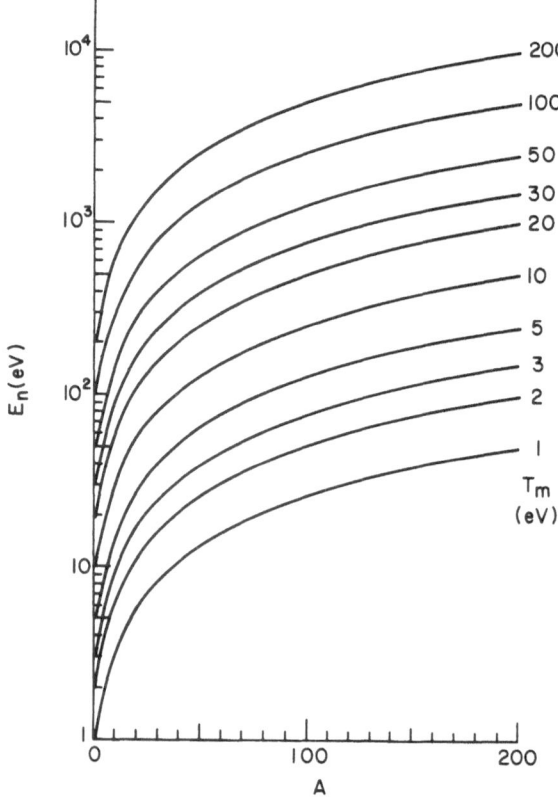

Fig. 70. Maximum recoil energy of an atom T_m versus atomic number A and incident neutron energy E_n.

the damage produced at near-maximum recoil energy it would be preferable to use neutrons since the resultant damage would not be greatly diluted by damage produced at low recoil energies. Since the atomic number of a nucleus is to a very good approximation given by M/M_n, with M the mass of the nucleus and M_n the mass of the neutron, we can rewrite the equation for the maximum recoil energy T_m in terms of the neutron energy E_n as

$$T_m = [4A/(A + 1)^2]E_n \qquad (80)$$

Figure 70 illustrates how the maximum recoil energy varies versus E_n and A. It may well be that studying processes which have a high T_d, e.g., multi-defect and spike creation processes, will be most readily carried out with neutron damage.

5. SURVEY OF DISPLACEMENT DAMAGE RESULTS

5.1. Introduction

In this section we will present the experimental details on damage results in semiconductors. Clearly we cannot present all of the data in full, but will be selective. We will include a discussion of the status of the experimental results in this section and refer a discussion of theory until Section 7.

Table V presents the threshold data for the materials we will discuss; silicon, germanium, diamond, graphite, and III–V and II–VI compounds. We include information on some of the oxides, although these compounds are frequently regarded as ionic, and are treated extensively in Vol. 1 of this Treatise. Similarly we discuss graphite, although it is not a semiconductor; as we will see, there are a good bit of data on graphite, and the fundamental bonding in graphite is primarily covalent so that a relevant comparison to semiconductors can be made. No threshold information is available on Te, Se, SiC, etc.

The threshold energy is, of course, only one parameter in the full physics of the displacement damage process. We will consequently discuss other experimental details, when available, more fully. In some systems we will find that a study of the temperature dependence of damage production has been made; in some systems a study of the doping dependence and the ionization dependence has been made; in some systems we will find that the anisotropy of the damage production with respect to the crystal and lattice has been studied; and in some systems we will find that an attempt

Table V. Value of Displacement Threshold Measured for Various Materials[a]

Material	T_d, eV	Temperature, °K	Ref.
Ge	~30	80	320, 321
	≤23	300	322
	22.3	300	323, 324
	14.5, ⟨18⟩	300	325
	15.5	300	326
	18.0	21	202
	14.5	79	202
	12.7	263	202
	14	78, 269	328
Si	20.9	300	322
	13	300	329
	14,[b] 21[c]	80, 300	334
	22	80	344
	11	300	345
	20	80	350
	~13	300	350
	⟨45⟩	300	342
Diamond	80	300	352, 353
Graphite	24.7	290	358
	60	15	189
	33	300	195
	31	300	196
	(23, 40, 0); (31, 30, −2)	6, 80	217
	(28, 42, 0)	285	217
InSb	5.7(In), 6.6(Sb)	78	359
	6.4(In), 8.5–9.9(Sb)	80	360
InAs	6.7(In), 8.3(As)	77	25, 361, 362
InP	6.7(In), 8.7(P)	77	25, 361, 362
GaAs	9.0(Ga), 9.4(As)	300	25, 361, 362
	15	77	365
GaSb	6.2(Ga), 7.5(Sb)	77	369
ZnSe	7.6(Zn), 8.2(Se)	85	371, 372
ZnTe	7.4(Zn)	10	373
	9.7(Zn), 6.7(Te)	10	375, 376
	4.2(Zn)	77	376, 376
	7.35(Zn)	10	377

Table V (*continued*)

Material	T_d, eV	Temperature, °K	Ref.
ZnS	9.9(Zn) or 20.2(S)	10	378
	9.9(Zn), 15(S)	10	379
CdS	8.7(S)	300	181, 380
	7.3(Cd)	77	381
CdSe	8.6(Se)	77	382
	8.1(Cd), 8.6(Se)	5	383
CdTe	7.8(Te)	77	384
	5.6(Cd), 7.8(Te)	5	385
MgO	60(O)	300	387
ZnO	30(Zn)	300	388
	57(Zn), 57(O)	300	389
BeO	76(O)	300	389

[a] See text for further discussion of the average values (shown by $\langle \rangle$), anisotropy, temperature dependence, etc.
[b] *n* type.
[c] *p* type.

has been made to relate the damage production process to a known defect, thereby more fully characterizing the nature of the damage production.

In no semiconductor will we find that all the parameters have been studied experimentally, and the results integrated into a comprehensive theoretical scheme, although in some systems substantial progress has been made.

5.2. Germanium

Klontz and Lark-Horovitz[320,321] performed the first experiment to determine a threshold for displacement damage in a semiconductor. (As we shall see, a number of techniques were first applied to germanium). They used electrical conductivity measurements on *n*-type samples irradiated at liquid nitrogen temperature to determine the onset of damage. They obtained the value of $\sim$30 eV.

Loferski and Rappaport[322] employed a *p–n* junction, a technique subsequently applied to a number of other systems, to study the threshold in bombardments at room temperature. Their preliminary result was that the threshold was $\leq$23 eV.

Vavilov *et al.*[323,324] used electrical resistivity measurements on *n*-type samples for irradiations carried out at room temperature. They found a threshold for damage production of approximately 22.3 eV.

Loferski and Rappaport,[325] in a more extensive report on their work, found that the onset of damage was at 14.5 eV. They argued that there was not a single well-defined threshold, advancing the following possible reasons: (1) orientation dependence of the threshold; (2) variation of the threshold near structural defects, e.g., dislocations; (3) variation in the recoil energy, and therefore in the threshold, with the nuclear mass (i.e., an isotope effect); and (4) the contribution of thermal vibrations to the recoil energy. They introduced a Gaussian distribution of thresholds to account for these effects and found that a *mean value* of the threshold of ∼18 eV fitted their data best.

Smirnov and Glazunov[326] performed the first "thick-sample" measurement of a threshold energy. They irradiated both *n*- and *p*-type samples which were thick compared to the depth of which damage is produced. They then used a microprobe technique to measure the electrical resistivity versus depth in the sample. Their irradiations were carried out at ∼30°C. The sort of data they obtained are shown in Fig. 71, which shows the carrier concentration versus depth into the sample. The data in Fig. 71 correspond to an irradiation with 700 keV. From such curves they evaluated the depth

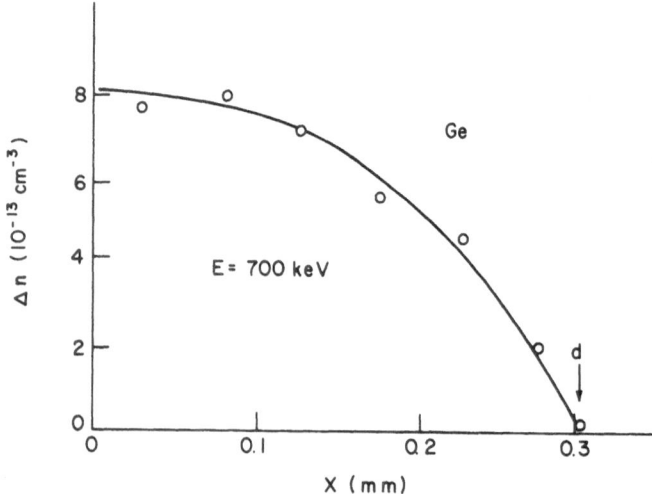

Fig. 71. Carrier removal rate Δn versus depth x for an irradiation of germanium by 700-keV electrons. The end of the penetration d is indicated. (After Smirnov and Glazunov.[326])

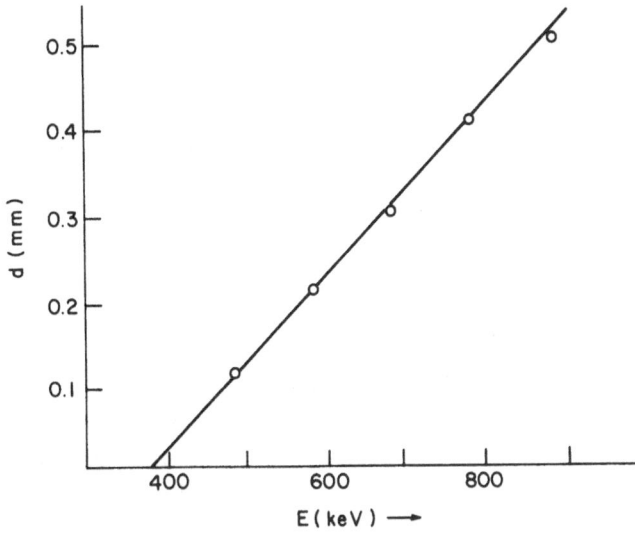

Fig. 72. The dependence of the depth of damage d upon the incident electron bombarding energy E. (After Smirnov and Glazunov.[326])

of damage and plotted this depth versus bombarding energy, as shown in Fig. 72. By extrapolating the energy dependence of the depth of damage production, they obtained a threshold of approximately 15.5 eV.

Brown and Augustyniak[202] performed the first measurements of the variation of the damage rate with respect to crystal orientation. They studied electrical conductivity in n-type samples irradiated at 79°K. Figure 73 shows some of their results (it should be noted that they studied thinner samples as well). We see that the onset of damage occurs *at the same value for all three principal crystallographic directions*: ⟨100⟩, ⟨110⟩, and ⟨111⟩. This result was not expected; it was anticipated that there would be a different threshold for different bombarding directions; we will return to this matter in Section 7. It is also to be noted that they observed subthreshold damage, i.e., damage, in the case of Fig. 73, below 400 keV. They suggested that this damage might be due to light impurity atoms in the lattice. They qualitatively confirmed the effect by irradiating a Ge–0.03% Si alloy. Naber and James[327] and Chen and MacKay[328] (see below) observed a similar effect in germanium grown in a hydrogen atmosphere. (We discuss this matter more fully in Section 6). Brown and Augustyniak[202] also studied the temperature dependence of the energy dependence of the damage rate. Their data are shown in Fig. 74. They found that the threshold varied with temperature: 18.0 eV for 21°K, 14.5 eV for 79°K, and 12.7 eV for 263°K; as we have discussed earlier, they recognized that thermal vibrations could contribute to the recoil energy and thereby change the threshold energy. There is also a variation in the energy dependence, depending upon whether annealing is carried out to 80°K or not. At the time their measurements were carried out, the rather more complete understanding of the defect processes in Ge which we now have were not available and could not be related to these results. Brown and Augustyniak noted that the damage production rate at 21°K was nonlinear; presumably they were observing the subsequently discovered radiation-annealing effect.

Whan[152] observed that the defect production in germanium is dependent upon the temperature at which the irradiation is carried out. She studied the production of infrared bands associated with defects in oxygen-doped Ge and found that the damage rate varied as shown in Fig. 75. We also show the earlier results which she obtained on infrared measurements on silicon. As we will discuss in Section 7, these results are described in terms of a metastable pair model.

Chen and MacKay[328] measured the energy dependence of damage

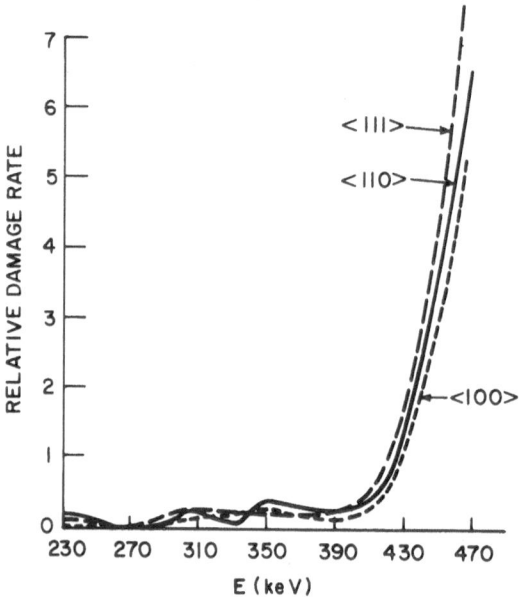

Fig. 73. Relative damage rate for electron irradiation of germanium along the three principal crystallographic axes versus electron energy E. (After Brown and Augustyniak.[202])

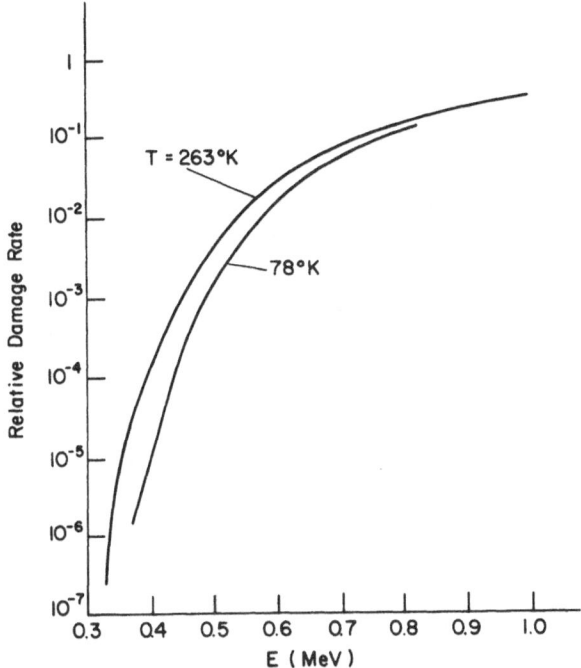

Fig. 74. Temperature dependence of the relative damage rate versus incident electron energy E for irradiation of germanium. (After Brown and Augustyniak.[202])

production at 78 and 269°K for irradiation along the $\langle 111 \rangle$ and $\langle 100 \rangle$ directions. The anisotropy they observed is shown in Fig. 76. They noted the trend toward an isotropic damage production mechanism near threshold and suggested that there exists a lower threshold corresponding to the first interstitial site along a $\langle 111 \rangle$ direction from a lattice atom. They suggested that the first site had $T_d \sim 6$ eV and for the second site $T_d \sim 14$ eV, with the probability of escape from the first site strongly energy dependent. But they noted that direct confirmation of the $T_d \sim 6$ eV value seemed beyond present experimental capabilities.

In summary, it is still true, of course, that we do not understand fully the defect recovery processes in germanium. Nonetheless it is unfortunate that the extensive studies of damage processes in germanium have not been pursued further in light of our present understanding. In particular, a study seems *much to be desired* which relates the work of Whan[151,152] to that of Brown and Augustyniak[202] and Chen and MacKay[328] and includes an investigation of the role of ionization.

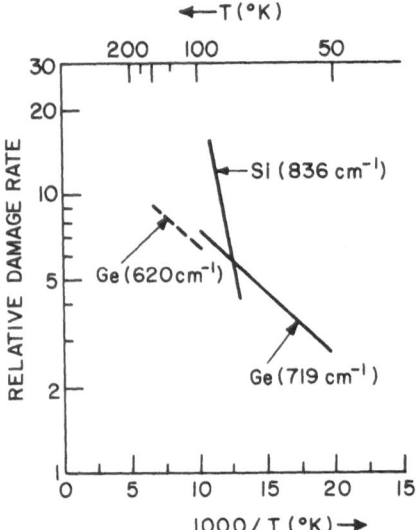

Fig. 75. Temperature dependence of the production rate of infrared bands in oxygen-doped
germanium and silicon. (After Whan.[152])

5.3. Silicon

Loferski and Rappaport[322] used the photovoltaic effect in *p–n* junctions
to study the threshold in room-temperature irradiations. They observed a
threshold for the onset of damage in silicon of 20.9 eV. Their measurements
were essentially a "thin-sample" experiment, but they noted their concern
for the thickness of their alloy junctions.

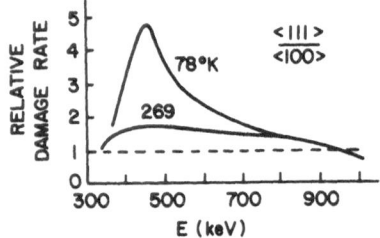

Fig. 76. Temperature dependence of the energy
dependence of the relative damage rate for
bombardment of germanium along the ⟨111⟩
and ⟨100⟩ directions. (After Chen and Mac-
Kay.[328])

Vavilov et al.[329] performed thick-sample measurements. They irradiated p-type silicon samples of different crystallographic orientations with 500-keV electrons and then used a microprobe technique to measure the electrical resistivity versus depth into the sample. They found that damage along the $\langle 111 \rangle$ orientation was larger than along $\langle 110 \rangle$, which is in turn larger than that along $\langle 100 \rangle$. They used the Yurkov[330,331] theory to calculate the distribution of damage versus penetration into the sample. They found that a threshold of 13 eV described their damage dependence versus depth better than a threshold of 28 eV.

Flicker, Loferski, and Scott-Monck[333] used the photovoltaic effect in very thin p–n junctions (which could be made using a technology improved over that in their earlier work) to measure the damage rate. They studied "n on p" and "p on n" junctions, using both floating-zone and pulled material and various dopings. The results among the various samples were in general agreement except that the commercial "p on n" cells had ~100 times larger damage rate, although they had roughly the same energy dependence. They found, however, that the energy dependence did not seem to have a sharp onset threshold, nor did the energy dependence follow the predictions for a step-function threshold. Novak[334] used electrical conductivity and Hall coefficient measurements to monitor the damage rate for irradiations at both room temperature and liquid nitrogen temperature. He found that in n-type material the threshold was ~14 eV, while in p-type material the threshold was ~21 eV. He also found the production rate to be Fermi level and temperature dependent. This probably accounts for the disagreement with the Vavilov[329] data. Flicker and Loferski[335] also disagree with the dependence of damage versus depth found by Vavilov et al.,[329] again a discrepancy probably due to the dependence on type of impurity, temperature, and Fermi level. Wikner, Horiye, and Harrity[336] measured carrier removal rate for an 80°K irradiation of n-type silicon. They irradiated with 5- and 45-MeV electrons. They found a rough agreement with the energy dependence characteristics of a step-function threshold with a threshold energy of 25 eV.

Next, a number of workers[337–344] investigated the orientation dependence of the energy dependence of the damage production process. There are a number of disagreements between these measurements. Figure 77 shows the results of Haddad and Banbury.[341] They converted these data into an anisotropy ratio as shown in Fig. 78. They also obtained data on p-type material, although they had more difficulty with these measurements. In the p-type results they found that at 0.4 MeV the carrier removal rate for $\langle 110 \rangle$ direction was greater than for the $\langle 111 \rangle$, which in turn was

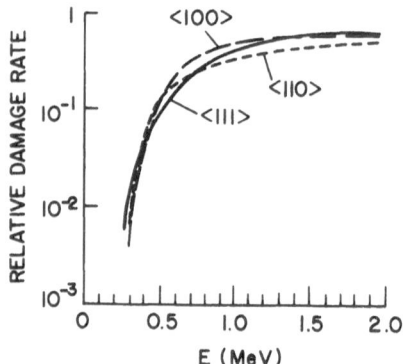

Fig. 77. Relative damage rate in silicon versus
bombarding energy E for irradiation along the
principal crystallographic directions. (After
Haddad and Banbury.[341])

greater than for the $\langle 100 \rangle$, while from 0.8 to 1.9 MeV the $\langle 100 \rangle$ was greater
than the $\langle 111 \rangle$, which was greater than the $\langle 110 \rangle$; at 2 MeV the $\langle 110 \rangle$
became the highest again. Their orientation dependence differed from the
finding of George and Gunnersen[338] and of Kryukova and Vavilov[337] on
n-type silicon and with those of Vavilov *et al.* for p-type material, but agreed
with the order of the orientation dependence found by Fang and Gdula[339]
at 0.3, 0.5, and 1.0 MeV in all respects, except that at 1 MeV Haddad and
Banbury[341] found that the $\langle 110 \rangle$ damage rate was less than the $\langle 111 \rangle$.
Grimshaw,[342] in work on junctions, found results in qualitative agreement
with the results of Haddad and Banbury; he found that the average damage

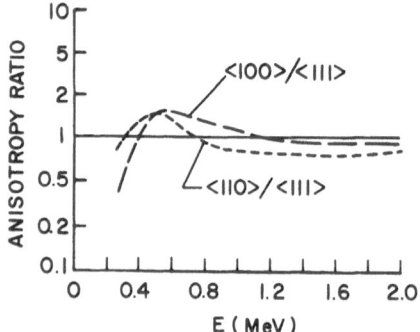

Fig. 78. The damage rate given in Fig. 77
expressed as a ratio to the damage rate along
the $\langle 111 \rangle$ direction. (After Haddad and Banbury.[341])

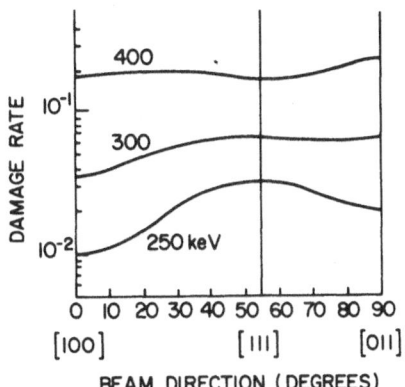

Fig. 79. Orientation dependence for the damage rate in silicon for a bombardment at 245°K. (After Hemment and Stevens.[343])

rate in n-type material over the range 0.35–2.0 MeV fit the primary displacement theory with an *average* threshold of 45 ± 5 eV. These unfortunate discrepancies among the results of various experimenters have never been fully accounted for; we note that the level of ionization may have something to do with the discrepancies; for example, Haddad *et al.* used a beam pulsed at 400 Hz while others used a dc beam.

Hemment and Stevens[343] reported measurements of the carrier removal rate in n-type silicon versus beam direction with respect to the crystalline axis. Their results for an irradiation temperature of $\sim$245°K are shown in Fig. 79 and their results for an irradiation at a temperature of $\sim$80°K are shown in Fig. 80. They also present data on the relative damage rate for an irradiation at $\sim$245°K as shown in Fig. 81. Banbury,[344] in considering similar data found in Ge by Chen and MacKay[328] (shown in Fig. 76), had noted that the anisotropy of the threshold seemed to disappear at lower energies. Banbury[344] had suggested that the true "threshold" was associated with the anisotropic damage process and that below that there occurred what he termed an isotropic "subthreshold" damage process. From such an argument he concluded that the true threshold was $\sim$22 eV rather than the more frequently quoted value of $\sim$14 eV. Hemment and Stevens[343] made this same argument based on the data in Fig. 81. We must note here that this "subthreshold" damage is *not what is conventionally termed subthreshold damage* (which we will discuss in Section 6). In part, these authors come to their conclusions on the basis of theories of the damage production process and its anisotropy; in Section 7 we will return

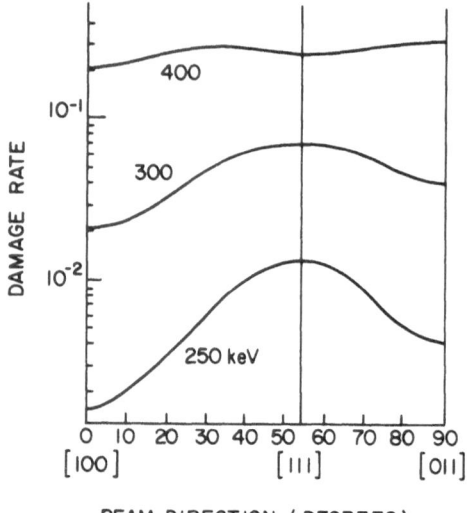

BEAM DIRECTION (DEGREES)

Fig. 80. Orientation dependence for the damage rate in silicon for a bombardment at 80°K. (After Hemment and Stevens.[343])

to a consideration of these data and their theories. We note here, however, that Hemment and Stevens[343] made measurements which are pertinent to this point of subthreshold damage. They measured the temperature dependence of the carrier concentration after irradiation. They concluded that they were observing both (vacancy + oxygen) and (vacancy + phosphorus) centers and at 200 keV both of these centers were produced at the same relative rate as at 400 keV! This clearly argues that they are observing the same type of damage at both 400 and 200 keV. But 200 keV is below their "threshold" energy of 230 keV. We note as well, as we will discuss more

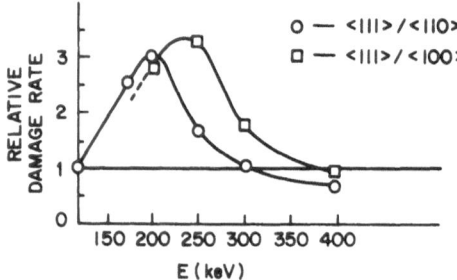

Fig. 81. Relative damage rate for the principal crystallographic directions for the data shown in Fig. 79.

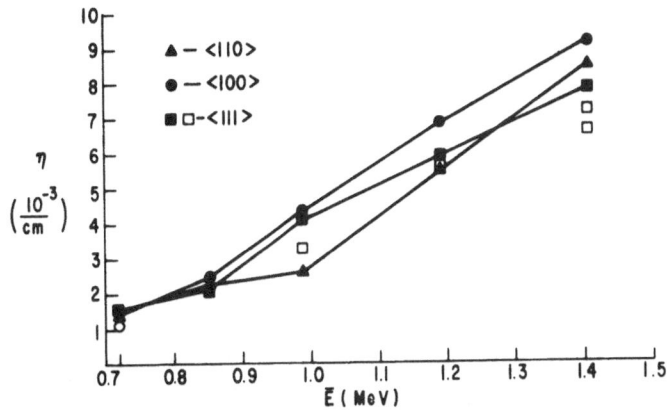

Fig. 82. Production rate η of the divacancy for irradiation along different crystallographic directions in silicon versus the mean bombarding energy $\bar{E}$. (After Corbett and Watkins.[121])

fully below, that Gerasimenko et al.[345] observed (vacancy + oxygen) centers for bombardment down to 125 keV, i.e., corresponding to a threshold of 11 eV!

Two experiments have used specific EPR-identified defects to study the damage production process. Corbett and Watkins[121,122] used EPR measurements of the (vacancy + oxygen) center and of the divacancy to study the damage process in n-type Si. They found over the range 0.7–56 MeV that their (vacancy + oxygen) data agreed with the data of Flicker, Loferski, and Scott-Monck.[333] Corbett and Watkins[121,122] used measurements of the divacancy to study the anisotropy of the energy dependence of the damage production rate, as shown in Fig. 82. Their measurements also lend themselves to a *direct* measurement of the anisotropy of divacancy production, i.e., of the number of divacancies along the beam direction for irradiation in the [111] direction versus the number produced in one of the other $\langle 111 \rangle$ directions. Figure 83 shows the energy dependence of the anisotropy R of the divacancy production along the beam [111] direction versus that along another $\langle 111 \rangle$ axis.

Gerasimenko et al.[345] used EPR measurements of (vacancy + oxygen) centers to study the threshold region of damage production. Their measurements were carried out at room temperature in n-type material. Their results are shown in Fig. 84. They observed a very interesting *ionization dependence* of the production rate above 400 keV, that is, the production rate of (vacancy + oxygen) centers depended upon the rate at which damage was created. They noted that extrapolation of such curves, in the case of

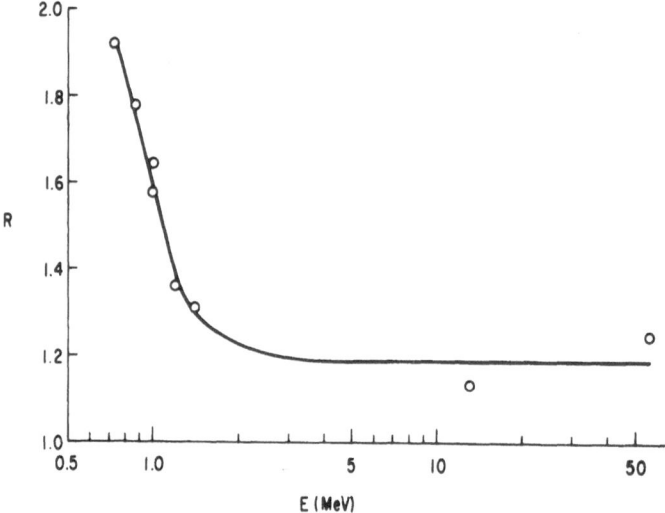

Fig. 83. Ratio of the divacancy production rate along the beam direction for irradiation in the [111] direction to the production rate in one of the other ⟨111⟩ directions versus electron bombarding energy E. (After Corbett and Watkins.[121])

poor-sensitivity measurements, would be ambiguous as to what was the threshold of the onset of damage. They were able to carry the measurements to a lower energy, where the ionization dependence did not occur and then found no ambiguity. They found no (vacancy + oxygen) centers were created below 125 keV, which corresponds to a threshold for the onset of damage of 11.0 eV.

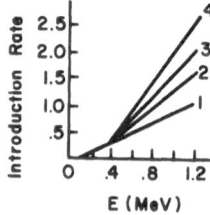

Fig. 84. Relative defect introduction rate as measured by the production rate of the (vacancy + oxygen) center versus electron bombarding energy E. The numbers indicate electron current density (μA/cm²): (1) 0.2; (2) 0.3; (3) 0.5; and (4) 20. (After Gerasimenko et al.[345])

In a study of the temperature dependence of the damage production rate, Whan and Vook[346] found that infrared bands associated with defects in silicon exhibited an exponential temperature dependence in the production rate at low temperatures. (See Fig. 75 in the section on germanium). Novak[347] observed a similar dependence in electrical measurements. Vook and Stein[348] extended these electrical measurements and studied the Fermi level dependence as well. Their results are shown in Fig. 85. These authors distinguish between ITD and ITI defects, i.e., defects whose production rates are irradiation temperature dependent (ITD) and irradiation temperature independent (ITI). Stein and Vook[349] showed (see Fig. 86) that the ITI defects fit the divacancy production rate, whereas the ITD defects fit roughly the energy dependence observed by Haddad and Banbury.[351] These authors presented an interpretation of these data in terms of a charge-state-dependent metastable pair model which we will discuss more fully in Section 7.

Kolomenskaya, Razumovskii, and Vitovkin[350] have made electrical measurements on n-type material to study the threshold for irradiation at 300 and 80°K. They found that the threshold is $\sim$13 eV at 300°K and is $\sim$20 eV at 80°K.

Panov and Smirnov[351] extended the study of the damage rate, using

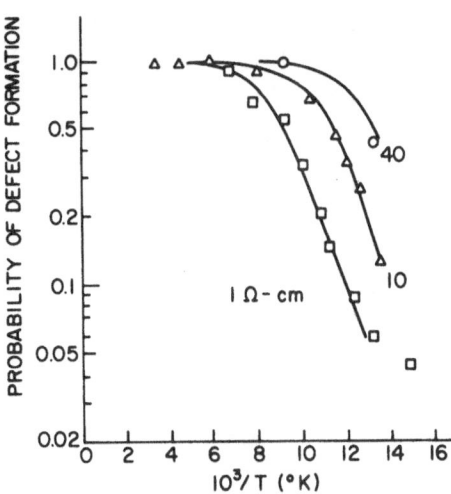

Fig. 85. Comparison of the charge-state-limited theory of defect formation (solid curves) to experimental points for irradiation of samples of silicon of various resistivities (as shown), versus temperature T. (After Vook and Stein.[348])

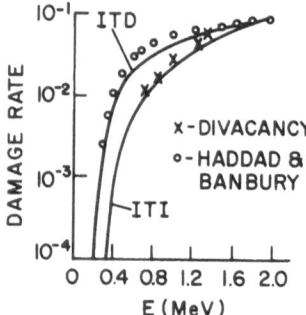

Fig. 86. Energy dependence of the ITD and ITI defects compared to the damage rate observed by Haddad and Banbury and to the divacancy production rate. (After Stein and Vook.[349])

the (vacancy + oxygen) EPR signal, to include a study of the temperature dependence of the threshold as shown in Fig. 87. Their measurements were carried out at a constant beam current of 0.5 $\mu A/cm^2$, and a study of the ionization dependence of their results has not been published. In their brief report they also did not go into the question of reconciling their data with those of Gerasimenko et al.,[345] who reported a threshold of 11.0 eV for a room-temperature irradiation, and those of Kolomenskaya et al.,[350] who found a 13-eV threshold at 300°K and a 20-eV threshold at 80°K; nor did they compare their results with those of Vook and Stein.[348] Clearly, more work in this area needs to be done, including studies which not only pursue the temperature dependence, but also the ionization and orientation dependences.

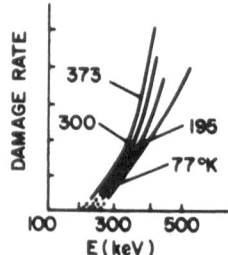

Fig. 87. Temperature dependence of the damage rate as determined by the introduction of (vacancy + oxygen) centers versus energy E. (After Panov and Smirnov.[351])

We recall that Gerasimenko *et al.*[345] found a strong ionization dependence of the damage rate. Aside from the dependence upon sample resistivity (Fig. 85), the ionization dependence of the energy dependence of the damage rate has not been studied.

5.4. Diamond and Graphite

There has been only one published measurement related to the threshold for damaged production in diamond. Clark, Kemmey, and Mitchell[352,353] studied both the optical properties of a type IIa diamond and the electrical properties of a semiconducting diamond. As a measure of the optical damage, they plotted the absorption at 2.0 eV. This absorption is at the peak of the multiphonon components of the set of optical bands usually designated as the GR1 system. The GR1 system has long been argued to be associated with the vacancy, an assignment supported recently by calulations of the line shape[354] and of the piezospectroscopic response of the zero-phonon lines.[355] The Clark, Kemmey, and Mitchell energy dependence data are shown in Fig. 88. As can be seen, the optical measurements favor

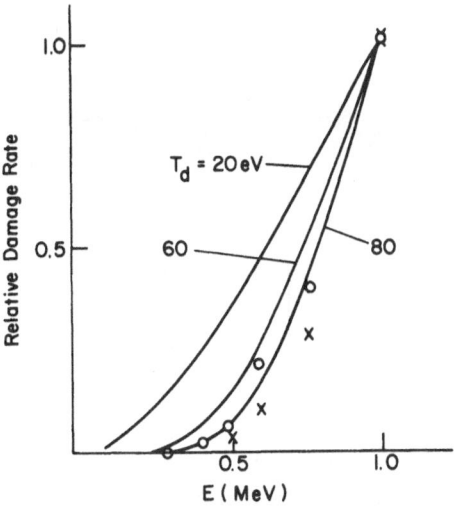

Fig. 88. Relative damage rate for electron irradiation of diamond as determined by optical measurements (circles) and electrical measurements (crosses) versus electron energy E. Theoretical curves for different choices of the displacement threshold T_d are given. (After Clark, Kemmey, and Mitchell.[353])

a threshold of 80 eV, while the electrical measurements favor an even higher threshold energy. The GR1 band persists to 950°C, so that if it is a simple intrinsic defect, it is a remarkably stable one. Threshold measurements have not been made following low-temperature radiations, although EPR measurements ($T_{irr} \geq 17°K$) by Lomer and Wild[154,156] and by Brosious, Bourgoin, and Corbett[155] ($T_{irr} \geq 77°K$) indicate that substantial annealing and probable long-distance (interstitial) migration occur below room temperature.

Graphite has, of course, a different crystallographic structure than diamond, as shown in Fig. 89, and is a semimetal. An approach viewing graphite from a covalent bonding point of view, however, is successful in treating the band structure of *both* diamond[356] and graphite,[357] as well as treating defect properties[82,85] theoretically. Hence we feel justified in including graphite for comparison to diamond.

Eggen[358] made the first measurement of the threshold energy of graphite, obtaining a value of 24.7 ± 0.9 eV for measurements at 290°K on polycrystalline graphite. Lucas and Mitchell[189] made electrical resistivity measurements on single crystals of graphite that were irradiated at 15°K. They studied the energy dependence of the damage process, and fitting their

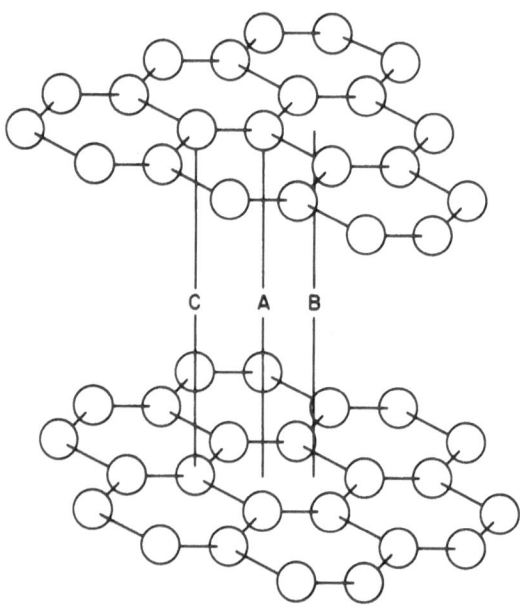

Fig. 89. Representation of the structure of graphite with several possible interstitial sites shown.

results to theory, obtained a displacement energy of 60 ± 10 eV (we note that this high value places a great deal of emphasis on their data at 300 keV). Montet[195] studied the damage process in natural graphite crystals through the observation of vacancy clusters in an electron microscope. He found a threshold energy along the c axis of 33 eV at room temperature. Further, he found that the threshold energy remains constant when the angle between the c axis and the electron beam was increased to 60°; above 60°, the threshold energy increased sharply, reaching a value of 60 eV at 80°. Montet suggested that this abrupt change in threshold energy was hard to understand on intuitive grounds. Montet and Myers,[196] in recent work, found a c-axis threshold of 31 eV.

Iwata and Nihira[217] have made extensive studies of the damage process in graphite. They studied the displacement process at 6, 80, and 285°K. They measured the a-axis electrical resistivity. Their data are shown in Fig. 90. They characterized the threshold energy as

$$T_d(\psi) = A \cos^2 \psi + B \sin^2 \psi + C(1 - \cos^4 \psi) \qquad (81)$$

and expressed their results in terms of the three parameters A, B, and C (in eV), and ψ, the angle between the c axis and the displacement direction. A and B are the threshold energies of the directions parallel and perpendicular to the x axis, respectively.

Graphite has four atoms per unit cell. Iwata and Nihira[217] did not argue that they can distinguish all four atoms per unit cell, but they did note that their data seemed to require two different threshold energies near $\psi = 0$, which they assumed corresponded to the displacement of two different types of atoms. For both the 6 and 80°K irradiations they obtained the sets of thresholds (23, 40, 0) and (31, 30, -2) (where all energies are to an accuracy of ± 2 eV). For the 285°K irradiation they obtain either a single set (28, 42, 0) or the sets (28, 42, 0) and (32, 42, -1). They also examined the multiple displacement probabilities and concluded that the 6 and 80°K irradiations seemed to involve a mixture of the Harrison–Seitz replacement model and the Kinchin–Pease model, whereas for the 285°K irradiation the cascade obeyed the Harrison–Seitz no-replacement model. We see that their c-axis threshold varies from 23 eV for the 6 and 80°K irradiations to 28 eV for the 285°K irradiation. Iwata, Nihira, and Matsuo[193,194] have studied the annealing stages which occur in low-temperature irradiated graphite. From their work we would conclude that the 6°K irradiation is below the temperature at which long-range migration of interstitial atoms seem to occur (which they assigned to the temperature range

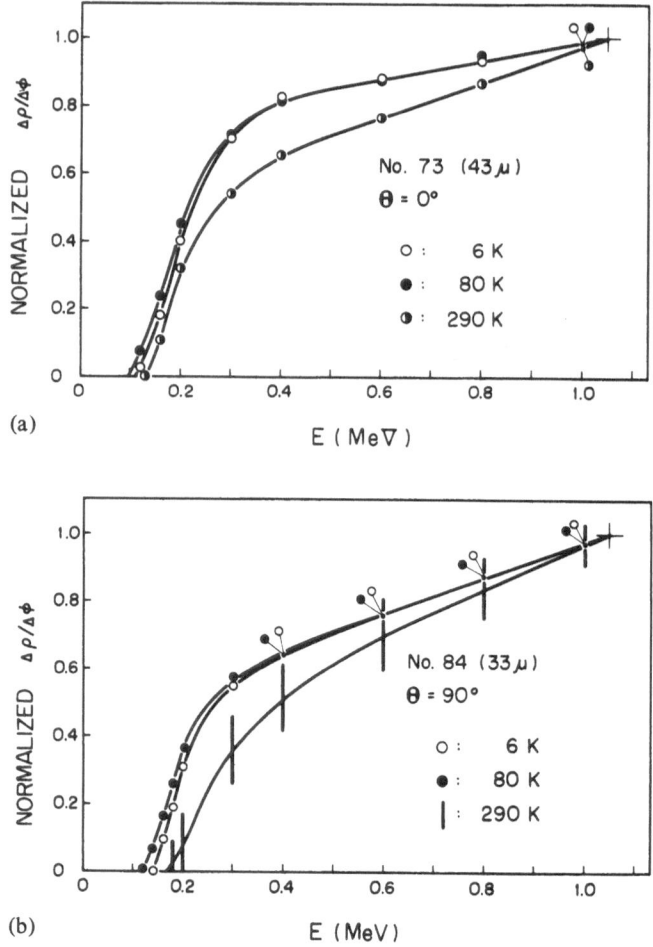

Fig. 90. Temperature dependence of the relative damage rate (expressed as the change in resistivity $\Delta\varrho$ per change in flux $\Delta\phi$) versus electron bombarding energy E. The sample number and thickness are indicated. (a) Irradiation along the c axis, $\theta = 0°$; (b) Irradiation perpendicular to the c axis, $\theta = 90°$. (From Iwata and Nihira.[217])

45–65°K), whereas of course the 80°K irradiation is above that temperature range.

We presume that ionization plays no role in the damage process in graphite (because it is a semimetal), and therefore must conclude that the studies of the energy dependence of the damage process in graphite are essentially complete, indeed more complete than in any other material we will consider. The studies in diamond are obviously in a much more primitive stage. This is quite unfortunate since, as we will note in Section 7, of the covalent bonded systems, these two are probably the most amenable to theoretical treatment.

5.5. III–V Compounds

5.5.1. *InSb, InAs, InP*

Eisen and Bickel[359] studied the carrier removal rate in indium antimonide for irradiation at 78°K. Their data are shown in Fig. 91. Bäuerlein[25] examined their data from the "thin sample" point of view, i.e., recognizing that near threshold the damage rate would rise quadratically with energy, as shown in Fig. 92. He singled out two thresholds, one at 240 keV and the other at 285 keV. (He did not consider the portion of the curve below 245 keV, recognizing that the mechanism for damage production below 240 keV had to be *of a fundamentally different character* than that occurring above the 240-keV threshold. There have subsequently been a number of experiments relating to this "subthreshold" damage in indium antimonide, but we will defer discussion of this until Section 6. Bäuerlein[25] assigned the 240-keV threshold to the displacement of In ($T_d = 5.7$ eV) and the 285-keV threshold to the displacement of Sb ($T_d = 6.6$ eV). Eisen[360] carried out experiments to test this assignment. He recognized a special feature of the structure of the III–V compounds. The diamond structure has, along the $\langle 111 \rangle$ direction, two lattice atoms in a row at tetrahedral sites, followed by two tetrahedral sites which are unoccupied, followed by two occupied sites, two empty sites, etc. In the case of InSb (the zinc-blende structure)

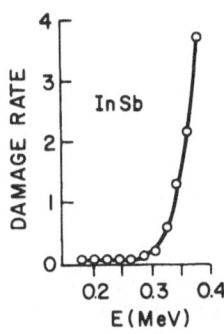

Fig. 91. Energy E dependence of the damage
rate in InSb. (After Eisen and Bickel.[359])

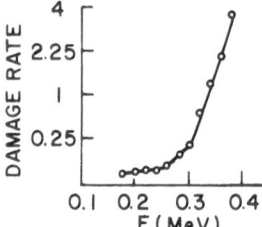

Fig. 92. Energy E dependence of the damage
rate shown in Fig. 91 plotted versus a squared
ordinate scale. (After Bäuerlein.[25])

in the [111] direction the order of the atoms is In, Sb, empty, empty, etc.
Eisen[360] reasoned that an irradiation in the [111] direction would find it
easier to displace an antimony atom, since it had empty spaces behind it,
whereas in an irradiation in the [$\bar{1}\bar{1}\bar{1}$] direction it would be easier to displace
the indium atoms. Eisen studied the energy dependence of the recovery
stages. He found that Stage I (centered at 83°K) had a higher threshold
than Stage II (centered at 128°K). His data are shown in Fig. 93. He associated Stage I with defects due to the displacement of antimony atoms to
which he assigned a threshold displacement energy of between 8.5 and
9.9 eV. He associated Stage II with the displacement of indium atoms with
a threshold of about 6.4 eV. As can be seen in Fig. 93, the Stage II data
exhibit the "breaks" which Bäuerlein had attributed to the displacement of
the different atoms—In and Sb. Eisen argued that the Bäuerlein interpretation is *not correct*. He argued from his studies of the recovery in Stage II
that the breaks are due to two different close-pair defects and he associated
the change in slope at 300 keV as the threshold for the production of the

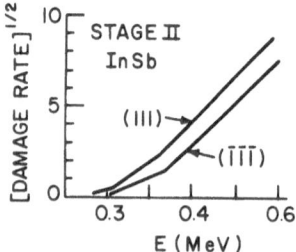

Fig. 93. Energy E dependence of the square
root of the damage rate for Stage II in electron
irradiation of InSb. The different curves are
for irradiation on the (111) and ($\bar{1}\bar{1}\bar{1}$) surfaces.
(After Eisen.[360])

second kind of these close pairs. He made no assignment to the change in slope at 300 keV. Clearly, more extensive recovery studies following up these suggestions should be made. We note in any event that these changes in slopes may simply represent the thresholds for new channels of producing the same defects, i.e., production at an angle different from the direction of bombardment.

Because of Eisen's work we must view with caution assignments of different thresholds for different atoms. We make this caveat here, although we will in our subsequent description of experiments report the assignments as the authors made them.

Bäuerlein[25,361,362] used the degradation of the photoresponse of diodes following electron irradiation at 77°K to study the threshold process in InAs diodes. He assigned a threshold of 6.7 eV to the displacement of indium atoms and 8.3 eV for arsenic atoms. In similar studies on InP diodes he assigned the displacement threshold of 6.7 eV for the displacement of indium and 8.7 eV for the displacement of phosphorus atoms.

Lindsay and Banbury[363] used Hall and conductivity measurements on n- and p-type InAs samples to study the energy dependence of the recovery stages: Stage I centered at 160°K, II at 250°K, and III at 330°K. Their results for Stages I and II are shown in Figs. 94 and 95 for an 80°K irradiation. They noted that their Stage III defects had a very similar energy dependence to those of Stage II, from which they suggested that both stages are caused by the same displacement event. They performed isothermal annealing experiments on each of the three stages, but failed to find simple kinetics. Employing the computer calculations used by Fisher and Banbury[364] for Si, now modified for InAs, they concluded that both Stage I and Stage II were due to the displacement of As atoms (and from our

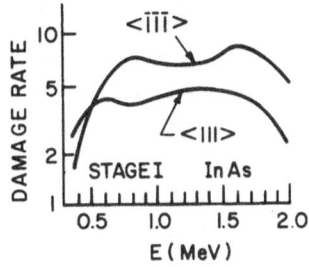

Fig. 94. Energy E dependence of the damage rate for Stage I observed in electron irradiation of InAs at 80°K. Data are shown for bombardment along the $\langle 111 \rangle$ and $\langle \bar{1}\bar{1}\bar{1} \rangle$ directions. (After Lindsay and Banbury.[363])

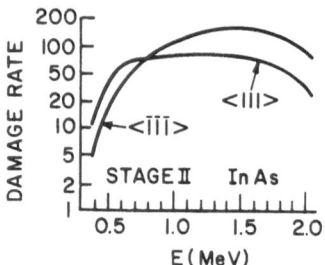

Fig. 95. Energy E dependence of the damage
rate for Stage II observed in electron irradiation of InAs at 80°K. Data are shown for
bombardment along the $\langle 111\rangle$ and $\langle\overline{1}\overline{1}\overline{1}\rangle$
directions. (After Lindsay and Banbury.[363])

comment above inferentially Stage III as well). The damage that persists
above 380°K, they suggested, is due to divacancies, the low-energy portion
being due primarily to As primaries and the higher energies being due to In
primaries.

5.5.2. GaAs, GaSb

Bäuerlein[25,361,362] studied the displacement threshold process using
GaAs photodiodes irradiated at 300°K. He assigned a threshold energy
of 9.0 eV (233 keV) to gallium and of 9.4 eV (256 keV) to arsenic. Grimshaw and Banbury[365] studied the electrical property changes in n-type
material irradiated at 77°K. They found no damage below ~370 keV,
corresponding to a threshold of approximately 15 eV. They found that
their energy dependence from 370 keV to 1.0 MeV could be best fit by an
average threshold of approximately 17 to 18 eV. They suggested that the
discrepancy between their measurements and Bäuerlein's measurements
was due to the fact that the damage cross section possesses a "tail" at lower
energies. Loferski and co-workers studied the radiation damage degradation
of optical properties in gallium arsenide. Loferski and Wu[366] studied n-type
material irradiated at 300°K and concluded that the damage threshold was
between 275 and 288 keV. Loferski et al.[367] studied p-type material irradiated
at 100°K and concluded that the damage threshold was between 250 and
300 keV. Thommen[368] studied the energy dependence of the recovery stages
in GaAs: Stage I, ~235°K; Stage II, ~280°K; and Stage III, ~520°K.
He studied electrical property changes in n-type GaAs irradiated near 5
and 77°K. He found that a small amount of irreversible ionization-induced
recovery was observed after irradiation at 5°K, and that changes which

occurred upon annealing below 200°K could be reversed by ionizing radiation. His results on the energy dependence of the recovery are shown in
Fig. 96. Arguing by analogy to the work in InSb, Thommen associated the
Stage III damage with gallium displacements. Unfortunately, he did not
carry his studies below 400 keV, as would be required to shed some light
on the discrepancies in the threshold energy. We note that if, again in
analogy with the work in InSb, the Stage I and II recovery is associated
with arsenic, then Thommen's data in Fig. 96 strongly disagree with Bäuerlein's assignment of a threshold of 9.4 eV to arsenic.

Thommen[369] studied the irradiation-induced changes in the electrical
properties of GaSb irradiated at 77°K. Using studies of the energy dependence of the recovery stages, he associated Stage I (centered at approximately 120°K) with the displacement of antimony atoms with a threshold
of 7.5 eV and Stage II (centered at 160°K) with the displacement of gallium
with a threshold of 6.2 eV.

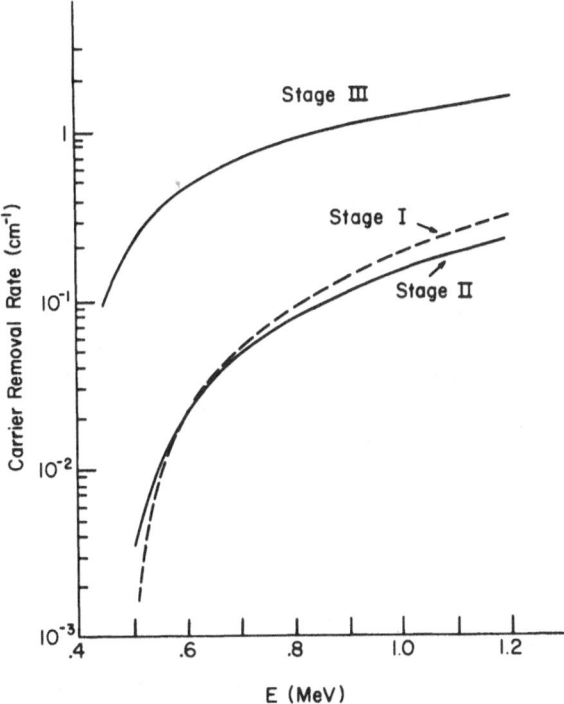

Fig. 96. Energy E dependence of the carrier removal rate
observed in three substages following electron irradiation of
GaAs. (After Thommen.[368])

In summary, the III–V compounds have some very interesting results. Particularly fascinating is the anisotropy of damage production, all the more so since the threshold energy is isotropic in elemental semiconductors. Much more work, however, is needed to fill out the full picture, i.e., the temperature, ionization, and orientation dependences and their relationship to known defects. Of course, the same aspects of the puzzling "subthreshold" damage (which will be discussed more fully in Section 6) must be elucidated.

5.6. II–VI Compounds

A few definitive defect identifications have been made using EPR studies on II–VI compounds.[115] The isolated chalcogenide vacancy has been reported in ZnS,[171] ZnO,[172,173] and BeO,[174] as has the isolated zinc vacancy in ZnSe[115] and the zinc vacancy–donor pair in ZnS and ZnSe.* In some cases these results have begun to be related to earlier optical measurements on II–VI compounds; it is these optical measurements that have largely been used in threshold studies. As yet, however, use of a firmly established defect has not been carried out systematically in threshold studies in II–VI compounds. A systematic study of the temperature dependence of damage production has not yet proven feasible, nor has a thoroughgoing study of the role of ionization been carried out. The II–VI compounds are, of course, considerably more ionic than the other materials we consider, yet the strong indications of the role of ionization are found in the other materials. We also must note that many of the II–VI compounds have significant impurity contents and often substantial stoichiometric defects as well. Working with stoichiometric defects, created by an excess vapor pressure of one component or the other, has permitted certain defect properties to be assigned to one vacancy or the other; defect properties established in this way have been used to study threshold effects.

There is a widespread view among workers in II–VI compounds that the observation of two thresholds implies that one threshold corresponds to the displacement of one component and the other threshold to the displacement of the other component. This may, of course, be correct. We must note, however, that the experience in the III–V compounds with two thresholds being observed for one sublattice and the lack of internal consistency in some of the II–VI work argues that care must be made in assignment of thresholds to sublattices.

---

* For a review of this work see Ref. 370.

5.6.1. *ZnSe, ZnTe, ZnS*

Kulp and Detweiler[371] found a threshold at 240 keV in ZnSe associated with the production at 85°K of a broad fluorescence band. Their data are shown in Fig. 97. Kulp[170] argued that this 240-keV threshold was due to a displacement of a selenium atom. Subsequently these authors[372] found a second threshold at 195 keV while bombarding at 10°K; this threshold they attributed to the displacement of Zn atoms ($T_d = 7.6$ eV). Their data for the 10°K irradiation are shown in Fig. 98. They found that upon annealing, the fluorescence spectrum produced at 10°K changes into that produced by the 80°K bombardment, from which they concluded that at least one and possibly more of the fluorescence centers observed following bombardment at 80°K is a "complex" center, i.e., it is composed of a defect of zinc or selenium in association with a foreign impurity. They attributed the 240-keV threshold to the displacement of selenium ($T_d = 8.2$ eV).

Bryant and Baker[373] and Meese[374] observed a threshold energy at approximately 185 keV which they associated with the displacement of Zn ($T_d = 7.4$ eV). Bryant and Baker used the bound-exciton emission (2.36–2.37 eV) and Meese used broadband emission (2.1–2.35 eV) to monitor the threshold energy.

Bryant and Baker[375] found that electron irradiation of *p*-type ZnTe at 77°K produced a sharp (bound exciton) emission at 2.362 eV. They used this emission to establish an energy threshold of 300 keV, which they assigned to the displacement of Te atoms ($T_d = 6.7$ eV). These same authors[376] confirmed the results of the production of a sharp line at 2.362 eV. They found another line at 2.358 eV which had a threshold at 110 keV for irradiation at 77°K and at 235 keV for irradiation at 10°K. This lower energy threshold they attributed to the displacement of Zn ($T_d = 4.2$ eV at 77°K and 9.7 eV at 10°K). Their data are shown in Fig. 99.

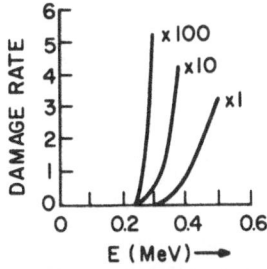

Fig. 97. Energy *E* dependence of the damage rate observed in the 85°K electron irradiation of ZnSe. (After Kulp and Detweiler.[371])

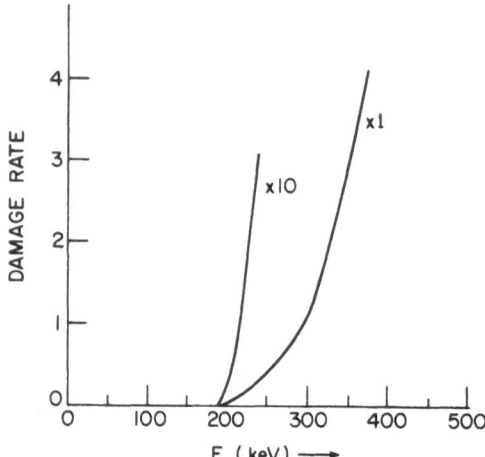

Fig. 98. Energy E dependence of the damage rate observed in the 10°K electron irradiation of ZnSe. (After Detweiler and Kulp.[372])

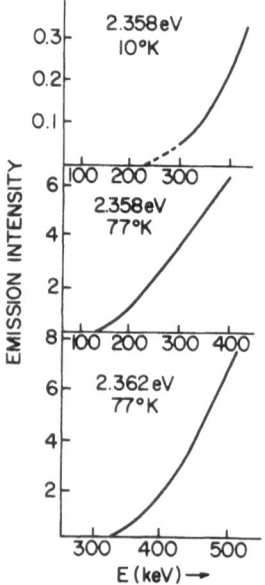

Fig. 99. Energy E dependence of the damage production, as observed by emission lines at the energies shown, for electron irradiation of ZnTe for electron irradiation at the temperatures shown. (After Bryant and Baker.[376])

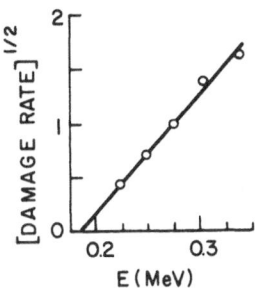

Fig. 100. Energy E dependence of the damage
rate (shown on a square root scale) for the
10°K electron irradiation of ZnTe. (After
Meese and Park.[377])

Meese and Park[377] used the zero-phonon lines at 2.320 and 2.234 eV
of two optical emission series to observe a damage threshold at 185 keV.
Their samples were irradiated at 10°K. They associated this threshold with
the displacement of Zn atoms ($T_d = 7.35$ eV). Their data are shown in
Fig. 100. They could not pinpoint the reason for the discrepancy between
their work (185 keV) and that of Bryant and Baker (235 keV).

Bryant and Cox[378] studied the thermoluminescence of ZnS and found
a damage threshold between 240 and 250 keV. This would correspond to
a $T_d = 20.2$ eV for displacement of a sulfur atom or 9.9 eV for displacement of a Zn atom. Bryant and Hamid[379] continued the study of thermoluminescence in ZnS following room-temperature irradiation and found
another threshold in the range 175–195 keV. They tentatively assigned the
185-keV threshold to the displacement of S ($T_d = 15.0$ eV) and the 240-keV
threshold to the displacement of Zn ($T_d = 9.9$ eV).

5.6.2. CdS, CdSe, CdTe

Kulp and Kelley[181,380] used the optical properties of CdS to study the
threshold for damage. They observed a threshold at 115 keV which they
associated with the displacement of the S atom ($T_d = 8.7$ eV). This is the
threshold for the production of green edge-emission centers and of centers
which give rise to a red fluorescence band with a maximum at about 7200 Å.
Some of their crystals show the edge-emission before bombardment, but
in these samples the edge-emission is removed by electron bombardment
($E = 2.5$–200 keV), i.e., a radiation annealing. They proposed a model of

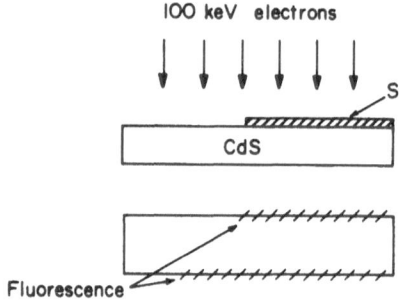

Fig. 101. Schematic representation of an overlay experiment used to demonstrate induced diffusion of sulfur. (After Kulp and Kelley.[181,380])

a sulfur interstitial as the center responsible for the edge-emission and a sulfur vacancy as the center responsible for the red emission band. They carried out a very interesting overlay experiment in support of their model. As shown in Fig. 101, they took a sample which had no green fluorescence and coated half its length with vacuum-evaporated sulfur. This crystal was then bombarded at $-100°C$ with 100-keV electrons, after which the half of the crystal under the sulfur turned a brilliant green, while the other half continued to fluoresce red under electron bombardment. Under UV stimulation the crystal fluoresced bright green only. They noted that the underside of the crystal glowed green *over a larger area* of the crystal than the top, indicating that electron-induced diffusion of interstitials occurred, i.e., a drift of the defect under the influence of the bombarding beam.

Kulp[381] observed a threshold at 290 keV for the production of two fluorescent bands in CdS under electron bombardment at 77°K. He associated this threshold with the displacement of cadmium atoms ($T_d = 7.3$ eV). As shown in Fig. 102, he carried out an overlay experiment to support the view that the fluorescence band at 6050 Å is due to cadmium interstitial atoms. He associated a band at 1.03 μm with cadmium vacancies since this band could also be produced by heating undoped cadmium sulfide crystals at 200°C in air, argon, or vacuum.

Kulp[382] used fluorescence emission to study the threshold for damage in CdSe. He irradiated at 77°K and obtained a threshold of 250 keV. His results are shown in Fig. 103. He felt he could not assign the threshold to either sublattice. Schulz and Kulp[383] irradiated CdSe at 5°K and found a

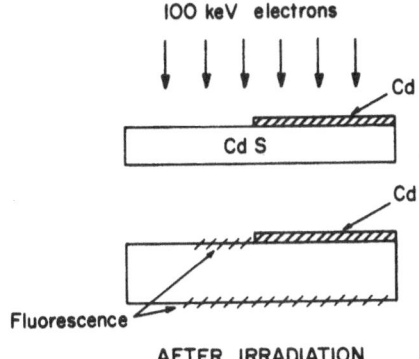

Fig. 102. Schematic representation of an overlay experiment demonstrating the induced motion of cadmium. (After Kulp.[381])

new emission band produced with a threshold of 320 keV. Since this emission band annealed below 77°K, they argued that they were not observing the temperature dependence of the threshold, but rather a separate displacement process. They assigned the 320-keV threshold to the displacement of Cd atoms ($T_d = 8.1$ eV) and the threshold at 250 keV to the displacement of Se ($T_d = 8.6$ eV).

Bryant and Webster[384] studied the luminescence emission spectrum in p-type CdTe following electron irradiation at 77°K. They found an energy threshold at 340 keV but did not assign the threshold to either sublattice. Their results are shown in Fig. 104. Bryant, Cox, and Webster[385] combined low-temperature luminescence measurements on electron-irradiated CdTe with luminescence data reported on nonstoichiometric specimens. They irradiated at liquid helium temperatures and found a displacement threshold at 235 keV which they associated with the displacement of cadmium

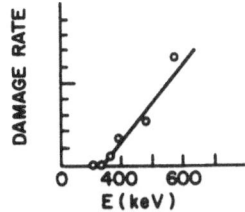

Fig. 103. Energy E dependence of the damage rate for a 77°K electron irradiation of CdSe. (After Kulp.[381])

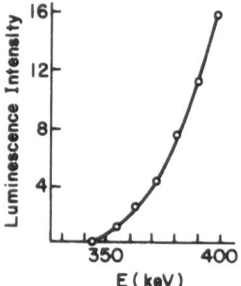

Fig. 104. Energy E dependence of the damage rate monitored by luminescence produced in a 77°K electron irradiation of CdTe. (After Bryant and Webster.[384])

atoms ($T_d = 5.64$ eV), and they associated the 340-keV threshold with the displacement of tellurium ($T_d = 7.79$ eV). Their results are shown in Fig. 105. They argued that some of their results correspond to a close pair, i.e., a cadmium vacancy with a nearby interstitial cadmium atom.

5.6.3. MgO, ZnO, BeO

Sibley and Chen[385] were apparently the first to realize that the highly ionic II–VI oxides, specifically MgO, require displacement damage, rather

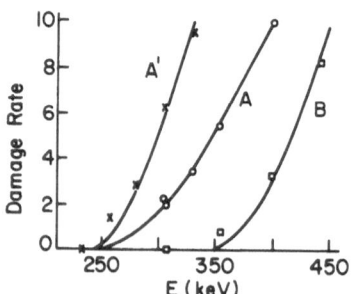

Fig. 105. Energy E dependence of the damage rate observed in the cathodoluminescence following electron irradiation of CdTe. Curve A is for emission at 8085 Å; curve A' is for emission at 8050 Å; curve B is for emission at 11000 Å. (After Bryant, Cox, and Webster.[385])

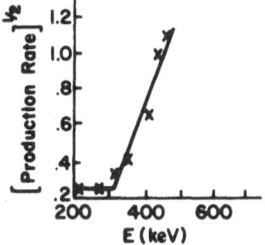

Fig. 106. Energy E dependence of the damage production rate (square root scale) as measured by the production of the F^+ center following electron irradiation of ZnO. (After Locker and Meese.[389])

than ionization damage. Chen *et al.*[387] studied optical absorption in MgO and observed a threshold at 330 keV which they associated with the displacement of oxygen ($T_d = 60$ eV), a *very* high displacement energy. Vehse *et al.*[388] studied the optical absorption of ZnO following 0.6- and 1.7-MeV electron irradiation. They correlated an absorption band which they observed with the yellow color centers produced by heat treatment in Zn vapor. They favored assignment of a 600-keV threshold with the displacement being that of Zn atoms ($T_d = 30$ eV). This work was followed up by Locker and Meese,[389] who studied the carrier removal rate following 40°C irradiation of ZnO crystals. They observed a threshold of 310 keV which they associated with oxygen displacement ($T_d = 57$ eV) and a threshold of 900 keV which they associated with Zn displacement ($T_d = 57$ eV), i.e., the same threshold energy for both the zinc and oxygen displacement. They also studied the EPR and optical properties of their samples. Their EPR data for the production rate of the F^+ center[173] are shown in Fig. 106. It will be noticed that there is a substantial subthreshold damage rate, which they found to be proportional to the electron beam power and suggested might be related to the generation of X-rays in the sample. For irradiations above 900 keV they found changes in their EPR spectra and observed the increase in the absorption band studied by Vehse *et al.*[388] Finally, Locker and Meese mention unpublished work on the displacement threshold in BeO by J. C. Pigg, A. K. Garrison, and S. Austerman; these authors observed a threshold at 405 keV which they associated with the displacement of oxygen ($T_d = 76$ eV). We see that the displacement energy in all these II–VI oxides appears to fall in the range of 50–80 eV.

6. SUBTHRESHOLD AND IONIZATION DAMAGE

In most of this review we have been concerned with displacement damage, i.e., the process whereby defects are created when energetic particles transmit to the atoms of the lattice an energy higher than a certain value called the threshold energy. But there are, as we encountered in Section 5, instances of damage production in semiconductors by so-called "subthreshold irradiation." There is substantial controversy in this area, specifically related to the question of whether ionization alone can create defects. Certainly ionization can alter the charge state of preexisting defects and dramatically alter the properties of a semiconductor. At issue is whether ionization can create new defects. We will first present the data and arguments in support of ionization creating new defects, and then turn to the evidence which argues against that point.

As Eisen[163] has reviewed, in InSb there are substantial data on the question of defect production in InSb by low-energy X-rays, subthreshold electrons, and light. Arnold and Vook[390] found that both the electrical and thermal conductivity of InSb were changed by irradiation with 100-keV X-rays. The samples they studied were p-type samples and samples which had been converted from n to p type by irradiation with MeV electrons. They argued that some form of ionization mechanism must be responsible for these changes because the 2-MeV electron irradiation should have saturated all charge-state changes, and thermal conductivity changes should not be affected by surface effects. Others have found similar changes in the electrical conductivity due to irradiation by X-rays[391] and electron bombardment with energies as low as 20 keV.[163] Vitovskii et al.[392] reported changes in the electrical conductivity of p-type InSb samples which had been irradiated with light. They found that the annealing of this damage occurred in the same temperature region ($\sim$100°K) as the annealing of damage in samples irradiated with gamma rays. They concluded that the same type of damage was created in both experiments. To some extent their results are similar to those of Kreutz,[393,394] who reported changes in both the electrical and thermal conductivities following X-ray irradiation. Kreutz also observed that the recovery in the electrical conductivity occurred at 100°K and observed the same activation energy as reported by Arnold and Vook[390] and Eisen.[163] Kreutz, however, concluded that at least the electrical conductivity changes were due to surface effects since the conductivity change depended upon the sample thickness and various surface treatments. He concluded that the thermal conductivity results could not be understood at that time. Concerning the congruity of the annealing behavior, Eisen[163]

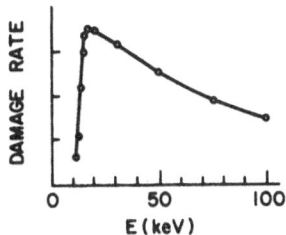

Fig. 107. Energy E dependence of the damage rate observed in an electron irradiation of a 50-μm silicon epitaxial film. (After Zaikovskaya, Kiv, and Niyazova.[396])

argued that there is a problem in distinguishing the damage produced by subthreshold mechanisms from that which ordinarily recovers in Stage II ($\sim$100°K). Although Eisen concluded that more work would be required to establish whether the damage created by light, X-rays, and low-energy electrons was really a surface effect or a real volume damage effect, he did note that the anisotropy experiments (i.e., the dependence of the damage upon the bombardment direction, [111] vs. [$\bar{1}\bar{1}\bar{1}$]) indicated that ionization mechanisms cannot play a significant role in the creation of this damage. Anisotropy experiments have not been reported on *sub*threshold damage in InSb.

Zaikovskaya, Kiv, and Niyazova[395-397] have studied irradiation-induced changes in the electrical resistance of silicon epitaxial films bombarded with electrons of energies up to 100 keV. They studied the dose, energy, thickness, and temperature dependences. Their resistance changes were substantial, i.e., in some cases a 30% resistance change for 10^{15} electrons/cm². The energy dependence of their relative damage rate is shown in Fig. 107 for a sample of 50 μm thickness. They found that the maximum in the energy dependence shifted to lower energies with decreasing film thickness. They concluded that the damage was due to defects which result from the ionization of the silicon K shell. Figure 108 shows the agreement of the energy dependence of the K-shell ionization cross section with the value of the relative damage rate for zero sample thickness, which they obtain from extrapolating their data. The temperature dependence of their damage rate is shown in Fig. 109, which includes data from 16- and 100-keV electron irradiations. In addition they investigated the dependence of the subthreshold defect production by 16-keV electron irradiation upon the boron concentration of the sample, finding a proportionality between the defect production rate and the hole concentration in *p*-type silicon.

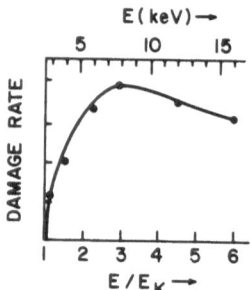

Fig. 108. Energy E dependence of the damage rate in a silicon epitaxial film extrapolated to zero thickness. The experimental points are shown as closed circles; the solid curve gives the K-shell ionization cross section with ionization energy E_K. (After Zaikovskaya, Kiv, and Niyazova.[396])

Pabst and Palmer[398] have studied Rutherford backscattering of Si. They argued that they have observed ionization-created damage in hydrogen or helium bombardment of Si because their observed defect introduction rate is substantially larger than expected. (We will return to their measurements later.) They also suggest a specific mechanism for ionization damage. They argued that the energy loss of the high-energy particles they used is high enough to occasionally yield the ionization of two *neighboring atoms*. The resultant Coulomb repulsion, they argued, is enough to cause the displacement of one or more atoms. This is an interesting model but it requires further quantitative elaboration, e.g., cross sections for formation of the ion pair, net damage cross section, etc. A question to be answered

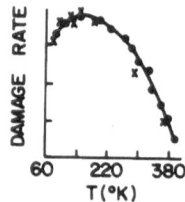

Fig. 109. Temperature T dependence of the damage rate in silicon epitaxial films. The filled circles correspond to 16-keV electron irradiation; the crosses correspond to 100-keV electron irradiation. (After Zaikovskaya, Kiv, and Niyazova.[396])

is why the two *holes* on the adjacent atoms do not simply repel one another rather than causing the *ions* to be repelled.

There have been a number of other suggestions as to mechanisms of subthreshold damage. One which has occurred commonly is the suggestion of displacement damage at "soft spots" in the lattice, e.g., at dislocations. Another suggestion is due to Oen,[399] who argued that subthreshold damage might occur if an electronic bond is broken in the same collision which imparts the recoil energy to the nucleus; the probability that a Rutherford collision simultaneously ionizes an atom is relatively high, but as Eisen[163] has noted, it seems difficult to envision that such a mechanism would lower the threshold more than about 25%. Other suggestions for damage by ionization have invoked mechanisms which have been proposed for ionic crystals, where ionization clearly does create lattice defects. One of these mechanisms is the Varley mechanism.[223-225] This mechanism requires the double ionization of an atom, as shown in Fig. 39, so that, say, an originally negative ion now finds itself in a positive charge state surrounded by positive ions, with the result that the central ion is expelled from its site—creating a Frenkel defect. This mechanism does not seem to be favored in the alkali halides and requires a high degree of ionicity which does not occur in most semiconductors. Balarin[400] suggested that the energetic photoelectrons created by high-energy ionizing particles could in turn impart energy through Coulomb collisions with a nucleus, i.e., the displacement damage mechanism we have already been discussing. Pooley,[226] in discussing Balarin's work, recognized that the photoionization of an electron would require that the atom recoil. He gave the maximum recoil energy T_m of an ion of mass M which would occur on the absorption of a photon of energy E_γ and the ejection of an electron from a level of ionization potential E_i as

$$T_m = (m/M)\{(E_\gamma - E_i) + E_\gamma [2(E_\gamma - E_i)/mc^2]^{1/2}\} \qquad (82)$$

with m the electron mass. This equation implies that the maximum recoil could be of the order of 0.2 eV for a 30-keV photon on the K shell of bromine. Again this process does not seem to be favored for damage production in the alkali halides, and while it could operate in the semiconductors, the magnitude of the recoil energy is small. Several authors[226-230] have suggested the model which seems to be the favored damage production mechanism in alkali halides. This mechanism recognizes that ionization can form, say, Cl_2^- in KCl with a very substantial displacement between the two chlorines, as shown in Fig. 40. The mechanism then argues that, when this Cl_2^- molecule captures an electron, sufficient strain energy is related to create an interstitial–vacancy pair, i.e., the release of the distortion along the $\langle 110 \rangle$ axis results in a substantial impulse along that direction. (See Fig. 41.) This specific model is not appropriate for semiconductors, in that the corresponding strain configuration does not occur; it may well be that we will eventually find that in the more ionic materials this type of damage production mechanism occurs, but for now there is no evidence of it in semiconductors. Platzman[401] suggested a mechanism which seems to operate in molecules. He argued that the ionization of a bond in a molecule permits the rupture of the molecule at that point in the case where there are no other bonds restraining the rupture. In solids there is other "bonding" and a single broken bond alone will not result in an atomic displacement.

Loferski and Rappaport[322] found experimentally that there was a low-energy "tail" in the energy dependence of their damage production curve, and they concluded that this was possibly due to the displacement of atoms near dislocations, i.e., a site where the threshold energy was less (the "soft spot" mentioned earlier). Brown and Augustyniak[202] found clear evidence of subthreshold damage in germanium, i.e., damage extending some 100 keV below a $\sim$370 keV threshold. They noted that the subthreshold damage was crystal dependent. They argued that the presence of substitutional silicon or carbon atoms in germanium, because of their light mass, might result in displacement damage at an energy below the threshold damage for the displacement of germanium atoms. They irradiated one alloy sample containing 0.03% silicon and found a noticeable increase in the carrier removal rate in the subthreshold region. Naber and James[327] suggested that a two-step process involving an intermediate hydrogen atom might give rise to the displacement of germanium atoms at an energy significantly less than the usual threshold energy. They envisioned that the hydrogen atom initially receives a recoil energy (for a given incident electron energy it can receive approximately four times as much energy as a germanium atom) and that the hydrogen atom then in turn collides with a germanium atom. They found that crystals grown in hydrogen exhibited subthreshold damage, whereas crystals grown in vacuum or helium did not. Chen and MacKay[402] studied subthreshold damage in germanium in detail. They studied the carrier removal rate at 78°K in crystals grown in vacuum, hydrogen, deuterium, nitrogen, or helium. They found subthreshold damage *only* in those crystals grown in hydrogen. They studied the nature of the defects, the orientation dependence, the energy dependence, and the annealing of the subthreshold damage. They found a threshold for subthreshold damage of $\sim$40 keV, with a peak in the damage at $\sim$150 keV. They found no orientation dependence in the damage production (see Fig. 110), whereas

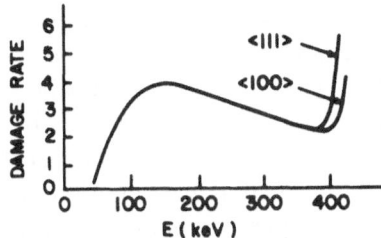

Fig. 110. Anisotropy of the energy E dependence of the subthreshold damage rate found in electron irradiation of germanium. (After Chen and MacKay.[402])

the lattice damage itself does exhibit an orientation dependence. They found that the recovery of the damage exhibits a special peak at $\sim$350°K. They concluded that the subthreshold damage did not correspond to a two-step displacement damage process, and in fact corresponded not to direct lattice damage itself, but rather to the relocation of hydrogen. Further, they found that a heat treatment in vacuum or a hydrogen atmosphere at $\sim$450°C for 24 hr rendered the crystal immune to subthreshold damage, although it still contains hydrogen. They suggested that the relocation of hydrogen occurs either by a direct knock-on process or by ionization of the hydrogen bond at its normal site. They favored the ionization explanation since they were not able to observe subthreshold damage in deuterium-grown crystals.

A final explanation which has been advanced for subthreshold damage is that it is an *artifact*, due to surface effects and/or the diffusion of impurities. For example, Norris[403] has studied irradiation effects in bulk silicon and thin silicon films at energies far below threshold and found no permanent damage. His experiments showed no permanent resistance changes in bulk samples at fluences of 20-keV electrons as high as 2×10^{18} electrons/cm² for irradiations carried out at 78°K. He also reported measurements made using EPR and optical techniques in an endeavor to detect the aluminum interstitial or (vacancy + oxygen) EPR centers or the divacancy IR absorption bands. Neither the EPR centers nor the optical bands were observed, which implied fewer than 10^{-4}–10^{-5} atomic displacements per 20-keV electon. He did detect electrical changes associated with a surface inversion layer related to the inherently oxidized silicon surface. He noted that a K-shell ionization damage theory is ruled out in that the resistance effects remain at energies well below the K-shell ionization potential. We note that Gerasimenko *et al.*,[345] in studying the threshold with EPR techniques, observed that there were no (oxygen + vacancy) centers created below 125 keV. As they noted, and Fig. 84 shows, above 400 keV their defect production rate was ionization sensitive, presumably related to the ionization changing the relative trapping cross sections of various defects.

We have argued[232] that much of the subthreshold damage could be attributed to ionization facilitating the diffusion of impurities; for example, in the Pabst and Palmer experiment[398] their techniques of Rutherford backscattering is extremely sensitive to the state of the surface layers, i.e., ionization-enhanced diffusion of impurities might then be the explanation for their enhanced damage rate.

InSb has had substantial work indicating subthreshold damage but, as we noted earlier, Eisen[162] observed that the anisotropy effect indicates that most of the damage is not ionization dependent. We note as well that Kreutz found his electrical conductivity changes to be dependent upon the thickness of the sample and upon the effects of various surface treatments, from which he concluded that at least the electrical conductivity changes were due to surface effects. We simply note again that impurity effects might very well fit into this pattern.

Mashovets, Vikhlii, and Vitovskii[167] have shown clearly that impurities can play a role in subthreshold damage in InSb; they found that in germanium-doped InSb there was an enhanced subthreshold damage, apparently due to ionization-dissociation of germanium pairs.

In summary, it is indisputable that subthreshold damage can occur. It can often be important in circumstances precisely relevant to devices. There is, however, substantial question as to the origin of this damage, and much work needs to be done on this subject.

7. THEORIES OF DISPLACEMENT DAMAGE PRODUCTION

In this section we will review the several theories which have been advanced to describe displacement damage production in semiconductors. It is useful to compare the relatively primitive situation in semiconductors with that in metals, where the theory of damage production is quite advanced. In metals the conduction electrons are generally treated as providing the cohesive energy, but are otherwise neglected in defect calculations. This simplifies matters. The lattice is modeled by repulsive spheres held together by the cohesive energy. The interatomic repulsion is described by an isotropic, two-body potential, for example, a Born–Mayer potential. This approach has yielded several dynamic calculations[404,405] as well as a number of static calculations[406-409] which give a good picture of the configuration of the defects and their migration energies. These dynamic calculations involve keeping track of the motion of approximately 1000 atoms during a simulated radiation damage event; such calculations have arrived at estimates of the displacement damage threshold and its anisotropy with regard to crystallographic orientation.

In semiconductors, of course, the valence electrons cannot be ignored. In fact from organic chemistry we know that there is not *one* interatomic potential between two carbon atoms, but several, depending upon what the electrons do; thus for a C–C single bond, i.e., a σ bond between the carbons, the equilibrium distance is ~1.54 Å; for a C=C bond, i.e., an additional π bond between them, the distance is about 1.33 Å, and for a C≡C, i.e., a σ and two π bonds, it is about 1.20 Å. As we discussed in Section 2, static calculations in semiconductors have been performed, including calculations which treat solely repulsive interactions, and calculations which include the bonding electrons. There are no dynamic calculations in semiconductors analogous to those in metals. In this section, however, we shall see a number of calculations treating the displacement damage process. Some of these calculations will consider only repulsive interactions, and others will attempt to consider the electronic bonding.

There have been two semiquantitative treatments of the displacement process in semiconductors. Very early, Kohn[410] considered that there were two contributions to the displacement energy T_d. The first is the potential energy of pushing a substitutional atom into an interstitial site; he estimated that the ion-core repulsions would be negligible for a displacement through the triangle of nearest-neighbor atoms along a ⟨111⟩ direction. That is, if a nearest-neighbor atom is in the [$\bar{1}\bar{1}\bar{1}$] direction from a lattice atom, it would then be easier to displace an atom in the [111] direction than in any other direction, i.e., he concluded that the displacement threshold would be *anisotropic*. The energy of displacement, Kohn argued, would be solely the bond-breaking energy; he assumed that in creating an interstitial one broke four sp^3 bonds. In this way he estimated the displacement energy in germanium to be ~15 eV, a value quite close to the experimental value. Bäuerlein[25,361,362] considered three contributions to T_d: The repulsive potential energy term, the bond-breaking energy, and a strain-energy term. He argued that the repulsive-energy term and the strain-energy term would be negligible and concurred with Kohn that the bond-breaking energy was dominant. Bäuerlein argued that the single bond energy is one-half the total bond energy E_B per atom, and hence the displacement energy is approximately $2E_B$. He took as E_B the heat of sublimation corrected for the fact that atoms sublime in the (s^2p^2) configuration rather than (sp^3). He used the total bonding energy since he believed that the electron would not undergo the (sp^3) transition during the time of the displacement. Thus he added the sublimation energy (which for silicon is ~ 4.6 eV) to the s–p promotion energy (3.2 eV), giving $E_B \sim 7.8$ eV or $T_d \sim 15.6$ eV, a value in good semiquantitative agreement with experiments.

Sometime later, but in spirit following Kohn and Bäuerlein, Bailly[411] sought to systematize threshold values by comparing them with bond energies. The results of his approach for the II–VI compounds are shown[376] in Fig. 111. The results of this approach for III–V compounds were reviewed by Eisen,[163] and are not quite as successful as in the II–VI compounds.

Pursuing this sort of phenomenological parametrization, we have found a rather good characterization of T_d as a function of the lattice parameter a_0. This is shown in Fig. 112, where we plot the mean displacement value for the data in Table V versus the reciprocal of the lattice parameter. The least-squares fit to these data gives

$$1.117T_d = (10/a_0)^{4.363} \tag{83}$$

with T_d in eV and a_0 in angstroms. As we will see later, the state of the theory is not such that these parameters can be directly interpreted, but presumably the dependence on a_0 reflects both the strength of binding and the repulsive interaction between atoms.

Banbury and co-workers, as well as Hemment and Stevens, have worked on the modeling of the anisotropy of the displacement process. Banbury and Haddad[412] began with a simple model. As implied by Kohn, they considered only displacements in the $\langle 111 \rangle$ directions, which they referred to as the $\langle 111 \rangle$ gaps. They assumed that an atom suffering an initial recoil

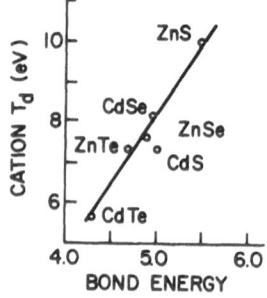

Fig. 111. Correlation of the cation damage threshold T_d with the bond energy of some II–VI compounds. (After Bailly.[411])

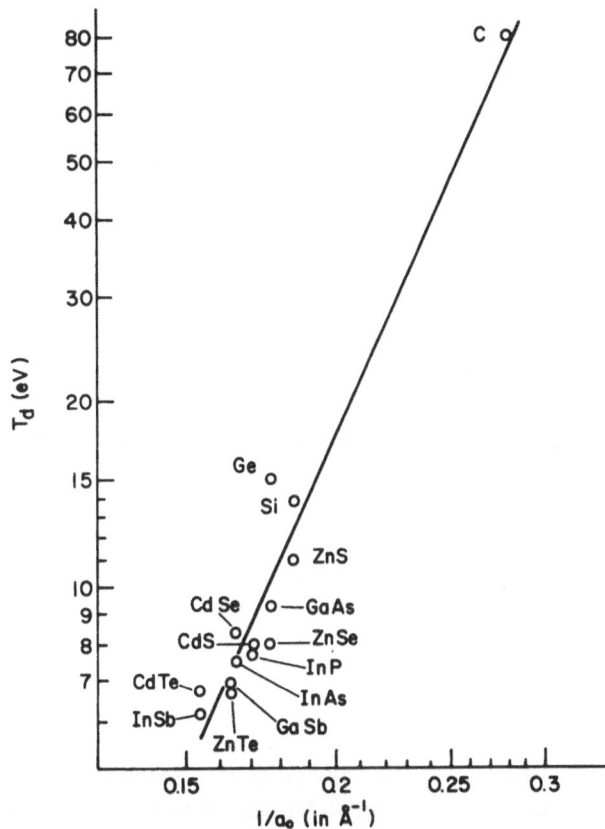

Fig. 112. Correlation of the mean damage threshold T_d with the lattice parameter of several elemental semiconductors and compound semiconductors.

in any other direction would encounter a substantially higher barrier and therefore not contribute to the damage. They utilized the fact that as the beam energy increases, the solid angle increases over which a minimum recoil energy is achieved. They considered the two atoms per unit cell in the diamond lattice (they labeled these atoms A and B), and tallied the number of accessible $\langle 111 \rangle$ gaps available to each type of atom for a given bombarding energy and bombarding direction. These results are shown in Table VI, including the relative order of damage rate. Relating the experimental data to the crossover values indicated in the table, they concluded that a value of T_d of 22 eV would permit their model to describe the anisotropy. This of course is higher than the commonly accepted value of 13–14 eV, and as we have discussed earlier (Section 5), they suggested that

Table VI. The Number of $\langle 111 \rangle$ Gaps Energetically Available to Recoiling Atoms of Energy E_R for the Atoms of Type A and B and the Consequent Anisotropies of Displacement Rate for Irradiation in Various Directions[a]

$E_R(\text{max})/E_d =$	1		1.5		3		9		$\rightarrow\infty$	
Recoil angle for $E_R = E_d$, deg	0		35		54		70		$\rightarrow 90$	
Irradiation direction	Number of $\langle 111 \rangle$ gaps available									
	A	B	A	B	A	B	A	B	A	B
$\langle 111 \rangle$	0	1	0	1	0	1	3	1	3	1
$\langle 110 \rangle$	0	0	1	1	1	1	1	1	3	3
$\langle 100 \rangle$	0	0	0	0	2	2	2	2	2	2
Relative order of displacement rate	$\langle 111 \rangle$ $\langle 110 \rangle$ $\left.\right\}$ $\langle 100 \rangle$		$\langle 110 \rangle$ $\langle 111 \rangle$ $\langle 100 \rangle$		$\langle 100 \rangle$ $\langle 110 \rangle$ $\langle 111 \rangle$		$\langle 111 \rangle$ $\left.\right\}$ $\langle 100 \rangle$ $\langle 110 \rangle$		$\langle 110 \rangle$ $\langle 111 \rangle$ $\left.\right\}$ $\langle 100 \rangle$	

[a] E_d is the displacement energy.

there may be another, isotropic damage mechanism at lower energies. They compared their model to equivalent data in germanium and came to similar conclusions. At the Santa Fe conference, Banbury[344] discussed the extension of their model using a computer; these results were subsequently presented by Fisher and Banbury.[413] They considered the threshold energy T_d to consist of an isotropic energy T_B which corresponds to the energy for breaking the bond, and a sum of two-body, central repulsive potentials $V(r)$ from fixed nearest neighbors evaluated at the closest approach for a particular recoil direction. They included secondary events in their model by using hard-sphere collisions based on a hard-sphere radius derived from the same repulsive potential. They neglected the beam dispersion and the energy loss on the grounds that the sample thickness they were dealing with was only ~ 20 μm. Using primary collisions alone, they found that the various channels for damage production lead to an abrupt change in the damage rate in the $\langle 111 \rangle$ direction, which does not coincide well with the experimental data. Including secondary collisions eliminates this abrupt step (or "knee"). By using their model to predict the anisotropy in the divacancy production rate obtained by Corbett and Watkins,[121] they found the best fit for the parameters in their model to be

$$T_B \simeq 21 \text{ eV} \quad \text{and} \quad V \simeq 5 \times 10^4 \exp(-R/0.2) \quad (84)$$

over the range of R from 1.0 to 2.3 Å.

At the Santa Fe conference, Hemment and Stevens[414] presented a computer modeling of the anisotropy of the damage rate in silicon, which included the angular dispersion of the electron beam and the energy loss of the bombarding electrons. They included the possibility of secondary displacement by using the Kinchin–Pease expression.[1] They considered first a "one-window" model in which displacements were only permitted within a 10° diameter of the $\langle 111 \rangle$ direction, i.e., the Banbury–Haddad model. They found that the crossovers were in good agreement, but the damage-rate ratios were badly in error. Increasing the angular size of the windows improved the fit of the damage rate ratios but agreement with the crossover energies was lost. To obtain a better fit, they extended their work to two- and to three-window models. They made the $\langle 111 \rangle$ direction the easiest direction for displacement, the $\langle 100 \rangle$ the next easiest, and the $\langle 110 \rangle$ direction the most difficult. Their "best fit" was for a two-window model with the parameters: $\langle 111 \rangle$, $T_d \simeq 22$ eV, window diameter $= 60°$; $\langle 100 \rangle$, $T_d \simeq 26$ eV, window diameter $\simeq 40°$. This "best fit" still had a pronounced "knee" in the curve for the $\langle 111 \rangle$ specimen. They then extended their calculations.[343,414,415] Using the parameters $T_d \langle 111 \rangle = 22$ eV, $T_d \langle 100 \rangle = 30$ eV, and $T_d \langle 110 \rangle = 35$ eV, they found the anisotropy ratio and the crossovers in the curves to be fairly accurately predicted, but the overall shape less satisfactory. They attributed this to their overly simple step-function displacement probability and the fact that they did not consider the anisotropy of the secondary displacements. As we have discussed, they also presented experimental results for the anisotropy of the energy dependence of the displacement process for irradiation along nonprincipal lattice directions. They found that their model could be improved by changing the sharp threshold to what they term a "graded" threshold (the linear displacement probability in Fig. 36b), namely a probability which goes from zero to unity as the recoil energy varies from $T_d - \Delta$ to $T_d + \Delta$. As shown in Fig. 113, this graded threshold, with $\Delta = 5$ eV, gives a relatively good fit to their data. In the process of their calculations they defined a T_d surface for each of the atoms in the unit cell (A and B) as shown in Fig. 114. For the type B atom there is a large window centered on the [111] direction, while there is a smaller window for the type A atom in that direction. The smaller window is interpreted as a replacement process in which the type A

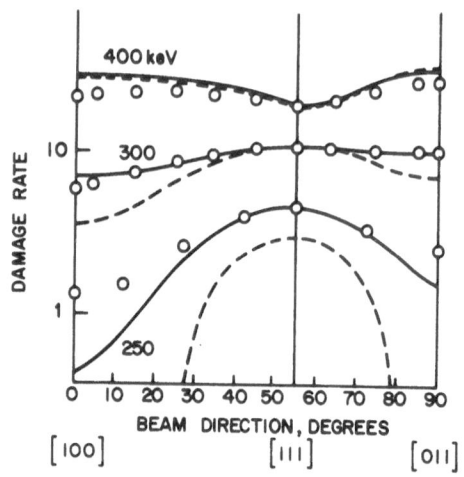

Fig. 113. Orientation dependence of the damage
rate for irradiation of silicon with electrons of
various energies. The dashed curve corresponds
to a model using a step-function probability;
the solid curves correspond to a model using a
graded threshold as shown in Fig. 36b, with
$\Delta = 5$ eV. (After Hemment and Stevens.[343])

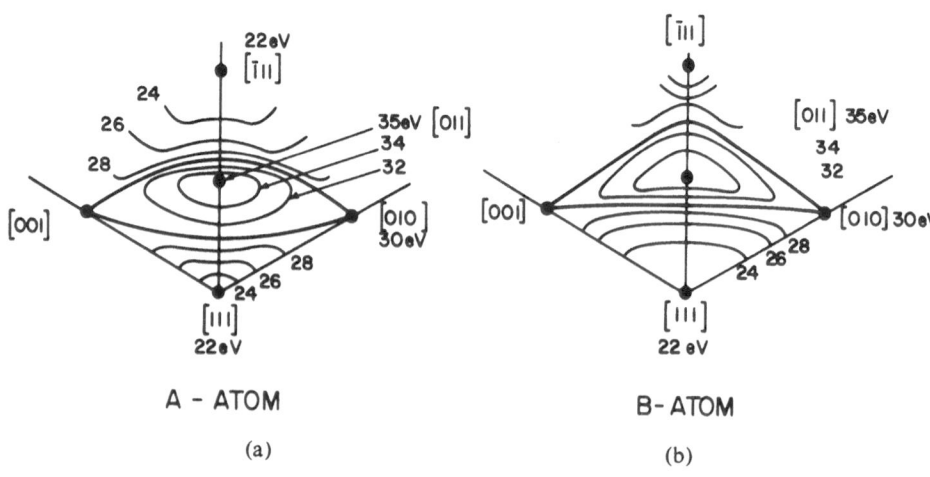

Fig. 114. Displacement energy surface for the two atoms (A and B) in the silicon unit cell.
(a) The surface for the A atom; (b) surface for the B atom. (After Hemment and
Stevens.[343])

atom can displace a type B atom. As the figures show, they concluded that the minimum threshold energy is 22 eV, considerably higher than the frequently quoted value of $\sim$14 eV. They, too, argued that the isotropic damage rate at low energies cannot be predicted by any form of anisotropic displacement energy surface, but we noted earlier (Section 6) that they found the same type of defect created below the 22-eV recoil energy that they had found above that energy.

As we discussed in Section 6, there have been a number of experimental investigations of the temperature dependence of the damage production process. Associated with their early study of the temperature dependence of the damage production process, Brown and Augustyniak[202] suggested that a possible explanation for the temperature dependence of the damage threshold was in thermal vibrations. They argued that if E_V is the energy of a vibration and E_C is the energy received in the collision with an atom at rest, the total energy E_T attained by an atom is

$$E_T = E_C + 2(E_C E_V)^{1/2} \tag{85}$$

This expression gave just about what was needed to superimpose their 78 and 263°K data. Panov and Smirnov[351] took a different view of the origin of the temperature dependence of the threshold energy. The form of their data suggested to them that the damage process gave rise to a localized excitation consisting of approximately 60 atoms. They then compared the mean-squared relative displacement of an atom at a given temperature to the mean-squared relative displacement at the melting temperature and from such consideration explained the temperature dependence which they observed (see Fig. 115). As we indicated in Section 5, there are some discrepancies between their data and some of the other published data.

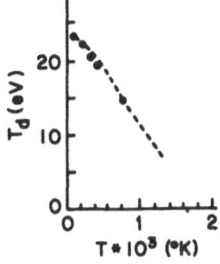

Fig. 115. Temperature T dependence of the damage threshold T_d in silicon as obtained by Panov and Smirnov. (After Panov and Smirnov.[351])

Whan, Vook, and Stein considered not the threshold energy, but the temperature dependence of the damage production rate, as discussed in Section 5. They found their irradiation-temperature-dependent (ITD) damage production process to be exponential in temperature at low temperatures. They discussed an early model proposed by Wertheim[416] which assumed a metastable close-pair configuration, as shown in Fig. 116a. The energy barrier to liberation E_L differs from the energy barrier for recombination E_R as shown. In this model the probability P of pair separation is given by

$$P = 1/\{1 + \lambda \exp[(E_L - E_R)/kT]\} \tag{85}$$

where λ is the ratio of the statistical weight of the jump that annihilates the pair to that of the jump that leads to liberation. On the basis of this

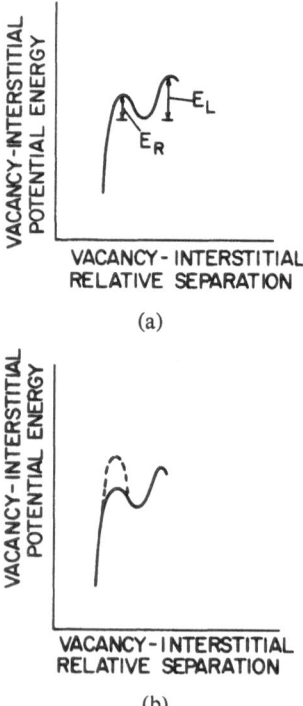

Fig. 116. Model potential for the vacancy–interstitial interaction versus their relative separation. E_L is the energy of liberation; E_R is the energy of recombination. (a) The simple metastable close-pair model; (b) the charge-state-dependent metastable model.

model, $E_L - E_R = 0.06$ eV for n-type silicon. MacKay and Klontz[417] suggested a charge-state modification of this model, as shown in Fig. 116b. Specifically following this model, Vook and Stein[348] assumed that the probability of liberation was unity for, say, the plus charge state and zero for the neutral charge state. In this case the probability of defect formation is then equal to the probability that the level is ionized. They assumed that the level position of the metastable pair with respect to the conduction band E_C is $E_C - E_M$. Then the probability of defect formation P is equal to

$$P = 1/\{1 + g \exp[(E_F - E_M)/kT]\} \qquad (86)$$

where E_F is the Fermi level (which of course depends upon the carrier concentration as well as the temperature) and g is the ratio of the degeneracies of the two states. As we have shown, this charge-state-dependent metastable pair model can explain not only the exponential irradiation-temperature dependence in n-type silicon, but also the resistivity dependence of the defect production. (See Fig. 85 in Section 5.)

We turn now to a different approach to the theory of defect production. As we have discussed earlier (Section 2), the LCAO molecular orbital calculations have provided a great deal of progress in our understanding of the static properties of defects. Unfortunately such an approach is not yet possible in dynamic calculations. Corbett, Bourgoin, and Weigel[65] considered the relevance of LCAO calculations for studies of the mechanisms of defect production. First they noted that the LCAO calculations favor the bonded interstitials over the tetrahedral T or hexagonal H interstitial sites which have been considered in the past. This has the consequence that Kohn and Bäuerlein erred in discussing the displacement process by choosing the wrong interstitial configuration. In the spirit of the Kohn–Bäuerlein arguments, we can estimate the threshold for a displacement of the split and bond-centered interstitials. As they did, we use only the "bond" energy. Both the T and H interstitials involve four broken bonds for the neutral interstitial. Following Bäuerlein, this implies a displacement energy of approximately 16 eV for silicon, i.e., 4 eV per bond. The bond-centered interstitial has, roughly speaking, sp bonding to the two adjacent neighbors with its two additional electrons in p orbitals. The split-$\langle 100 \rangle$ configuration is more involved since it has two atoms in the configuration; both of these atoms make three sp^2 bonds to their neighbors and have one nonbonding p electron. If we make the approximation that the sp, sp^2, and the sp^3 bonds (and their mixtures) have the same bond energy[418] and that the electrons not involved in σ bonds are equivalent, then intercomparison is easy; both the bond-centered interstitial and the split-$\langle 100 \rangle$ interstitial involve two broken bonds. Thus where Kohn and Bäuerlein obtain for T (and H) a T_d of ~ 16 eV, the revised estimate would give a T_d of ~ 8 eV for silicon.

We did not discuss it earlier, but the Kohn–Bäuerlein estimates were embarrassingly in agreement with experiment. The reason is that their estimates are for an *adiabatic* displacement process, i.e., a process in which no energy is radiated into the lattice in the form of heat. But all of the dynamic calculations in metals indicate that a great deal of energy is lost in the displacement process, e.g., in copper three times as much energy is radiated as is stored in the lattice. In the spirit of the Kohn–Bäuerlein estimate, of course, there should be no energy loss, because they assume all of the energy goes into breaking electronic bonds. The revised estimate, on the basis of the split-interstitials, provides that as much energy is lost in heating the lattice as is stored in the lattice. Corbett, Bourgoin, and Weigel have pursued the LCAO calculations to test further the assumptions made by Kohn and Bauerlein. Specifically, they have calculated the potential energy for the displacement of a carbon lattice atom in various directions, as shown in Fig. 117. The energy they evaluated is for the displacement of a lattice atom with all of the other lattice atoms remaining fixed, and hence *no* energy can be radiated into the lattice. The most striking point of their calculations is that the energy of displacement is largely isotropic and there is no $\langle 111 \rangle$ gap! We see also that the energy rises quite quickly, so that even before the atom is displaced into an interstitial site greater than 10 eV is required. They carried out the equivalent calculation for the silicon lattice, shown in Fig. 118. We see again the isotropy and the fact that greater than 10 eV will be required to displace an atom into the nearest interstitial site. We hasten to add that those calculations were not pursued to the point of carrying the displaced atom to a stable site. Such calculations remain to be done.

The isotropy found by Corbett, Bourgoin, and Weigel is particularly fascinating. Presumably the energy they are evaluating is to be construed as the bond-breaking energy. In any event at present it seems quite likely that this isotropy will prove to be related to the isotropy which, experiment after experiment, has been found in the near-threshold energy regime.

The LCAO calculations introduced another complication. These authors considered the electronic energy levels as the atom is displaced from its lattice site. Figure 119 shows the electron energy levels in the vicinity of the forbidden gap for the displacements shown in Fig. 117 for diamond. Analysis shows that the energy levels coming down from the

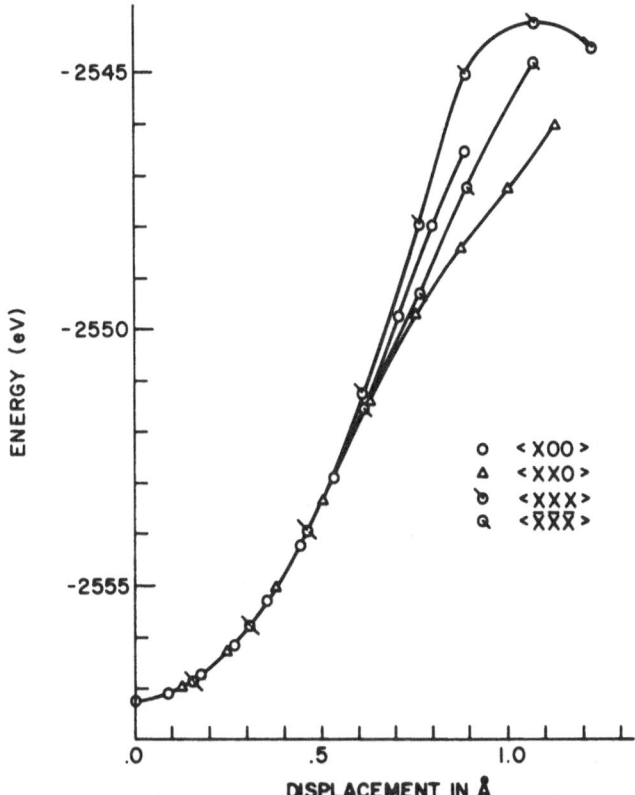

Fig. 117. Potential energy for the displacement of a carbon lattice atom (the central one in a rigid 35-carbon-atom cluster) in various directions. (After Corbett, Bourgoin, and Weigel.[65])

conduction band are those associated with the interstitial, while the energy levels rising from the valence band are those primarily for the vacancy. At the bottom of Fig. 119 is plotted the charge on the interstitial atom as a function of its displacement from the lattice site. As is seen, the interstitial *loses an electron*. That means that rather than having the reaction

$$\text{lattice} + T_d \rightarrow I^0 + V^0 \tag{87}$$

for the creation of the Frenkel pair, we have

$$\text{lattice} + T_d \rightarrow I^+ + V^- \tag{88}$$

A charge transfer automatically takes place and the neutral Frenkel pair has charged constituents. We hasten to add that the theoretical technique

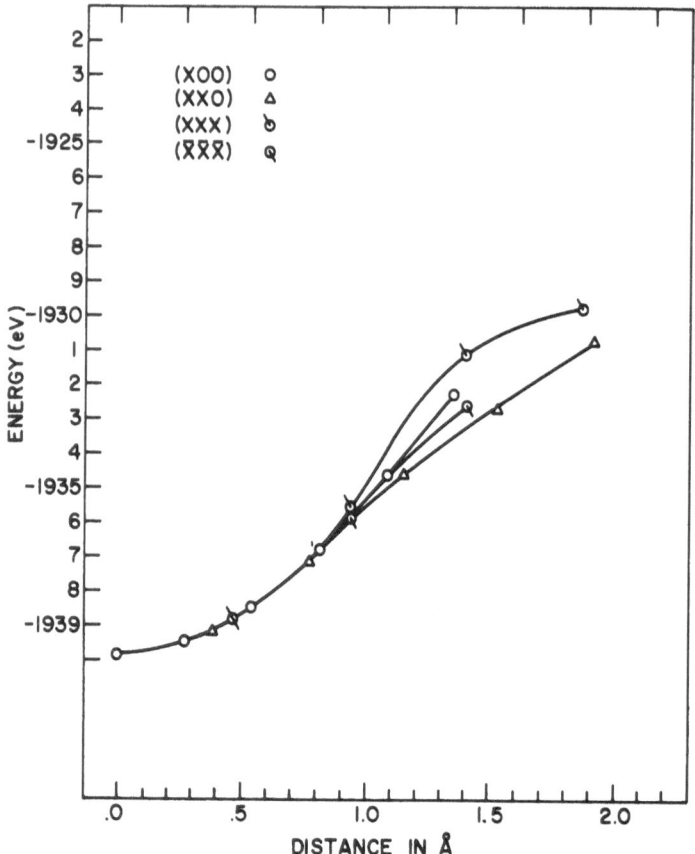

Fig. 118. Potential energy for the displacement of a silicon lattice atoms (the central one in a rigid 35-silicon-atom cluster) in various directions. (After Corbett, Bourgoin, and Weigel.[65])

which they used is *not* a technique capable of handling charge transfer in a self-consistent manner. But we view their result as being more general than the specific computational technique. Presumably nature does not have the highly unlikely situation in which the energy levels of the vacancy are identical to those of the interstitial. If the levels are different, a charge transfer is very natural when the defects are close.

The fact that the constituents of the neutral Frenkel pair are charged means that there will be an additional contribution to the displacement energy due to the attractive Coulomb interaction. The time scale of a recoil of a 20-eV carbon atom is such that it will traverse a bond distance in about 10^{-14} sec. This velocity is large, but the atomic electrons can keep up. Put in another way, 10^{14} sec^{-1} is an optical frequency and the dielectric dispersion function is well defined. Hence the Coulomb interaction will be diminished by the polarization of the solid. The dielectric constant K enters directly in the denominator of the interaction energy and so for materials with a low dielectric constant may contribute to a relatively high displacement energy T_d. This may be the explanation for the high threshold ($T_d \sim 80 \text{ eV}$) in diamond, where K is ~ 5, compared with silicon, $K \simeq 12$, $T_d \sim 14 \text{ eV}$. We note as well that the Coulomb interaction term is isotropic and would contribute to an isotropic displacement energy.

The charge transfer shown in Fig. 119 would give rise to an attractive potential of $\sim 10 \text{ eV}$, i.e., a value comparable to the other contributions in the displacement energy.

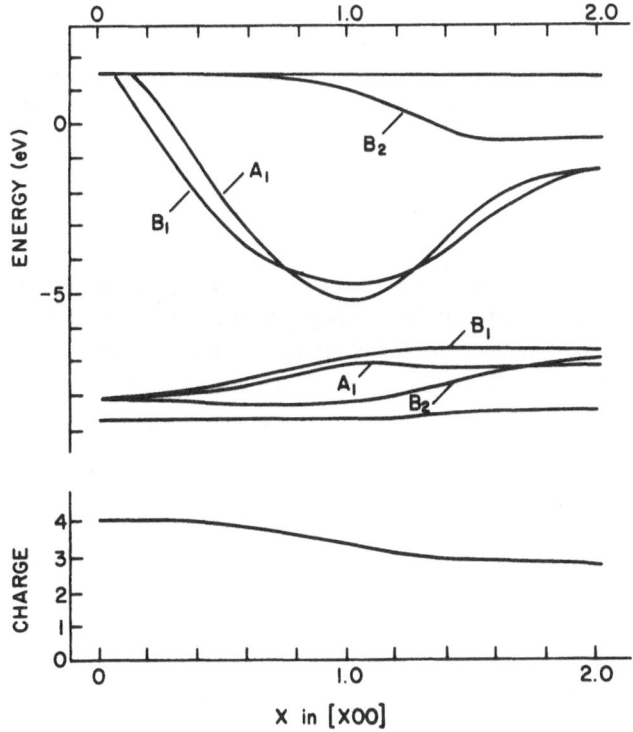

Fig. 119. Top: Electron energy levels in the vicinity of the forbidden gap for atom displacements shown in Fig. 117. The interstitial levels are those coming down from the conduction band, the vacancy levels are those coming up from the valance band. Bottom: The charge on in the interstitial atom versus displacement X. (After Corbett, Bourgoin, and Weigel.[65])

Hopefully, computers will soon permit a complete and sophisticated calculation to be carried out. What is clearly needed is a calculation which considers a larger model lattice than was considered by Corbett, Bourgoin, and Weigel and treats the charge transfer self-consistently. The results obtained thus far suggest that the displacement energy obtained will probably be larger than the experimental value. Consequently, we think that the calculation should consider a relaxation of the neighboring atoms of the recoiling atom so that the split-interstitial will be created directly.

As Corbett, Bourgoin, and Weigel[65] pointed out, the results of their calculations suggest an alternate explanation of the ITD defect production rate observed by Whan, Vook, and Stein. The calculations suggest that the assumption of an intermediate, metastable close-pair stage is not necessary. Specifically, they argued[65] that the recoil atom naturally becomes charged vis-à-vis the vacancy and hence has a Coulomb barrier for damage production. If the temperature is high enough and the attendant energy level is shallow enough, then the charge state can change during the recoil, with the result that the Coulomb attraction will change and liberation will be facilitated. (They noted that the phrase "during the recoil" need not be restricted to mean just passage beyond an initial barrier, but the full time span in which the interstitial is in interaction with the vacancy.)

In summary, at this stage we see that the state of the theoretical calculations leaves much to be desired. There is an emerging understanding of the origin of the anisotropy in the damage production rate and perhaps of the temperature dependence of the damage production rate. Techniques are also emerging which have the promise of permitting more reliable calculations. Time (specifically computer time) and effort seem to be what remains.

8. SUMMARY

In this review we have been concerned with the creation of damage in semiconductors. The definition of damage is somewhat arbitrary. Damage can be taken to mean any deleterious change in a material. Certainly in some device applications the creation of a large, transient change in the electronic state of the system constitutes substantial damage, e.g., changing the state of a computer logic element. We have endeavored to confine our attention to the relatively more permanent damage associated with defect production. But the division between purely "electronic" and purely "defect" damage has not proven simple. In the past "defect" damage has been associated with displacement damage. We have reviewed the displacement damage phenomenology in Section 3, and surveyed the results of the studies of the physics of displacement damage in Section 5. These experimental studies of displacement damage have been performed under a variety of conditions; for that reason we have provided a survey of what is known about defect properties in the various systems in Section 2 against which the displacement damage studies can be assessed.

But the role of ionization cannot be neatly excised from our considerations. For example, defect motion apparently occurs in some systems at the very lowest temperatures by an athermal migration process facilitated by ionization. And then there is the question of whether ionization alone can produce defects as we have discussed in Section 6.

Theory, which we reviewed in Section 7, has made a great deal of progress in recent years, but still has not provided the incisive help which is needed.

In fact it is possible to summarize the theoretical and experimental situation quite briefly. There has been a great deal of progress in the understanding of damage processes in the past several years. Most of that progress, however, has been of the character of asking questions which were heretofore either unasked or not fully appreciated. As we have discussed in this review, those questions primarily relate to the role of ionization and charge state in the damage process. As we have reviewed, there has been some progress in answering those questions. While much remains to be answered, the momentum of the past several years argues that we will see much progress in this area in the next several years.

REFERENCES

1. G. H. Kinchin and R. S. Pease, *Rept. Progr. Phys.* **18,** 1 (1955).
2. A. Seeger, in *Handbuch der Physik*, Ed. by S. Flügge (Springer Verlag, Berlin, 1955), Vol. VII, Part 1.
3. F. Seitz and J. S. Koehler, *Solid State Phys.* **2,** 307 (1956).
4. J. H. Crawford, Jr. and J. W. Cleland, in *Progress in Semiconductors*, Ed. by A. F. Gibson (Wiley, New York, 1957), Vol. 2, p. 67.
5. G. J. Dienes and G. H. Vineyard, *Radiation Effects in Solids* (Wiley, New York, 1957).
6. "Proceedings Gatlinburg Conference on Radiation Effects in Semiconductors," *J. Appl. Phys.* **30,** 1117–1322 (1959).
7. H. G. Van Bueren, *Imperfections in Crystals* (North-Holland, Amsterdam, 1960).
8. D. S. Billington and J. H. Crawford, Jr., *Radiation Damage in Solids* (Princeton Univ. Press, Princeton, 1961).
9. *Radiation Damage in Solids* (IAEA, Vienna, 1962), Vol. III.

10. J. H. Crawford, Jr., in *Radiation Damage in Solids*, Ed. by D. S. Billington (Academic Press, New York, 1962), p. 333.
11. V. S. Vavilov, *Dveistviye Izluchenii na Poluprovodniki* (Fizmatgiz, Moscow, 1963) [English transl.: *Effects of Radiation on Semiconductors*, Consultans Bureau, New York, 1965].
12. "Proceedings International Conference on Crystal Lattice Defects," *J. Phys. Soc. Japan* **18**, Suppl. II and III (1963).
13. P. Baruch, ed., *Radiation Damage in Semiconductors* (Dunod, Paris, 1965).
14. V. S. Vavilov, *Phys. Stat. Sol.* **11**, 447 (1965).
15. J. W. Corbett, *Electron Radiation Damage in Semiconductors and Metals* (Academic Press, New York, 1966).
16. F. Cambou, ed., *Radiation Effects on Semiconductor Components* (Journées d'Electronique, Toulouse, 1967), Vols. I and II.
17. *Conference on Calculations of Properties of Vacancies and Interstitials, Skyland, Virginia, 1966* (U. S. Natl. Bur. Std., Washington, 1967), Misc. Publ. 287.
18. R. R. Hasiguti, ed., *Lattice Defects in Semiconductors* (Univ. Tokyo Press, Tokyo, 1968).
19. F. L. Vook, ed., *Radiation Effects in Semiconductors* (Plenum Press, New York, 1968).
20. R. Bäuerlein, in *Festkörperprobleme* (Akademie Verlag, Berlin, 1968), Vol. 8, p. 1.
21. A. Sosin and W. Bauer, in *Studies in Radiation Effects in Solids*, Ed. by G. J. Dienes (Gordon and Breach, New York, 1969), Vol. 3, pp. 153–328.
22. V. S. Vavilov and N. A. Ukhin, *Radiatsionniye Effekty v Poluprovodnikakh* (Atomizdat, Moscow, 1969).
23. J. W. Corbett and G. D. Watkins, eds., *Radiation Effects in Semiconductors* (Gordon and Breach, New York, 1971).
24. *Radiation Damage and Defects in Semiconductors* (Inst. of Phys., London, 1973).
25. R. Bäuerlein, in *Radiation Damage in Solids*, Ed. by D. S. Billington (Academic Press, New York, 1962), p. 358.
26. E. W. J. Mitchell, in *Radiation Damage in Semiconductors*, Ed. by P. Baruch (Dunod, Paris, 1965), p. 367.
27. J. B. Wagner, to appear in a later volume of this treatise.
28. H. C. Casey, Jr. and G. L. Pearson, this volume, Chapter 2.
29. A. D. Franklin, in *Point Defects in Solids*, Ed. by J. H. Crawford, J. and L. M. Slifkin (Plenum Press, New York, 1972), Vol. 1, p. 1.
30. C. Kittel and A. H. Mitchell, *Phys. Rev.* **96**, 1488 (1954).
31. W. Kohn and J. M. Luttinger, *Phys. Rev.* **97**, 1721 (1955).
32. W. Kohn and J. M. Luttinger, *Phys. Rev.* **98**, 915 (1955).
33. W. Kohn, in *Solid State Physics*, Ed. by F. Seitz and D. Turnbull (Academic Press, New York, 1957), Vol. 5, p. 258.
34. C. A. Coulson and M. J. Kearsley, *Proc. Roy. Soc. (Lond.) A* **241**, 433 (1957).
35. T. Yamaguchi, *J. Phys. Soc. Japan* **17**, 1359 (1962).
36. B. S. Gourary and A. E. Fein, *J. Appl. Phys.* **33** (Suppl. 1), 331 (1962).
37. T. Yamaguchi, *J. Phys. Soc. Japan* **18**, 368 (1963).
38. T. Yamaguchi, *J. Phys. Soc. Japan* **18**, 923 (1963).
39. T. Yamaguchi, in *Radiation Damage in Semiconductors*, Ed. by P. Baruch (Dunod, Paris, 1965), p. 323.
40. C. Doggett, *Proc. Phys. Soc.* **86**, 393 (1965).
41. A. M. Stoneham, *Proc. Phys. Soc. (Lond.)* **88**, 135 (1966).

42. P. Goosens and P. Phariseau, *Physica* **32**, 1713, 1724 (1966).
43. J. Friedel, M. Lannoo, and G. Leman, *Phys. Rev.* **164**, 1056 (1967).
44. A. B. Lidiard and A. M. Stoneham, in *Science and Technology of Industrial Diamonds* (Industrial Diamond Information Bureau, London, 1967), Vol. I, p. 1.
45. J. Friedel, in *Radiation Effects on Semiconductor Components*, Ed. by F. Cambou (Journées d'Electronique, Toulouse, 1967), Vol. I, paper A-11.
46. M. Lannoo and G. Leman, *J. de Phys.* **28** (Suppl. 5-6), C3-168 (1967).
47. M. Lannoo and A. M. Stoneham, *J. Phys. Chem. Solids* **29**, 1987 (1968).
48. M. Lannoo, G. Leman and J. Friedel, in *Radiation Effects in Semiconductors*, Ed. by F. L. Vook (Plenum Press, New York, 1968), p. 37.
49. M. Lannoo and A. M. Stoneham, in *Radiation Effects in Semiconductors*, Ed. by F. L. Vook (Plenum Press, New York, 1968), p. 43.
50. M. Lannoo and P. Lenglart, *J. Chem. Phys. Solids* **30**, 2409 (1969).
51. R. P. Messmer and G. D. Watkins, *Phys. Rev. Lett.* **25**, 656 (1970).
52. R. P. Messmer and G. D. Watkins, in *Radiation Effects in Semiconductors*, Ed. by J. W. Corbett and G. D. Watkins (Gordon and Breach, London, 1971), p. 23.
53. G. D. Watkins and R. P. Messmer, *Phys. Rev. B* **4**, 2066 (1971).
54. E. B. Moore, Jr. and C. M. Carlson, *Phys. Rev. B* **4**, 2063 (1971).
55. C. A. Coulson and F. P. Larkins, *J. Phys. Chem. Solids* **32**, 2245 (1971).
56. F. P. Larkins and A. M. Stoneham, *J. Phys. C: Solid State Phys.* **4**, 143, 154 (1971).
57. F. P. Larkins, *J. Phys. Chem. Solids* **32**, 2123 (1971).
58. A. M. Stoneham, *Rad. Eff.* **9**, 165 (1971).
59. D. Rouhani, M. Lannoo, and P. Lenglart, *Rad. Eff.* **9**, 173 (1971).
60. G. D. Watkins, in *Radiation Damage and Defects in Semiconductors* (Inst. of Phys., London, 1973), p. 228.
61. A. B. Lidiard, in *Radiation Damage and Defects in Semiconductors* (Inst. of Phys., London, 1973), p. 238.
62. C. A. Coulson, in *Radiation Damage and Defects in Semiconductors* (Inst. of Phys., London, 1973), p. 249.
63. R. P. Messmer and G. D. Watkins, in *Radiation Damage and Defects in Semiconductors* (Inst. of Phys., London, 1973), p. 255.
64. G. D. Watkins, R. P. Messmer, C. Weigel, D. Peak, and J. W. Corbett, *Phys. Rev. Lett.* **27**, 1573 (1971).
65. J. W. Corbett, J. C. Bourgoin, and C. Weigel, in *Radiation Damage and Defects in Semiconductors* (Inst. of Phys., London, 1973), p. 1.
66. A. Seeger and W. Frank, in *Radiation Damage and Defects in Semiconductors* (Inst. of Phys., London, 1973), p. 262.
67. C. Weigel, D. Peak, J. W. Corbett, G. D. Watkins, and R. P. Messmer, *Phys. Rev. B* **8**, 2906 (1973).
68. R. A. Swalin, *J. Phys. Chem. Solids* **18**, 290 (1961).
69. A. Scholz, *Phys. Stat. Sol.* **3**, 42 (1963).
70. A. Seeger and A. Scholz, *Phys. Stat. Sol.* **3**, 1480 (1963).
71. K. H. Bennemann, *Phys. Rev.* **137**, A1497 (1965).
72. A. Scholz and A. Seeger, in *Radiation Damage in Semiconductors*, Ed. by P. Baruch (Dunod, Paris, 1965), p. 315.
73. K. H. Bennemann, in *Conference on Calculations of Properties of Vacancies and Interstitials, Skyland, Virginia, 1966* (U. S. Natl. Bur. Std., Washington, 1967), Misc. Publ. 287, p. 127.

74. F. Bailly, in *Lattice Defects in Semiconductors*, Ed. by R. R. Hasiguti (Univ. of Tokyo Press, Tokyo, 1968), p. 231.
75. K. H. Bennemann, in *Lattice Defects in Semiconductors*, Ed. by R. R. Hasiguti (Univ. of Tokyo Press, Tokyo, 1968), p. 30.
76. C. A. Coulson and F. P. Larkins, *J. Phys. Chem. Solids* **30**, 1963 (1969).
77. G. Schmid, K. P. Chik, and A. Seeger, in *Radiation Effects in Semiconductors*, Ed. by F. L. Vook (Plenum Press, New York, 1968), p. 60.
78. R. R. Hasiguti, in *Lattice Defects in Semiconductors*, Ed. by R. R. Hasiguti (Univ. of Tokyo Press, Tokyo, 1968), p. 131.
79. R. Bogdanović, *Phys. Stat. Sol.* **22**, 603 (1967).
80. P. R. Wallace, *Sol. State Comm.* **4**, 521 (1966).
81. C. Baker and A. Kelly, *Nature* **193**, 235 (1962).
82. C. A. Coulson, M. A. Herraez, M. Leal, E. Santos, and S. Senent, *Proc. Roy. Soc. A* **274**, 461 (1963).
83. E. B. Moore, Jr. and C. M. Carlson, *Sol. State Comm.* **4**, 47 (1965).
84. T. Iwata, F. E. Fujita, and H. Suzuki, *J. Phys. Soc. Japan* **16**, 197 (1961).
85. P. A. Thrower and R. T. Loader, *Carbon* **7**, 467 (1969).
86. K. Lark-Horovitz, in *Semi-Conducting Materials*, Ed. by H. K. Henisch (Butterworth, London, 1951), p. 47.
87. H. M. James and K. Lark-Horovitz, *Z. Phys. Chem. (Leipzig)* **198**, 107 (1951).
88. J. Callaway and A. J. Hughes, *Phys. Rev.* **156**, 860 (1967).
89. J. Callaway and A. J. Hughes, in *Lattice Defects in Semiconductors*, Ed. by R. R. Hasiguti (Univ. of Tokyo Press, Tokyo, 1968), p. 36.
90. D. Rouhani, M. Lannoo, and P. Lenglart, *J. de Phys.* **31**, 597 (1970).
91. F. P. Larkins, *J. Phys. Chem. Solids* **32**, 965 (1971).
92. F. P. Larkins, *Rad. Eff.* **9**, 5 (1971).
93. C. Weigel, D. Peak, J. W. Corbett, G. D. Watkins, and R. P. Messmer, to be published.
94. J. Callaway and A. J. Hughes, in *Radiation Effects in Semiconductors*, Ed. by F. L. Vook (Plenum Press, New York, 1968), p. 27.
95. J. Callaway and A. J. Hughes, *Phys. Rev.* **164**, 1043 (1967).
96. A. Seeger and K. P. Chik, *Phys. Stat. Sol.* **29**, 455 (1968).
97. J. Callaway, *Phys. Rev. B* **3**, 2556 (1971).
98. R. R. Hasiguti, *J. Phys. Soc. Japan* **21**, 1927 (1966).
99. R. R. Hasiguti, in *Conference on Calculations of Properties of Vacancies and Interstitials, Skyland, Virginia, 1966* (U. S. Natl. Bur. Std., Washington, 1967), Misc. Publ. 287, p. 27.
100. R. R. Hasiguti, *J. Phys. Soc. Japan* **21**, 1927 (1966).
101. E. I. Blount, *Phys. Rev.* **113**, 995 (1959).
102. C. J. Hwang and L. A. K. Watt, *Phys. Rev.* **171**, 958 (1968).
103. R. R. Hasiguti and S. Motomiya, *Rad. Eff.* **9**, 25 (1971).
104. J. C. Bourgoin and J. W. Corbett, *Phys. Lett.* **38A**, 135 (1972).
105. J. W. Corbett and J. C. Bourgoin, *Trans. IEEE* **NS-18**(6), 11 (1971).
106. B. J. Masters, *Sol. State Comm.* **9**, 283 (1971).
107. J. Hornstra, *Philips Res. Lab. Rept.* No. 3497 (1959).
108. K. Weiser, *Phys. Rev.* **126**, 1427 (1962).
109. G. D. Watkins, in *Radiation Effects on Semiconductor Components*, Ed. by F. Cambou (Journées d'Electronique, Toulouse, 1967), Vol. I, paper A1.

110. M. Cherki and A. H. Kalma, *Phys. Rev. B* **1**, 647 (1970).

111. J. Hornstra, *J. Phys. Chem. Solids* **5**, 129 (1958).

112. S. Amelinckx, *The Direct Observation of Dislocations* (Academic Press, New York, 1964).

113. J. P. Hirth and J. Lothe, *Theory of Dislocations* (McGraw-Hill, New York, 1968).

114. J. S. Kasper and S. M. Richards, *Acta Cryst.* **17**, 752 (1964).

115. G. D. Watkins, in *Radiation Effects in Semiconductors*, Ed. by J. W. Corbett and G. D. Watkins (Gordon and Breach, New York, 1971), p. 301.

116. G. D. Watkins, *J. Phys. Soc. Japan* **18** (Suppl. II), 2 (1963).

117. G. D. Watkins, in *Radiation Damage in Semiconductors*, Ed. by P. Baruch (Dunod, Paris, 1965), p. 97.

118. G. D. Watkins, in *Radiation Effects in Semiconductors*, Ed. by F. L. Vook (Plenum Press, New York, 1968), p. 67.

119. G. D. Watkins, *Trans. IEEE* **NS-16**(6), 13 (1969).

120. G. D. Watkins and J. W. Corbett, *Phys. Rev.* **138**, A543 (1965).

121. J. W. Corbett and G. D. Watkins, *Phys. Rev.* **138**, A555 (1965).

122. J. W. Corbett and G. D. Watkins, *Phys. Rev. Lett.* **7**, 314 (1961).

123. G. D. Watkins and J. W. Corbett, *Disc. Faraday Soc.* No. 31, 86 (1961).

124. K. L. Brower, *Rad. Eff.* **8**, 213 (1971).

125. Y. H. Lee and J. W. Corbett, *Phys. Rev. B* **8**, 2810 (1973).

126. G. D. Watkins and J. W. Corbett, *Phys. Rev.* **134**, A1359 (1964).

127. G. D. Watkins, *Phys. Rev.* **155**, 802 (1967).

128. E. L. Elkin and G. D. Watkins, *Phys. Rev.* **174**, 881 (1968).

129. G. D. Watkins and J. W. Corbett, *Phys. Rev.* **121**, 1001 (1961).

130. J. W. Corbett, G. D. Watkins, R. M. Chrenko, and R. S. McDonald, *Phys. Rev.* **121**, 1015 (1961).

131. A. Brelot and J. Charlemagne, *Rad. Eff.* **9**, 65 (1971).

132. A. Brelot, in *Radiation Damage and Defects in Semiconductors* (Inst. of Phys., London, 1973), p. 191.

133. A. R. Bean, R. C. Newman, and R. S. Smith, *J. Phys. Chem. Solids* **31**, 739 (1969).

134. A. R. Bean and R. C. Newman, *Sol. State Comm.* **8**, 175 (1970).

135. R. C. Newman and R. S. Smith, *J. Phys. Chem. Solids* **30**, 1493 (1969).

136. J. W. MacKay and E. E. Klontz, *J. Appl. Phys.* **30**, 1269 (1959).

137. I. Arimura and J. W. MacKay, in *Radiation Effects in Semiconductors*, Ed. by F. L. Vook (Plenum Press, New York, 1968), p. 204.

138. S. Ishino and E. W. J. Mitchell, in *Lattice Defects in Semiconductors*, Ed. by R. R. Hasiguti (Univ. of Tokyo Press, Tokyo, 1968), p. 185.

139. J. Zizine, in *Radiation Effects in Semiconductors*, Ed. by F. L. Vook (Plenum Press, New York, 1968), p. 186.

140. W. D. Hyatt and J. S. Koehler, in *Radiation Effects in Semiconductors*, Ed. by J. W. Corbett and G. D. Watkins (Gordon and Breach, New York, 1971), p. 59.

141. J. W. MacKay, E. E. Klontz, and G. W. Gobeli, *Phys. Rev. Lett.* **2**, 146 (1959).

142. E. E. Klontz and J. W. MacKay, *J. Phys. Soc. Japan* **18** (Suppl. III), 216 (1963).

143. R. E. Penczer and H. M. DeAngelis, *Phys. Rev.* **171**, 862 (1968).

144. H. M. DeAngelis and R. E. Penczer, *J. Appl. Phys.* **39**, 5842 (1968).

145. J. Bourgoin and F. Mollot, *Phys. Lett.* **30A**, 264 (1969).

146. J. Bourgoin and F. Mollot, in *Radiation Effects in Semiconductors*, Ed. by J. W. Corbett and G. D. Watkins (Gordon and Breach, New York, 1971), p. 63.

147. G. K. Wertheim, *Phys. Rev.* **115**, 568 (1959).
148. J. Zizine, in *Radiation Effects on Semiconductor Components*, Ed. by F. Cambou (Journées d'Electronique, Toulouse, 1967), paper A-23.
149. J. M. Meese and J. W. MacKay, in *Radiation Effects in Semiconductors*, Ed. by J. W. Corbett and G. D. Watkins (Gordon and Breach, New York, 1971), p. 51.
150. J. W. MacKay and E. E. Klontz, in *Radiation Effects in Semiconductors*, Ed. by J. W. Corbett and G. D. Watkins (Gordon and Breach, New York, 1971), p. 41.
151. R. E. Whan, *Phys. Rev.* **140**, A690 (1965).
152. R. E. Whan, in *Radiation Effects in Semiconductors*, Ed. by F. L. Vook (Plenum Press, New York, 1968), p. 195.
153. J. A. Baldwin, Jr., *J. Appl. Phys.* **36**, 2079 (1965).
154. J. N. Lomer and A. M. A. Wild, *Phil. Mag.* **24**, 273 (1971).
155. P. Brosious, J. C. Bourgoin, and J. W. Corbett, to be published.
156. J. N. Lomer and A. M. A. Wild, *Rad. Eff.* **17**, 37 (1973).
157. J. H. E. Griffiths, J. Owen, and I. M. Ward, in *Defects in Solids* (Physical Soc., London, 1955), p. 81.
158. E. A. Faulkner and J. N. Lomer, *Phil. Mag.* **7**, 1995 (1962).
159. J. Owen, in *Physical Properties of Diamonds*, Ed. by R. Berman (Clarendon Press, Oxford, 1965), Chapter 10.
160. C. D. Clark and J. Walker, *Proc. Roy. Soc. (Lond.), A* **334**, 241 (1973).
161. L. W. Aukerman, in *Semiconductors and Semimetals*, Ed. by R. K. Willardson and A. C. Beer (Academic Press, New York, 1968), Vol. 4, p. 343.
162. F. H. Eisen, *Phys. Rev.* **123**, 736 (1961).
163. F. H. Eisen, in *Radiation Effects in Semiconductors*, Ed. by J. W. Corbett and G. D. Watkins (Gordon and Breach, New York, 1971), p. 273.
164. R. Bauerlein, *Z. Naturforsch.* **16A**, 1002 (1961).
165. F. H. Eisen, in *Radiation Damage in Semiconductors*, Ed. by P. Baruch (Dunod, Paris, 1965), p. 163.
166. M. U. Jeong, J. Shirafuji, and Y. Inuishi, in *Radiation Effects in Semiconductors*, Ed. by J. W. Corbett and G. D. Watkins (Gordon and Breach, New York, 1971), p. 287.
167. T. V. Mashovets, G. A. Vikhlii, and N. A. Vitovskii, *Phys. Stat. Sol. A* **14**, 439 (1972).
168. J. Blanc, R. H. Bube, and L. R. Weisberg, *Phys. Rev. Lett.* **9**, 252 (1962).
169. B. Goldstein and N. Almeleh, *Appl. Phys. Lett.* **2**, 130 (1963).
170. B. A. Kulp, in *Radiation Damage in Semiconductors*, Ed. by P. Baruch (Dunod, Paris, 1967), p. 173.
171. J. Schneider and A. Rauber, *Sol. State Comm.* **5**, 779 (1967).
172. J. M. Smith and W. H. Vehse, *Phys. Lett.* **31A**, 147 (1970).
173. D. R. Locker and J. M. Meese, *IEEE Trans.* **NS-19(6)**, 237 (1972).
174. R. C. DuVarney, A. K. Garrison, and R. H. Thorland, *Phys. Rev.* **188**, 657 (1969).
175. D. Galland and A. Herve, *Phys. Lett.* **33A**, 1 (1970).
176. A. L. Taylor, G. Filipovich, and G. K. Linberg, *Sol. State Comm.* **8**, 1359 (1970).
177. A. Herve and B. Maffeo, *Phys. Lett. A* **32**, 247 (1970).
178. B. Maffeo, A. Herve, and R. Cox, *Sol. State Comm.* **8**, 2169 (1970).
179. A. L. Taylor, G. Filipovich, and G. K. Linberg, *Sol. State Comm.* **9**, 945 (1971).
180. M. A. Kaneev, *Soviet Phys.—Solid State* **10**, 726 (1968).
181. B. A. Kulp and R. H. Kelley, *J. Appl. Phys.* **31**, 1057 (1960).

182. L. A. de S. Balona and J. H. M. Loubser, *J. Phys. C: Sol. State Phys.* **3**, 2344 (1970).
183. E. Gmelin, R. Stapf, P. Klemt, G. Landwehr, W. Lichtenberg, and A. Przybylski, in *Radiation Effects in Semiconductors*, Ed. by J. W. Corbett and G. D. Watkins (Gordon and Breach, New York, 1971), p. 337.
184. E. Gmelin, R. Saffert, G. Landwehr, W. Lichtenberg, and P. Klemt, in *Radiation Damage and Defects in Semiconductors* (Inst. of Phys., London 1973), p. 394.
185. H. Kronmüller, J. Jaumann, and K. Seiler, *Z. Naturforsch.* **11b**, 243 (1956).
186. C. Weigel, Ph.D. thesis, Univ. of Würzburg (1974).
187. S. B. Austerman and J. E. Hove, *Phys. Rev.* **100**, 1214 (1955).
188. C. E. Klabunde, T. H. Blewitt, and R. R. Coltman, *Bull. Am. Phys. Soc.* **6**, 129 (1961).
189. M. W. Lucas and E. W. J. Mitchell, *Carbon* **1**, 345 (1964).
190. T. Iwata and H. Suzuki, in *Radiation Damage in Reactor Materials* (IAEA, Vienna, 1963), p. 565.
191. W. Bailey, to be published.
192. L. Bochirol and E. Bonjour, *Carbon* **6**, 661 (1968).
193. T. Iwata, T. Nihara, and H. Matsuo, *Phys. Lett.* **28A**, 146 (1968).
194. T. Iwata, T. Nihara, and H. Matsuo, *J. Phys. Soc. Japan* **33**, 1060 (1972).
195. G. L. Montet, *Carbon* **5**, 19 (1967).
196. G. L. Montet and G. E. Myers, *Carbon* **9**, 179 (1971).
197. T. Tsuzuku and S. Arai, *Japan J. Appl. Phys.* **10**, 580 (1971).
198. F. Seitz, *Disc. Faraday Soc.* **5**, 271 (1949).
199. J. B. Sampson, H. Hurwitz, Jr., and E. F. Clancy, *Phys. Rev.* **99**, 1657 (1955).
200. J. W. Corbett, J. M. Denney, M. D. Fiske, and R. M. Walker, *Phys. Rev.* **108**, 954 (1957).
201. A. E. Fein, *Phys. Rev.* **109**, 1076 (1958).
202. W. L. Brown and W. M. Augustyniak, *J. Appl. Phys.* **30**, 1300 (1959).
203. P. G. Lucasson and R. M. Walker, *Phys. Rev.* **127**, 485 (1962).
204. C. H. Sherman, L. F. Lowe, and E. A. Burke, *Phys. Rev.* **145**, 568 (1966).
205. C. M. Jimenez, L. F. Lowe, E. A. Burke, and C. H. Sherman, *Phys. Rev.* **153**, 735 (1967).
206. G. W. Isler, H. I. Dawson, A. S. Mehner, and J. W. Kauffman, *Phys. Rev.* **146**, 468 (1966).
207. H. Wollenberger and J. Wurm, *Phys. Stat. Sol.* **9**, 601 (1965).
208. G. H. Kinchin and R. S. Pease, *J. Nucl. Energy* **1**, 200 (1955).
209. W. S. Snyder and J. Neufeld, *Phys. Rev.* **97**, 1636 (1955).
210. W. A. Harrison and F. Seitz, *Phys. Rev.* **98**, 1530 (1955).
211. J. Neufeld and W. S. Snyder, *Phys. Rev.* **99**, 1326 (1955).
212. W. S. Snyder and J. Neufeld, *Phys. Rev.* **103**, 862 (1956).
213. D. Heinrich and R. Lenk, *Phys. Stat. Sol.* **3**, 676 (1963).
214. M. T. Robinson, in *Radiation-Induced Voids in Metals*, Ed. by J. W. Corbett and L. C. Ianniello (U. S. AEC, Washington, D.C., 1972), Symposium 26, p. 397.
215. J. A. Brinkman, *J. Appl. Phys.* **25**, 961 (1954).
216. J. A. Brinkman, *Am. J. Phys.* **24**, 246 (1956).
217. T. Iwata and T. Nihara, *J. Phys. Soc. Japan* **31**, 1761 (1971).
218. H. Wollenberger, in *Vacancies and Interstitials in Metals*, Ed. by A. Seeger, D. Schumacher, W. Schilling, and J. Diehl (North-Holland, Amsterdam, 1970), p. 215.
219. F. Dessauer, *Z. Phys.* **38**, 12 (1923).

220. J. B. Gibson, A. N. Goland, M. Milgrim, and G. H. Vineyard, *Phys. Rev.* **120**, 129 (1960).
221. R. L. Fleischer, P. B. Price, and R. M. Walker, *J. Appl. Phys.* **36**, 3645 (1965).
222. E. Sonder and W. A. Sibley, in *Point Defects in Solids*, Ed. by J. H. Crawford, Jr. and L. M. Slifkin (Plenum Press, New York, 1972), Vol. 1, p. 201.
223. J. H. O. Varley, *Nature* **174**, 886 (1954).
224. J. H. O. Varley, *J. Nucl. Energy* **1**, 130 (1954).
225. J. H. O. Varley, *J. Phys. Chem. Solids* **23**, 985 (1962).
226. D. Pooley, *Proc. Phys. Soc.* **87**, 245, 257 (1966).
227. D. Pooley and W. A. Runciman, *Sol. State Comm.* **4**, 351 (1966).
228. J. D. Konitzer and H. N. Hersh, *J. Phys. Chem. Solids* **27**, 771 (1966).
229. H. N. Hersh, *Phys. Rev.* **148**, 928 (1966).
230. C. B. Lushchik, G. K. Vale, E. R. Ilmas, N. S. Roose, A. A. Elango, and M. A. Elango, *Opt. and Spect.* **21**, 377 (1966).
231. J. C. Bourgoin, J. W. Corbett, and H. L. Frisch, *J. Chem. Phys.* (in press).
232. J. C. Bourgoin and J. W. Corbett, *Rad. Eff.* (in press).
233. H. B. Huntington, in *Encyclopedia of Chemical Technology* (Wiley, New York, 1971), 2nd ed., Suppl., p. 278.
234. A. M. Barnett, in *Semiconductors and Semimetals*, Ed. by R. K. Willardson and A. C. Beer (Academic Press, New York, 1970), Vol. 6, p. 141; E. M. Conwell, *High Field Transport in Semiconductors* (Academic Press, New York, 1967).
235. F. C. Frank and D. Turnbull, *Phys. Rev.* **104**, 617 (1956).
236. M. Yoshida, *Japan. J. Appl. Phys.* **8**, 1211 (1969).
237. M. Lax, *Phys. Rev.* **119**, 1502 (1960).
238. G. Duesing, N. Sassin, W. Schilling, and H. Heinmerich, *Crystal Lattice Defects* **1**, 55 (1969).
239. H. E. Schepp, F. Dworschak, H. Schuster, and H. Wollenberger, *Rad. Eff.* **7**, 219 (1971).
240. D. Meissner and W. Schilling, *Z. Naturforsch* **26a**, 502 (1971).
241. F. Seitz, *Adv. Phys.* **1**, 43 (1952).
242. P. Haasen and A. Seeger, in *Halbleiterprobleme*, Ed. by W. Schottky (Vieweg Verlag, Braunschweig, 1958), Vol. IV, p. 68).
243. W. C. Ellis and E. S. Greiner, *Phys. Rev.* **92**, 1061 (1953).
244. A. G. Tweet, *Phys. Rev.* **99**, 1245 (1955).
245. W. Schröter, *Phys. Stat. Sol.* **21**, 211 (1967).
246. W. Schröter and R. Labusch, *Phys. Stat. Sol.* **36**, 539 (1969).
247. H. Alexander, R. Labusch, and W. Sander, *Sol. State Comm.* **3**, 357 (1965).
248. F. D. Wöhler, H. Alexander, and W. Sander, *J. Phys. Chem. Solids* **31**, 1381 (1970).
249. V. A. Grazhulis and Yu. A. Osip'yan, *Soviet Phys.–JETP* **31**, 677 (1970).
250. V. A. Grazhulis and Yu. A. Osip'yan, *Soviet Phys.—JEPT* **33**, 623 (1971).
251. F. D. Wöhler, *Z. Metallkunde* **62**, 240 (1971).
252. B. Reppich, P. Haasen, and B. Ilscher, *Acta Met.* **12**, 1283 (1964).
253. H. Alexander and P. Haasen, in *Solid State Physics*, Ed. by F. Seitz, D. Turnbull, and H. Ehrenreich (Academic Press, New York, 1965), Vol. 22, p. 27.
254. J. R. Patel and A. R. Chandhuri, *Phys. Rev.* **143**, 601 (1966).
255. H. R. Frisch and J. R. Patel, *Phys. Rev. Lett.* **18**, 784 (1967).
256. J. R. Patel and P. E. Freeland, *Phys. Rev. Lett.* **18**, 833 (1967).
257. J. R. Patel and H. L. Frisch, *Appl. Phys. Lett.* **13**, 32 (1968).

258. R. I. Garber, I. I. Soloshenko, and O. A. Khaldei, *Soviet Phys.—Solid State* **7**, 2147 (1966).

259. R. I. Garber, I. I. Soloshenko, and I. A. Charkina, *Soviet Phys.—Solid State* **12**, 2766 (1971).

260. A. Seeger and M. L. Swanson, in *Lattice Defects in Semiconductors*, Ed. by R. R. Hasiguti (Univ. of Tokyo Press, Tokyo, 1966), p. 93.

261. A. Seeger, *Rad. Eff.* **9**, 15 (1971).

262. B. L. Kendall and D. B. DeVries, in *Semiconductor Silicon*, Ed. by R. R. Haberect (Electro-Chem. Soc., New York, 1969), p. 358.

263. A. Seeger, in *Radiation Effects in Semiconductors*, Ed. by J. W. Corbett and G. D. Watkins (Gordon and Breach, New York, 1971), p. 29.

264. R. F. Peart, *Phys. Stat. Sol.* **15**, K119 (1966).

265. R. N. Goshtagore, *Phys. Rev. Lett.* **16**, 890 (1966).

266. J. W. Corbett, J. C. Bourgoin, and H. L. Frisch, to be published.

267. G. Bemski and C. A. Dias, *J. Appl. Phys.* **35**, 293 (1964).

268. L. Elstner and W. Kamprath, *Phys. Stat. Sol.* **22**, 541 (1967).

269. M. L. Swanson, *Phys. Stat. Sol.* **33**, 721 (1969).

270. V. M. Malovetskaya, G. M. Galkin, and V. S. Vavilov, *Soviet Phys.—Solid State* **4**, 1008 (1962).

271. V. S. Vavilov, G. M. Galkin, V. M. Malovetskaya, and A. S. Plotnikov, *Soviet Phys.—Solid State* **4**, 1442 (1963).

272. H. J. Stein and R. Gerreth, *J. Appl. Phys.* **39**, 2890 (1968).

273. G. Swenson, *Phys. Stat. Sol. A* **2**, 803 (1970).

274. S. Mayburg, *Phys. Rev.* **95**, 38 (1954).

275. R. A. Logan, *Phys. Rev.* **101**, 1455 (1956).

276. A. Hiraki and T. Suita, *J. Phys. Soc. Japan* **18** (Suppl. III), 254 (1963).

277. A. Hiraki and Y. Inuishi, in *Lattice Defects in Semiconductors*, Ed. by R. R. Hasiguti (Univ. of Tokyo Press, Tokyo, 1968), p. 138.

278. A. Hiraki, *J. Phys. Soc. Japan* **21**, 34 (1966).

279. B. Samuelson, *Ark. Fys.* **35**, 321 (1968).

280. V. D. Tkachev and V. I. Urenev, *Soviet Phys.—Semicond.* **4**, 1880 (1971).

281. S. N. Abdurakhmanova, M. A. Vitovskii, M. Maksimov, and T. V. Mashovets, *Soviet Phys.—Semicond.* **4**, 1979 (1971).

282. M. Cardona, in *Semiconductors and Semimetals*, Ed. by R. K. Willardson and A. C. Beer (Academic Press, New York, 1967), Vol. 3, p. 125.

283. W. G. Spitzer, in *Semiconductors and Semimetals*, Ed. by R. K. Willardson and A. C. Beer (Academic Press, New York, 1967), Vol. 3, p. 17.

284. E. Storm and H. I. Israel, "Photon Cross-Sections from 0.001 to 100 MeV for Elements 1 through 100," Report of Los Alamos Scientific Laboratory, Los Alamos, New Mexico, LA-3753 (1967).

285. L. G. Parratt, *Phys. Rev.* **50**, 1 (1936).

286. R. D. Richtmeyer, *Phys. Rev.* **49**, 1 (1936).

287. V. P. Sachenko and V. F. Demekhin, *Soviet Phys.—JETP* **22**, 532 (1966).

288. V. P. Sachenko and E. V. Burtsev, *Bull. Acad. Sci. USSR, Phys. Ser.*, **31**, 980 (1968).

289. F. B. Marion and F. C. Young, *Nuclear Reaction Analysis* (North-Holland, Amsterdam, 1968).

290. H. A. Bethe and J. Ashkin, in *Experimental Nuclear Physics*, Ed. by E. Segré (Wiley, New York, 1953), Vol. I, p. 166.

291. C. Williamson and J. P. Boujot, *Tables of Energy Loss* (Saclay, 1962).
292. N. Bohr, *Kgl. Danske Vidensk. Selk, Biol. Medd.* **18**, 9 (1948).
293. G. Leibfried, *Bestrahlungseffekte in Festkörpern* (B. G. Teubner, Stuttgart, 1965).
294. J. Lindhard, *Kgl. Danske Vidensk. Selsk., Mat.-Fys. Medd.* **28**, No. 8 (1954).
295. J. Lindhard and M. Scharff, *Phys. Rev.* **124**, 128 (1961).
296. J. Lindhard, M. Scharff, and H. E. Schiott, *Kgl. Danske Vidensk. Selsk., Mat.-Fys. Medd.* **33**, No. 10 (1963).
297. J. Lindhard, V. Nielsen, and M. Scharff, *Kgl. Danske Vidensk. Selsk., Mat.-Fys. Medd.* **36**, No. 10 (1968).
298. O. B. Firsov, *Soviet Phys.—JETP* **36**, 1076 (1959).
299. E. S. Parilis and L. M. Kishinevskii, *Soviet Phys.—Solid State* **3**, 885 (1960).
300. L. M. Kishinevskii, *Bull. Acad. Sci. USSR, Phys. Ser.* **26**, 1433 (1962).
301. D. I. Porat and K. Ramavatarum, *Proc. Phys. Soc. (Lond.)* **77**, 97 (1961).
302. I. M. Cheshire and J. M. Poate, in *Atomic Collision Phenomena in Solids*, Ed. by D. W. Palmer, M. W. Thompson, and P. D. Townsend (North-Holland, Amsterdam, 1970), p. 351.
303. C. P. Bhalla, J. N. Bradford, and G. Reese, in *Atomic Collision Phenomena in Solids*, Ed. by D. W. Palmer, M. W. Thompson, and P. D. Townsend (North-Holland, Amsterdam, 1970), p. 361.
304. M. T. Robinson and O. S. Oen, *Phys. Rev.* **132**, 2385 (1963).
305. J. M. Hansteen and O. P. Mosebekk, *Z. Phys.* **234**, 281 (1970).
306. P. C. Gehlen, J. R. Beeler, Jr., and R. R. Jaffee, eds., *Interatomic Potentials and Simulation of Lattice Defects* (Plenum Press, New York, 1972).
307. E. Rutherford, *Phil. Mag.* **21**, 669 (1911).
308. R. H. Silsbee, *J. Appl. Phys.* **28**, 1246 (1957).
309. E. J. Williams, *Proc. Roy. Soc. A* **125**, 420 (1929).
310. L. D. Landau, *J. Phys. (USSR)* **8**, 201 (1944).
311. O. Blunck and S. Liesegang, *Z. Physik* **128**, 500 (1950).
312. P. F. Hebbard and P. R. Wilson, *Austral. J. Phys.* **8**, 90 (1955).
313. W. Heitler, *Quantum Theory of Radiation* (Oxford Univ. Press, London (1954), Sect. 37.
314. C. G. Darwin, *Phil. Mag.* **25**, 301 (1913).
315. N. F. Mott, *Proc. Roy. Soc. A* **124**, 426 (1929).
316. N. F. Mott, *Proc. Roy. Soc. A* **135**, 429 (1932).
317. W. A. McKinley, Jr. and H. Feshbach, *Phys. Rev.* **74**, 1759 (1948).
318. O. S. Oen, Oak Ridge National Laboratory, Oak Ridge, Tennessee, private communication.
319. B. T. Kelly, *Irradiation Damage to Solids* (Pergamon Press, Oxford, 1966).
320. E. E. Klontz and K. Lark-Horovitz, *Phys. Rev.* **82**, 763 (1951).
321. E. E. Klontz and K. Lark-Horovitz, *Phys. Rev.* **86**, 643 (1952).
322. J. J. Loferski and P. Rappaport, *Phys. Rev.* **98**, 1861 (1955).
323. V. S. Vavilov, L. S. Smirnov, G. N. Galkin, A. V. Spitsyn, and V. M. Patskevich, *Soviet Phys.—Tech. Phys.* **1**, 1805 (1957).
324. V. S. Vavilov, L. S. Smirnov, G. N. Galkin, A. V. Spitsyn, and V. M. Patskevich, *Soviet Phys.—Tech. Phys.* **3**, 894 (1958).
325. J. J. Loferski and P. Rappaport, *Phys. Rev.* **111**, 432 (1958).
326. L. S. Smirnov and P. A. Glazunov, *Soviet Phys.—Solid State* **1**, 1262 (1960).
327. J. A. Naber and H. M. James, *Bull. Am. Phys. Soc.* **6**, 303 (1961).

328. Y. Chen and J. W. MacKay, *Phil. Mag.* **19**, 357 (1969).
329. V. S. Vavilov, V. M. Patskevich, B. Ya. Yurkov, and P. Ya. Glazunov, *Soviet Phys.—Solid State* **2**, 1301 (1961).
330. B. Ya. Yurkov, *Soviet Phys.—Solid State* **1**, 633 (1959).
331. B. Ya. Yurkov, *Soviet Phys.—Solid State* **2**, 2412 (1961).
332. B. Ya. Yurkov, *Soviet Phys.—Solid State* **4**, 626 (1962).
333. H. Flicker, J. J. Loferski, and J. Scott-Monck, *Phys. Rev.* **128**, 2557 (1962).
334. R. L. Novak, *Bull. Am. Phys. Soc.* **8**, 235 (1963).
335. H. Flicker and J. J. Loferski, *J. Appl. Phys.* **34**, 2146 (1963).
336. E. G. Wikner, H. Horiye, and J. W. Harrity, *J. Phys. Soc. Japan* **18**(Suppl. III), 222 (1963).
337. I. V. Kryukova and V. S. Vavilov, *Soviet Phys.—Solid State* **6**, 266 (1964).
338. G. G. George and E. M. Gunnersen, in *Radiation Damage in Semiconductors*, Ed. by P. Baruch (Dunod, Paris, 1965), p. 385.
339. P. H. Fang and W. Gdula, *Bull. Am. Phys. Soc.* **10**, 321 (1965).
340. I. N. Haddad, P. C. Banbury, and J. A. Grimshaw, *Phil. Mag.* **12**, 1203 (1965).
341. I. N. Haddad and P. C. Banbury, *Phil. Mag.* **14**, 829 (1966).
342. J. A. Grimshaw, *Phys. Lett.* **22**, 372 (1966).
343. P. L. F. Hemment and P. R. C. Stevens, in *Atomic Collision Phenomena in Solids*, Ed. by D. W. Palmer, M. W. Thompson, and P. D. Townsend (North-Holland, Amsterdam, 1970), p. 217.
344. P. C. Banbury, in *Radiation Effects in Semiconductors*, Ed. by F. L. Vook (Plenum, New York, 1968), p. 280.
345. N. N. Gerasimenko, A. Z. Dvurechenskii, V. I. Panov and L. S. Smirnov, *Soviet Phys.—Semicond.* **5**, 1439 (1972).
346. R. E. Whan and F. L. Vook, *Phys. Rev.* **153**, 814 (1967).
347. R. L. Novak, Ph.D. Thesis, Univ. of Penn., 1964, unpublished.
348. F. L. Vook and H. J. Stein, in *Radiation Effects in Semiconductors*, Ed. by F. L. Vook (Plenum Press, New York, 1968), p. 99.
349. H. J. Stein and F. L. Vook, in *Radiation Effects in Semiconductors*, Ed. by F. L. Vook (Plenum Press, New York, 1968), p. 115.
350. T. I. Kolomenskaya, S. M. Razumovskii, and S. E. Vitovkin, *Soviet Phys.—Semicond.* **1**, 652 (1967).
351. V. I. Panov and L. S. Smirnov, *Fiz. Tekh. Poluprov.* **7**, 212 (1973).
352. C. D. Clark, P. J. Kemmey, and E. W. J. Mitchell, in *Proc. Int. Conf. on Semiconductor Physics, Prague, 1960* (Czech. Acad. Sci., Prague, 1961), p. 316.
353. C. D. Clark, P. J. Kemmey, and E. W. J. Mitchell, *Disc. Faraday Soc.* No. 31, 96 (1961).
354. J. T. Ritter, *Sol. State Comm.* **8**, 773 (1970).
355. C. D. Clark and J. Walker, *Proc. Roy. Soc.*, to be published.
356. R. P. Messmer, *Chem. Phys. Lett.* **11**, 589 (1971).
357. R. P. Messmer, B. McCarroll, and C. M. Singal, *J. Vacuum Sci. Tech.* **9**, 891 (1972).
358. D. T. Eggen, referred to by G. R. Hennig and J. H. Hove, in *Proc. Int. Conf. on Peaceful Uses of Atomic Energy* (United Nations, Geneva, 1955), Vol. 7, p. 666.
359. F. H. Eisen and P. W. Bickel, *Phys. Rev.* **115**, 345 (1959).
360. F. H. Eisen, *Phys. Rev.* **135**, A1394 (1964).
361. R. Bäuerlein, *Z. Naturforsch.* **14a**, 1069 (1959).
362. R. Bäuerlein, *Z. Physik* **176**, 498 (1963).

363. D. J. C. Lindsay and P. C. Banbury, in *Radiation Damage and Defects in Semiconductors* (Inst. of Phys., London, 1973), p. 34.
364. S. B. Fisher and P. C. Banbury, in *Atomic Collision Phenomena in Solids*, Ed. by D. W. Palmer, M. W. Thompson, and P. D. Townsend (North-Holland, Amsterdam, 1970), p. 232.
365. J. A. Grimshaw and P. C. Banbury, *Proc. Phys. Soc.* **84**, 151 (1964).
366. J. J. Loferski and N. H. Wu, in *Radiation Damage in Semiconductors*, Ed. by P. Baruch (Dunod, Paris, 1965), p. 213.
367. J. J. Loferski, H. Flicker, R. M. Esposto, and N. H. Wu, in *Lattice Defects in Semiconductors*, Ed. by R. R. Hasiguti (Univ. of Tokyo Press, Tokyo, 1968), p. 355.
368. K. Thommen, *Rad. Eff.* **2**, 201 (1970).
369. K. Thommen, *Phys. Rev.* **174**, 938 (1968); erratum, *Phys. Rev.* **179**, 920 (1969).
370. J. Schneider, in *II–VI Semiconducting Compounds*, Ed. by D. G. Thomas (Benjamin, New York, 1967), p. 40.
371. B. A. Kulp and R. M. Detweiler, *Phys. Rev.* **129**, 2422 (1963).
372. R. M. Detweiler and B. A. Kulp, *Phys. Rev.* **146**, 513 (1966).
373. F. J. Bryant and A. T. J. Baker, *Phys. Lett.* **35A**, 457 (1971).
374. J. M. Meese, *Appl. Phys. Lett.* **19**, 86 (1971).
375. F. J. Bryant and A. T. J. Baker, *J. Phys. C: Solid State Phys.* **5**, 2283 (1972).
376. F. J. Bryant and A. T. J. Baker, in *Radiation Damage and Defects in Semiconductors*, Ed. by J. E. Whitehouse (Inst. of Phys., London, 1973), p. 42.
377. J. M. Meese and V. F. Park, in *Radiation Damage and Defects in Semiconductors*, Ed. by J. E. Whitehouse (Inst. of Phys., London, 1973), p. 51.
378. F. J. Bryant and A. F. J. Cox, *Phys. Lett.* **20**, 108 (1966).
379. F. J. Bryant and S. A. Hamid, *Phys. Rev. Lett.* **23**, 304 (1969).
380. B. A. Kulp and R. H. Kelley, *J. Appl. Phys.* **32**, 1290 (1961).
381. B. A. Kulp, *Phys. Rev.* **125**, 1865 (1962).
382. B. A. Kulp, *J. Appl. Phys.* **37**, 4936 (1966).
383. H. J. Schulz and B. A. Kulp, *Phys. Rev.* **159**, 603 (1967).
384. F. J. Bryant and E. Webster, *Phys. Stat. Sol.* **21**, 315 (1967).
385. F. J. Bryant, A. F. J. Cox, and E. Webster, *J. Phys. C* **1**, 1737 (1968).
386. W. A. Sibley and Y. Chen, *Phys. Rev.* **160**, 712 (1967).
387. Y. Chen, D. L. Trueblood, O. E. Schow, and H. T. Tohver, *J. Phys. C* **3**, 2501 (1970).
388. W. E. Vehse, W. A. Sibley, F. J. Keller, and Y. Chen, *Phys. Rev.* **167**, 828 (1968).
389. D. R. Locker and J. M. Meese, *IEEE Trans.* NS-19(6), 237 (1972).
390. G. W. Arnold and F. L. Vook, *Phys. Rev.* **137**, A1839 (1965).
391. B. Chelustra, R. Yu. Khansevarov, T. V. Mashovets, and I. R. Kozlova, *Soviet Phys.—Solid State* **9**, 253 (1967).
392. N. A. Vitovskii, G. A. Vikhlii, V. V. Galavanov, T. V. Mashovets, and R. Yu. Khansevarov, *Soviet Phys.—Semicond.* **3**, 106 (1969).
393. E. W. Kreutz, *Z. Ang. Phys.* **27**, 244 (1969).
394. E. W. Kreutz, H. Pagnia, and W. Waidelich, *Phys. Stat. Sol.* **27**, K111 (1968).
395. M. A. Zaikovskaya, A. E. Kiv, and O. R. Niyazova, *Fiz. Tekh. Poluprov.* **1**, 1131 (1967).
396. M. A. Zaikovskaya, A. E. Kiv, and O. R. Niyazova, *Phys. Stat. Sol. a* **3**, 99 (1970).
397. M. A. Zaikovskaya, A. E. Kiv, and O. R. Niyazova, *Phys. Stat. Sol. a* **8**, K1333 (1971).

398. H. J. Pabst and D. W. Palmer, in *Proc. 5th Int. Conf. on Ion Interactions with Solids* (1973), to be published.
399. O. S. Oen, in *Radiation Effects in Semiconductors*, Ed. by F. L. Vook (Plenum, New York, 1968), p. 264.
400. M. Balarin, *Kernenergie* **7**, 434 (1964).
401. R. L. Platzman, in *Symposium on Radiobiology*, Ed. by J. J. Nickson (Wiley, New York, 1952), p. 97.
402. Y. Chen and J. W. MacKay, *Phys. Rev.* **167**, 828 (1968).
403. C. B. Norris, *J. Appl. Phys.* **43**, 4060 (1972).
404. J. B. Gibson, A. N. Goland, M. Milgram, and G. H. Vineyard, *Phys. Rev.* **120**, 1229 (1960).
405. C. Erginsoy, G. H. Vineyard, and A. Englert, *Phys. Rev. A* **133**, 595 (1964).
406. K. H. Bennemann, *Phys. Rev.* **124**, 669 (1961).
407. A. Seeger, E. Mann, and R. v. Jan, *J. Phys. Chem. Solids* **23**, 639 (1962).
408. R. A. Johnson and E. Brown, *Phys. Rev.* **127**, 446 (1962).
409. R. A. Johnson, *Phys. Rev.* **134A**, 1329 (1964).
410. W. Kohn, *Phys. Rev.* **94A**, 1409 (1954).
411. F. Bailly, *J. Phys. Radium* **27**, 335 (1966).
412. P. C. Banbury and I. N. Haddad, *Phil. Mag.* **14**, 841 (1966).
413. S. B. Fisher and P. C. Banbury, in *Atomic Collision Phenomena in Solids*, Ed. by D. W. Palmer, M. W. Thompson, and P. D. Townsend (North-Holland, Amsterdam, 1970), p. 232.
414. P. L. F. Hemment and P. R. C. Stevens, in *Radiation Effects in Semiconductors*, Ed. by F. L. Vook (Plenum Press, New York, 1968), p. 290.
415. P. L. F. Hemment and P. R. C. Stevens, *J. Appl. Phys.* **40**, 4893 (1969).
416. G. K. Wertheim, *Phys. Rev.* **115**, 568 (1959).
417. J. W. MacKay and E. E. Klontz, *J. Appl. Phys.* **30**, 1269 (1959).
418. R. S. Mulliken, *J. Chem. Phys.* **19**, 900 (1951).

Chapter 2

DIFFUSION IN SEMICONDUCTORS

H. C. Casey, Jr.

Bell Telephone Laboratories, Inc.
Murray Hill, New Jersey

and

G. L. Pearson

Stanford University
Stanford, California

1. INTRODUCTION

The desired physical and electrical properties of electronic materials are generally obtained by processing at elevated temperatures, and an understanding of the changes that occur during these processes is based on a knowledge of diffusion. One of the most widespread processes is the preparation of *p–n* junctions in semiconductors by diffusion of impurities from the vapor. These diffused *p–n* junctions are the building blocks for most transistors and integrated circuits. The detailed steps used in commercial diffusion processes for *p–n* junction preparation are generally worked out by trial and error. However, a combination of accurate experimental data and a well-founded understanding of basic diffusion mechanisms permits selection of the most promising conditions. The study of diffusion not only leads to useful commercial processes, but also provides fundamental knowledge on the behavior of both native defects and impurities in crystals.

There are numerous texts that treat the mathematical solutions of the diffusion equations, describe the basic mechanisms of diffusion, and list a summary of pertinent experimental data. A thorough treatment of the mathematics of the partial differential diffusion equations can be found in Crank's well-known text.[1] Two representative texts on the mechanisms of diffusion are by Shewmon[2] and Manning.[3] The diffusion data are generally given for a particular field of study, for example, metals, oxides, organic crystals, halides, or semiconductors. The books on diffusion in semiconductors by Boltaks[4] and Sharma[5] give extensive compilations of diffusion data, and the book edited by Shaw[6] presents analyses of the observed behavior. With several books devoted to diffusion in semiconductors, a single chapter obviously cannot completely cover the subject. Therefore, this chapter is limited to the development of the basic fundamental concepts necessary for a general understanding and appreciation of the problems associated with diffusion in semiconductors. However, extensive experimental diffusivity data are given in graphical and tabular form. The vacancy mechanisms will be illustrated by impurity and self-diffusion in Ge and Si, the interstitial–substitutional mechanism is illustrated by Zn diffusion in GaAs, and self-diffusion in compounds is illustrated with selected II–VI materials.

The treatment of diffusion can be undertaken from two principal approaches. One is the *atomistic method* which considers the actual movements of the atoms in the lattice. The second is the *continuum method* which is concerned with obtaining an appropriate mathematical description of the diffusion. The mathematical method generally becomes a problem in the solution of the flux and continuity equations (Fick's laws) for the appropriate boundary conditions. The atomistic approach will first be briefly considered in order to establish a description of the various diffusion mechanisms and to permit definition of such terms as energy of formation, energy of motion, and jump frequency. The major part of this chapter will be concerned with the continuum approach for both analysis and interpretation of experimental data.

2. ATOMIC THEORY OF DIFFUSION

2.1. Diffusion Mechanisms

The atomistic theory of diffusion is used to describe the processes by which an atom gets from one part of a crystal to another. The lattice sites in a crystal are generally taken as the fixed locations of the atoms making

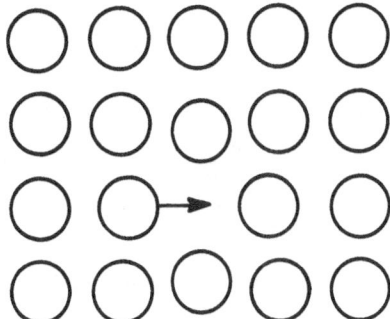

Fig. 1. The vacancy diffusion mechanism. The nearest-neighbor atoms are shown to be displaced toward the vacancy.

up the crystal. The study of specific heats clearly demonstrates that atoms oscillate around these lattice sites, which are their equilibrium positions. These oscillations lead to a finite probability that an atom will move from its lattice site to another position in the crystal. There are several mechanisms by which atoms can move from one site in the crystal to another. These mechanisms are described below.

One of the most important processes to consider for semiconductors in the vacancy mechanism. Thermodynamic considerations require that some of the lattice sites be vacant, and that the number of vacant lattice sites generally is a function of temperature. When a lattice atom moves into an adjacent vacant site, this process is called the *vacancy diffusion mechanism*. The vacancy mechanism is illustrated in Fig. 1. The atom that exchanges position with the vacancy can be either an atom of the host crystal (self-diffusion) or an impurity atom (impurity diffusion). In some cases, isolated vacancies can be bound together so that adjacent lattice sites are vacant. Diffusion can proceed by these bound pairs and this process is termed *divacancy diffusion*. In most cases, single-vacancy diffusion, rather than divacancy diffusion, is the dominate vacancy diffusion mechanism.

In addition to occupying lattice sites, atoms can reside in the space between the lattice sites. These interstitial atoms can readily move to adjacent interstitial sites without displacing the lattice atoms as shown in Fig. 2. The interstitial atoms can be impurity atoms or atoms of the host lattice, but in either case they are generally present only in very dilute amounts. These atoms, however, can be highly mobile and are the dominant diffusion mechanism in certain cases. The interstitial mechanism will be

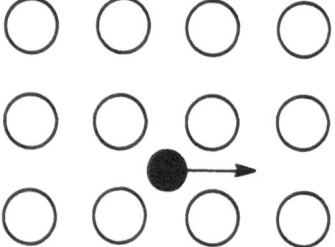

Fig. 2. The interstitial diffusion mechanism.

discussed further in the analysis of Zn diffusion in GaAs and for self-diffusion in the II–VI compounds.

A mechanism related to interstitial diffusion is the *interstitialcy mechanism*. In this process, an interstitial atom moves into a lattice site by displacing the atom on that site into an adjacent interstitial site. This process is illustrated in Fig. 3.

There are several other diffusion mechanisms that are frequently discussed in the atomistic theory of diffusion.[2,3] These include the exchange mechanism, the ring mechanism, the crowdion mechanisms, and the relaxation mechanism. Since these mechanisms do not appear to have a significant role in the diffusion of semiconductors, they will not be considered here. There are also paths of high diffusivity such as grain boundaries, surfaces, and dislocation pipes. These diffusion mechanisms operate in regions in which the lattice structure breaks down and the diffusion is rapid because of the open nature of these regions. The diffusion behavior is strongly dependent on the particular atomic arrangement in the immediate vicinity of the high diffusivity path, which prevents detailed analytical treatment (Ref. 2, pp. 164–187). Diffusion studies in semiconductors generally avoid high-diffusivity paths and will not be considered further.

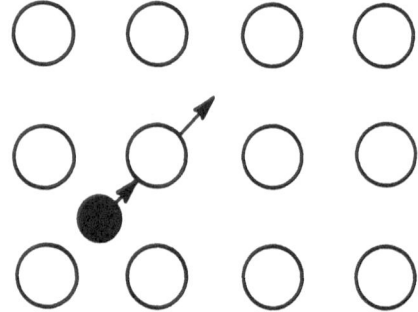

Fig. 3. The interstitialcy diffusion mechanism.

2.2. The Flux Equation and Diffusivity

2.2.1. *The Diffusive Flux*

The number of atoms which cross a unit area in unit time is known as the flux. In the one-dimensional case, the atoms only move to the right or left when they change position along the x axis as indicated in Fig. 4. The atoms in this simple case are taken to be located in planes located at x_0 and $x_0 + a_0$. The flux J is simply the concentration C times the velocity v, so that

$$J_x = Cv \tag{1}$$

The net flux is the difference between the flux to the right and from the left:

$$J_x = \tfrac{1}{2}v(C_{x_0} - C_{x_0+a_0}) \tag{2}$$

where C_{x_0} and $C_{x_0+a_0}$ are the concentrations at $x = x_0$ and $x = x_0 + a_0$, respectively. The factor of one-half occurs in Eq. (2) because at any one plane, half of the atoms move in the $+x$ direction and the other half move in the $-x$ direction. When a_0 approaches zero,

$$a_0 \frac{C_{x_0} - C_{x_0+a_0}}{a_0} = -a_0 \frac{dC}{dx} \tag{3}$$

and Eq. (2) becomes

$$J_x = -\tfrac{1}{2}va_0 \, dC/dx \tag{4}$$

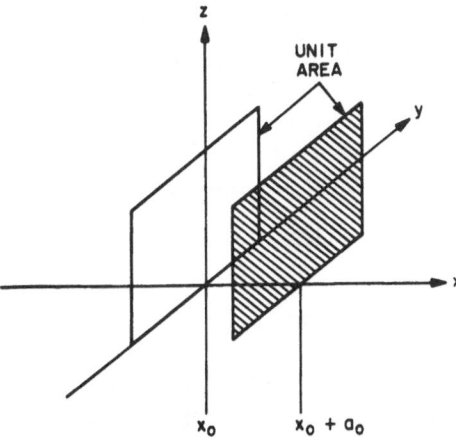

Fig. 4. Flux in the x direction through the unit area A in unit time. The planes of unit area are located at $x = x_0$ and $x = x_0 + a_0$.

For motion by discrete jumps between planes which are separated by a distance a_0, the velocity is the number of jumps/sec, Γ, times the distance a_0 of each jump. Equation (4) can now be written as

$$J_x = -\tfrac{1}{2}a_0^2\Gamma\, dC/dx \qquad (5)$$

and the quantity $\tfrac{1}{2}a_0^2\Gamma$ is called the diffusivity or diffusion coefficient D, where

$$D = \tfrac{1}{2}a_0^2\Gamma \qquad (6)$$

When the flux equation is written in terms of the diffusivity, we have

$$J_x = -D\, dC/dx \qquad (7)$$

This equation is called *Fick's first law*.

2.2.2. Jump Frequency

Equation (6) shows that for diffusion by a particular mechanism, calculation of the diffusivity is reduced to an evaluation of the jump frequency Γ. The jump frequency by the vacancy mechanism is

$$\Gamma = X_V\omega \qquad (8)$$

where ω is the frequency at which an atom exchanges position with an adjacent vacancy. Since the jump frequency also depends on the probability that the adjacent site is vacant, ω is multiplied by the ratio of vacant to occupied sites, which is called the vacancy atom fraction X_V.

For an elemental crystal, the formation reaction for a vacancy V is given by the reaction

$$0 \rightleftarrows V \qquad (9)$$

The Gibbs free energy change for this vacancy formation reaction is

$$\Delta G_f = \Delta H_f - T\,\Delta S_f = -kT \ln X_V \qquad (10)$$

or

$$X_V = \exp(\Delta S_f/k)\exp(-\Delta H_f/kT) \qquad (11)$$

where ΔS_f and ΔH_f are the entropy and enthalpy of formation. The enthalpy change ΔH_f is taken as the energy of formation.

The frequency ω at which an atom jumps into an adjacent vacancy is much more difficult to derive. In fact, the inability to evaluate ω from fundamentals has led to the use of the continuum approach rather than the atomistic approach in the analysis of diffusion in semiconductors. Nevertheless, a discussion of the factors that enter into the evaluation of ω provides useful insight into the quantities that affect the diffusivity.

In self-diffusion by the vacancy mechanism, a lattice atom moves from a normal lattice site to a vacancy. As shown in Fig. 5, the atom must move from the normal lattice site (a) to the saddle-point position (b) to reach the vacancy at (c). The energy at the saddle point is greater than at the equilibrium lattice sites, and atoms must be sufficiently "activated" in order to move first to (b) and then to (c). The fraction of the lattice atoms activated to the saddle point is related to the Gibbs free energy change between positions (a) and (b). In the same manner as for the atom fraction of vacancies, the atom fraction of activated atoms is

$$X_m = \exp(\Delta S_m/k)\exp(-\Delta H_m/kT) \qquad (12)$$

where ΔS_m and ΔH_m are called the entropy and enthalpy of motion.

The rate at which the atoms at the saddle point go to the vacant site is the frequency ν, which is of the order of the mean vibrational frequency of an atom about its equilibrium site. Therefore, $X_m\nu$ atoms will jump to a neighboring vacancy per second and

$$\omega = X_m\nu = \nu\exp(\Delta S_m/k)\exp(-\Delta H_m/kT) \qquad (13)$$

The frequency ν is generally not known and is usually taken as the Debye frequency

$$\nu_D = k\theta_D/h \qquad (14)$$

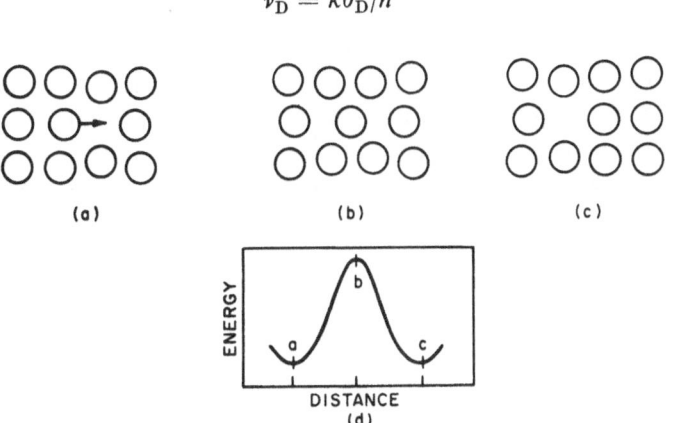

Fig. 5. The sequence (a)–(c) shows the movement of an atom from a normal lattice site to an adjacent vacancy. (d) The variation of free energy as the atom moves from (a) to (c). (Ref. 2, p. 58.)

where θ_D is the Debye temperature, and ν_D is of the order of 10^{13} sec^{-1}. From Eqs. (8), (11), and (13), the jump frequency for vacancy self-diffusion is

$$\Gamma = \nu \exp[(\Delta S_f + \Delta S_m)/k] \exp[-(\Delta H_f + \Delta H_m)/kT] \qquad (15)$$

2.2.3. Diffusivity and Activation Energy

Experimentally, it is found that diffusivity is given by the Arrhenius expression

$$D = D_0 \exp(-Q/kT) \qquad (16)$$

where Q is termed the activation energy. By combining Eqs. (6) and (15), it can be shown that the diffusivity is

$$D = \tfrac{1}{2}a_0^2\nu \exp[(\Delta S_f + \Delta S_m)/k] \exp[-(\Delta H_f + \Delta H_m)/kT] \qquad (17)$$

By comparing Eqs. (16) and (17), it is seen that the prefactor D_0 is

$$D_0 = \tfrac{1}{2}a_0^2\nu \exp[(\Delta S_f + \Delta S_m)/k] \qquad (18)$$

and the activation energy is

$$Q = \Delta H_f + \Delta H_m \qquad (19)$$

The numerical prefactor of one-half in Eq. (18) will be somewhat different for an actual three-dimensional crystal (see Refs. 2 and 3), but has been neglected here.

These equations are difficult to evaluate numerically, but they do demonstrate the quantities that make up the diffusivity and activation energy. The diffusivity is basically the product of the lattice vibrational frequency, the vacancy concentration, and the activated lattice concentration. Also, the activation energy for vacancy diffusion depends upon the energy necessary to form the vacancy and to move the lattice atom into an adjacent vacant site.

3. CONTINUUM THEORY OF DIFFUSION

3.1. Fick's Laws

As presented in Section 2.2.1, Eq. (7), Fick's first law for one-dimensional diffusion is

$$J_x = -D \, dC/dx$$

The minus sign appears because the flux is in the direction of decreasing concentration. In three dimensions, the flux is the vector

$$\mathbf{J} = -D\,\nabla C \tag{20}$$

For crystals with cubic symmetry, D is the same in all directions. However, for noncubic crystals, D can be nonisotropic and is then written as a second-order tensor. The analysis of diffusion in semiconductors presented here will only be concerned with the one-dimensional flux equation and an isotropic diffusion coefficient.

Fick's second law is the continuity equation. In one dimension, this equation is

$$\frac{\partial C}{\partial t} = \frac{\partial}{\partial x}\left(D\frac{\partial C}{\partial x}\right) \tag{21}$$

and for the three-dimensional case it is

$$\partial C/\partial t = -\nabla\cdot\mathbf{J} \tag{22}$$

A majority of the semiconductor diffusion analyses are based on Eqs. (7) and (21).

3.2. The Diffusion Profile

In most cases, the problem of interest is a determination of the diffusion profile for a diffusion species whose concentration in the host crystal is initially zero. The diffusion profile will depend on the initial and boundary conditions. When the diffusion coefficient is constant, the diffusion equation in one dimension is

$$\partial C/\partial t = D(\partial^2 C/\partial x^2) \tag{23}$$

Two different solutions to Eq. (23) are frequently encountered. These are designated by the boundary condition. The first is the case where the total amount of the diffusing species is fixed, and the second is for an infinite source that maintains a constant concentration at the sample surface during the diffusion. In either case, the sample is generally of sufficient thickness to require only consideration of diffusion from one surface.

Many commercial diffusion processes in Si are based on a two-step diffusion. First, a very thin layer having a high concentration of the dopant is obtained by using a short diffusion time at high temperature. This initial impurity distribution can be reasonably approximated by a constant concentration C_0 over a distance δ, and gives a concentration per unit area of $\alpha = C_0\delta$. The desired diffusion profile is obtained by using the initial distribution as the impurity source for a more lengthy diffusion. Under these conditions, Eq. (23) is solved for the case of a constant amount of diffusant (often called the plated source case). The initial condition at the boundary is taken as a delta function of α atoms per unit area, and the diffusion profile is given by[1,5]

$$C(x, t) = [\alpha/(\pi Dt)^{1/2}] \exp(-x^2/4Dt) \qquad (24)$$

which is a Gaussian profile. As the diffusion proceeds, the surface concentration decreases with time as

$$C(0, t) = \alpha/(\pi Dt)^{1/2} \qquad (25)$$

The variation of the diffusion profile with time is illustrated in Fig. 6. In the preparation of p–n junctions, it is of interest to know the junction depth. For a background impurity concentration of C_b, the junction depth x_j

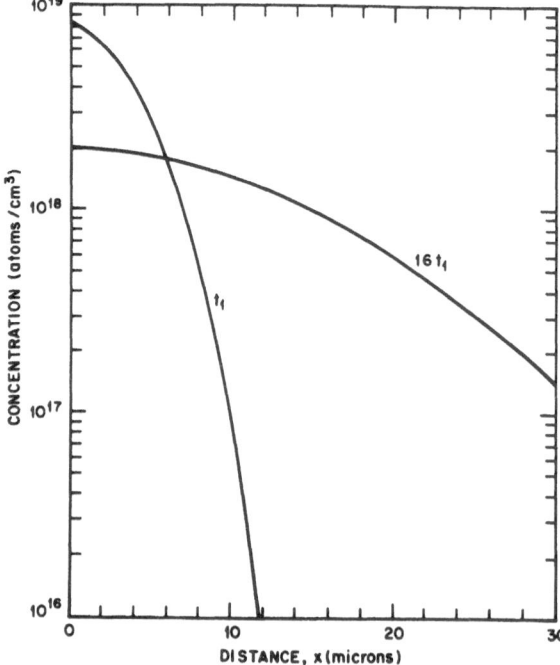

Fig. 6. The Gaussian diffusion profile for a constant amount of diffusant at diffusion times of t_1 and $16t_1$.

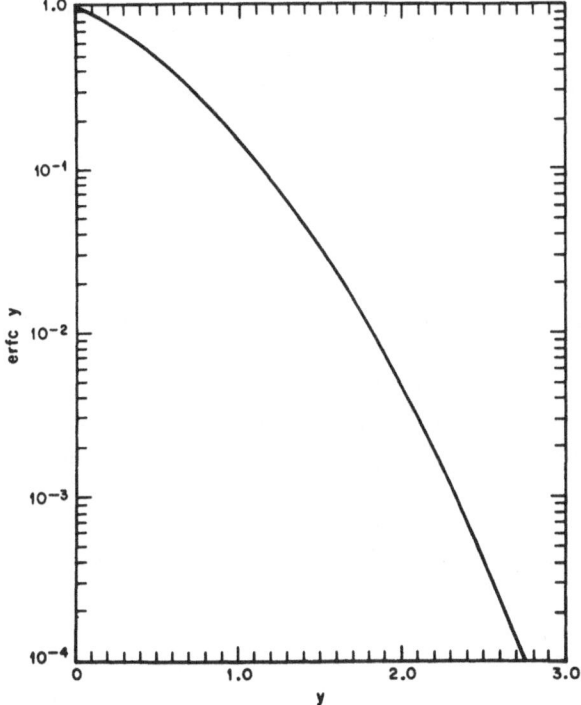

Fig. 7. Normalized erfc plot with $x/2(Dt)^{1/2} = y$.

varies as

$$x_j = \{4Dt \ln[\alpha/C_b(\pi Dt)^{1/2}]\}^{1/2} \tag{26}$$

It should also be noted, as indicated in Fig. 6, that the concentration gradient at a fixed value of concentration depends on the diffusion time.

The other common diffusion process is for a constant surface concentration C_0 during the entire diffusion. Solution of Eq. (23) with $C(0, t) = C_0$ gives[1,5]

$$C(x, t) = C_0 \operatorname{erfc}[x/2(Dt)^{1/2}] \tag{27}$$

where erfc is the complementary error function. This function is most readily evaluated from a normalized plot as given in Fig. 7. It is frequently convenient to consider an average diffusant penetration depth $\langle x \rangle$ as

$$\langle x \rangle = \tfrac{1}{2}(\pi Dt)^{1/2} \tag{28}$$

Experimentally, the diffusivity is often obtained by fitting Eq. (27) to the experimental diffusion profile.

3.3. Impurity Diffusion for a Moving Boundary

3.3.1. *Experimental Conditions for a Moving Boundary*

The previous section treated the most commonly encountered solutions to the diffusion equation, which are for planar diffusion into a semiinfinite sample. Most *p–n* junctions, transistors, and integrated circuits are prepared by impurity diffusion into epitaxial layers. These epitaxial layers are generally thin, lightly doped layers that are grown on heavily doped, single-crystal substrates. In either vapor-phase or liquid-phase epitaxy, the growth temperatures, although below the melting point of the crystal, are high enough to permit appreciable impurity diffusion in both the substrate and the epitaxial layer. Since the sample surface moves during epitaxial growth, this condition requires solution of the diffusion equation for a moving boundary.

Another moving boundary problem occurs when the semiconductor surface evaporates during diffusion. For most impurity and self-diffusion studies in the III–V compounds, the diffusion temperature is within 200–300°C of the host lattice melting point and long diffusion times are necessary to achieve measurable penetration. When the rate of evaporation is comparable to the rate of diffusion, it is necessary to quantitatively account for the rate of evaporation in the assignment of the diffusion coefficient from experimental data. This section briefly discusses the change of variables in the diffusion equation that is necessary for obtaining a solution with a moving boundary and then gives the expressions for the diffusion profiles.

3.3.2. *Diffusion into a Growing Layer*

The diffusion of impurities during epitaxial growth has been treated by Rice[7] and Grove *et al.*[8] Hu[9] has also considered the redistribution of impurities during additional high-temperature diffusions when forming junctions. The geometry of the growing layer is shown in Fig. 8. As the epitaxial layer grows, the sample surface recedes from the substrate at the rate v and is at x' rather than $x = 0$. Solution of the diffusion equation requires transformation of Eq. (23) by change of variables

$$x' = x + vt \tag{29}$$

and

$$t = t' \tag{30}$$

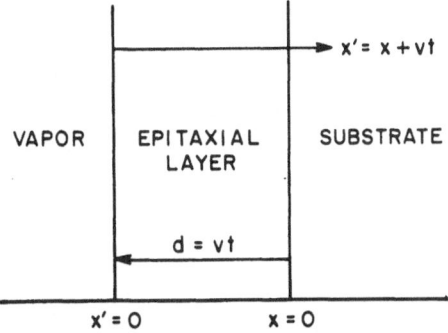

Fig. 8. Diffusion during epitaxial growth. The original surface is at $x = 0$ and an epitaxial layer is grown with a constant growth rate v for a time t to give a layer thickness $d = vt$.

which gives the diffusion equation

$$\frac{\partial C}{\partial t} = D \frac{\partial^2 C}{\partial x'^2} - v \frac{\partial C}{\partial x'} \tag{31}$$

If the substrate has an original concentration C_1 and the layer has a concentration C_2 added during epitaxial growth, the solution when the rate of loss of C_1 at the surface can be neglected is given by[7,10]

$$C(x', t) = \frac{C_2}{2} \left\{ \mathrm{erfc}\left[\frac{x' - vt}{2(D_2 t)^{1/2}}\right] + \exp\left(\frac{vx'}{D_2}\right) \mathrm{erfc}\left[\frac{x' + vt}{2(D_2 t)^{1/2}}\right] \right\}$$
$$+ \frac{C_1}{2} \left\{ 1 + \mathrm{erf}\left[\frac{x' - vt}{2(D_1 t)^{1/2}}\right] \right\} \tag{32}$$

In Eq. (32), the error function erf is related to the complementary error function erfc by

$$\mathrm{erfc}(y) = 1 - \mathrm{erf}(y) \tag{33}$$

and

$$\mathrm{erf}(-y) = -\mathrm{erf}(y) \tag{34}$$

An example given by Rice[7] for Ge illustrates the behavior of Eq. (32). For the conditions $T = 900°C$, $vt = 3.8$ μm, $t = 3.3 \times 10^2$ sec, $C_1 = 9 \times 10^{17}$ cm^{-3}, $D_1 = 6 \times 10^{-13}$ cm^2/sec, $C_2 = 10^{16}$ cm^{-3}, and $D_2 = 4 \times 10^{-11}$ cm^2/sec, the resulting distributions of C_1 and C_2 are shown in Fig. 9. It is seen that the distribution of C_1 is displaced from the substrate into the epitaxial layer by a significant amount.

3.3.3. *Diffusion into an Evaporating Surface*

The analysis for the evaporating surface is very similar to that of the growing surface. Kucher[11] has solved Eq. (23) for the condition of a surface evaporating at the rate v as represented in Fig. 10. The change of variable is similar to the case of diffusion during epitaxial growth, namely

$$x' = x - vt \tag{35}$$

which gives

$$\frac{\partial C}{\partial t} = D\frac{\partial^2 C}{\partial x'^2} + v\frac{\partial C}{\partial x'} \tag{36}$$

For a constant surface concentration C_0, the solution of the diffusion equation with an evaporating surface is[11]

$$C(x', t) = \frac{C_0}{2}\left\{\mathrm{erfc}\left[\frac{x' + vt}{2(Dt)^{1/2}}\right] + \exp\left[\frac{-vx'}{D}\right]\mathrm{erfc}\left[\frac{x' - vt}{2(Dt)^{1/2}}\right]\right\} \tag{37}$$

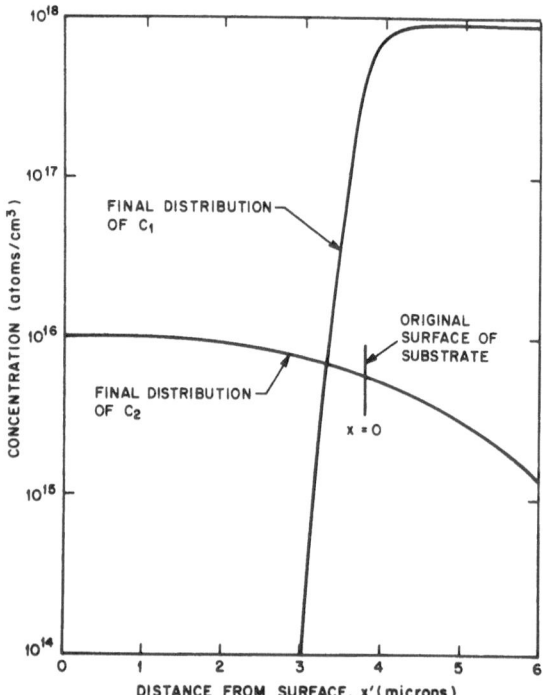

Fig. 9. Concentration distribution of C_1 and C_2 as given by Eq. (32) for the parameters given in the text, from Ref. 7.

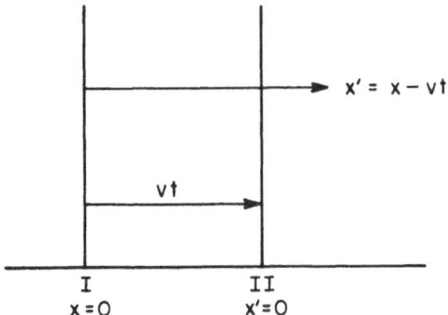

Fig. 10. Diffusion into an evaporating surface. The original surface I is at $x = 0$, and evaporates to surface II at the rate v. Distances from the new surface at $x = vt$ are x'.

There are two limiting cases of Eq. (37). If the evaporation rate goes to zero ($v = 0$), then Eq. (37) reduces to $C = C_0 \, \text{erfc}[x/2(Dt)^{1/2}]$, which is identical to Eq. (27). At the other extreme of rapid evaporation, when

$$v \geq 2.5(D/t)^{1/2} \tag{38}$$

the reduced form is

$$C(x', t) \approx C_0 \exp(-vx'/D) \tag{39}$$

This limiting case can readily be found for the steady-state solution[12] ($\partial C/\partial t = 0$) of Eq. (36). Families of curves for solutions between these two conditions are given in Ref. 11. In evaluation of the diffusion coefficient from experimental data, the rate of evaporation should be determined to see if Eq. (37) is applicable.

3.4. The Boltzmann–Matano Analysis

All of the solutions of the diffusion equation considered in the previous sections have been for a constant diffusion coefficient. In many cases, the impurity diffusion coefficient has been found to vary with the impurity concentration. Therefore the diffusion coefficient will have a different value at each position along the diffusion profile and be a function of the independent variable x. The one-dimensional diffusion equation [Eq. (21)] now becomes

$$\frac{\partial C}{\partial t} = \frac{\partial}{\partial x}\left(D\frac{\partial C}{\partial x}\right) = D\frac{\partial^2 C}{\partial x^2} + \frac{\partial D}{\partial x}\frac{\partial C}{\partial x} \tag{40}$$

Because of the $\partial D/\partial x$ term, Eq. (40) is an inhomogeneous equation and is very difficult to solve even in the simplest cases.

A technique that has become known as the Boltzmann–Matano analysis allows determination of the concentration-dependent diffusion coefficient from the experimental diffusion profile, thus avoiding a solution of Eq. (40). Boltzmann[13] showed that Eq. (40) can be reduced to an ordinary differential equation when D is a function of C only by defining a new variable

$$\eta = x/t^{1/2} \tag{41}$$

The partial derivative of C with respect to time becomes

$$\frac{\partial C}{\partial t} = \frac{\partial \eta}{\partial t}\frac{\partial C}{\partial \eta} = -\frac{1}{2}xt^{-3/2}\frac{\partial C}{\partial \eta} \tag{42}$$

while the partial derivative with respect to distance is

$$\frac{\partial C}{\partial x} = \frac{\partial \eta}{\partial x}\frac{\partial C}{\partial \eta} = t^{-1/2}\frac{\partial C}{\partial \eta} \tag{43}$$

Substitution of Eqs. (42) and (43) into Eq. (40) gives

$$-\frac{1}{2}xt^{-3/2}\frac{dC}{d\eta} = \frac{\partial}{\partial x}\left(Dt^{-1/2}\frac{dC}{d\eta}\right) \tag{44}$$

or

$$-\frac{1}{2}xt^{-3/2}\frac{dC}{d\eta} = \frac{\partial \eta}{\partial x}\frac{d}{d\eta}\left(Dt^{-1/2}\frac{dC}{d\eta}\right) \tag{45}$$

With $\partial \eta/\partial x = t^{-1/2}$, Eq. (45) becomes the ordinary differential equation

$$\frac{-\eta}{2}\frac{dC}{d\eta} = \frac{d}{d\eta}\left(D\frac{dC}{d\eta}\right) \tag{46}$$

and because Eq. (46) is in terms of total differentials, multiplying through by $d\eta$ gives

$$\frac{-\eta}{2}dC = d\left(D\frac{dC}{d\eta}\right) \tag{47}$$

Equation (47) can be integrated between $C = 0$ and $C = C(x_0)$ at some distance x_0 from the surface to give

$$D\frac{dC}{d\eta}\bigg|_{x=x_0} = -\frac{1}{2}\int_0^{C(x_0)}\eta\,dC \tag{48}$$

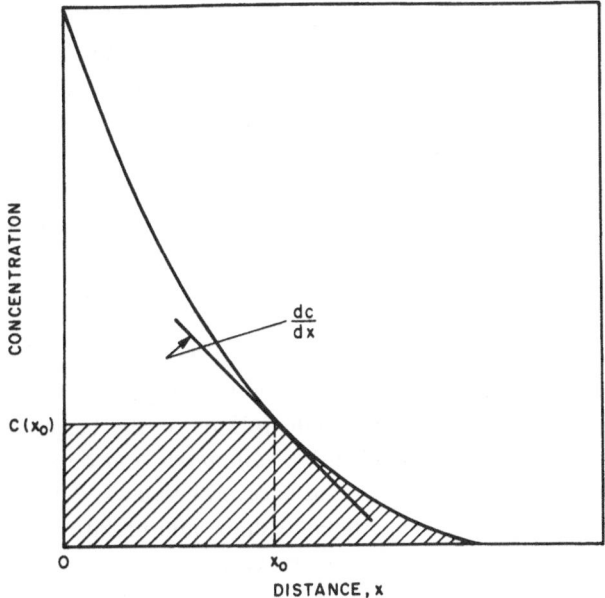

Fig. 11. Determination of D at x_0 by the Boltzmann–Matano method. The cross-hatched area represents the integral in Eq. (49), while dC/dx is given by the slope of the diffusion profile at $x = x_0$.

with $dC/d\eta = 0$ at $C = 0$. Substitution of Eqs. (41) and (43) into Eq. (48) gives

$$D = - \left(2t \left. \frac{dC}{dx} \right|_{x=x_0} \right)^{-1} \int_0^{C(x_0)} x \, dC \qquad (49)$$

Matano[14] used this method to determine the concentration-dependent diffusion coefficient as illustrated in Fig. 11. The integral is represented by the cross-hatched area. The Boltzmann–Matano analysis will be used in a later section to obtain the concentration dependence of As diffusion in Si and Zn diffusion in GaAs.

4. DIFFUSION IN Ge AND Si

4.1. Introduction

The fundamentals developed in the previous sections will now be applied to diffusion in the group IV semiconductors. Some of the properties of the group IV semiconductors Si, Ge, diamond, and SiC are summarized in Table I. A detailed summary of the properties of these semiconductors is given in Ref. 28. Very little diffusion data are available for diamond and SiC because of the very high temperatures necessary to obtain measurable diffusion. Discussion of diffusion in the group IV semiconductors will be limited to the elemental semiconductors Ge and Si.

Because of the many commercial applications of Ge and Si, the diffusion behavior in these two elemental semiconductors has been vigorously studied and the published literature is very extensive. Almost all of the $p–n$ junction rectifiers and bipolar transistors are made by diffusion of impurities into Si. Many integrated circuits contain these diffused devices, but field-effect transistors (FET's) are also widely used. Silicon is probably the most intensively investigated element during the last 20 years. The discussion of diffusion given here is limited to illustrations of the basic concepts and techniques of analysis. Comprehensive summaries and detailed listings of the original publications can be found in several reviews.[4,5,10,29–31] Although there is a great wealth of experimental data, interpretation in terms of a definitive mechanism is lacking in most cases.

In this section on diffusion in Ge and Si, the experimental self-diffusion data are given, and these experimental results are then interpreted in terms of the theory for vacancy self-diffusion. Next, experimental results of the slowly diffusing group III and V impurities are presented. Concentration-dependent diffusion coefficients and the formation of neutral donor–vacancy complexes will be illustrated for As in Si. Diffusion of many of the other

Table I. Properties of the Group IV Semiconductors

Group IV semiconductor	Crystal structure	Lattice constant, Å	Density, g/cm^3	Melting, point, °C	Energy gap at 300°K, eV
Si	Diamond	$a = 5.430951$[15]	2.32902[16]	1412[b]	1.120[18]
Ge	Diamond	$a = 5.646133$[19]	5.32674[16]	937[c]	0.663[18]
C(diamond)	Diamond	$a = 3.56683$[20]	3.515[21]	4100[22]	5.47[23]
SiC[a]	Wurtzite	$a = 3.0865$[24] $c = 15.11738$[24]	3.211[25]	2830[26,d]	2.996[27]

[a] There are numerous polymorphous modifications of SiC. The cubic structure is called β-SiC while the others are hexagonal and are called α-SiC. The lattice constant given here is for the hexagonal 6H-SiC.
[b] Ref. 17, Supplement November 1965.
[c] Ref. 17, Supplement March 1965.
[d] Decomposes.

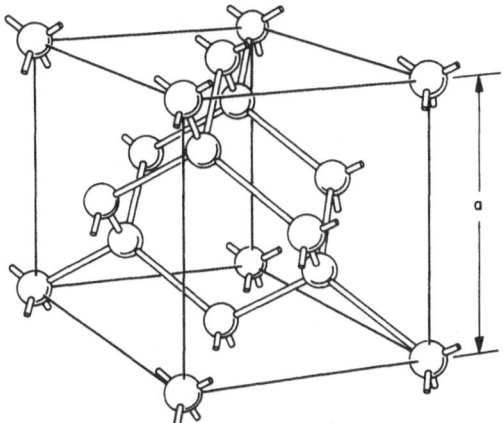

Fig. 12. The diamond structure for Ge and Si with lattice constant a.

impurities must be analyzed in terms of the dissociative mechanisms which consider the interaction between the substitutional and interstitial species. The relatively open diamond (cubic) lattice illustrated in Fig. 12 permits the interstitial atoms to dominate the diffusive flux for certain impurities. A detailed example of the interstitial–substitutional diffusion is given in the next section for Zn in GaAs. Therefore the dissociative mechanism for Si and Ge will not be discussed in detail.

4.2. Self-Diffusion in Ge and Si

Self-diffusion in Ge and Si has been studied through the use of radioactive isotopes. In Ge, Letaw et al.[32] diffused radioactive ^{71}Ge (half-life = 11 days) into single-crystal Ge at temperatures between 766 and 928°C. The self-diffusion in intrinsic Ge can be represented by[32]

$$D_{\text{self}}(\text{Ge}) = 7.8 \exp(-2.97/kT) \text{ cm}^2/\text{sec} \qquad (50)$$

with an uncertainty of ± 3.4 cm²/sec for D_0 and ± 0.04 eV for Q. At the melting point of Ge ($T_{\text{mp}} = 937°\text{C}$), $D_{\text{self}} = 3.4 \times 10^{-12}$ cm²/sec. For silicon, Masters and Fairfield[33,34] diffused radioactive ^{31}Si (half-life = 2.6 hr) into single-crystal Si at temperatures between 1110 and 1300°C. Within this temperature range, the self-diffusion in intrinsic Si can be represented by[33,34]

$$D_{\text{self}}(\text{Si}) = 9.0 \times 10^3 \exp(-5.13/kT) \text{ cm}^2/\text{sec} \qquad (51)$$

Because of the relatively narrow temperature range, the preexponential was considered to be an order-of-magnitude estimate. The uncertainty in Q was given as 0.07 eV. Measurements by others[35,36] have resulted in D_0 values of 1.2×10^3 to 1.8×10^3 and Q of ~ 4.75 eV. These values for Si are substantially larger than those for Ge. The diffusivity at the melting point ($T_{mp} = 1412°C$) as given by Eq. (51) is $D_{self} = 3.7 \times 10^{-12}$ cm²/sec.

Determination of the self-diffusion mechanism has been based on a comparison of experimental and theoretical values of the activation energy. As discussed in the derivation of Eq. (17), the activation energy is the sum of the vacancy formation and the vacancy migration energies. The energy of monovacancy formation ΔH_{f_1} was calculated by Swalin[37] to be 2.07 eV for Ge and 2.32 eV for Si. The energy of monovacancy motion ΔH_{m_1} was calculated by Swalin[37] to be 0.95 eV for Ge and 1.06 eV for Si.* These theoretical values give an activation energy of 3.02 eV for Ge and 3.38 eV for Si. Comparison with the experimental values gives for Ge

$$Q_{Ge}(\text{exp: } 2.97 \text{ eV}) \approx Q_{Ge}(\text{theor: } 3.02 \text{ eV}) \tag{52}$$

and for Si

$$Q_{Si}(\text{exp: } 5.13 \text{ eV}) > Q_{Si}(\text{theor: } 3.38 \text{ eV}) \tag{53}$$

On the basis of the good agreement between the experimental and theoretical values of the activation energy in Ge, self-diffusion in this material has been taken to proceed by the monovacancy mechanism. Because the experimental activation energy for Si is significantly larger than the theoretical value for the monovacancy, the divacancy mechanism has been proposed.[30,35,38]

For an elemental crystal, the association of lattice vacancies □ to form divacancies ⊟ is given by the reaction

$$\square + \square \rightleftarrows \boxminus \tag{54}$$

The Gibbs free energy change for the divacancy reaction is

$$\Delta G_{f_2} = \Delta H_{f_2} - T \Delta S_{f_2} = -kT \ln[X_{V_2}/(X_{V_1})^2] \tag{55}$$

or

$$X_{V_2} = (X_{V_1})^2 \exp(\Delta S_{f_2}/k) \exp(-\Delta H_{f_2}/kT) \tag{56}$$

where X_{V_2} is the atom fraction of divacancies and X_{V_1} is the atom fraction of monovacancies. From the reaction for divacancy formation it can be

---

* Calculated values in a recent paper[233] give ΔH_{f_1} as 2.53~2.63 eV for Ge and 2.74~2.84 eV for Si, and ΔH_{m_1} as 0.31~0.40 eV for Ge and 0.38~0.50 eV for Si.

seen that ΔH_{f_2} represents the enthalpy difference between the associated and dissociated pair, and is called the divacancy binding energy E_B. For an attractive potential, $\Delta H_{f_2} = -E_B$. The monovacancy atom fraction given by Eq. (11) permits Eq. (56) to be rewritten as

$$X_{V_2} = \exp[(2\Delta S_{f_1} + \Delta S_{f_2})/k]\,\exp[-(2\Delta H_{f_1} - E_B)/kT] \qquad (57)$$

In the same manner as for the monovacancy in Eq. (17), the diffusion coefficient for self-diffusion by the divacancy is

$$
\begin{aligned}
D = \tfrac{1}{2}a_0{}^2 v \, &\exp[(2\,\Delta S_{f_1} + \Delta S_{f_2} + \Delta S_{m_2})/k] \\
&\times \exp[-(2\,\Delta H_{f_1} - E_B + \Delta H_{m_2})/kT]
\end{aligned} \qquad (58)
$$

where ΔS_{m_2} and ΔH_{m_2} are the entropy and enthalpy of motion. The pre-factor $1/2$ is actually 2.0 for the diamond lattice, and v is the vibration frequency discussed in conjunction with Eq. (13).

Examination of Eq. (58) demonstrates why the divacancy was suggested for self-diffusion in Si. The activation energy for divacancy self-diffusion is

$$Q(V_2) = 2\,\Delta H_{f_1} - E_B + \Delta H_{m_2} \qquad (59)$$

while from Eq. (19),

$$Q(V_1) = \Delta H_{f_1} + \Delta H_{m_1}$$

The entropy terms in D_0 are

$$\Delta S(V_2) = 2\,\Delta S_{f_1} + \Delta S_{f_2} + \Delta S_{m_2} \qquad (60)$$

and

$$\Delta S(V_1) = \Delta S_{f_1} + \Delta S_{m_1} \qquad (61)$$

Therefore

$$Q(V_2) > Q(V_1) \qquad (62)$$

if

$$\Delta H_{f_1} + \Delta H_{m_2} > \Delta H_{m_1} + E_B \qquad (63)$$

Since it is reasonable that the monovacancy formation energy is greater than the divacancy binding energy and the motion energy is greater for a divacancy than a monovacancy, $Q(V_2)$ certainly is expected to exceed $Q(V_1)$. Evaluation of Eqs. (59), (19), and (63) with the available experimental values shows that this conclusion is indeed correct. Watkins and Corbett[39] give $E_B \geq 1.6\,\text{eV}$ and $\Delta H_{m_2} = 1.3\,\text{eV}$, and Watkins[40] assigned ΔH_{m_1} as

0.33 eV. The monovacancy formation energy was given by Elstner and Kamprath[41] as 2.5 eV. These values give

$$Q(V_2) = 4.7 \text{ eV} > 2.83 \text{ eV} = Q(V_1) \tag{64}$$

Note that Swalin's[37] theoretical values were $\Delta H_{f_1} = 2.32$ eV and $\Delta H_{m_1} = 1.06$ eV. These various values are compared in Table II and show that the activation energy for divacancy self-diffusion approaches the values observed experimentally.

The agreement between the observed and expected activation energies is not the only factor to consider. In fact,

$$\exp[-Q(V_2)/kT] \ll \exp[-Q(V_1)/kT] \tag{65}$$

which requires that

$$D_0(V_2) \gg D_0(V_1) \tag{66}$$

in order for the divacancy mechanism to dominate. The requirement of Eq. (66) also suggests that the experimental behavior is dominated at high temperature by divacancies. The value of D_0 for monovacancy diffusion in most metals is of the order of unity, while the experimental value of D_0

Table II. Comparison of the Experimental and Theoretical Values for the Activation Energy for Self-Diffusion by Monovacancies and Divacancies in Si

	Theoretical value, eV	Experimental value, eV
ΔH_{f_1}	2.32[87]	2.5[41]
ΔH_{m_1}	1.06[87]	0.33[40]
$2\Delta H_{f_1}$	4.64	5.0
ΔH_{m_2}	1.3[40,a]	1.3[39]
E_B	0.93[42]	≥ 1.6[39]
Q(monovacancy) $= \Delta H_{f_1} + \Delta H_{m_1}$	3.38	2.83
Q(divacancy) $= 2\Delta H_{f_1} - E_B + \Delta H_{m_2}$	4.99	4.7
Q(experimental)		5.1[33,34] 4.7[35,36]

[a] Experimental value.

for Si is $\sim 10^3$–10^4 cm^2/sec. Comparison of the entropy terms in D_0 [see Eqs. (60) and (61)] shows that $D_0(V_2)$ must be significantly larger than $D_0(V_1)$. Thus, both a large prefactor and a large activation energy are characteristic of divacancy diffusion. For the divacancy self-diffusion to be greater than monovacancy self-diffusion, we must have

$$D_0(V_2) \exp[-Q(V_2)/kT] > D_0(V_1) \exp[-Q(V_1)/kT] \qquad (67)$$

and because $Q(V_2) > Q(V_1)$, it is possible for $D(V_2)$ to dominate at high temperature and $D(V_1)$ to dominate at lower temperatures. The large value of $Q(V_2)$ requires that $D_0(V_2) \geq 10^5 D_0(V_1)$ for divacancy self-diffusion to dominate at $T > 1100°C$. This requirement of a $D_0(V_2)$ so much larger than $D_0(V_1)$ has caused the divacancy mechanism to be seriously questioned. In the diamond lattice, a divacancy cannot move continuously through the lattice. It must first dissociate as shown in Fig. 13(b) and then associate as shown in Fig. 13(c) to move the divancy from the original location of Fig. 13(a). Note that the Si lattice atom moves two lattice positions.

In order to reconcile this discrepancy between the vacancy migration energy ΔH_{m_1} of 0.33 eV inferred from the low-temperature electron paramagnetic resonance studies of radiation-induced defects[40] and the theoretical estimate of 1.06 eV,[37] Masters[43] has suggested the semivacancy pair. In the semivacancy configuration, two adjacent lattice sites are unoccupied and a Si atom is in the interstitial site between the two vacant sites. For this model, ΔH_{m_1} of 0.33 eV is taken as the energy for the conversion of monovacancies to a more stable semivacancy pair. Recently, Van Vechten[44] has attempted to demonstrate that the monovacancy diffusion dominates by suggesting that the enthalpy and entropy of monovacancy migration increase above a transition temperature. In this model, both ΔH_{m_1} and ΔS_{m_1} at the high diffusion temperatures are significantly greater than the low-temperature values. Because of the difficulties in assignment of the dominant vacancy, a definite assignment of the self-diffusion mechanism in Si cannot be made at the present time. The analysis presented here is intended to illustrate the basic concepts of monovacancy and divacancy self-diffusion, and to demonstrate the uncertainties that presently exist in the studies of self-diffusion.

In the discussion of self-diffusion and impurity diffusion, it is often convenient to have some estimate of the vacancy concentration. Hiraki[45] gives the neutral monovacancy concentration in Ge as

$$C_{V_1}(T) = 1.85 \times 10^{23} \exp(-1.9/kT) \text{ cm}^{-3} \qquad (68)$$

which results in $C_{V_1} = 2.3 \times 10^{15}$ cm^{-3} at the melting point. Kendall and

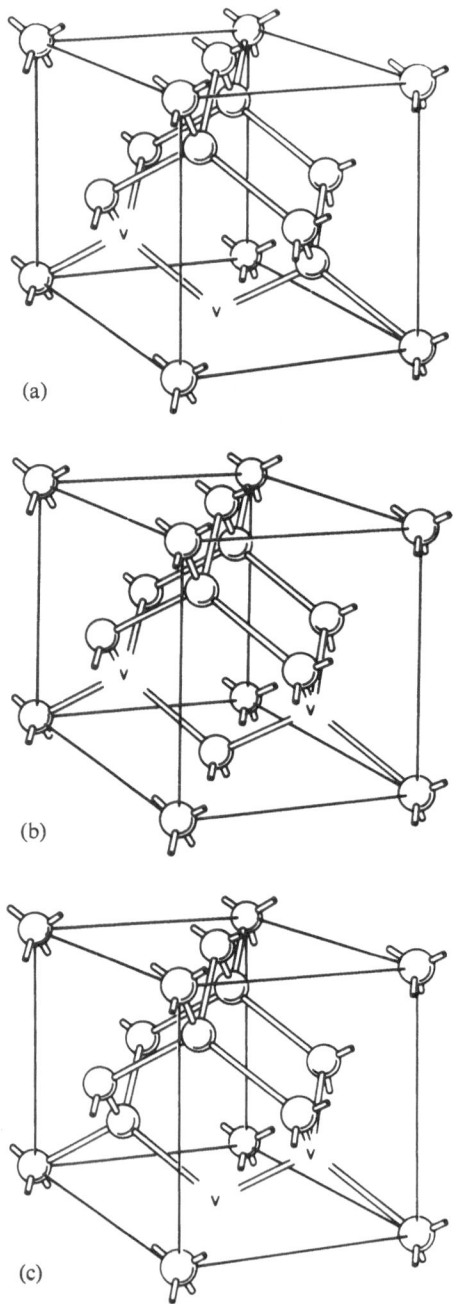

Fig. 13. Dissociation of the divacancy in the diamond lattice in order to change position.
(a) Initial divacancy position. (b) Divacancy dissociation. (c) Reassociation of divacancy.

De Vries[30] represent the neutral Si monovacancy concentration by

$$C_{V_1}(T) = 5 \times 10^{22} \exp(-2.53/kT) \, \text{cm}^{-3} \tag{69}$$

giving $C_{V_1} = 1.3 \times 10^{15} \, \text{cm}^{-3}$ at the melting point. These expressions show that it is reasonable to take the vacancy concentration in Ge or Si as $\sim 10^{15}$ cm^{-3} near the melting point.

Although it has not been discussed here, low-temperature electron paramagnetic resonance spectra have led to assignments of charge states for the monovacancy and divacancy.[39,40] The properties of lattice defects have been reviewed by Hu.[31] In addition to a neutral charge state, the monovacancy has been suggested to be a deep donor, and a singly or doubly charged acceptor. The divacancy has been assigned a neutral state, one donor level and two acceptor levels. The position of the Fermi level will determine whether the neutral or ionized vacancy is the dominant lattice defect. These possible charge states further complicate self-diffusion under extrinsic conditions, and therefore only the intrinsic case has been considered in the above analysis.

4.3. Diffusion of the Group III and V Impurities in Ge and Si

4.3.1. Low-Concentration Case

The most commonly used dopants in Ge and Si are the group III and V elements. These impurities are among the most soluble in these host crystals and also give the shallowest donor and acceptor levels. These elements are,

Table III. Diffusion Coefficients of Group III and Group V Elements in Ge

Impurity	D_0, cm^2/sec	Q, eV	Ref.
B	1.1×10^7	4.54	46
Al	1.6×10^2	3.24	46
Ga	40	3.15	47
In	33	3.03	48
P	2.5	2.49	47
As	10.3	2.51	49
Sb	3.2	2.42	48

Table IV. Diffusion Coefficients of Group III and Group V Elements in Si

Impurity	D_0, cm²/sec	Q, eV	Ref.
B	5.1	3.70	50
Al	8.0	3.47	51
Ga	3.6	3.51	51
In	16.5	3.91	51
P	10.5	3.69	51
As	60.0	4.20	52
Sb	12.9	3.98	53
Bi	1.03×10^3	4.63	51

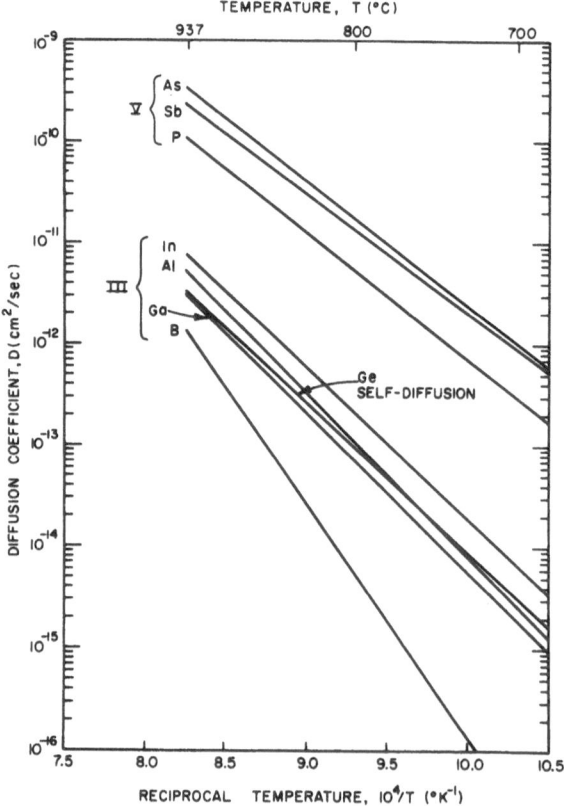

Fig. 14. The diffusion coefficient as a function of temperature for group III and group V elements in Ge.

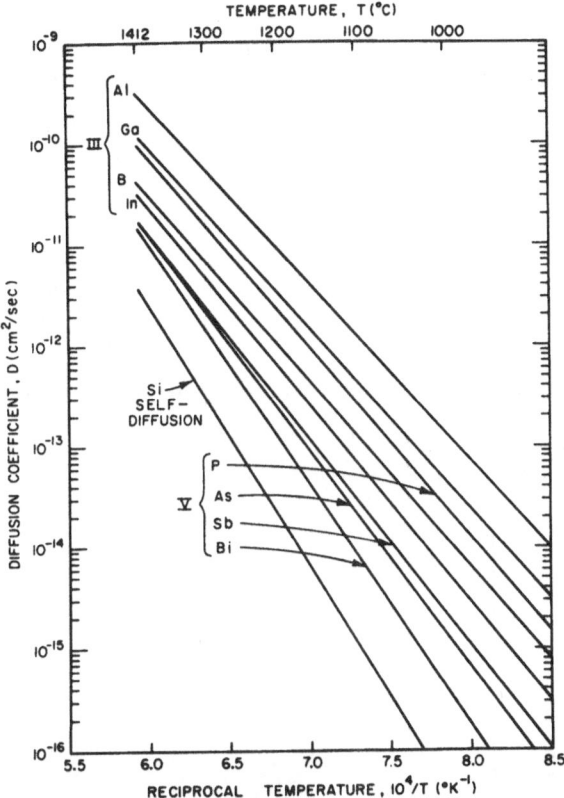

Fig. 15. The diffusion coefficient as a function of temperature
for group III and group V elements in Si.

however, among the slowest diffusing impurities. Boron is the most commonly used acceptor because it is the only group III element that can readily
be masked by SiO_2 layers for selected area diffusion. The donors P and As
can readily be masked and are extensively utilized in selected area diffusion.
The diffusivities of the group III and V elements are summarized in Tables
III and IV. Only representative values of D_0 and Q are given.[31] These
diffusivities are taken as the values at low concentrations. As discussed
later in this section, the diffusivity of these elements has been observed
to be concentration dependent at high concentrations. The diffusion coefficients of these elements are plotted as a function of reciprocal temperature for Ge in Fig. 14 and for Si in Fig. 15. Note that the group V elements
diffuse more rapidly than the group III elements in Ge, while in Si the
group III elements diffuse faster.

The diffusion profiles of the group III and group V elements at low concentrations are well behaved and can be fitted to solutions of the diffusion equation for the appropriate boundary conditions. There is, however, considerable controversy over the proper surface conditions to obtain the bulk diffusivity. The P diffusivities obtained by Ghostagore[54] for several different P source and surface conditions are shown in Fig. 16. The most commonly used condition is given by the upper curve in Fig. 16 for diffusion from an oxide source in a controlled oxidizing atmosphere and results in the values given in Table IV and Fig. 15. For diffusion in a hydrogen atmosphere with a doped vapor source or the doped Si layer (polycrystalline or epitaxial), Fig. 16 shows that the diffusivity is less than for the oxide source. At the present time, these differences are not understood

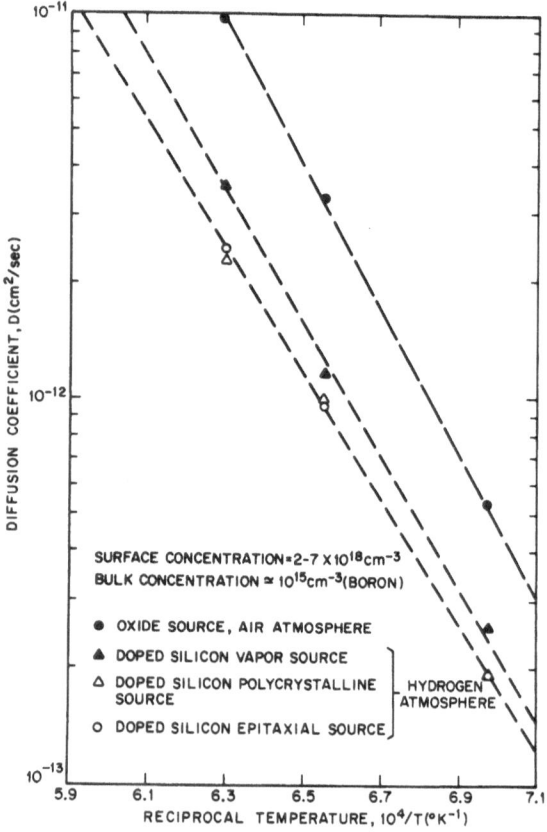

Fig. 16. Phosphorus diffusivity in Si as a function of temperature for the indicated source conditions.[54]

and there are no rigorous explanations for the diffusion mechanism of the group III and V elements that are generally accepted.

Part of the difficulty in assignment of a diffusion mechanism results from the fact that group V elements diffuse more rapidly than group III elements in Ge, while the reverse is true for Si (see Figs. 14 and 15). There is agreement, however, that diffusion is by a substitutional mechanism. Whether the impurity depends upon exchange with monovacancies or divacancies is unclear.

4.3.2. High-Concentration Case

a. *Phosphorus in Si.* At low concentrations for a constant surface concentration, the diffusion is well behaved and the diffusion profile is a complementary error function. The diffusion profiles are not complementary error functions at high impurity levels because the diffusing impurity can greatly enhance the ionized vacancy concentration, which results in enhanced diffusion. The results reported by Tannenbaum[55] with radioactive P are shown in Fig. 17. In addition to the departure from a complementary error function, it was also found that only a portion of the P atoms are electrically active at concentrations above 10^{20} atoms/cm^3.

Departure of the diffusion profile from the complementary error function is representative of a concentration-dependent diffusion coefficient. As described in Section 3.4, the concentration-dependent diffusion coefficient can be found by the Boltzmann–Matano analysis as given by Eq. (40).

A high impurity concentration has also been observed to result in the generation of diffusion-induced dislocations.[56] As the impurity concentration increases, the difference in atomic size between the diffusing impurity and the host crystal results in macroscopic strain in the lattice which can be relieved by the generation of dislocations. Because the atomic size mismatch between B and Si is greater than that for P and Si, the diffusion-induced dislocations occur at a lower impurity concentration in B-doped crystals.[56] Departures from complementary error function diffusion profiles and diffusion-induced dislocations generally occur at P concentrations of $\sim 10^{20}$ atoms/cm^3 and higher.

The difference between the total and the electrically active P concentrations has been ascribed to precipitation by Jaccodine.[57] Observations of SiP precipitates have been made by transmission electron microscopy.[58] As the precipitates grow, strain in the system also increases, which results in dislocation loops about the precipitates. It has been postulated[57] that these

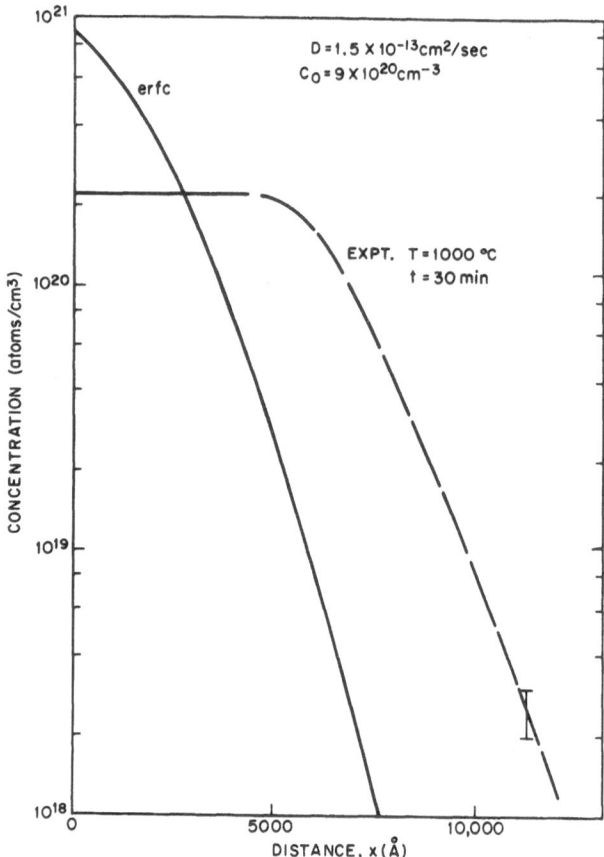

Fig. 17. Comparison of the complementary error function with
the experimental P diffusion profile.[55]

precipitates are the beginning stages of the dislocation networks discussed
in the preceding paragraph.

b. *Arsenic in Si.* At high As concentrations, the diffusion profiles
are similar to those for P. The diffusion profiles are not complementary
error functions and considerable discrepancy occurs between the total As
and the electrically active As. Although the initial investigations of diffusion
at high concentrations were for P, the later results for As more clearly
describe the basic diffusion behavior. The total As diffusion profile and the
electrically active As profile are shown in Fig. 18.[59] The total As concentration $C_{As(total)}$ was obtained by neutron activation, while the electrically

active As concentration C_{As}^+ was obtained from incremental resistivity measurements. The resistivity measurements are related to concentration with the curves given by Irvin.[60] The concentration-dependent diffusion coefficient for As in Si can be obtained by the Boltzmann–Matano analysis as given by Eq. (49). The concentration-dependent As diffusion coefficients obtained by this method, based on the data of Fair and Weber[59] and Kennedy and Murley,[61] are shown in Fig. 19. In the analysis given below, the following features of the concentration-dependent diffusion are explained: (1) The linear increase in the diffusivity between 10^{19} and 10^{20} atoms/cm³ is due to an enhancement of the ionized (acceptor) vacancy concentration by the As donor; (2) the difference in the total As concentration and electrically active As is due to the formation of electrically inactive

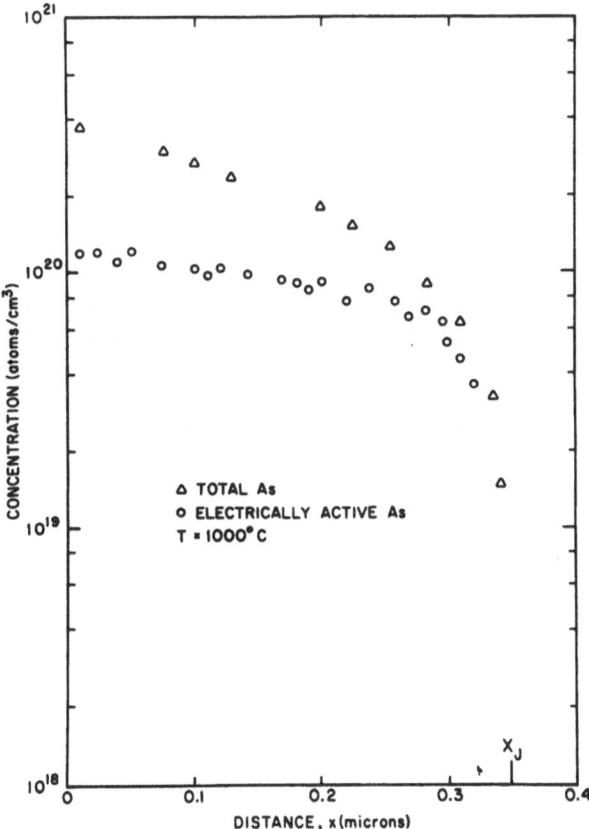

Fig. 18. Total As and electrically active As in Si following a 60-min diffusion at 1000°C. The diffusion source was As-doped SiO₂.[59]

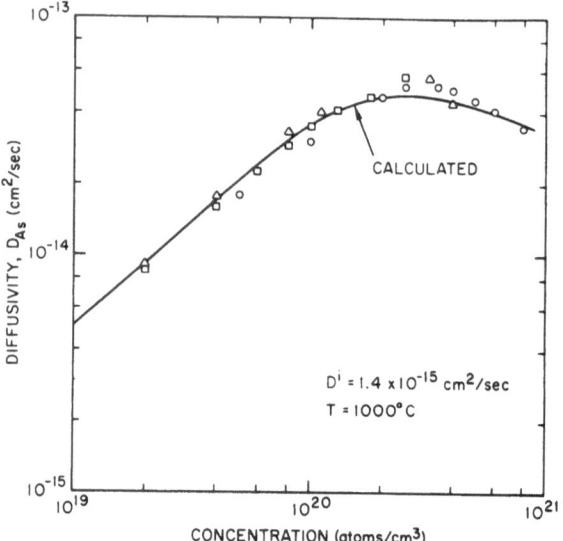

Fig. 19. The effective diffusivity of As in Si at 1000°C. The calculated curve is based on Eq. (84).[59] △, □, Ref. 59; ○, Ref. 61.

As complexes; and (3) the decrease in the diffusivity above 10^{20} atoms/cm³ occurs because only the electrically active As atoms remain mobile.

For a substitutional diffusion mechanism, Eqs. (6) and (8) show that the diffusivity is proportional to the vacancy concentration. The Si monovacancy has been reported to be an electron acceptor.[40] The ionization reaction is given by

$$V_{Si}^0 \rightleftarrows V_{Si}^- + e^+ \tag{70}$$

and the equilibrium relation is

$$K_1(T) = \frac{X_{V_{Si}^-} p}{X_{V_{Si}^0}} \tag{71}$$

where p is the hole concentration, $X_{V_{Si}^0}$ is the atom fraction of neutral Si vacancies V_{Si}^0, and $X_{V_{Si}^-}$ is the atom fraction of ionized acceptor vacancies V_{Si}^-. For intrinsic material

$$K_1(T) = \frac{X_{V_{Si}^-} p}{X_{V_{Si}^0}} = \frac{X_{V_{Si}^-}^i n_i}{X_{V_{Si}^0}^i} \tag{72}$$

where the superscript i is used to represent the intrinsic case. Since the neutral vacancy concentration is the same for intrinsic or extrinsic conditions,

$$X_{V_{Si}^-} = X_{V_{Si}}^i \frac{n_i}{p} = X_{V_{Si}}^i \frac{n}{n_i} \tag{73}$$

with the electron–hole mass action relations

$$e^- + e^+ \rightleftharpoons 0 \tag{74}$$

and

$$np = K_i(T) = n_i^2 \tag{75}$$

The diffusing As will be an ionized donor and control the electron concentration n which is equal to C_{As}^+. If the diffusivity for As in intrinsic Si is D^i and is proportional to the ionized vacancy concentration, then at high As concentrations, $X_{V_{Si}^-}$ is enhanced and the diffusivity is given by

$$D = D^i n / n_i \tag{76}$$

Equation (76) gives the linear concentration dependence shown in Fig. 19 at concentrations between 10^{19} and 10^{20} atoms/cm^3. Also, Fig. 18 shows that all the As is electrically active up to 10^{20} atoms/cm^3 where D_{As} then begins to decrease.

Above concentrations of 10^{20} atoms/cm^3, it is necessary to consider the formation of electrically inactive As complexes. Although there appears to be general agreement that electrically inactive As complexes are responsible for the difference between $C_{As(total)}$ and C_{As}^+, the actual structure of the complex is not understood. Several reasonable possibilities exist. The annealing results of Schwenker et al.[62] are consistent with a model consisting of two As atoms associated with one or more vacancies. In Fig. 20, the relationship between $C_{As(total)}$ and C_{As}^+ can be represented by

$$C_{As(total)} = C_{As}^+[1 + \beta(C_{As}^+)^3] \tag{77}$$

where β is a fitting parameter. These data were taken from the As diffusion profiles,[59] as well as from measurements on grown crystals.[63] The scatter in the onset of the sublinear dependence in Fig. 20 appears to result from uncertainties in analysis of such shallow diffusion profiles and uncertainties in the assignment of the electrically active As concentration.

A possible model which is consistent with these observations is a complex composed of two As atoms and a vacancy in the configuration shown in Fig. 21. This complex can propagate through the lattice in a structure that is similar to the pyrite-like structure SiAs$_2$.[59] The reaction

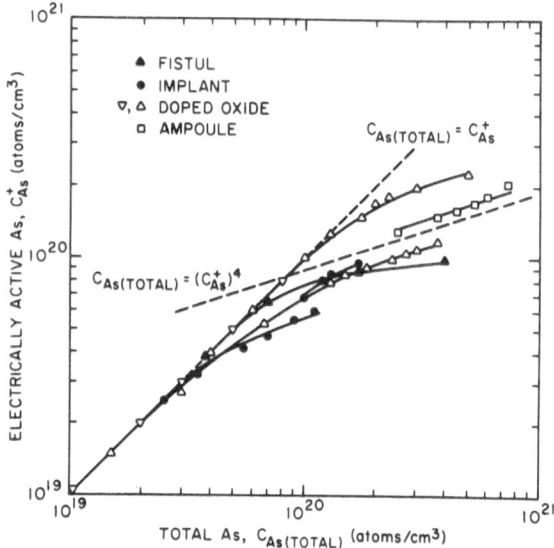

Fig. 20. Electrically active As as a function of the total As concentration.[59]

for formation of the complex of Fig. 21 is

$$2As^+ + V_{Si}^- + e^- \rightleftarrows V_{SiAs_2}$$ (78)

which gives the equilibrium relation

$$K_2(T) = C_{V_{SiAs_2}}/(C_{As}^+)^2 X_{V_{Si}^-} n$$ (79)

where $C_{V_{SiAs_2}}$ is the complex concentration, and $X_{V_{Si}^-}$ is the ionized Si

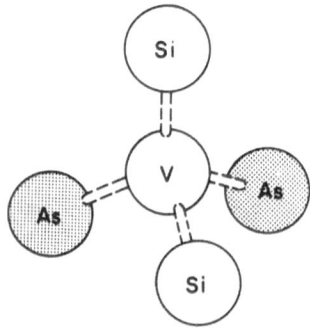

Fig. 21. Structure of the As–Si vacancy complex.

vacancy atom fraction. The total As concentration is the sum of the As in complexes plus the ionized As:

$$C_{As(total)} = C_{As}^+ + 2K_1(T)K_2(T)X_{V_{Si}^0}(C_{As}^+)^4/n_i^2 \qquad (80)$$

where C_{As}^+ has been taken as equal to n, and Eqs. (72) and (73) have been used to substitute for $X_{V_{Si}^-}$. Equation (80) is the same as the empirical result expressed by Eq. (77) with

$$\beta = 2K_1(T)K_2(T)X_{V_{Si}^0}/n_i^2 \qquad (81)$$

This result shows that β is a function of temperature only.

An effective diffusion coefficient can readily be derived for conditions of complex formation where only the electrically active As remains mobile. Equation (7) can be written as

$$J_{As} = -D\frac{dC}{dx} = -D\frac{\partial C_{As}^+}{\partial C_{As(total)}}\frac{\partial C_{As(total)}}{\partial x} \qquad (82)$$

from which the effective diffusion coefficients D_{As} is defined as

$$D_{As} = -J_{As}\left(\frac{\partial C_{As(total)}}{\partial x}\right)^{-1} = D\frac{\partial C_{As}^+}{\partial C_{As(total)}} \qquad (83)$$

The built-in electric field due to the p–n junction has been ignored in this simplified analysis, but it was included in the original analysis of Fair and Weber.[59] Their results demonstrate that the effects of the built-in electric field can be neglected for the example given here. Detailed treatment of this effect is given in Section 5.4.2 for Zn in GaAs.

The effective diffusion coefficient can now be found by differentiation of Eq. (77), and D is taken from Eq. (76) to give

$$D_{As} = \frac{D^i(C_{As}^+/n_i)}{1 + 4\beta(C_{As}^+)^3} \qquad (84)$$

The parameter β can be obtained from the difference in the total As and the electrically active As. The calculated diffusivity in Fig. 19 was obtained by use of Eq. (84).

4.4. Interstitial Diffusion of the Alkali Elements and Inert Gases

A diffusion behavior distinctly different from the substitutional diffusion of the group III and group V elements is observed for the alkali elements

Table V. Diffusion Coefficients of the Alkali Elements and Inert Gases in Ge

Impurity	D_0, cm²/sec	Q, eV	D at 937°C, cm²/sec	Ref.
Li	2.5×10^{-3}	0.52	1.7×10^{-5}	64
He	6.1×10^{-3}	0.70	3.8×10^{-5}	65

and inert gases. These elements are interstitial impurities in Ge and Si and diffuse by the interstitial mechanism illustrated in Fig. 2. The diffusivities of these elements in Ge and Si are summarized in Tables V and VI. Note that the activation energy is less than 1 eV for interstitial diffusion and that the diffusion coefficient is approximately 10^{-5} cm²/sec at the melting point of Ge and Si.

4.5. Interstitial–Substitutional Diffusion of the Transition Elements

4.5.1. Preliminary Considerations

In the two preceding sections, the elements have been grouped into either substitutional or interstitial impurities. Both interstitial and substitutional species must be considered for the transition elements. For some transition elements the substitutional species C_S dominates, while for others the interstitial species C_I is the controlling species. Hall and Racette[67] found that the ratio of substitutional to interstitial Cu is

$$C_S/C_I = 6.0 \tag{85}$$

Table VI. Diffusion Coefficients of the Alkali Elements and Inert Gases in Si

Impurity	D_0, cm²/sec	Q, eV	D at 1412°C, cm²/sec	Ref.
Li	2.3×10^{-3}	0.66	2.4×10^{-5}	64
Na	1.65×10^{-3}	0.72	1.2×10^{-5}	66
K	1.1×10^{-3}	0.76	5.8×10^{-6}	66
H	9.4×10^{-3}	0.48	3.4×10^{-4}	65
He	0.11	1.26	1.8×10^{-5}	65

in Ge and

$$C_S/C_I \approx 10^{-4} \tag{86}$$

in Si at 700°C in intrinsic semiconductors. In the case for Si, the diffusion is determined by the interstitial diffusion coefficient, which is

$$D = 4.7 \times 10^{-3} \exp(-0.43/kT) \tag{87}$$

This interstitial diffusivity for Cu in Si has a D_0 and a Q comparable to those of the alkali elements and inert gases as summarized in Table VI. When $C_S > C_I$, the interstitial and substitutional species can interact and give a very complex diffusion behavior in that the interstitial atom can react with a vacancy to give a substitutional atom, or the substitutional atom can become interstitial. The dissociative interstitial–substitutional diffusion has been reviewed by Hu.[31] Three categories of interstitial–substitutional diffusion can be designated as outlined below.

4.5.2. Low Solubility and Slow Interstitial Diffusion

The first case is for low solubility and slow interstitial diffusion. The interstitial concentration is below its equilibrium value while the vacancy

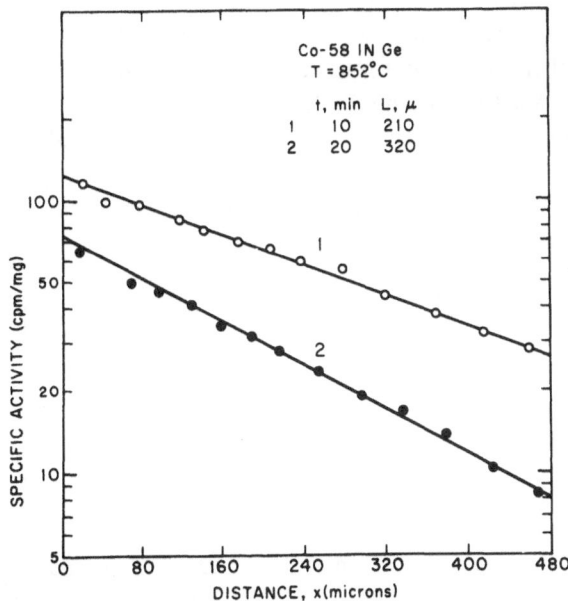

Fig. 22. Penetration profiles of Co-diffused Ge.[68] The concentration varies as $A(t) \exp(-x/L)$.

concentration maintains its equilibrium concentration. The diffusing interstitial is trapped by oxygen in the crystal or other trapping complexes. This type of mechanism gives exponential rather than complementary error function profiles as shown in Fig. 22 for Co in Ge.[68]

4.5.3. Intermediate Solubility and Rapid Interstitial Diffusion

The second case is for intermediate solubility and rapid interstitial diffusion. The interstitial concentration maintains its equilibrium value, but the diffusion rate is controlled by vacancy diffusion. An example of this case is Au in Si.[69,70] Gold is an important impurity in Si because it is sometimes diffused into Si to reduce the minority-carrier lifetime for high-frequency transistors. The interstitial Au diffusion coefficient in Si was found to be[69]

$$D = 2.4 \times 10^{-4} \exp(-0.38/kT) \qquad (88)$$

The substitutional Au concentration is much greater than the interstitial concentration.

4.5.4. High Solubility and Slow Interstitial Diffusion

The third case is for high solubility and slow interstitial diffusivity. These conditions result in a two-stream diffusion where the interstitial and substitutional species do not interact significantly. Kosenko[71] observed this behavior for Ag, In, Zn, and Te in Ge. Although In, Zn, and Te are not transition elements, they all appear to diffuse by the double-stream mechanism. The diffusion profile for Te in Ge is shown in Fig. 23.[71] The substitutional branch has a higher solubility and lower diffusivity ·than the interstitial branch.

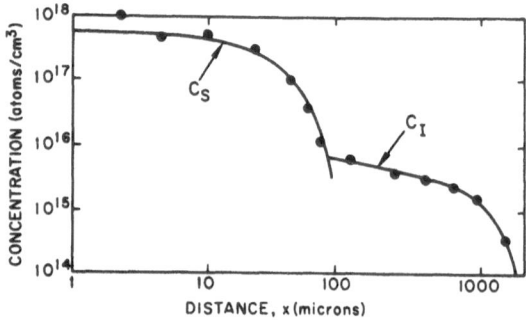

Fig. 23. Diffusion of Te in Ge by the double-stream mechanism.[71]

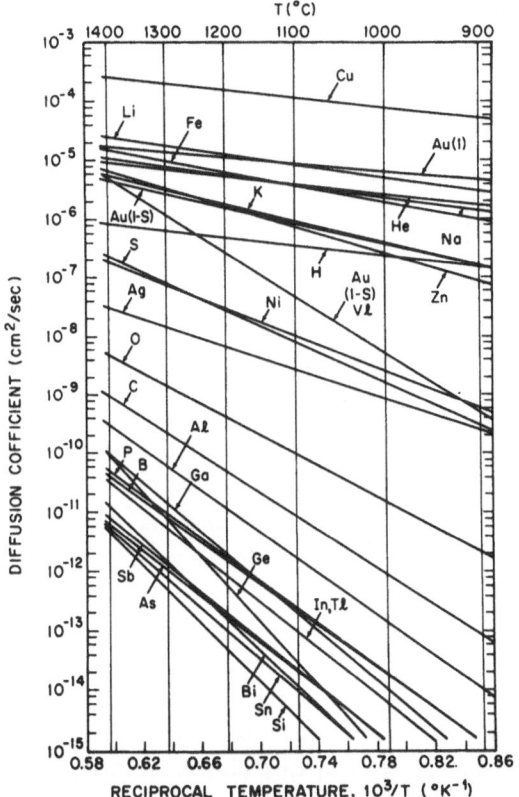

Fig. 24. Diffusion coefficients of active impurities in Si.[30]

4.6. Graphical Summary of the Diffusion Coefficients in Si

The diffusion coefficients in Si are plotted as a function of temperature in Fig. 24. The diffusivities of some elements not discussed in the preceding sections are also included. This figure was taken from the review paper of Kendall and De Vries.[30]

5. DIFFUSION IN THE III–V COMPOUNDS

5.1. Introduction

Of the various compound semiconductors, the III–V compounds have properties most similar to the group IV elemental semiconductors. Like Si and Ge, the III–V compounds (except GaN) can readily be doped n or p

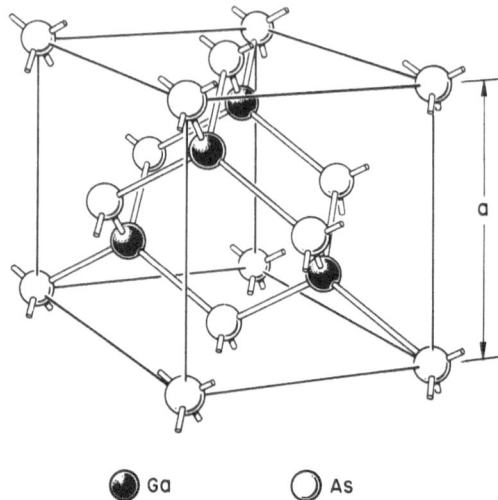

● Ga   ○ As

Fig. 25. The zinc-blende structure for GaAs. The
lattice constant is *a*.

type to form *p–n* junctions, the most useful and widespread application of
semiconductors. They are the one-to-one chemical compounds of the group
III elements B, Al, Ga, or In with the group V elements N, P, As, or Sb.
The III–V compounds are tetrahedrally coordinated, and the majority
crystallize in the zinc-blende structure illustrated in Fig. 25 for GaAs. The
zinc-blende structure is the diamond lattice of Si or Ge, but with group III
and V atoms occupying adjacent lattice sites. Some of the properties of the
III–V compound semiconductors are summarized in Table VII. A detailed
summary of properties for these semiconductors is given by Neuberger.[111]
The varied band structures and large range of energy gaps possible with
the III–V compounds have led to many potential applications.

The greatest impact of the III–V compound semiconductors has not
been in areas dominated by Si and Ge, such as the bipolar and field effect
transistors, but rather in applications that depend on the unique properties
of these compounds. For example, one rapidly evolving technology where
Si, Ge, and other semiconductors are unable to compete with III–V compounds is electroluminescence. Electroluminescence is the emission (visible
and near infrared) associated with the application of a small forward dc
bias to a *p–n* junction. In the preparation of *p–n* junctions, whether by
diffusion from the vapor phase, growth by liquid- or vapor-phase epitaxy,
or growth from solution, it is necessary to consider the behavior of impurities in the lattice, especially movement in the solid by diffusion.

Table VII. Properties of the III–V Compounds

Compound	Crystal structure	Lattice constant, Å	Melting point, °C	Pressure at melting point, atm	Energy gap at 300°K, eV	Ionicity[75]
BN	Zinc-blende	$a = 3.615$[72]	~3000[73]	(a)	>5[h]	0.256
BP	Zinc-blende	$a = 4.538$[76]	~3000[77]	(b)	2.0[79]	0.006
BAs	Zinc-blende	$a = 4.7778$[80]	Unknown	(c)	Unknown	0.002
AlN	Wurtzite	$a = 3.111$[81]; $c = 4.978$[81]	>2400[82]	(d)	5.9[83]	0.449
AlP	Zinc-blende	$a = 5.451$[84]	2530 ± 50[85]	Unknown	2.45[86]	0.307
AlAs	Zinc-blende	$a = 5.6607$[87]	1740 ± 20[88]	1.0[88]	2.16[86]	0.274
AlSb	Zinc-blende	$a = 6.1355$[89]	1060[90]	(e)	1.5[91,92 i]	0.426
GaN	Wurtzite	$a = 3.189$[93]; $c = 5.185$[93]	1500[82]	(f)	3.39[94]	0.500
GaP	Zinc-blende	$a = 5.4511$[95]	1465[96]	32[97]	2.261[98]	0.374
GaAs	Zinc-blende	$a = 5.6532$[99]	1238[96]	0.976[100]	1.424[101]	0.310
GaSb	Zinc-blende	$a = 6.09593$[102]	712[103]	(e)	0.72[104]	0.261
InN	Wurtzite	$a = 3.53$[93]; $c = 5.693$[93]	1200[82]	(g)	2.4[h]	0.578
InP	Zinc-blende	$a = 5.86875$[89]	1068 ± 2[105]	18 ± 5[105]	1.351[106]	0.421
InAs	Zinc-blende	$a = 6.0584$[80]	943 ± 3[107]	0.33[107]	0.35[108]	0.357
InSb	Zinc-blende	$a = 6.47937$[102]	525[109]	(e)	0.180[110]	0.321

a Begins to dissociate in vacuum at 2700°C.[73]
b Above 1100°C, BP loses P and B_6P is formed.[78]
c At a temperature of about 1100°C, BAs goes to a more stable lower arsenide with an orthorhombic structure.[78]
d Begins to dissociate in vacuum at 1750°C.[82]
e Although the pressures for the Sb compounds are not reported, they are known to be low.
f Begins to dissociate in vacuum at 1050°C.[82]
g Begins to dissociate in vacuum at 620°C.[82]
h Taken from Table 2 of Ref. 74. This original reference was not cited.
i The band structure is very complicated and considerable uncertainty presently exists.

Most of the discussion given here on impurity diffusion in the III–V compound semiconductors will apply to the diffusion of Zn in GaAs. This choice permits a fundamental and quantitative presentation of the interstitial–substitutional diffusion mechanism which can be substantiated by comparison with experimental results. The results derived for Zn in GaAs are directly applicable to Zn in GaP. Zinc diffusion in other III–V compounds also shows the strong concentration dependence that is characteristic of interstitial–substitutional diffusion. Although a fundamental analysis for donor diffusion in the III–V compounds is not presently possible, the empirical results will be presented. As an example of the diffusivities in a III–V compound, the experimental diffusion coefficients for both impurity and self-diffusion in GaAs are presented in graphical form as D versus $1/T$. Additional descriptions of diffusion behavior and compilations of diffusivities in the other III–V compounds can be found in reviews by Kendall[112] and Casey.[113] In the discussion that follows, only diffusions in sealed ampoules are considered since open or flowing vapor systems require the maintenance of constant vapor pressures in the flowing gas and represent control problems outside the scope of this presentation.

5.2. Ternary Considerations

An unpublished report by Allen and Pearson[114] emphasized that an understanding of diffusion requires that the diffusion conditions and system compositions be related to the ternary phase diagram. The relationship of the ternary phase diagram to the diffusion analysis was illustrated for Zn in GaAs by Casey and Panish.[115] If a diffusion system is to be interpreted in terms of a fundamental diffusion model, the diffusion source must reach an equilibrium composition as defined by the components in the system and the temperature. For example, in the Zn–GaAs system the *starting* source material for Zn diffusion in sealed ampoules has generally been either elemental Zn, dilute solutions of Zn in Ga, or various combinations of Zn and As. The *actual* diffusion source in any ternary system must either consist of an equilibrium ternary mixture or must approach this mixture in composition. The composition of the ternary mixture is determined by the total amount of each element present, the temperature, and the volume of the system. The actual diffusion source composition will determine the partial pressures of the impurity, Ga, and As. These partial pressures control the impurity surface concentrations in the III–V compound semiconductor. Any diffusion mechanism for impurities that reside substitutionally in the lattice depends upon the vacancy concentration and hence depends on the actual source composition. Empirical behavior can be obtained by simply introducing a known amount of the elemental impurity in the diffusion system, but the assignment of the actual source conditions and compositions is usually not possible. Thus, neglect of ternary considerations in diffusion experiments has prevented the development of fundamental mechanisms for the diffusion of most impurities in III–V compounds.

Figure 26 shows the projected ternary diagram of Panish.[116] The position of the boundaries in the high-As portion of the diagram is not known, and the dashed parts simply represent the situation schematically. The possible diffusion sources and diffusion behavior depend greatly on the temperature. Above 1015°C, the only solid is GaAs doped with Zn, which is in equilibrium with the liquid composition represented by the liquidus line. Between 1015 and 950°C, both GaAs doped with Zn and Zn_3As_2 doped with Ga are present. A representative isothermal section between 1015 and 950°C is shown in Fig. 27. Below 950°C the ternary isotherms become more complicated. The isothermal sections continue to change as the temperature is lowered and are rather complex (see Ref. 115). The system becomes simpler again at lower temperatures since no liquid exists below 723°C in the As–Zn binary except at the Zn-rich end.

The relationship between the *starting* source composition and the *actual* source composition is illustrated by considering Fig. 27. In this figure, any overall diffusion source composition within the liquidus curves *ab* and *cd*, represented as regions *B* and *C*, involves a liquid in equilibrium with one

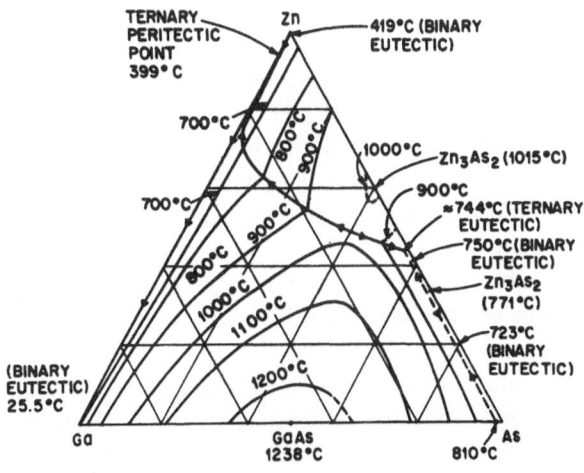

Fig. 26. The Ga–As–Zn ternary phase diagram.[116]

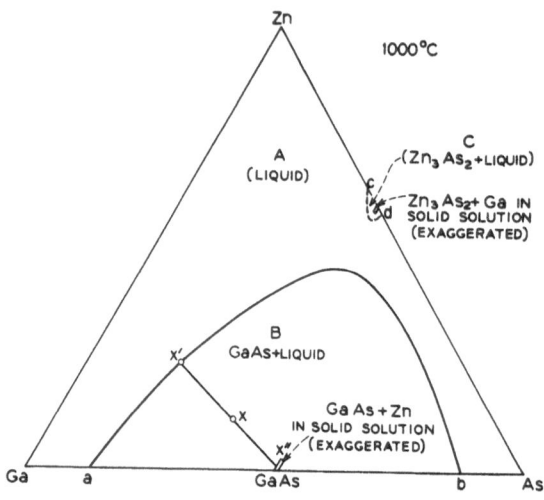

Fig. 27. The Ga–As–Zn ternary phase diagram at 1000°C.[115]

solid. The solid phase in region B is GaAs doped with Zn, while Zn_3As_2 doped with Ga is the solid phase in equilibrium with the liquid in region C. The composition of the liquid phase is somewhere on curve ab or cd; for example, a source of overall composition X in region B of Fig. 27 leads to a composition X' in the liquid and X'' in the solid. For an isothermal system in contact with the vapor, the number of degrees of freedom F allowed in various source composition ranges is given by the phase rule[117]

$$F = 3 - P \qquad (89)$$

where P is the number of condensed phases. Therefore, in this example, the phase rule specifies a single degree of freedom which is a univariant region. The single degree of freedom means that the composition of the liquid can be varied along curve ab or cd by varying the overall composition while still maintaining the desired solid phase. This liquid composition then determines the solid solubility.

Since no solid phase exists within region A, starting compositions within this region are not desirable diffusion sources. With elemental Zn or Ga–Zn sources the system starts with a liquid somewhere in region A. To establish equilibrium, the source must react with GaAs until the solid is consumed or the boundary ab is reached by the liquid composition. For an elemental Zn source the liquid can form only on the sample surface since the vapor pressure of Zn is much greater than the Ga vapor pressure.

With the Ga–Zn source the ternary liquid will form both in the source and on the sample surface. Arsenic must vaporize from the sample and transport to the source, where it must distribute itself to maintain the same concentration in both the liquid in the source and the new liquid on the sample surface. For both the elemental and Ga–Zn sources, there are equilibrium As, Zn, and Ga pressures within the ampoule that are supplied from the phases present and completely determined by the liquid composition at equilibrium.

5.3. Concentration Gradient Diffusion of Zn in GaAs

In diffusion studies of Zn, the limiting Zn concentration at the diffused surface is often taken as the equilibrium solid solubility but, in general, the conditions of the diffusion are not sufficiently defined to permit an unambiguous determination of the equilibrium conditions during diffusion. When the equilibrium conditions can be determined, as was done by Chang and Pearson[118] and Shih, Allen, and Pearson,[119] the surface concentrations were in agreement with solubilities obtained by solution growth.[120] It therefore appears that for Zn in GaAs, the limiting Zn concentration at the surface is the equilibrium solid solubility. Complete analysis of the Zn solid solubility can be found in published papers[113,120] and will not be repeated here.

The non-complementary error function shape of the Zn concentration versus depth curves in GaAs that were obtained by Cunnell and Gooch[121] initiated extensive studies of Zn diffusion in the III–V compound semiconductors. Their diffusion profiles obtained at 1000°C are shown in Fig. 28. Diffusion profiles of this shape were verified by extensive, although unpublished, studies by Kendall and Jones.[122] The profiles are characterized by a very abrupt drop in Zn concentration at the diffusion front.

As discussed in Section 3.4, the diffusion coefficient can be obtained from the experimentally derived profiles of C_{Zn} versus x in Fig. 28 by the Boltzmann–Matano analysis. Using this technique, Cunnell and Gooch[121] obtained a diffusion coefficient that varied as the square of the Zn concentration. The diffusion coefficient reached a maximum and decreased slightly at the highest concentrations.

Longini,[123] in an attempt to explain the deterioration of GaAs tunnel diodes, suggested that Zn diffused interstitially as a donor although it is a substitutional acceptor. Only plausibility arguments were given. Certain aspects of the model have been disputed by various investigators, but the basic concept of interstitial–substitutional diffusion is now well established.

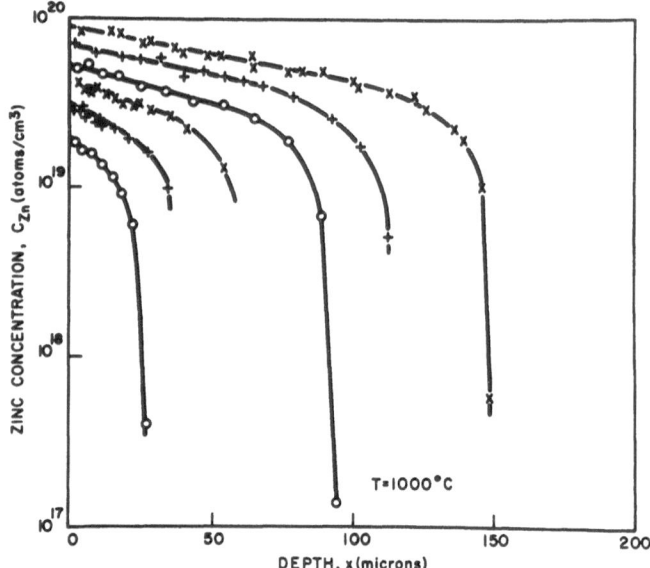

Fig. 28. Diffusion profiles of Zn in GaAs after annealing at 1000°C for 10⁴ sec. The different surface concentrations were obtained by maintaining the Zn source at temperatures in the range of 600 to 800°C.[121]

Weisberg and Blanc[124] applied computer solutions to the calculation of diffusion profiles for concentration-dependent diffusion coefficients. A very good fit to the data shown in Fig. 28 was obtained by assuming that $D \propto C_{\mathrm{Zn}}^2$. Present concepts are based on their demonstration of the agreement between the interstitial–substitutional diffusion model and the data obtained by Cunnel and Gooch at 1000°C.

5.4. Interstitial–Substitutional Diffusion

5.4.1. *The Flux Equation*

The derivation of the effective diffusion coefficient to be presented here is taken, in part, from the analysis of Casey, Panish, and Chang.[125] The total Zn flux J_{Zn} is the sum of the interstitial and substitutional flux, each of which consists of a flux term due to the concentration gradient and a flux term for the built-in field. The one-dimensional expression for the total flux along x is

$$J_{\mathrm{Zn}} = -D_I(\partial C_I/\partial x) \pm B_I C_I \mathscr{E} - D_S(\partial C_S/\partial x) \pm B_S C_S \mathscr{E} \qquad (90)$$

where D_I is the diffusion coefficient for the interstitial species and C_I its concentration, D_S and C_S are the substitutional diffusion coefficient and concentration, respectively, and $\mathscr{E}$ is the built-in field. The proper sign for the field term is determined by both the direction of the field and the sign of the ionized impurity. The mobility B is related to the diffusion coefficient by the Einstein relation

$$B = qD/kT \tag{91}$$

where q is the electronic charge.

5.4.2. *The Built-in Field*

In order to proceed with the determination of the effective diffusion coefficient, it is necessary to evaluate the built-in field of Eq. (90). This is given by the condition that at equilibrium, there can be no net flow of electric current. Therefore, in inhomogeneous material a built-in field must exist to give a drift current that is equal and opposite to the diffusive current. Since current is proportional to the mobility–concentration product, the most mobile of the species present in the largest concentration will dominate the expression for current; i.e., the hole current. Fick's first law as defined in Eq. (7) states that the diffusive flux (current density is flux times q) is proportional to the concentration gradient. However, the more general approach of the phenomenological equations, which do not result from a diffusion model, gives the flux from the observed conditions of equilibrium.[2] These equations indicate that the flux is proportional to the gradient of the chemical potential. This more general approach is necessary in order to derive the built-in field.

The phenomenological equation for the diffusive hole current density i_p is

$$i_p(\text{diffusive}) = -qM\,\partial[\mu(p) - \mu^\circ]/\partial x \tag{92}$$

where M is the proportionality constant, $\mu(p)$ is the chemical potential for holes, and μ° is the reference potential. The chemical potential of any species can be written as[126]

$$\mu = \mu^\circ + kT \ln \gamma p A_0 \tag{93}$$

where γ is the activity coefficient and A_0 is the activity of the reference state. From the reaction for the creation of a hole–electron pair

$$0 \leftrightarrows e^- + e^+$$

the sum of the electron chemical potential $\mu(n)$ and the hole chemical potential is zero:

$$\mu(n) + \mu(p) = 0 \tag{94}$$

From statistical mechanics, it can be shown that[127]

$$\mu(n) = E_{\mathrm{F}} \tag{95}$$

and therefore

$$\mu(p) = -E_{\mathrm{F}} \tag{96}$$

where E_{F} is the Fermi level.

The hole activity coefficient is unity and Boltzmann statistics apply for dilute concentrations. For these conditions, the band theory of semiconductors gives the hole concentration as[128]

$$p = N_v \exp[(E_v - E_{\mathrm{F}})/kT] \tag{97}$$

where N_v is the effective density of states and E_v is the valence band edge. Rewriting Eq. (97) as

$$-E_{\mathrm{F}} = -E_v + kT \ln(p/N_v) \tag{98}$$

shows that μ° in Eq. (93) is $-E_v$ and A_0 is $1/N_v$. Therefore, when the hole activity coefficient γ_p is unity,

$$\frac{\partial[\mu(p) - \mu^\circ]}{\partial x} = \frac{kT}{p}\frac{dp}{dx} \tag{99}$$

Equation (92) for the diffusive hole current reduces to

$$i_p(\text{diffusive}) = -\frac{qMkT}{p}\frac{dp}{dx} = -qD_p\frac{dp}{dx} \tag{100}$$

where the second expression is simply Fick's law for hole current of diffusivity D_p. Therefore

$$M = D_p p/kT \tag{101}$$

or by Eq. (91)

$$M = \mu_p p/q \tag{102}$$

with μ_p the hole mobility, and Eq. (92) is

$$i_p(\text{diffusive}) = -q\mu_p p\,\frac{1}{q}\frac{\partial[\mu(p) - \mu^\circ]}{\partial x} \tag{103}$$

The drift hole current density is, by definition,

$$i_p(\text{drift}) = q\mu_p p \mathscr{E} \tag{104}$$

so that the total current density becomes

$$i_p = -q\mu_p p\left(\frac{1}{q}\frac{\partial[\mu(p) - \mu^\circ]}{\partial x} - \mathscr{E}\right)$$ (105)

At equilibrium $i_p = 0$ and the built-in field is

$$\mathscr{E} = \frac{1}{q}\frac{\partial[\mu(p) - \mu^\circ]}{\partial x} = \frac{kT}{q}\left(\frac{1}{p}\frac{\partial p}{\partial x} + \frac{1}{\gamma_p}\frac{\partial\gamma_p}{\partial x}\right)$$ (106)

with $\mu(p)$ given by Eq. (93). A detailed derivation of the built-in field, which includes γ_p, has been given because such a treatment is not available in the published literature, and is crucial to the derivation that follows.

For a constant activity coefficient, Eq. (106) is simply

$$\mathscr{E} = \frac{kT}{q}\frac{1}{p}\frac{\partial p}{\partial x}$$ (107)

The electric field can be obtained by expressing p in terms of the electrical neutrality condition

$$p + C_D{}^+ = n + C_A{}^-$$ (108)

where $C_D{}^+$ and $C_A{}^-$ are the ionized donor and acceptor concentrations. The hole concentration then reduces to

$$p = \tfrac{1}{2}\{(C_A{}^- - C_D{}^+) + [(C_A{}^- - C_D{}^+)^2 + 4n_i{}^2]^{1/2}\}$$ (109)

Then

$$\mathscr{E} = \frac{kT}{q}\frac{1}{[(C_A{}^- - C_D{}^+)^2 + 4n_i{}^2]^{1/2}}\frac{dC_A{}^-}{dx}$$ (110)

can be used in the flux Eq. (90). Several treatments[129,130] of diffusion have considered the influence of $\mathscr{E}$ as expressed by Eq. (110).

5.4.3. The Effective Diffusion Coefficient

The built-in field given by Eq. (106) and the Einstein relation of Eq. (91) permit one to write the Zn flux Eq. (90) as

$$J_{\text{Zn}} = -D_I\left[\frac{\partial C_I}{\partial C} \pm C_I\left(\frac{1}{p}\frac{\partial p}{\partial C} + \frac{1}{\gamma_p}\frac{\partial\gamma_p}{\partial C}\right)\right]\frac{\partial C}{\partial x}$$
$$-D_S\left[\frac{\partial C_S}{\partial C} \pm C_S\left(\frac{1}{p}\frac{\partial p}{\partial C} + \frac{1}{\gamma_p}\frac{\partial\gamma_p}{\partial C}\right)\right]\frac{\partial C}{\partial x}$$ (111)

where the total Zn concentration $C = C_I + C_S = C_{Zn} = p$, and $(\partial/\partial x)$ has been replaced by $(\partial/\partial C)(\partial C/\partial x)$. An effective diffusion coefficient $D_{Zn} = -J_{Zn}(\partial C/\partial x)^{-1}$ can be identified as

$$D_{Zn} = D_I \frac{\partial C_I}{\partial C} \pm D_I C_I \left(\frac{1}{p}\frac{\partial p}{\partial C} + \frac{1}{\gamma_p}\frac{\partial \gamma_p}{\partial C}\right)$$
$$+ D_S \frac{\partial C_S}{\partial C} \pm D_S C_S \left(\frac{1}{p}\frac{\partial p}{\partial C} + \frac{1}{\gamma_p}\frac{\partial \gamma_p}{\partial C}\right) \qquad (112)$$

The proper designation of the sign for the interstitial and substitutional field terms depends on the charge state of the interstitial and substitutional species. Ions of charge state identical to that of the dominant free carrier (the hole) will be retarded and have a minus sign in Eq. (112), while oppositely charged ions will be aided and enter Eq. (112) with a plus sign.

5.4.4. Interstitial–Substitutional Equilibrium

In order to evaluate Eq. (112), the relationship between C_I and C must be obtained from the equilibrium reaction. At equilibrium the interstitial Zn atom, which is assumed to be a singly ionized donor, reacts with a neutral Ga vacancy to form an ionized substitutional Zn acceptor and two holes according to the relation

$$Zn_I^+ + V_{Ga} \rightleftarrows Zn_{Ga}^- + 2e^+ \qquad (113)$$

For these charge states, the built-in field will retard the interstitial Zn_I^+ diffusion and aid the substitutional Zn_{Ga}^- diffusion. The equilibrium relation for this reaction is

$$K_1(T) = C_S(\gamma_p p)^2/C_I X_{V_{Ga}} \qquad (114)$$

where $X_{V_{Ga}}$ is the atom fraction of Ga vacancies. From the decomposition reaction,

$$GaAs(s) \rightleftarrows Ga(l) + \tfrac{1}{2}As_2(g) \qquad (115)$$

one obtains the Ga activity

$$a_{Ga(l)} = K_2(T)/p_{As_2}^{1/2} \qquad (116)$$

The solid, liquid, and gas phases are represented by s, l, and g.
The reaction for the formation of Ga vacancies is

$$Ga_{Ga} \rightleftarrows Ga(l) + V_{Ga} \qquad (117)$$

and

$$X_{V_{Ga}} = K_3(T)/a_{Ga(l)} \tag{118}$$

Combining Eqs. (116) and (118) gives

$$X_{V_{Ga}} = \frac{K_3(T)}{K_2(T)} p_{As_2}^{1/2} \tag{119}$$

so that Eq. (114) can be written as

$$K_4(T) = \frac{K_1(T)K_3(T)}{K_2(T)} = \frac{C_S(\gamma_p p)^2}{C_I p_{As_2}^{1/2}} \tag{120}$$

and

$$C_I = C_S(\gamma_p p)^2/K_4(T)p_{As_2}^{1/2} \tag{121}$$

with the activity coefficients of interstitial and substitutional Zn taken as constants. At hole concentrations above 5×10^{18} cm^{-3}, p is equal to C_{Zn},[131,132] and Eq. (121) can be written

$$C_I = \gamma_p^2 C_S^3/K_4(T)p_{As_2}^{1/2} \tag{122}$$

5.4.5. Analysis of Experimental Data

Under the condition $C_I \ll C_S$, which makes $C_S = C = p$, and with the elimination of C_I from Eq. (112) by substituting Eq. (122), the effective diffusion coefficient becomes

$$D_{Zn} = \frac{D_I}{K_4(T)p_{As_2}^{1/2}} C_S^2 \gamma_p^2 \left[\left(3 + 2 \frac{C_S}{\gamma_p} \frac{d\gamma_p}{dC_S} \right) - \left(1 + \frac{C_S}{\gamma_p} \frac{d\gamma_p}{dC_S} \right) \right]$$
$$+ D_S \left[1 + \left(1 + \frac{C_S}{\gamma_p} \frac{d\gamma_p}{dC_S} \right) \right] \tag{123}$$

Equation (123) is written in the above form to emphasize that the term $[1 + (C_S/\gamma_p)(d\gamma_p/dC_S)]$ represents the effect of the built-in field.

As seen from the results of Cunnell and Gooch,[121] where the Zn concentrations are low enough to assign $\gamma_p = 1$, the effective diffusion coefficient can vary as the square of the Zn concentration only if substitutional diffusion is neglected. Also, substitutional diffusion requires divacancies to permit movement from one Ga site to another. For these reasons, D_S in Eq. (123) is neglected, and the effective diffusion coefficient becomes

$$D_{Zn} = \frac{2D_I C_S^2}{K_4(T)p_{As_2}^{1/2}} \gamma_p^2 \left(1 + \frac{C_S}{2\gamma_p} \frac{d\gamma_p}{dC_S} \right) \tag{124}$$

At low concentrations where $\gamma_p = 1$, $D_{Zn} \propto C_S^2$. At higher concentrations where $\gamma_p < 1$, the derivative $d\gamma_p/dC_S$ is negative and Eq. (124) leads to a decrease in D_{Zn}.

Weisberg and Blanc[124] analyzed the data of Cunnell and Gooch[121] shown in Fig. 28. For these diffusion profiles, the Zn concentrations are low enough to assume that $\gamma_p = 1$. Under these conditions, Eq. (124) reduces to

$$D_{Zn} = 2D_I C_S^2/K_4(T)p_{As_2}^{1/2} \qquad (125)$$

Their computer-generated diffusion profiles for $D_{Zn} \propto C_S^2$ were in excellent agreement with the experimental data shown in Fig. 28. This agreement strongly supports the interstitial–substitutional mechanism.

Diffusions at high Zn concentrations, where $\gamma_p < 1$, were made by Casey, Panish, and Chang.[125] The approximate liquidus compositions for the diffusion sources are shown in the ternary phase diagram of Fig. 29. The diffusion profiles were obtained with radioactive ^{65}Zn and are shown in Fig. 30. Diffusion time was 1.5 hr for the 800, 900, and 1000°C profiles and 168 hr for the 700°C profile. The 700°C profile was normalized to 1.5 hr according to the $x/t^{1/2}$ relation. The detailed experimental procedure for obtaining the diffusion profiles is described in the literature.[133]

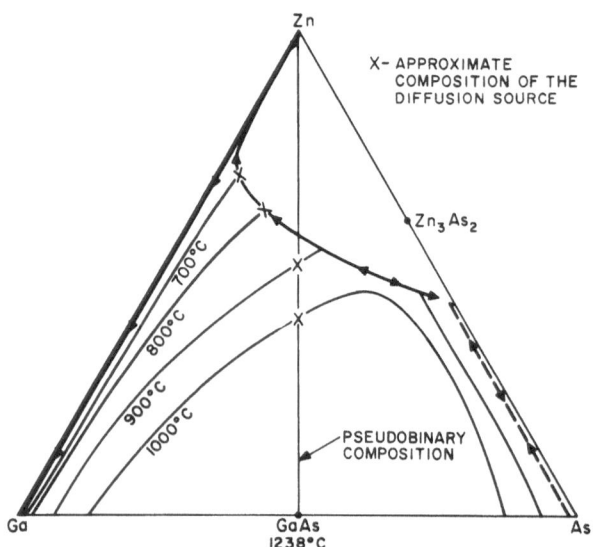

Fig. 29. The Ga–As–Zn ternary phase diagram illustrating the approximate compositions of the diffusion sources used for the profiles of Fig. 30.[125]

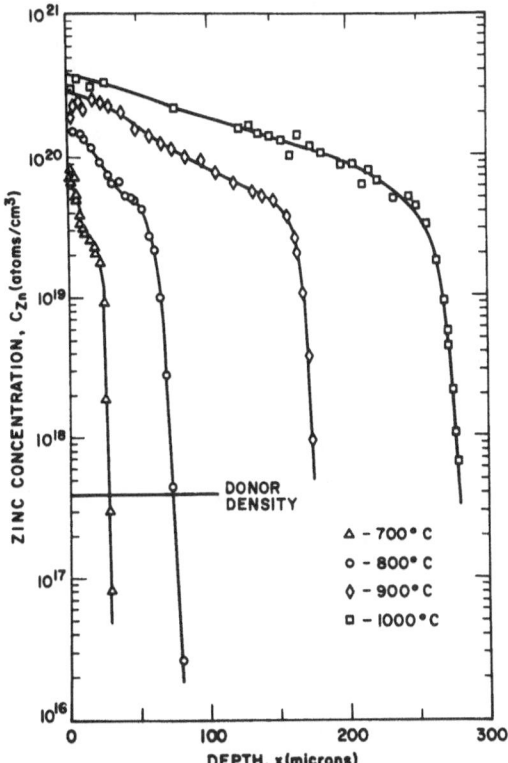

Fig. 30. Diffusion profiles of Zn in GaAs at 700, 800, 900, and 1000°C. Diffusion time is normalized to 1.5 hr.[125]

The concentration-dependent diffusion coefficients can be obtained for these diffusion profiles by the Boltzmann–Matano analysis based on Eq. (49). The experimental data have their greatest errors at low concentrations, and therefore both the slope and $x \, dC$ in the integral of Eq. (49) are difficult to evaluate. The slope evaluation is subject to significant errors at high concentrations. The greatest errors in the Boltzmann–Matano analysis thus occur at the lowest and highest concentrations. The concentration-dependent diffusion coefficients obtained in this manner are shown in Fig. 31 and cover the range of C_{Zn} where these errors are least significant. Iso-concentration diffusion values at 900°C which were taken from Refs. 133 and 134 are shown for comparison. Isoconcentration diffusion is discussed in the next part of this section.

It should be noted in Fig. 31 that, for each temperature, the D versus C_{Zn} curve varies as C_{Zn}^2 at the lower concentration, goes successively through

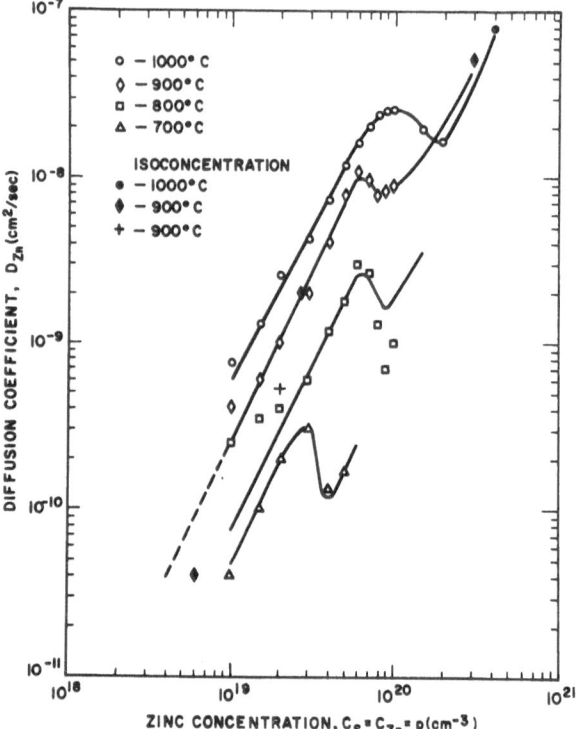

Fig. 31. Diffusion coefficient of Zn in GaAs versus Zn concentration at specified temperatures as derived from a Boltzmann–Matano analysis.[125] ●, ♦, Ref. 133; +, Ref. 134.

a peak and a valley, and then increases. These features will now be considered in terms of the effective diffusion coefficient given in Eq. (124). At high concentrations, γ_p is expected to decrease from unity.[120] Since γ_p enters Eq. (124) as $[\gamma_p^2 + (C_S\gamma_p/2)(d\gamma_p/dC_S)]$, an explicit expression for γ_p cannot be obtained and the values of γ_p as a function of Zn concentration in the solid must be found by trial and error. The relationship between p, γ_p, and E_F as obtained from Eqs. (93) and (98) is

$$p = (N_v/\gamma_p) \exp[(E_v - E_F)/kT] \tag{126}$$

The possible values of γ_p are further limited by the restriction that the Fermi level is a single-valued, monotonic function of p. This restriction means that in the region where $\gamma_p < 1$, there must be a minimum possible value of γ_p at any given C_S for E_F to decrease smoothly. The γ_p curves shown in Fig. 32 were obtained in this manner from the curves of Fig. 31. At the Zn concentrations considered in this diffusion analysis, the Zn is fully ionized so that $C_S = C_{Zn} = p$. The initial departure from the C_S^2 dependence indicates the concentration where γ_p becomes less than unity. The valley in the D_{Zn} versus C_S curve occurs approximately at the inflection point of a linear γ_p versus C_S plot. At this point, the quantity $(C_S/2\gamma_p)(d\gamma_p/dC_S)$ is negative and approaching unity. At still higher concentrations, the C_S^2 dependence of D_{Zn} indicates that γ_p is again approaching a constant value, but for $\gamma_p < 1$. Even if the shape of the curve in the region of decreasing D_{Zn} in Fig. 31 is not entirely due to γ_p, solubility results given in Ref. 120 and theoretical considerations[135,136] require that $\gamma_p < 1$. The γ_p versus p curves should be essentially the same as for D_{Zn} versus C_S curves that increase monotonically and do not have a minimum at high concentrations.

5.4.6. Isoconcentration Diffusion

For isoconcentration diffusion,[122,133] nonradioactive Zn (preferably a ternary composition) is first diffused into the sample for a sufficiently long time to obtain a uniform Zn concentration throughout the entire sample.

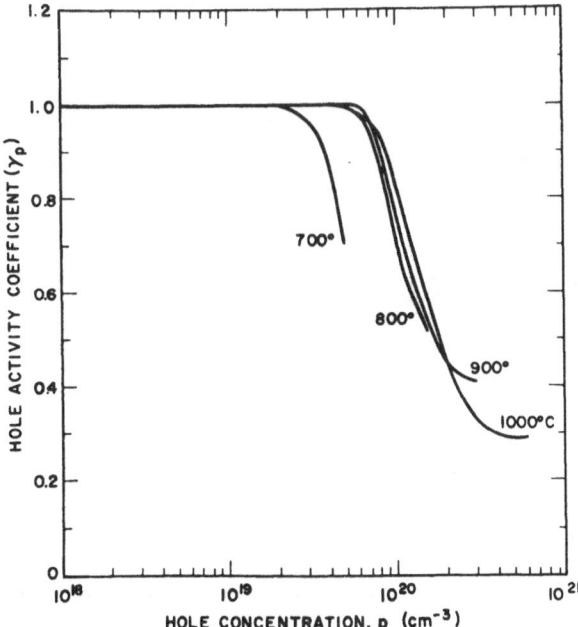

Fig. 32. Variation of the hole activity coefficient with hole concentration at specified temperature.[125]

The sample is then diffused with radioactive ^{65}Zn from a source of the same composition so that nonradioactive Zn diffuses out while radioactive Zn diffuses in. The total Zn concentration throughout remains constant at all times.

The isoconcentration profiles obtained by Ting and Pearson[137] are shown in Fig. 33. The ^{65}Zn profiles are complementary error functions which readily permit determination of the diffusivity at the fixed Zn concentration of $\sim 2\times10^{19}$ cm^{-3}. The temperature was varied between 600 and 1015°C. The resulting diffusion coefficients, as shown in Fig. 34, are relatively temperature independent.

It should be noted that the diffusion coefficient for concentration gradient diffusion shown in Fig. 31 varies between 2×10^{-10} and 2×10^{-9} cm^2/sec at a concentration of 2×10^{19} atoms/cm^3. These values tend to bracket the isoconcentration value of 9×10^{-10} cm^2/sec. In addition to the numerical difficulties of obtaining the diffusivity from the experimental diffusion profile by the Boltzmann–Matano analysis, there are reasons for the concentration-gradient and isoconcentration diffusivities to differ slightly. In the isoconcentration diffusion, there is no gradient in either the hole concentration or in the total Zn concentration but only in the nonradioactive and radioactive Zn concentrations. In addition, there is no built-in electric field. Therefore the effective diffusion coefficient for isoconcentration diffusion becomes[113]

$$D_{Zn} = D_I C_S{}^2 \gamma_p{}^2 / K_4(T) p_{As_2}^{1/2} \qquad (127)$$

At concentrations where $\gamma_p = 1$, concentration-gradient and isoconcentra-

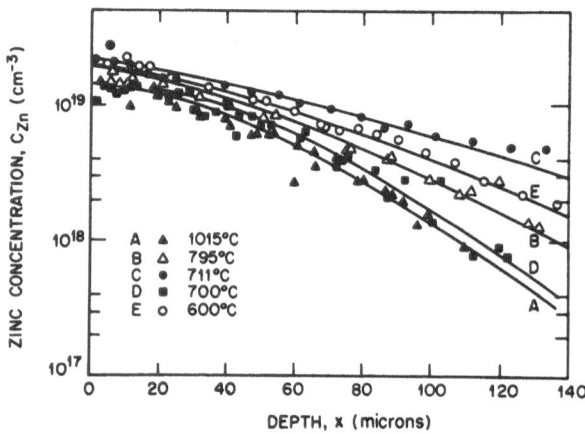

Fig. 33. Isoconcentration-diffusion profiles of Zn in GaAs at various temperatures with fixed surface concentrations.[137]

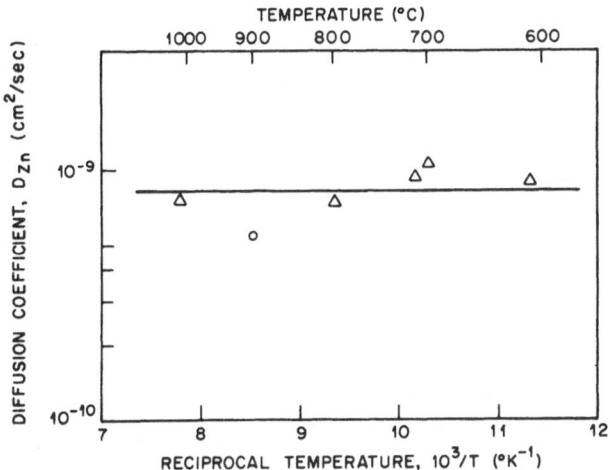

Fig. 34. Effective diffusion coefficient of Zn in GaAs versus $1/T$ as obtained from isoconcentration diffusions at 2×10^{19} cm^{-3}.[137] $\triangle$, Ref. 137; $\bigcirc$, Ref. 133.

tion diffusivities only differ by the prefactor of 2.0 [see Eq. (125)] and the difference in $p_{As_2}^{1/2}$. These relatively temperature-independent isoconcentration diffusion coefficients further demonstrate the validity of the interstitial–substitutional model for Zn diffusion in GaAs.

5.5. Effects of Arsenic Pressure

The effective diffusion coefficient for interstitial–substitutional diffusion, as expressed by Eq. (124), varies as $p_{As_2}^{-1/2}$ at a fixed temperature and Zn concentration. Numerous papers have reported the use of excess As to obtain more planar p–n junction interfaces, but most of these studies concern only the empirical behavior. However, the work of Shih, Allen, Pearson[119,138] provided a quantitative understanding of the effect of arsenic pressure on the Zn diffusion. In their work, the diffusion source compositions were varied from the pseudobinary composition to the As-rich side of the liquidus isotherm. In the binary Ga–As system, As$_4$ dominates on the As-rich side, and therefore the arsenic pressure measurements of Shih et al.[138] on the As-rich side of the liquidus isotherm are expected to measure only the As$_4$ pressure. For this reason, it is more convenient to express Eq. (124) in terms of $p_{As_4}^{1/4}$ as

$$D_{Zn} = \frac{2D_I C_S^2}{K_4'(T) p_{As_4}^{1/4}} \gamma_p^2 \left(1 + \frac{C_S}{2\gamma_p} \frac{d\gamma_p}{dC_S}\right) \qquad (128)$$

Since p_{As_2} and p_{As_4} are related in a known manner, as shown in Fig. 35, either expression can be used. Shih *et al.*[138] assumed that As_4 was the dominant species under all conditions of their measurements, and As_2 was neglected. The recent measurements of Arthur[140] showed that As_2 was the dominant species in the Ga-rich region. The pressure measurements of Shih *et al.*[138] are shown with Jordan's[139] calculated pressures in Fig. 35. For the diffusion compositions used in Ref. 119, As_4 is the dominant arsenic vapor species.

The diffusion profiles obtained with As-rich diffusion sources of increasing $X_{As(l)}$, and hence greater p_{As_4}, became progressively shallower. The diffusion coefficients were obtained from these profiles by the Boltzmann–Matano analysis. When these diffusion coefficient values for constant Zn concentrations are plotted as a function of As_4 pressure as shown in

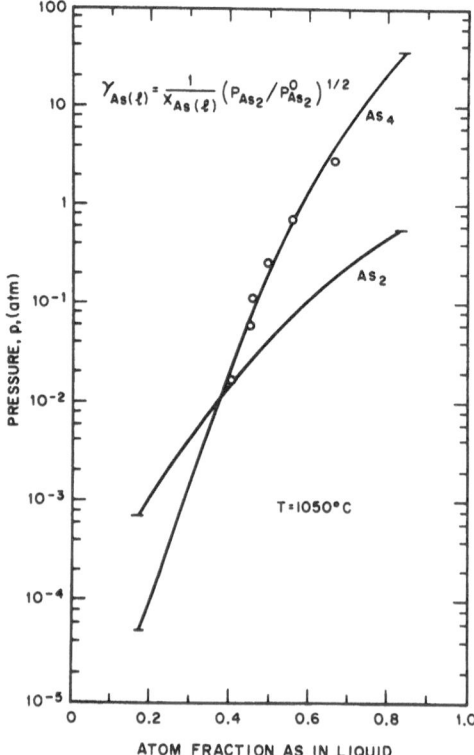

Fig. 35. The As_2 and As_4 partial pressures versus $X_{As(l)}$ along the 1050°C liquidus isotherm in the Ga-As-Zn system.[139] $\bigcirc$, Ref. 138.

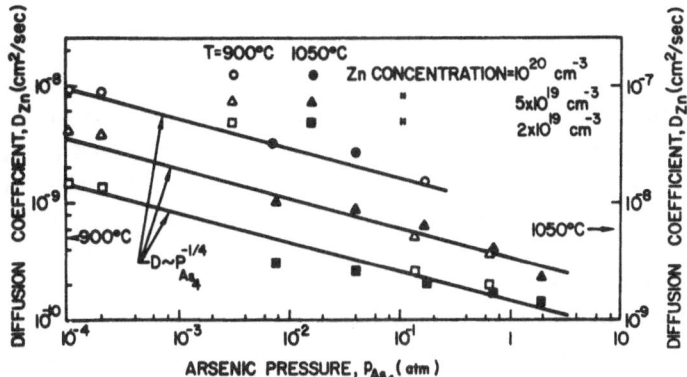

Fig. 36. Diffusion coefficient at a given Zn concentration as a function of p_{sA_4} at 900 and 1050°C.[119]

Fig. 36, a reasonable $p_{As_4}^{-1/4}$ dependence is obtained. Data from profiles of material which had been diffused at 900°C are also included in this figure. As the arsenic pressure increases, the profiles become shallower and accurate diffusion coefficients become more difficult to obtain, so that the fit at each concentration to $p_{As_4}^{-1/4}$ is not exact. The results are quite reasonable when one considers the difficulties of obtaining the diffusion profiles and the arsenic pressure.

The results of Shih, Allen, and Pearson demonstrate two very basic concepts. First, the assignment of an interstitial–substitutional diffusion model is verified by the observed dependence of the effective diffusion coefficient on arsenic pressure. Second, their results illustrate that in the interpretation of diffusion in the III–V compound semiconductors it is necessary to consider the ternary nature of the system and that results derived from treatment of the compound as a single element in a binary system can be very misleading.

5.6. Departure from Equilibrium

The analyses given in the preceding sections assumed equilibrium conditions. However, several of the experimental observations can be related to nonequilibrium conditions. One is the irregular junction interface that is frequently observed when low-arsenic-pressure sources are used. Another is the generation of dislocations in the diffused layer. In addition, it is necessary to consider the generation rate and mechanism of vacancy supply.

Numerous publications have dealt with diffusion-induced dislocations, precipitation, and complex formation. For example, P, B, and As in Si were discussed in Section 4.3. It appears that for a given Zn surface concentration, there is an incubation period of diffusion without diffusion-induced dislocations.[141] Rhines and Stevenson[142] have studied precipitation during Zn diffusion in GaAs. Their diffusion times were sufficient to exceed the incubation period for diffusion-induced dislocations. By using transmission electron microscopy, these precipitates, which nucleate upon diffusion induced dislocations, were found to be tetrakaidecahedra (14 sides) and did not exceed 0.1 μm in diameter. The precipitate compound is unknown and contains a small fraction of the total Zn. A complete quantitative description is not presently available. However, it should be emphasized that a number of commonly used techniques permit preparation of p–n junctions by diffusion without diffusion-induced dislocations.

The question of vacancy equilibrium is particularly difficult to resolve. There are really two aspects that must be considered. One is the achievement of the proper stoichiometry required by the ternary partial pressures and the other is the maintenance of local vacancy equilibrium as high substitutional Zn concentrations fill the Ga vacancies. These vacancies can be supplied by internal sources such as dislocations and do not depend on supply from the surface. Shaw and Showan have considered Zn diffusion when the rate of supply of Ga vacancies dominates the diffusion behavior.[143] However, at the present time, there are rather formidable difficulties in a quantitative approach to this problem.

The question of nonequilibrium should not detract from the basic concepts of interstitial–substitutional diffusion. Such effects may be contributing factors, but certainly are not dominant in controlling diffusion. These concepts of departure from equilibrium are briefly mentioned to give a complete representation of the status of the behavior of Zn in GaAs. Further discussion of this problem has been given by Casey.[113]

5.7. Compilation of Diffusion Coefficients in the III–V Compounds

The reported self- and impurity-diffusion coefficients are compiled in Tables VIII–XIV. These tables have been taken from Ref. 113. Table IX for GaAs is presented in graphical form in Fig. 37. One of the diffusants, the rare earth Tm, is shown to have a diffusion coefficient that decreases with increasing temperature. The effective diffusion coefficient for that case, $D = D_I C_I / C_S$, showed that the inverse temperature dependence results from the fact that the interstitial-to-substitutional concentration ratio de-

Table VIII. Self- and Impurity-Diffusion Coefficients in AlSb

Diffusant	D_0, cm²/sec	Q, eV	Ref.	Comments
Al	7.3×10^{-1}	1.8	144	X-ray measurements on
Sb	2.9×10^{-1}	1.6	144	the growth of AlSb in sandwich specimens of Al and Sb layers
Zn	(a)	—	145	Radioactive profiles
Cd	(a)	—	146	Radioactive profiles
Cu	3.5×10^{-3}	0.36	147	Radioactive profiles

a Concentration dependent.

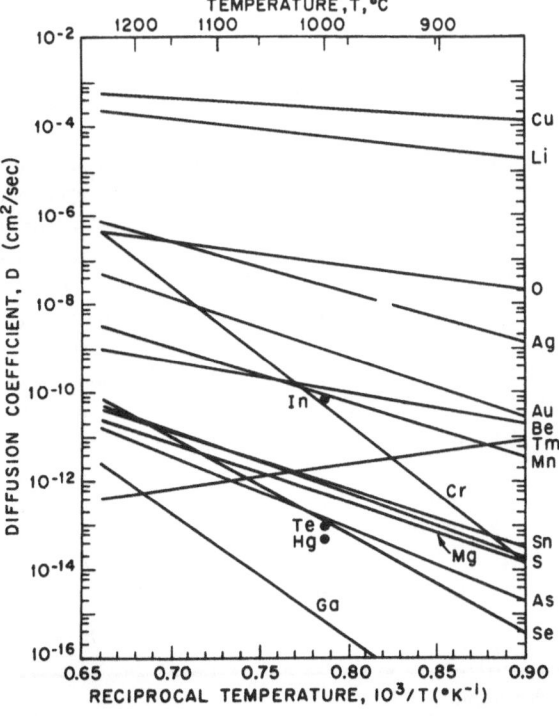

Fig. 37. Summary of the self- and impurity-diffusion coefficients in GaAs. See Table IX for the literature references.

Table IX. Self- and Impurity-Diffusion Coefficients in GaAs

Diffusant	D_0, cm²/sec	Q, eV	Ref.	Comments
As	7×10^{-1}	3.2	134	Radioactive As
Ga	1×10^7	5.6	148	Radioactive Ga
Ag	2.5×10^1	2.27	149	Radioactive profiles
Au	2.9×10^1	2.64	149	Radioactive profiles
Be	7.3×10^{-6}	1.2	150	Incremental sheet resistivities
Cd	(^{a})	—	151	Radioactive profiles
Cr	4.3×10^3	3.4	149	Radioactive profiles, data scatter
Cu	3×10^{-2}	0.53	67	Radioactive profiles
Hg	5×10^{-14} ^{b}	—	152	Radioactive profiles
In	7×10^{-11} ^{b}	—	153	Radioactive profiles
Li	5.3×10^{-1}	1.0	154	Chemical analysis
Mg	2.6×10^{-2}	2.7	155	Sheet resistance and p–n junction measurements
Mn	6.5×10^{-1}	2.49	156	Radioactive profiles with excess arsenic pressure
O	2×10^{-3}	1.1	157	Estimated from mass-spectrometric data on O out-diffusion
S	1.85×10^{-2}	2.6	158	Radioactive profiles at high arsenic pressure
Se	3.0×10^3	4.16	148	Radioactive profiles, surface alloying
Si	(^{c})	—	159	Junction depth measurement
Sn	3.8×10^{-2}	2.7	160	Radioactive profiles and junction depth
Te	10^{-13} ^{b} 2×10^{-12} ^{d}	—	161	Radioactive profiles diffusion through SiO_2 films
Tm	2.3×10^{-16}	$(-)1.0$	162	Radioactive profiles
Zn	(^{a})	—	—	See text

^{a} Concentration dependent.
^{b} $D(1000°C)$.
^{c} Junction depth at 900 and 1000°C only.
^{d} $D(1100°C)$.

Table X. Impurity-Diffusion Coefficients in GaP

Diffusant	D_0, cm²/sec	Q, eV	Ref.	Comments
Zn	$(^a)$	—	163	Radioactive profiles
S	3.2×10^3	4.7	158	Radioactive profiles
Be	$(^a)$	—	164	Chemical analysis

^{a} Concentration dependent.

creases rapidly at higher temperatures and overcomes the increase in the interstitial diffusion coefficient.[162] The data are expressed by the usual Arrhenius relation $D = D_0 \exp(-Q/kT)$. These diffusion data for impurities in the III–V compounds, except for a few cases, have been obtained by introducing a small amount of elemental impurity into the diffusion ampoule. As demonstrated by the detailed treatment of Zn in GaAs, it is difficult to determine the *actual* ternary diffusion conditions for the reported data. Different group V partial pressures can lead to significantly different diffusion behavior. Also, to correctly determine diffusion behavior, radio-tracer diffusion profiles should be obtained by sectioning. Often *p–n* junction depth measurements are used with an assumed erfc profile to determine the diffusion coefficient. No discussion of these diffusion data will be given here. It should be emphasized that these data have been compiled only for completeness.

Table XI. Self- and Impurity-Diffusion Coefficients in GaSb

Diffusant	D_0, cm²/sec	Q, eV	Ref.	Comments
Ga	3.2×10^3	3.15	165	Radioactive profiles
Sb	3.4×10^4	3.45	165	Radioactive profiles
Cd	1.5×10^{-6}	0.72	166	Junction depth measurements
In	1.2×10^{-7}	0.53	167	Radioactive profiles
Sn	2.4×10^{-5}	0.80	167	Radioactive profiles
Te	3.8×10^{-4}	1.20	167	Radioactive profiles

Table XII. Self- and Impurity-Diffusion Coefficients in InAs

Diffusant	D_0, cm^2/sec	Q, eV	Ref.	Comments
As	3×10^7	4.45	168	[a]
In	6.0×10^5	4.0	168	[a]
Ag	7.3×10^{-4}	0.26	169	[a]
Au	5.8×10^{-4}	0.65	170	[a]
Cd	7.4×10^{-4}	1.15	171	[a]
Cu	3.6×10^{-3}	0.52	172	[a]
Hg	1.45×10^{-5}	1.32	173	[a]
Ge	3.74×10^{-6}	1.17	174	[b]
Mg	1.98×10^{-6}	1.17	174	[b]
S	6.78	2.20	174	[b]
Sn	1.49×10^{-6}	1.17	174	[b]
Se	12.6	2.20	174	[b]
Te	3.43×10^{-5}	1.28	174	[b]
Zn	[c]	—	175	[a]

[a] Radioactive profiles.
[b] Measurement of p–n junction depth.
[c] Concentration dependent.

Table XIII. Self- and Impurity-Diffusion Coefficients in InP

Diffusant	D_0, cm^2/sec	Q, eV	Ref.	Comments
In	1×10^5	3.85	148	Radioactive profiles
P	7×10^{10}	5.65	148	Radioactive profiles
Ag	3.6×10^{-4}	0.59	176	Radioactive profiles
Au	1.32×10^{-5}	0.48	177	Radioactive profiles
Cd	1.8	1.9	178	Radioactive profiles
Cu	3.8×10^{-3}	0.69	179	Radioactive profiles
Zn	[a]	—	180	Radioactive profiles

[a] Concentration dependent.

Table XIV. Self- and Impurity-Diffusion Coefficients in InSb

Diffusant	D_0, cm^2/sec	Q, eV	Ref.	Comments
In	1.76×10^{13}	4.3	181	Radioactive profiles
Sb	3.1×10^{13}	4.3	181	Radioactive profiles
Ag	$\sim 1.0 \times 10^{-7}$	~ 0.25	182	Radioactive profiles
Au	7.0×10^{-4}	0.32	183	Radioactive profiles
Co	$\sim 1.0 \times 10^{-7}$	~ 0.25	182	Radioactive profiles
Cd	1.3×10^{-4}	1.2	184	Measurement of p–n junction depth
Cu	9.0×10^{-4}	1.08	185	Radioactive profiles
Fe	$\sim 1.0 \times 10^{-7}$	~ 0.25	182	Radioactive profiles
Hg	4.0×10^{-6}	1.17	186	Radioactive profiles
Li	7.0×10^{-4}	0.28	187	Electrical measurements
S	0.09	1.40	188	Electrical measurements
Se	1.6	1.87	189	Measurement of profile by capacitance voltage
Sn	5.5×10^{-8}	0.75	190	Radioactive profiles
Te	1.7×10^{-7}	0.57	191	Radioactive profiles
Zn	(a)	—	134	Radioactive profiles

a Concentration dependent.

6. DIFFUSION IN THE II–VI COMPOUNDS

6.1. Introduction

The II–VI compounds are one-to-one chemical compounds of the group II and group VI elements. These compounds include wide-energy-gap insulators such as BeO and the narrow-energy-gap semiconductor HgTe. The most commonly encountered group are the semiconducting compounds formed between Zn or Cd and S, Se, or Te. These six Zn and Cd chalcogenides have energy gaps between 1.44 and 3.6 eV. In this section on diffusion in II–VI compounds, only CdS, CdSe, CdTe, ZnS, ZnSe, and ZnTe will be considered, and hereafter, reference to the II–VI compounds will mean only these six materials.

These II–VI compounds are different from the III–V compounds in several important respects. Perhaps the most significant difference is that

Table XV. Properties of the II–VI Compounds

Compound	Crystal structure	Lattice constant, Å	Melting point,[198] °C	Pressure at melting point,[193] atm	Energy gap at 300°K, eV	Ionicity[75]	Conductivity type
CdS	Wurtzite	$a = 4.1368^{192}$ $c = 6.7163^{192}$	1475	3.8	2.41^{194}	0.685	n
CdSe	Wurtzite	$a = 4.2975^{195}$ $c = 6.9892^{195}$	1239	0.41	1.670^{196}	0.699	n
CdTe	Zinc-blende	$a = 6.481^{197}$	1092	0.23	1.44^{198}	0.675	n and p
ZnS	Zinc-blende Wurtzite	$a = 5.4093^{199}$ $a = 3.820^{199}$ $c = 6.260^{199}$	1830	3.7	3.6^{200}	0.623	n
ZnSe	Zinc-blende	$a = 5.6687^{192}$	1520	0.53	2.7^{201}	0.676	n, p, high resistivity
ZnTe	Zinc-blende	$a = 6.1037^{202}$	1295	0.64	2.26^{203}	0.546	p

this group of compounds, with the exception of CdTe, has only been obtained as either *n* or *p* type, but not as both conductivity types. This property means that it has not been possible to prepare *p–n* junctions (except heterojunctions) with these compounds.

Although it is not possible to change the conductivity type by impurity doping as commonly utilized for Si, Ge, and the III–V compounds, there are several very important commercial applications of II–VI compounds. Both CdS and CdSe are extensively used as photoconductive radiation detectors from the near infrared to gamma radiation. When doped with the proper impurities, ZnS is an efficient visible luminescent material in cathode ray tube applications. Another application utilizes the strong piezoelectric coupling of CdS and CdSe for acoustoelectric devices, such as ultrasonic amplifiers and thin-film transducers.

The properties of the II–VI compounds are summarized in Table XV. These compounds are tetrahedrally coordinated and crystallize in the cubic zinc-blende structure, illustrated in Fig. 25 for GaAs, or the hexagonal wurtzite structure, illustrated in Fig. 38 for CdS. The melting points are rather high and the ionicity, which is in excess of 0.5, indicates that the bonding tends to be slightly more ionic than covalent.

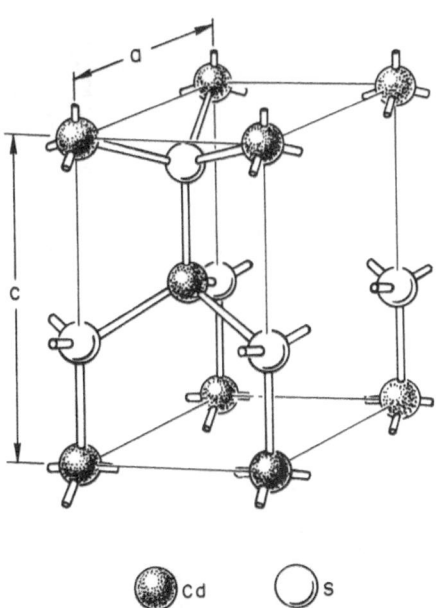

Fig. 38. The wurtzite structure for CdS.

The discussion given here will be limited to the consideration of self-diffusion. Extensive discussion of diffusion in the II–VI compounds can be found in several recent reviews.[5,204–206] In this group of compounds, many of the observed properties are dominated by ionized vacancies. Control of the vacancy concentration requires consideration of the partial pressures of the components in equilibrium with a crystal, and this discussion of self-diffusion will begin with the solid–liquid–vapor equilibria of ZnTe. This compound was selected to illustrate the basic principles since very complete data are available.[207,208] Self-diffusion in CdS will next be used to illustrate the concepts, terminology, and experimental techniques that are unique to these compounds. Detailed reference to the many publications on impurity and self-diffusion can be found in Refs. 5 and 204–206.

6.2. Solid–Liquid–Vapor Equilibrium

6.2.1. *Component Partial Pressures*

The relationships between the nonstoichiometric compound, the liquid, and the vapor are illustrated in Figs. 39 and 40 for ZnTe. The liquidus curve in Fig. 39 was taken from Jordan,[207] whose analysis is based on the

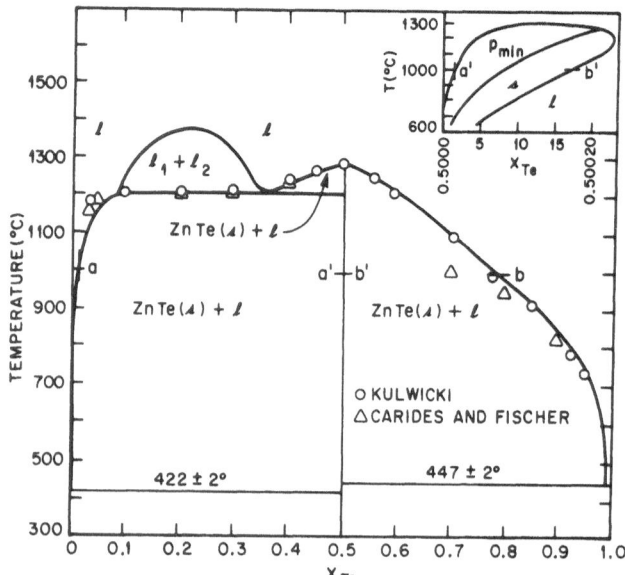

Fig. 39. Liquidus diagram for the Zn–Te system.[207] The insert is the calculated solidus curve of ZnTe.[208]

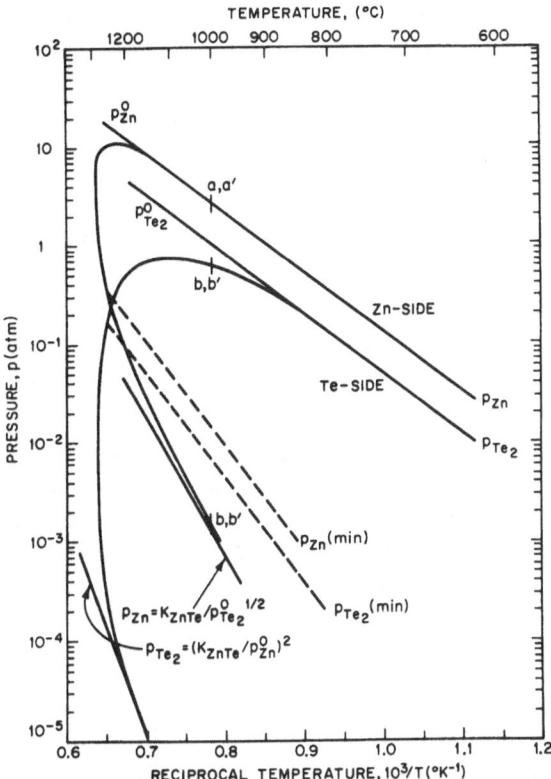

Fig. 40. The Zn and Te partial pressures versus reciprocal
temperature for the Zn–Te system.[208]

data of Kulwicki[209] and Carides and Fischer.[210] The solidus curve shown
as the insert in Fig. 39 and the partial pressures of the vapor species that
are shown in Fig. 40 are taken from Jordan and Zupp.[208] The points a, a'
in Fig. 39 represent the equilibrium of the liquidus and solidus on the Zn-
rich side, while b, b' represent the liquidus and solidus on the Te-rich side.
The corresponding Zn and Te_2 partial pressures are indicated by a, a' and
b, b' in Fig. 40.

The decomposition of ZnTe at high temperature is given by the reaction[211]

$$ZnTe(s) \rightleftarrows Zn(g) + \tfrac{1}{2}Te_2(g) \tag{129}$$

The equilibrium relation for this reaction is

$$K_{ZnTe} = p_{Zn} p_{Te_2}^{1/2} \tag{130}$$

where p_{Zn} and p_{Te_2} are the partial pressures of Zn and Te_2, and K_{ZnTe} is the equilibrium constant. The equilibrium constant can be evaluated by measurement of the partial pressures or by calculation from the standard enthalpy and entropy of formation of the elements by the relation

$$-RT \ln K_{ZnTe} = \Delta H° - T \Delta S° \qquad (131)$$

These heats of formation and entropies for the II–VI compounds have been summarized by Lorenz.[193] As shown in Fig. 40, the partial pressure of Zn approaches p_{Zn}^0, the pressure over pure Zn, at lower temperatures on the Zn-rich side of the liquidus. The same is true for Te_2 on the Te-rich side. Therefore, by Eq. (130), the Zn partial pressure on the Te-rich side approaches $K_{ZnTe}/p_{Te_2}^{1/2}$ as shown in Fig. 40. The Te_2 partial pressure on the Zn-rich side is similarly given by $(K_{ZnTe}/p_{Zn}^0)^2$. These curves indicate the range of partial pressure variation that can be achieved at a given temperature. For example, at 1000°C the Zn partial pressure can be varied from 2.5 atm on the Zn-rich side of the liquidus to 1.6×10^{-3} atm on the Te-rich side. The Te_2 partial pressure can vary from 6.5×10^{-1} to about 10^{-6} atm. The pressures have a ratio of 6.4×10^2 for Zn and the square of this ratio, 4.1×10^5, for Te_2.

6.2.2. Congruent Evaporation and Minimum Pressure

Other useful concepts to consider are congruent evaporation and minimum pressure. The total minimum pressure can be found from Eq. (130) by writing the total pressure as

$$p_t = p_{Zn} + p_{Te_2} = p_{Zn} + (K_{ZnTe}/p_{Zn})^2 \qquad (132)$$

The minimum total pressure can be obtained by setting $\partial p_t/\partial p_{Zn} = 0$, which gives

$$K_{ZnTe}^2 = p_{Zn}^3/2 \qquad (133)$$

Substitution of Eq. (133) into Eq. (130) gives

$$p_{Te_2} = \tfrac{1}{2}p_{Zn} \qquad (134)$$

for minimum pressure. At p_{min} the compositions of the solid and vapor phases are always the same, which is defined as congruent evaporation. The compositions of the solid and vapor phases will not in general correspond to the stoichiometric composition. The minimum pressures[208] for p_{Zn} and p_{Te_2} are shown in Fig. 40.

Congruent evaporation has several properties which must be considered when heating crystals during self-diffusion studies. For a crystal within the solidus shown by the insert in Fig. 39, heating in a large-volume container results in a change of the composition toward the composition indicated by p_{min} and evaporation occurs without formation of the liquid phase. Congruent evaporation at elevated temperatures is typical of the II–VI compounds but only occurs at temperatures below 700°C for III–V compounds[140] such as GaAs.

6.2.3. *The Solidus Curve*

The solidus curve of Jordan and Zupp[208] was based on the resistivity measurements of Thomas and Sadowski[212] at temperatures between 700 and 950°C with p_{Zn} from 1.3×10^{-2} to 0.52 atm. A Zn pressure was maintained around the crystal at the elevated temperature while the resistivity was measured.[212] The hole concentration, in the absence of intentionally added impurities, varies as $p_{Zn}^{-1/3}$ and was interpreted as due to the presence of double-ionized acceptor Zn vacancies. It was found that the hole concentration is

$$p = 4.0 \times 10^{22} p_{Zn}^{-1/3} \exp(-1.31/kT) \tag{135}$$

where p_{Zn} is in atm. A similar pressure dependence was also observed by Smith in a recent and more extensive study.[213] Small numerical differences for the hole concentrations (factors of two or less) exist between Refs. 212 and 213 because Thomas and Sadowski extrapolated mobility data to the high temperature used in their resistivity measurements, in order to calculate the hole concentration, whereas the measured mobility of Smith is less than the extrapolated value.

The result that the hole concentration depends on pressure led to the assignment of the acceptor as a native defect rather than an impurity.[212,213] The equilibrium between Zn atoms on the Zn sublattice Zn_{Zn}, Zn vacancies V_{Zn}, holes $h^{\cdot}$, and Zn vapor is given by

$$Zn_{Zn} \rightleftarrows V_{Zn}'' + 2h^{\cdot} + Zn(g) \tag{136}$$

where V_{Zn}'' represents the double negative charged acceptor vacancy. The ionization energy of this level was reported as 0.20 eV.[213] Since each vacancy gives two holes, the vacancy concentration as given by Eq. (135) is

$$[V_{Zn}''] = p/2 = 2.0 \times 10^{22} p_{Zn}^{-1/3} \exp(-1.31/kT) \tag{137}$$

where p is the hole concentration and $[V''_{Zn}]$ represents the concentration of V''_{Zn}.

It should be noted that the notation used to designate donors and acceptors in the II–VI compounds is considerably different from that for Si, Ge, and the III–V compounds. A summary of the notation used with the II–VI compounds is given in Table XVI. This difference in notation results from the early workers in the II–VI compounds beginning with insulator crystals. Even though the II–VI compounds are semiconductors, this notation for insulators is commonly used.

If V''_{Zn} is taken as the dominant native defect, which includes the assumption that the Te vacancy concentration is negligibly small, then Eq. (137) permits one to calculate the solidus curve. The Te atom fraction in the solid $X_{Te}(s)$ is given by

$$X_{Te}(s) = [Te_{Te}]/([Te_{Te}] + [Zn_{Zn}]) \qquad (138)$$

Table XVI. Notation for Defects in the II–VI Compounds

Defect	Symbol
Metal atom	M
Chalcogen atom	X
Interstitial metal atom	M_i
Interstitial chalcogen atom	X_i
Metal vacancy	V_M
Chalcogen vacancy	V_X
Metal atom on metal site	M_M
Chalcogen atom on chalcogen site	X_X
Free electron	e′
Free hole	h·

Superscript notation

Condition	Symbol
Neutral state	×
Singly ionized acceptor (negative charge)	′
Doubly ionized acceptor	″
Singly ionized donor (positive charge)	·
Doubly ionized donor	··

where $[Te_{Te}]$ and $[Zn_{Zn}]$ represent the Te and Zn concentrations in the solid. The number of Te or Zn sites per cm³ is given by

$$N = \frac{\text{Avogadro's number} \times \text{density}}{\text{molecular weight}} \qquad (139)$$

and amounts to 1.75×10^{22} Te sites/cm³ and 1.75×10^{22} Zn sites/cm³. In Eq. (138), all the Te sites are occupied so that $[Te_{Te}] = 1.75 \times 10^{22}$ cm⁻³, while for Zn the number of occupied sites $[Zn_{Zn}]$ is $1.75 \times 10^{22} - [V''_{Zn}]$. Therefore Eq. (138) becomes

$$X_{Te}(s) = 1/(2 - [V''_{Zn}]/1.75 \times 10^{22}) \qquad (140)$$

The p_{Zn} curve in Fig. 40 permits calculation of $[V''_{Zn}]$ from Eq. (137) and then the solidus curve is calculated from Eq. (140). These results show that ZnTe is nearly a stoichiometric compound, but is always slightly Te-rich. The composition for p_{min} is also shown in Fig. 39.

Although ZnTe has been treated in detail, these concepts of solid–liquid–vapor equilibrium also apply to the other II–VI compounds. The dominant native defect is not necessarily the same, however, and the defect behavior must be examined separately for each compound. Doubly charged native donor centers have been reported for CdS,[214–216] CdSe,[217,218] CdTe,[219,220] and ZnS,[221] while for ZnSe the native donor is singly charged.[222]

6.3. Self-Diffusion

6.3.1. General Comments

It is difficult to visualize a substitutional self-diffusion mechanism in a compound because such a model requires significant group II atoms in group VI sites and vice versa. Therefore, a self-diffusion mechanism permitting transport through the lattice would imply a significant concentration of antistructure defects, but presently no results suggest this model. Self-diffusion by movement through an interstitial site in reaching an adjacent vacancy on its own sublattice is a reasonable vacancy mechanism.

Several interstitial mechanisms can be suggested for self-diffusion. Lattice vacancies are not required in the interstitialcy mechanism (Fig. 3). The interstitial atom moves into a normal lattice site and the atom which originally occupied this site is pushed into an interstitial position. Another possibility is movement as an interstitial atom which reacts with a lattice vacancy to form the substitutional atom. Interstitial atoms are highly mobile but their concentrations are expected to be very small so that slow self-diffusion would result for this interstitial–substitutional mechanism. This self-diffusion process is described below in order to illustrate the manner in which an experimental radioactive tracer can be treated.

6.3.2. Effective Diffusion Coefficient

The radioactive flux for either group II or group VI atoms can be written as

$$J_I = D_I \, \partial C_I^* / \partial x \tag{141}$$

where D_I is the interstitial diffusion coefficient, and C_I^* is the interstitial radioactive tracer concentration. Equation (141) represents a *concentration gradient in the radioactive tracer concentration and not the total concentration,* which is constant with distance. This equation can be written

$$J_I = -D_I (\partial C_I^* / \partial C^*) \, \partial C^* / \partial x \tag{142}$$

where $C^* = C_L^* + C_I^*$ is the total radioactive tracer concentration in lattice sites C_L and interstitial sites C_I. By analogy with Eq. (141), the effective diffusion coefficient can be written

$$D_e = -J_I / (\partial C^* / \partial x) = D_I \, \partial C_I^* / \partial C^* \tag{143}$$

It should be noted that for self-diffusion with a radioactive tracer, the total concentration of atoms C is the sum of radioactive atoms C^* and nonradioactive atoms C° or $C = C^\circ + C^*$. The nonradioactive atoms are the sum of the lattice site and interstitial atoms, $C^\circ = C_L^\circ + C_I^\circ$. The ratio of interstitial radioactive atoms to the total radioactive atom concentration is the same as the ratio of interstitial nonradioactive atoms to the total nonradioactive atom concentration, $C_I^*/C^* = C_I^\circ/C^\circ$, and $C_I^\circ > C_I^*$ and $C^\circ > C^*$. When $C_I^*/C^* = C_I/C$, it is useful to write Eq. (143) as

$$D_e = D_I C_I / C \tag{144}$$

As shown in Fig. 39 for ZnTe, the number of lattice vacancies is small, and C in units of atom fraction is nearly 0.5. With C_I also in atom fraction, Eq. (144) can be written as

$$D_e = 2 D_I C_I \tag{145}$$

Thus, even though $D_I = 10^{-5} \, \text{cm}^2/\text{sec}$ and $C_I = 10^{-5}$, D_e would be $2 \times 10^{-10} \, \text{cm}^2/\text{sec}$ for an interstitial–substitutional self-diffusion mechanism.

6.3.3. Self-Diffusion in CdS

a. *The Observed Native Defects.* A number of studies have reported self-diffusion in CdS.[214,216,223] The discussion given here will follow the results of Kumar and Kröger.[216] In this work the Cd and S diffusivities were measured in undoped and donor-doped (In) CdS single crystals as a function of temperature, Cd or S_2 partial pressure, and In concentration. The diffusion results also considered the high-temperature Hall effect measurements in Cd vapor by Hershman and Kröger[215] in order to be consistent with both the diffusive and electrical behavior. These results permit assignment of the Cd and S diffusion mechanisms and the determination of the double-ionized S vacancy $V_S^{\cdot\cdot}$ (double donor) as the native defect that controls the carrier concentration at high temperature.[216] At high Cd pressures above 700°C, S diffuses by a vacancy mechanism via the doubly ionized S vacancy $V_S^{\cdot\cdot}$, while Cd diffuses by the singly ionized (donor) Cd interstitial. For high S_2 pressure (low p_{Cd}), S diffuses by the interstitial mechanism via the neutral interstitial $S_i^\times$, while Cd diffusion depends upon doubly ionized Cd (V_{Cd}'' acceptor) vacancies.

b. *The CdS Partial Pressures.* As illustrated for ZnTe, the range of partial pressures is limited. The Cd partial pressure as given by Kumar and Kröger[216] based on the data of Shiozawa and Jost[224] is shown in Fig. 41(a), while Fig. 41(b) gives the total pressure on the S-rich side ($p_t \approx p_{S_2}$), the total pressure on the Cd-rich side ($p_t \approx p_{Cd}$), and the total minimum pressure ($p_{S_2} = \frac{1}{2}p_{Zn}$). Diffusion profiles were obtained[216] for various partial pressures and temperatures. Since $p_{Cd}p_{S_2}^{1/2} = K_{CdS}$ [Eqs. (129) and (130)], low Cd pressures can be obtained by establishing high p_{S_2} pressures. From the results of Shiozawa and Jost,[224] the total S partial pressure is p_{S_2}.

c. *Self-Diffusion at High p_{Cd}.* The Cd self-diffusion profiles were obtained[216] by using radioactive ^{109}Cd, while the S profiles were obtained with radioactive ^{35}S. Sectioning was achieved by removal of 2–5-μm-thick layers on fine emory paper. Penetration profiles, as illustrated in Fig. 42, permitted assignment of the diffusivity by fitting with a complementary error function.

The variation of the Cd self-diffusion at high Cd pressures is shown in Fig. 43. The data of Kumar and Kröger[216] obtained at 800 and 900°C are shown along with the results of Shaw and Whelan[214] at 850°C. At both temperatures, D_{Cd} is proportional to $p_{Cd}^{2/3}$. From electrical measurements $n \propto p_{Cd}^{1/3}$.[215] The defect controlling the Cd diffusion cannot, therefore, be the majority defect controlling charge neutrality. This result leads to assign-

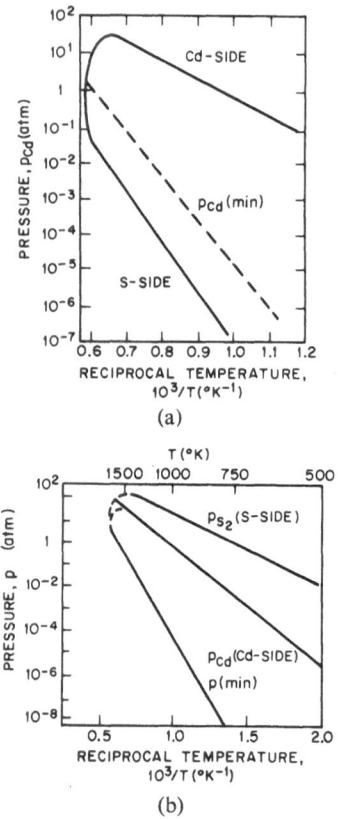

Fig. 41. The CdS partial pressures.[216] (a) The
Cd partial pressure. (b) The total pressure on
the S-rich side is p_{S_2} and on the Cd-rich side
is p_{Cd}.

ment of a doubly ionized S vacancy $V_S^{\cdot\cdot}$ as the dominant electrically active
defect. This assignment was verified[216] by S diffusion as a function of p_{Cd}.
The results are shown in Fig. 44 and illustrate that both n and D_S vary
as $p_{Cd}^{1/3}$.

The reactions and equilibrium relations that give $[V_S^{\cdot\cdot}] \propto p_{Cd}^{1/3}$ begin
with the reaction of Cd in the vapor with a neutral Cd vacancy to form a
Cd atom in the lattice according to the relation

$$Cd(g) + V_{Cd}^x \rightleftarrows Cd_{Cd}^x \tag{146}$$

and the Schottky reaction

$$0 = V_{Cd}^x + V_S^x \tag{147}$$

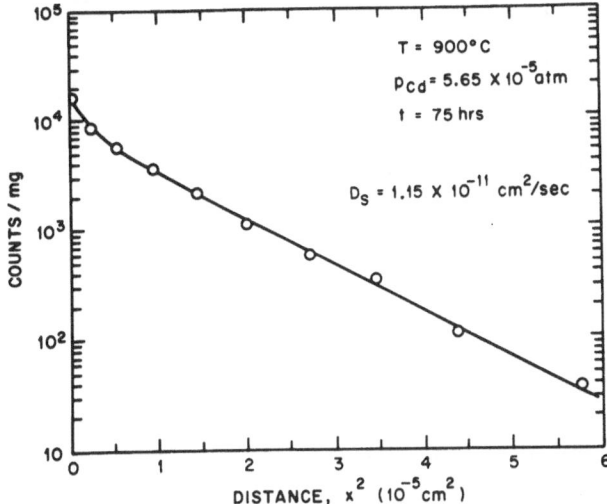

Fig. 42. Penetration profile of a typical tracer diffusion experiment in CdS. The curve represents a complementary error function and the points indicate experimental data for the diffusion of S tracer at 900°C and $p_{Cd} = 5.6 \times 10^{-5}$ atm.[216]

Addition of Eq. (146) and (147) gives

$$Cd(g) \rightleftarrows Cd_{Cd}^{x} + V_S^{x} \tag{148}$$

and the double ionization of the neutral S vacancy is given by

$$V_S^{x} \rightleftarrows V_S^{\cdot\cdot} + 2e' \tag{149}$$

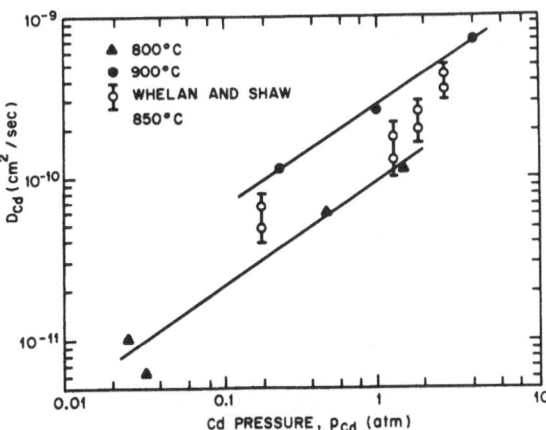

Fig. 43. Variation of D_{Cd} with p_{Cd} in CdS at 800 and 900°C[216]; 850°C data by Shaw and Whelan are also given.[214]

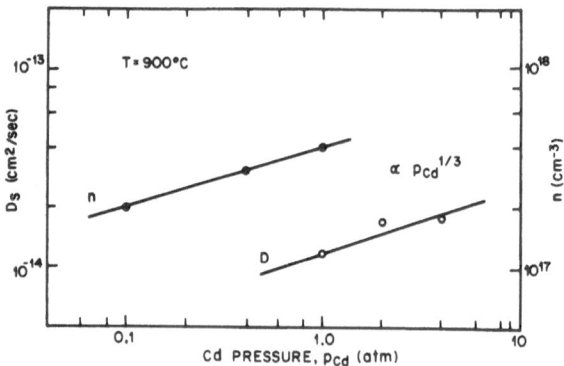

Fig. 44. Variation of D_S and n with p_{CD} in CdS at 900°C.[216]

Equation (148) can now be rewritten as

$$Cd(g) \rightleftarrows Cd_{Cd}^x + V_S^{\cdot\cdot} + 2e' \tag{150}$$

which gives the equilibrium relation

$$K_{SV}'' = [V_S^{\cdot\cdot}]n^2/p_{Cd} \tag{151}$$

Since the electrical neutrality condition is

$$n = 2[V_S^{\cdot\cdot}] \tag{152}$$

Eq. (151) reduces to

$$n/2 = [V_S^{\cdot\cdot}] = (K_{SV}''/4)^{1/3}p_{Cd}^{1/3} \tag{153}$$

which is the $p_{Cd}^{1/3}$ dependence for both n and D_S ($D_S \propto V_S$).

The reactions and equilibrium relations that give $Cd_i \propto p_{Cd}^{2/3}$ for $D_{Cd} \propto p_{Cd}^{2/3}$ are based on the relation

$$Cd(g) \rightleftarrows Cd_i^x \rightleftarrows Cd_i^{\cdot} + e' \tag{154}$$

for the singly ionized Cd interstitial, which has the equilibrium relation

$$K_{Cd_i} = [Cd_i^{\cdot}]n/p_{Cd} \tag{155}$$

Substitution for n from Eq. (153) gives

$$[Cd_i^{\cdot}] = K_{Cd_i}(2K_{SV}'')^{-1/3}p_{Cd}^{2/3} \tag{156}$$

which demonstrates that a two-thirds power dependence is expected when

$[Cd_i^{\cdot}]$ is the dominant diffusing species. Comparison of Figs. 43 and 44 shows that $D_{Cd} \gg D_S$, and therefore, rapidly moving $Cd_i^{\cdot}$ should be the species that brings about the change in stoichiometry when the partial pressure or temperature is changed.

Expressions for $[V_S^{\cdot\cdot}]$, n, and D_{Cd} can be obtained as a function of temperature and pressure from the electrical measurements and self-diffusion studies in the high-Cd partial pressure region. The equilibrium constant $K_{SV}^{\prime\prime}$ in Eq. (151) can be evaluated by Eq. (153) from the electrical measurements of Hershman and Kröger.[215] A plot of n versus $1/T$ for $p_{Cd} = 1$ atm gives[216]

$$K_{SV}^{\prime\prime} = 1.6 \times 10^{60} \exp(-1.75/kT) \, \text{cm}^{-3} \, \text{atm}^{-1} \qquad (157)$$

With this value for $K_{SV}^{\prime\prime}$, the concentration of the major doubly ionized native donor is

$$[V_S^{\cdot\cdot}] = 7.4 \times 10^{19} \exp(-0.58/kT) \, p_{Cd}^{1/3} \, \text{cm}^{-3} \qquad (158)$$

The temperature dependence of D_{Cd} for $p_{Cd} = 1$ atm is shown in Fig. 45 and permits the diffusivity of Cd at high Cd pressure to be written as[216]

$$D_{Cd}(\text{high } p_{Cd}) = 7.3 \times 10^{-5} \exp(-1.26/kT) \, p_{Cd}^{2/3} \qquad (159)$$

Equations (158) and (159) summarize the most important results in the high-Cd-pressure region.

d. *Self-Diffusion at Low* p_{Cd}. It was found that going from a high-p_{Cd} to a low-p_{Cd} (high-p_{S_2}) region results in a change in the diffusion mechanism for both Cd and S. The diffusivity of S at low p_{Cd} (high p_{S_2}) is shown in Figs. 46 and 47. Figure 46 demonstrates the temperature dependence at

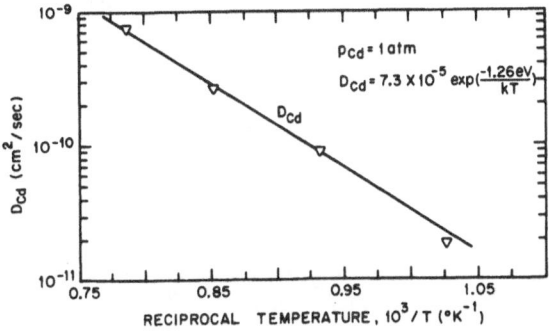

Fig. 45. Temperature variation of D_{Cd} in CdS at high p_{Cd}.[216]

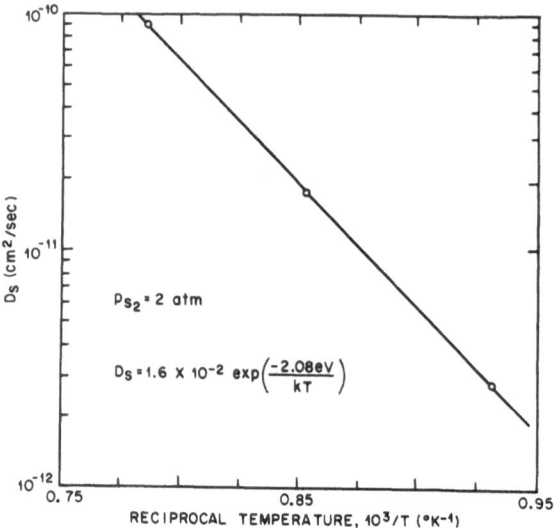

Fig. 46. Temperature variation of D_S in CdS at high p_{S_2}.[216]

a fixed p_{S_2}, while Fig. 47 shows that D_S (high p_{S_2}) varies as p_{Cd}^{-1}. The $D_S \propto p_{Cd}^{-1}$ relation suggests that S diffuses by the neutral S interstitial. The neutral condition of the S interstitial was confirmed by demonstrating that D_S did not change when the donor In was added to the crystal. From Figs. 46 and 47, the diffusivity of S for high p_{S_2} can be represented by[216]

$$D_S(\text{high } p_{S_2}) = 1.1 \times 10^{-2} \exp(-2.08/kT)\, p_{S_2}^{1/2} \tag{160}$$

and, since $p_{Cd} p_{S_2}^{1/2} = K_{CdS}$, $D_S \propto p_{Cd}^{-1}$.

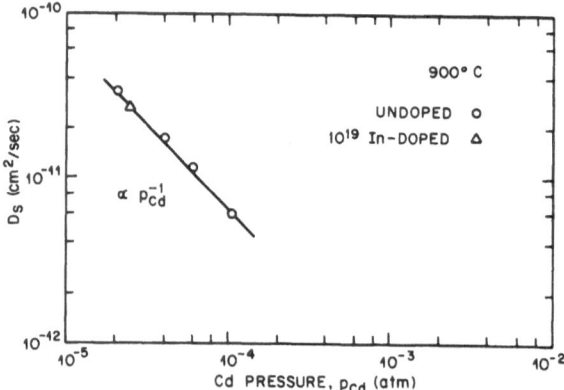

Fig. 47. Sulfur diffusivity for pure and In-doped CdS at 900°C.[216]

e. *The Defect Equilibrium Diagram.* In discussions of the II–VI compounds, it is frequently useful to plot the defect concentrations as a function of the partial pressure. This type of plot is called an equilibrium or Brouwer diagram.[225] In the case of CdS, the important defects are the electrons e' and holes h, the doubly ionized S vacancy $V_S^{\cdot\cdot}$, the singly ionized interstitial $Cd_i^{\cdot}$, and the doubly ionized Cd vacancy V_{Cd}''. The assignment of the Cd vacancy as a doubly ionized acceptor was obtained from studies with In-doped (donor) CdS.[216,226]

There are two electrical neutrality conditions to be considered for CdS. The first is the intrinsic case when $n = p$, which will be designated region I, and the extrinsic case in which $n = 2[V_S^{\cdot\cdot}]$. The equilibrium relation was given for $[V_S^{\cdot\cdot}]$ in Eq. (151), and for $[Cd_i^{\cdot}]$ in Eq. (156). The Schottky relation

$$0 = V_{Cd}'' + V_S^{\cdot\cdot} \tag{161}$$

permits an evaluation of V_{Cd}''. The corresponding equilibrium relation is

$$K_S'' = [V_{Cd}''][V_S^{\cdot\cdot}] \tag{162}$$

In Ref. 226, the Schottky constant K_S'' was evaluated as

$$K_S'' = 2.4 \times 10^{47} \exp(-4.09/kT) \tag{163}$$

The $[V_{Cd}'']$ can be obtained from Eq. (162) with the value of K_S'' given in Eq. (163). The hole concentration p is given by the reaction

$$e' + h^{\cdot} \rightleftarrows 0$$

This corresponds to the equilibrium relation

$$K_i = np = n_i^2$$

where the equilibrium constant K_i is the intrinsic carrier concentration squared n_i^2. The intrinsic carrier concentration n_i is given by[128]

$$n_i = (N_c N_v)^{1/2} \exp(-E_g/2kT) \tag{164}$$

The effective density of states for the valence band is

$$N_v = 2.5 \times 10^{19} (m_p/n_0)^{3/2} (T/300)^{3/2} \tag{165}$$

Table XVII. Defect Formation Reactions and the Corresponding Equilibrium Relations

Reaction	Equilibrium relation region I intrinsic $(n = p = n_i)$	Equilibrium relation region II extrinsic $(n > n_i)$
1. $Cd(g) \rightleftharpoons Cd_{Cd}^x + V_S^{\cdot\cdot} + 2e'$	$K_{SV}'' = [V_S^{\cdot\cdot}]n_i^2/p_{Cd}$	$K_{SV}'' = [V_S^{\cdot\cdot}]n^2/p_{Cd}$
2. $Cd(g) \rightleftharpoons Cd_i^{\cdot} + e'$	$K_{Cd_i} = [Cd_i^{\cdot}]n_i/p_{Cd}$	$K_{Cd_i} = [Cd_i^{\cdot}]n/p_{Cd}$
3. $0 \rightleftharpoons V_{Cd}'' + V_S^{\cdot\cdot}$	$K_S'' = [V_{Cd}''][V_S^{\cdot\cdot}]$	$K_S'' = [V_{Cd}''][V_S^{\cdot\cdot}]$
4. $0 \rightleftharpoons e' + h^{\cdot}$	$K_i = np = n_i^2$	$K_i = np = n_i^2$
5. Charge neutrality	$n = p = n_i$	$n = 2[V_S^{\cdot\cdot}]$

and the effective density of states for the conduction band is

$$N_c = 2.5 \times 10^{19}(m_n/m_0)^{3/2}(T/300)^{3/2} \tag{166}$$

where m_p is the effective hole mass, m_n is the effective electron mass, and m_0 is the free-electron mass. The energy gap is E_g. Values[227] for CdS have been given as $m_p = 0.7m_0$ and $m_n = 0.2m_0$. The energy gap is much less certain and the value of n_i at 900°C was given by Kumar and Kröger[216] as $\sim 5 \times 10^{15}$ cm^{-3}. The remaining quantity $[Cd_i^{\cdot}]$ requires knowledge of K_{Cd_i}, which is unknown. Therefore the value of $[Cd_i^{\cdot}]$ can only represent the $p_{Cd}^{2/3}$ dependence and its absolute magnitude is unknown.

The reactions and equilibrium relations for the intrinsic region I and the extrinsic region II are given in Table XVII, and the values of the equilib-

Table XVIII. Values of the Equilibrium Constants Given in Table XVII

Equilibrium constant	Value at 900°C
1. $K_{SV}'' = 1.6 \times 10^{60} \exp(-1.75/kT)$ cm^{-3} atm^{-1}	4.9×10^{52}
2. K_{Cd_i}, unknown	Estimate
3. $K_S'' = 2.4 \times 10^{47} \exp(-4.09/kT)$	6.5×10^{29}
4. $\sqrt{K_i} = (N_v N_c)^{1/2} \exp(-E_g/2kT)$	$\sim 5 \times 10^{15}$

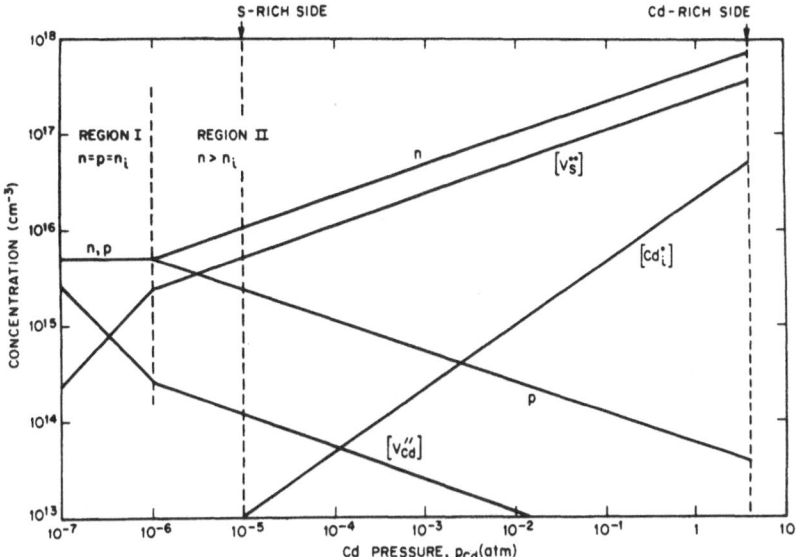

Fig. 48. Native defect concentrations at 900°C in undoped CdS.[216]

rium constants are given in Table XVIII. These equilibrium constants have been used to draw the equilibrium diagram shown in Fig. 48. For the range of p_{Cd} between the S-rich side and the Cd-rich side, the crystal is always extrinsic and dominated by $V_S^{\cdot\cdot}$. Note that since $V_S^{\cdot\cdot}$ dominates, the solid is always Cd-rich. As for ZnTe, it appears that the component with the highest vapor pressure will also have the greatest vacancy concentration. The equilibrium diagram has sufficient information to enable one to draw the solidus curve as was done for ZnTe in Fig. 39.

6.3.4. Compilation of Self-Diffusion Coefficients in the II–VI Compounds

The self-diffusion behavior in the other II–VI compounds will now be briefly summarized. Diffusion of the metal component in CdTe, ZnTe, and ZnSe is independent of component pressure over the entire phase field, while in CdSe and CdS the diffusivity increases with increasing metal partial pressure. A summary of metal component diffusivity, taken from Stevenson,[206] is shown in Fig. 49. The self-diffusion of the chalcogen component D_X is summarized in Fig. 50. The same behavior is observed for the II–VI compounds at the chalcogen boundary and can be represented by

$$D_X = K p_M^{-1} \tag{167}$$

which indicates that the diffusivity varies as the inverse of the metal partial pressure. Approach to the metal-rich compositions leads to a change of the dependence upon component pressure. As pointed out for CdS, the chalcogen component diffusivity is much slower than the metal diffusivity.

In the treatment given here of diffusion in the II–VI compounds, it has been necessary to omit discussion of most of the extensive publications on this subject. Many questions relating to diffusion in these compounds remain to be resolved. The discussion given here illustrates the concepts necessary for interpretation of diffusive behavior in II–VI compounds. Further details of diffusion in the individual compounds and extensive literature references can be found in review papers on this subject.[5,204–206]

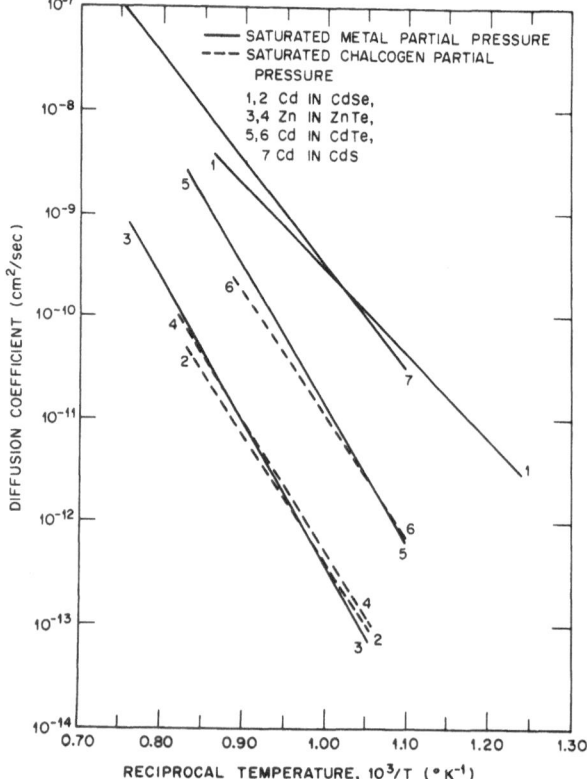

Fig. 49. Some representative curves for Cd and Zn diffusion coefficients in II–VI compounds. These curves show the dependence upon temperature and partial pressure.[206] 1, 2, Ref. 228; 3, 4, Ref. 229; 5–7, Ref. 231.

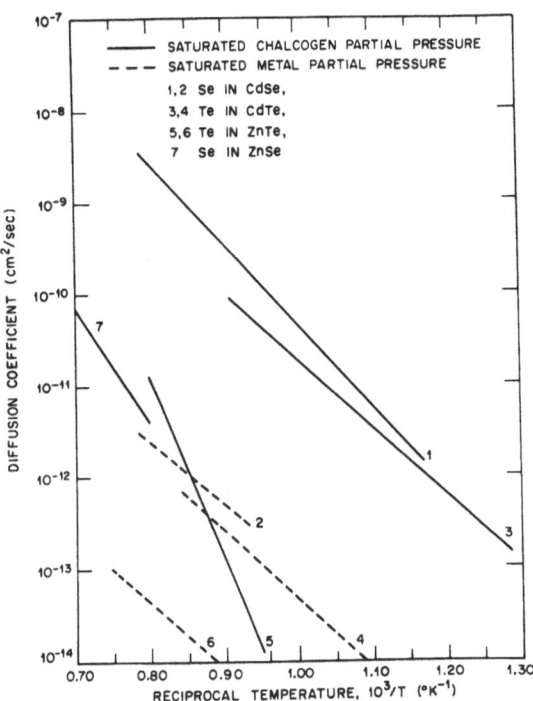

Fig. 50. Some representative curves for Se and Te diffusion in II–VI compounds. These curves show the dependence upon temperature and partial pressure.[206] 1, 2, 7, Ref. 232; 3, 4, Ref. 230; 5, 6, Ref. 229.

7. SUMMARY AND CONCLUSIONS

The purpose of this chapter has been to provide the theory, terminology, and experimental results of diffusion in semiconductors. An exhaustive compilation of all the available data has not been given, but representative results have been included. The basic theory of diffusion has been introduced by consideration of both the mechanism of atomic motion and the mathematical treatment by Fick's laws. The Boltzmann–Matano analysis, which is necessary for treatment of concentration-dependent diffusion, is given and its use illustrated. Sufficient basic concepts are presented so that recourse to general texts on diffusion is not necessary for those without previous background in diffusion theory. Three groups of semiconductors are considered: the group IV semiconductors Si and Ge, the III–V compounds, and the II–VI compounds.

For self-diffusion in Si and Ge, the monovacancy mechanism appears to dominate for Ge, while for Si the dominant mechanism is still subject to considerable controversy. The commonly used group III and V impurities in Si diffuse by a vacancy mechanism. These impurities result in concentration-dependent diffusion profiles at concentrations in excess of approximately 10^{19} impurities/cm^3. The concentration-dependent diffusion is still the subject of extensive investigation. However, it is clear that electrically inactive impurity–vacancy complexes play a significant role in diffusion at high concentrations.

In the III–V compounds, proper interpretation of experimental diffusion profiles requires consideration of the diffusing impurity in the compound as a ternary system. The example of Zn in GaAs demonstrates the relationship between the ternary phase diagram, the species partial pressures, and the resulting diffusion profiles. Diffusion in this case is by an interstitial–substitutional mechanism. Further studies are necessary in order to understand the establishment and maintenance of vacancy equilibria during diffusion.

Only self-diffusion was considered for the II–VI compounds. The relationships between the nonstoichiometric compound, the liquid, and the vapor were illustrated for ZnTe. Self-diffusion in CdS was given as a detailed example of self-diffusion in a II–VI compound. The diffusion behavior in conjunction with high-temperature electrical measurements permits assignment of the dominant lattice defects and the diffusion mechanism. For the II–VI compounds, as well as Si and Ge, and the III–V compounds, a detailed understanding of the vacancy behavior is necessary in order to properly interpret impurity diffusion and self-diffusion.

ACKNOWLEDGMENTS

The authors would like to thank several colleagues for helpful discussions on the selection of topics to be covered and the interpretation of various results. C. F. Gibbon provided a useful perspective on diffusion in Si. M. B. Panish, C. D. Thurmond, and A. S. Jordan contributed many suggestions for the treatment of the III–V and II–VI compounds.

REFERENCES

1. J. Crank, *Mathematics of Diffusion* (Oxford University Press, London, 1956).
2. P. G. Shewmon, *Diffusion in Solids* (McGraw-Hill, New York, 1963).
3. J. R. Manning, *Diffusion Kinetics for Atoms in Crystals* (Van Nostrand, 1968).

4. B. I. Boltaks, *Diffusion in Semiconductors* (Academic Press, New York, 1963).
5. B. L. Sharma, *Diffusion in Semiconductors* (Trans. Tech. Publications, Rocky River, Ohio, 1970).
6. D. Shaw, ed., *Atomic Diffusion in Semiconductors* (Plenum, London, 1973).
7. W. Rice, *Proc. IEEE* **52**, 284–295 (1964).
8. A. S. Grove, A. Roder, and C. T. Sah, *J. Appl. Phys.* **36**, 802–810 (1965).
9. S. M. Hu, *J. Appl. Phys.* **39**, 3844–3849 (1968).
10. W. R. Runyan, *Silicon Semiconductor Technology* (McGraw-Hill, New York, 1965), p. 127.
11. T. I. Kucher, *Fiz. Tverd. Tela* **3**, 547–552 (1961) [English transl. *Soviet Phys.—Solid State* **3**, 401–404 (1961)].
12. R. L. Batdorf and F. M. Smits, *J. Appl. Phys.* **30**, 259–264 (1959).
13. L. Boltzmann, *Ann. Physik* **53**, 948–964 (1894).
14. C. Matano, *Japan J. Phys. (Trans.)* **8**, 109–113 (1933).
15. R. L. Barns, *Mater. Res. Bull.* **2**, 273–282 (1967).
16. A. Smakula and V. Sils, *Phys. Rev.* **99**, 1744–1746 (1955).
17. R. Hultgren, R. L. Orr, P. D. Anderson, and K. K. Kelly, in *Selected Values of Thermodynamic Properties of Metals and Alloys* (Wiley, New York, 1963).
18. T. P. McLean, *Prog. Semiconductors* **5**, 53–102 (1960).
19. A. S. Cooper, *Acta Cryst.* **15**, 578–582 (1962).
20. W. Kaiser and W. L. Bond, *Phys. Rev.* **115**, 857–863 (1959).
21. V. A. Bochko and Yu. L. Orlov, *Doklady Akad. Nauk SSSR* **191**, 341–344 (1970) [English transl., *Soviet Phys.—Doklady* **15**, 204–207 (1970)].
22. F. P. Bundy, *J. Chem. Phys.* **38**, 631–643 (1963).
23. C. D. Clark, P. J. Dean, and P. V. Harris, *Proc. Roy. Soc. (Lond.)* A **277**, 312–329 (1964).
24. A. Taylor and R. M. Jones, in *Conf. on SiC, Boston, 1959*, Ed. by J. R. O'Connor and J. Smiltens (Pergamon, New York, 1960), pp. 147–154.
25. A. H. G. de Mesquita, *Acta Cryst.* **23**, 610–617 (1967).
26. R. I. Scace and G. A. Slack, in *Conf. on SiC, Boston, 1959*, Ed. by J. R. O'Connor and J. Smiltens (Pergamon, New York, 1960), pp. 24–30.
27. W. J. Choyke, in *Int. Conf. Silicon Carbide*, 1968, Ed. by H. K. Henisch and R. Roy (Pergamon, New York, 1969), pp. 5141–5152.
28. M. Neuberger, in *Handbook of Electronic Materials*, Vol. 5, *Group IV Semiconducting Materials* (IFI/Plenum, New York, 1971).
29. A. Seeger and K. P. Chik, *Phys. Stat. Sol.* **29**, 455–542 (1968).
30. D. L. Kendall and D. B. De Vries, in *Semiconductor Silicon*, Ed. by R. R. Haberecht and E. L. Kern (The Electrochem. Soc., New York, 1969), pp. 358–421.
31. S. M. Hu, in *Atomic Diffusion in Semiconductors*, Ed. by D. Shaw (Plenum, New York, 1973).
32. H. Letaw, W. M. Portnoy, and L. Slifkin, *Phys. Rev.* **102**, 636–639 (1956).
33. B. J. Masters and J. M. Fairfield, *Appl. Phys. Lett.* **8**, 280–281 (1966).
34. J. M. Fairfield and B. J. Masters, *J. Appl. Phys.* **38**, 3148–3154 (1967).
35. R. F. Peart, *Phys. Stat. Sol.* **15**, K119–K122 (1966).
36. R. N. Ghoshtagore, *Phys. Rev. Lett.* **16**, 890–892 (1966).
37. R. A. Swalin, *J. Phys. Chem. Solids* **18**, 290–296 (1961).
38. R. N. Ghoshtagore, *Phys. Stat. Sol.* **20**, K89–K94 (1967).
39. G. D. Watkins and J. W. Corbett, *Phys. Rev.* **138**, A543–A560 (1965).

40. G. D. Watkins, in *Radiation Damage in Semiconductors* (1964 Conf. Proceedings) (Dunod, Paris, 1965), pp. 97–113.
41. L. Elstner and W. Kamprath, *Phys. Stat. Sol.* **22**, 541–547 (1967).
42. R. R. Haisguti, in *Lattice Defects in Semiconductors*, Ed. by R. R. Haisguti (Univ. of Tokyo Press, Tokyo, and Pennsylvania State Univ. Press, Univ. Park, 1968), pp. 131–137.
43. B. J. Masters, *Solid State Comm.* **9**, 283–286 (1971).
44. J. A. Van Vechten, *Phys. Rev.* **B 10**, 1482–1506 (1974).
45. A. Hiraki, *J. Phys. Soc. Japan* **21**, 34–41 (1966).
46. W. Meer and D. Pommerrenig, *Z. Angew. Phys.* **23**, 369–372 (1967).
47. W. C. Dunlap, Jr., *Phys. Rev.* **94**, 1531–1540 (1954).
48. P. V. Pavlov, *Fiz. Tverd. Tela* **8**, 2977–2981 (1966) [English transl., *Soviet Phys.—Solid State* **8**, 2377–2380 (1967)].
49. N. Isawa, *Japan J. Appl. Phys.* **7**, 81–82 (1968).
50. M. Okamura, *Japan J. Appl. Phys.* **8**, 1440–1448 (1969).
51. C. S. Fuller and J. A. Ditzenberger, *J. Appl. Phys.* **27**, 544–553 (1956).
52. B. J. Masters and J. M. Fairfield, *J. Appl. Phys.* **40**, 2390–2394 (1969).
53. J. J. Rohan, N. E. Pickering, and J. Kennedy, *J. Electrochem. Soc.* **106**, 705–709 (1959).
54. R. N. Ghostagore, *Phys. Rev. Lett.* **25**, 856–858 (1970).
55. E. Tannenbaum, *Solid-State Electronics* **2**, 123–132 (1961).
56. H. J. Queisser, *J. Appl. Phys.* **32**, 1776–1780 (1961).
57. R. J. Jaccodine, *J. Appl. Phys.* **39**, 3105–3108 (1968).
58. P. F. Schmidt and R. Stickler, *J. Electrochem. Soc.* **111**, 1188–1189 (1964).
59. R. B. Fair and G. R. Weber, *J. Appl. Phys.* **44**, 273–279 (1973).
60. J. C. Irvin, *Bell Syst. Tech. J.* **41**, 387–410 (1962).
61. D. P. Kennedy and P. C. Murley, *Proc. IEEE* **59**, 335–336 (1971).
62. R. O. Schwenker, E. S. Pan, and R. F. Lever, *J. Appl. Phys.* **42**, 3195–3200 (1971).
63. V. I. Fistul, in *Heavily Doped Semiconductors* (Plenum Press, New York, 1969), pp. 245–280.
64. C. S. Fuller and J. C. Severiens, *Phys. Rev.* **96**, 21–24 (1954).
65. A. Van Wieringen and N. Warmoltz, *Physica* **22**, 849–865 (1956).
66. L. Svob, *Solid-State Electronics* **10**, 991–996 (1967).
67. R. N. Hall and J. H. Racette, *J. Appl. Phys.* **35**, 379–397 (1964).
68. L. Y. Wei, *J. Phys. Chem. Solids* **18**, 162–174 (1961).
69. W. R. Wilcox and T. J. Chappelle, *J. Appl. Phys.* **35**, 240–246 (1964).
70. J. Martin, E. Haas, and E. Raithel, *Solid-State Electronics* **9**, 83–85 (1966).
71. V. E. Kosenko, *Fiz. Tverd. Tela* **3**, 2102–2104 (1961) [English transl., *Soviet Phys.—Solid State* **3**, 1526–1528 (1962)].
72. R. H. Wentorf, Jr., *J. Chem. Phys.* **26**, 956 (1957).
73. I. E. Campbell, C. F. Powell, D. H. Nowicki, and B. W. Gonser, *J. Electrochem. Soc.* **96**, 318–333 (1949).
74. V. V. Sobolev, *Fiz. Tverd. Tela* **6**, 3124–3130 (1964) [English transl., *Soviet Phys.—Solid State* **6**, 2488–2493 (1965)].
75. J. C. Phillips, *Rev. Mod. Phys.* **42**, 317–356 (1970).
76. J. A. Perri, S. La Placa, and B. Post, *Acta Cryst.* **11**, 310 (1958).
77. F. V. Williams, in *Compound Semiconductors: Preparation of III–V Compounds*,

Vol. 1, Ed. by R. K. Willardson and H. L. Goering (Reinhold, New York, 1962), p. 171.

78. F. V. Williams and R. A. Ruehrwein, *J. Am. Chem. Soc.* **82**, 1330–1332 (1960).
79. R. J. Archer, R. Y. Koyama, E. E. Loebner, and R. C. Lucas, *Phys. Rev. Lett.* **12**, 538–540 (1964).
80. S. M. Ku, *J. Electrochem. Soc.* **113**, 813–816 (1966).
81. G. A. Jeffrey and G. S. Parry, *J. Chem. Phys.* **23**, 406 (1955).
82. T. Renner, *Z. Anorg. Allgem. Chem.* **298**, 22–33 (1959).
83. G. A. Cox, D. O. Cummins, K. Kawabe, and R. H. Tredgold, *J. Phys. Chem. Solids* **28**, 543–548 (1967).
84. A. Addamiano, *J. Am. Chem. Soc.* **82**, 1537–1540 (1960).
85. W. Koschio, *J. Inorg. Nucl. Chem.* **27**, 750–751 (1965).
86. M. R. Lorenz, R. J. Chicotka, G. D. Pettit, and P. J. Dean, *Solid State Comm.* **8**, 693–697 (1970).
87. W. M. Yim, *J. Appl. Phys.* **42**, 2854–2856 (1971).
88. W. Kischio, *Z. Anorg. Allgem. Chem.* **328**, 187–193 (1964).
89. G. Giesecke and H. Pfister, *Acta Cryst.* **11**, 369–371 (1958).
90. H. Weker, *Z. Naturforsch.* **8a**, 248–251 (1953).
91. C. A. Mead and W. G. Spitzer, *Phys. Rev. Lett.* **11**, 358–360 (1963).
92. R. F. Blunt, H. P. R. Frederikse, J. H. Becker, and W. R. Hosler, *Phys. Rev.* **96**, 578–580 (1954).
93. H. P. Maruska and J. J. Tietjen, *Appl. Phys. Lett.* **15**, 327–329 (1969).
94. R. Dingle, D. D. Sell, S. E. Stokowski, and M. Ilegems, *Phys. Rev. B* **4**, 1211–1218 (1971).
95. R. L. Barns, private communication.
96. C. D. Thurmond, *J. Phys. Chem. Solids* **26**, 785–802 (1965).
97. S. F. Nygren, C. M. Ringel, and H. W. Verleur, *J. Electrochem. Soc.* **118**, 306–312 (1971).
98. M. R. Lorenz, G. D. Pettit, and R. C. Taylor, *Phys. Rev.* **171**, 876–881 (1968).
99. M. E. Straumanis and C. D. Kim, *Acta Cryst.* **19**, 256–259 (1965).
100. J. R. Arthur, *J. Phys. Chem. Solids* **28**, 2257–2267 (1967).
101. D. D. Sell, H. C. Casey, Jr., and K. W. Wecht, *J. Appl. Phys.* **45**, 2650–2657 (1974).
102. M. E. Straumanis and C. D. Kim, *J. Appl. Phys.* **36**, 3822–3825 (1965).
103. N. N. Sirota, in *Semiconductors and Semimetals*, Vol. 4, *Physics of III–V Compounds*, Ed. by R. K. Willardson and A. C. Beer (Academic, New York, 1968), p. 61.
104. E. J. Johnson, I. Filinski, and H. Y. Fan, in *Proc. Int. Conf. on Physics of Semiconductors, Exeter, 1962* (Inst. of Phys. and Phys. Soc., London, 1962), p. 375.
105. M. B. Panish and J. R. Arthur, *J. Chem. Thermo.* **2**, 299–318 (1970).
106. W. J. Turner and W. E. Reese, in *Radiative Recombination in Semiconductors, Paris, 1964* (Dunod, Paris, 1965), pp. 59–64.
107. J. van den Boomgaard and K. Schol, *Philips Res. Repts.* **12**, 127–140 (1957).
108. J. R. Dixon and J. M. Ellis, *Phys. Rev.* **123**, 1560–1566 (1961).
109. T. S. Liu and C. A. Perretti, *Trans. Am. Soc. Metals* **44**, 539–548 (1952).
110. S. Zwerdling, B. Lax, and L. M. Roth, *Phys. Rev.* **108**, 1402–1408 (1957).
111. M. Neuberger, in *Handbook of Electronic Materials*, Vol. 2, *III–V Semiconducting Compounds* (IFI/Plenum, New York, 1971).
112. D. L. Kendall, in *Semiconductors and Semimetals*, Vol. 4, *Physics of III–V Com-

pounds, Ed. by R. K. Willardson and A. C. Beer (Academic, New York, 1968), pp. 163–259.

113. H. C. Casey, Jr., in *Atomic Diffusion in Semiconductors*, Ed. by D. Shaw (Plenum, New York, 1973).
114. J. W. Allen and G. L. Pearson, NASA Cr-438, Stanford University, Stanford, California, April 1966.
115. H. C. Casey, Jr. and M. B. Panish, *Trans. Met. Soc. AIME* 242, 406–412 (1968).
116. M. B. Panish, *J. Electrochem. Soc.* 113, 861 (1966).
117. J. E. Ricci, in *The Phase Rule and Heterogeneous Equilibrium* (van Nostrand, Princeton, New Jersey, 1951), p. 1.
118. L. L. Chang and G. L. Pearson, *J. Phys. Chem. Solid* 25, 23–30 (1964).
119. K. K. Shih, J. W. Allen, and G. L. Pearson, *J. Phys. Chem. Solids* 29, 379–386 (1968).
120. M. B. Panish and H. C. Casey, Jr., *J. Phys. Chem. Solids* 28, 1673–1684 (1967).
121. F. A. Cunnell and C. H. Gooch, *J. Phys. Chem. Solids* 15, 127–133 (1960).
122. D. L. Kendall and M. E. Jones, AIEE–IRE Device Research Conference, Stanford (1961).
123. R. L. Longini, *Solid-State Electronics* 5, 127–130 (1962).
124. L. R. Weisberg and J. Blanc, *Phys. Rev.* 131, 1458–1552 (1963).
125. H. C. Casey, Jr., M. B. Panish, and L. L. Chang, *Phys. Rev.* 162, 660–668 (1967).
126. G. N. Lewis and M. Randall, in *Thermodynamics* (McGraw-Hill, New York, 1961), p. 250.
127. R. A. Swalin, in *Thermodynamics of Solids* (Wiley, New York, 1962), p. 253.
128. J. S. Blakemore, in *Semiconductor Statistics* (Pergamon, New York, 1962), p. 80.
129. D. Shaw and A. L. J. Wells, *Br. J. Appl. Phys.* 17, 999–1004 (1966).
130. C. Van Opdorp, *J. Appl. Phys.* 38, 5411–5412 (1967).
131. B. Goldstein, *Phys. Rev.* 118, 1024–1027 (1960).
132. F. Ermanis and K. B. Wolfstirn, *J. Appl. Phys.* 37, 1963–1966 (1966).
133. L. L. Chang and G. L. Pearson, *J. Appl. Phys.* 35, 1960–1965 (1964).
134. D. L. Kendall, unpublished.
135. J. R. Brews and C. J. Hwang, *J. Chem. Phys.* 8, 3263–3268 (1971).
136. C. J. Hwang, and J. R. Brews, *J. Phys. Chem. Solids* 32, 837–845 (1971).
137. C. H. Ting and G. L. Pearson, *J. Appl. Phys.* 42, 2247–2251 (1971).
138. K. K. Shih, J. W. Allen, and G. L. Pearson, *J. Phys. Chem. Solids* 29, 367–377 (1968).
139. A. S. Jordan, *Met. Trans.* 2, 1965–1970 (1971).
140. J. R. Arthur, *J. Phys. Chem. Solids* 28, 2257–2267 (1967).
141. C. H. Ting and G. L. Pearson, *J. Electrochem. Soc.* 118, 1454–1458 (1971).
142. W. D. Rhines and D. A. Stevenson, *J. Electron. Mat.* 2, 341–358 (1973).
143. D. Shaw and S. R. Showan, *Phys. Stat. Sol.* 32, 109–118 (1969).
144. B. Ya. Pines and E. F. Chaikovskii, *Fiz. Tverd. Tela* 1, 946–952 (1959) [English transl., *Soviet Phys.—Solid State* 1, 864–869 (1959)].
145. S. R. Showan and D. Shaw, *Phys. Stat. Sol.* 32, 97–108 (1969).
146. D. Shaw and S. R. Showan, *Phys. Stat. Sol.* 34, 475–482 (1969).
147. R. H. Wieber, H. C. Gorton, and C. S. Peet, *J. Appl. Phys.* 31, 608 (1960).
148. B. Goldstein, *Phys. Rev.* 121, 1305–1311 (1961).
149. K. B. Wolfstirn, unpublished.
150. E. A. Poltoratskii and V. M. Stuchebnikov, *Fiz. Tverd. Tela* 8, 963–965 (1966) [English transl., *Soviet Phys.—Solid State* 8, 770–771 (1966)].

151. S. R. Showan and D. Shaw, *Phys. Stat. Sol.* **35**, K97–K99 (1969).

152. J. A. Kanza, unpublished.

153. D. L. Kendall, *Appl. Phys. Lett.* **4**, 67–68 (1964).

154. C. S. Fuller and K. B. Wolfstirn, *J. Appl. Phys.* **33**, 2507–2514 (1962).

155. R. G. Moore, Jr., M. Belasco, and H. Strack, *Bull. Am. Phys. Soc.* **10**, 731 (1965).

156. M. S. Seltzer, *J. Phys. Chem. Solids* **26**, 243–250 (1965).

157. J. Rachmann and R. Biermann, *Solid State Comm.* **7**, 1771–1775 (1969).

158. A. B. Y. Young and G. L. Pearson, *J. Phys. Chem. Solids* **31**, 517–527 (1970).

159. G. R. Antell, *Solid-State Electronics* **6**, 383–387 (1963).

160. R. W. Fane and A. J. Goss, *Solid-State Electronics* **6**, 383–387 (1963).

161. J. F. Osborne, K. G. Heinen, and H. Riser, unpublished.

162. H. C. Casey, Jr., and G. L. Pearson, *J. Appl. Phys.* **35**, 3401–3407 (1964).

163. L. L. Chang and G. L. Pearson, *J. Appl. Phys.* **35**, 374–378 (1964).

164. M. Ilegems and W. C. O'Mara, *J. Appl. Phys.* **43**, 1190–1197 (1972).

165. F. H. Eisen and C. E. Birchenall, *Acta Met.* **5**, 265–274 (1957).

166. J. Bougnot, L. Szepessy, and S. F. DaCunka, *Phys. Stat. Sol.* **26**, K127–K129 (1968).

167. B. I. Boltaks and Ya. A. Gutorov, *Fiz. Tverd. Tela* **1**, 1015–1021 (1960) [English transl., *Soviet Phys.—Solid State* **1**, 930–935 (1960)].

168. H. Kato, M. Yokozawa, R. Kohara, Y. Okabayashi, and S. Takayanagi, *Solid-State Electronics* **12**, 137–139 (1969).

169. B. I. Boltaks, S. I. Remberza, and B. L. Sharma, *Fiz. Tekh. Poluprov.* **1**, 247–254 (1967) [English transl., *Soviet Phys.—Semicond.* **1**, 196–203 (1967)].

170. S. I. Rembeza, *Fiz. Tekh. Poluprov.* **1**, 615–617 (1967) [English transl., *Soviet Phys.—Semicond.* **1**, 516–518 (1967)].

171. K. A. Arseni, B. I. Boltaks, and S. I. Rembeza, *Fiz. Tverd. Tela* **8**, 2809–2811 (1966) [English transl., *Soviet Phys.—Solid State* **8**, 2248–2249 (1967)].

172. C. S. Fuller and K. B. Wolfstirn, *J. Electrochem. Soc.* **114**, 856–861 (1967).

173. B. L. Sharma, R. K. Purohit, and S. N. Makerjee, *J. Phys. Chem. Solids* **32**, 1397–1399 (1971).

174. E. Schillman, in *Compound Semiconductors: Preparation of III–V Compounds*, Vol. 1, Ed. by R. K. Willardson and H. L. Goering (Reinhold, New York, 1962), p. 358.

175. M. G. Buehler and G. L. Pearson, unpublished.

176. K. A. Arseni and B. I. Boltaks, *Fiz. Tverd. Tela* **10**, 2783–2789 (1968) [English transl., *Soviet Phys.—Solid State* **10**, 2190–2196 (1969)].

177. S. I. Rembeza, *Fiz. Tekh. Poluprov.* **3**, 612–613 (1969) [English transl., *Soviet Phys.—Semicond.* **3**, 519–520 (1969)].

178. K. A. Arseni, B. I. Boltaks, V. L. Gordin, and Ya. A. Ugai, *Izv. Akad. Nauk SSSR, Neorg. Mater.* **3**, 1679–1681 (1967) [English transl., *Inorg. Mat.* **3**, 1465–1467 (1967)].

179. K. A. Arseni, *Fiz. Tverd. Tela* **10**, 2864–2866 (1968) [English transl., *Soviet Phys.—Solid State* **10**, 2263–2265 (1969)].

180. L. L. Chang and H. C. Casey, Jr., *Solid-State Electronics* **7**, 481–485 (1964).

181. D. L. Kendall and R. A. Huggins, *J. Appl. Phys.* **40**, 2750–2759 (1969).

182. L. A. K. Watt and W. S. Chen, *Bull. Am. Phys. Soc.* **7**, 89 (1962).

183. B. I. Boltaks and V. I. Sokolov, *Fiz. Tverd. Tela* **6**, 771–775 (1964) [English transl., *Soviet Phys.—Solid State* **6**, 600–603 (1964)].

184. R. B. Wilson and E. L. Heasell, *Proc. Phys. Soc. (Lond.)* **79**, 403–408 (1962).

185. H. J. Stocker, *Phys. Rev.* **130**, 2160–2169 (1963).

186. I. A. Gusev and A. N. Murin, *Fiz. Tverd. Tela* **6**, 1563 (1964) [English transl., *Soviet Phys.—Solid State* **6**, 1229 (1964)].

187. G. I. Rekalova, U. Kebe, and L. A. Megrina, *Fiz. Tekh. Poluprov.* **5**, 776–778 (1971) [English transl., *Soviet Phys.—Semicond.* **5**, 685–686 (1971)].

188. T. Takabatake, H. Ikari, and Y. Uyeda, *Japan J. Appl. Phys.* **5**, 839–840 (1966).

189. G. I. Rekalova, A. A. Shakov, and V. V. Gaurushko, *Fiz. Tekh. Poluprov.* **2**, 1744–1747 (1969) [English transl., *Soviet Phys.—Semicond.* **2**, 1452–1455 (1969)].

190. S. M. Sze and L. Y. Wei, *Phys. Rev.* **124**, 84–89 (1961).

191. B. I. Boltaks and G. S. Kulikov, *Zh. Tekh. Fiz.* **27**, 82–84 (1957) [English transl., *Soviet Phys.—Tech. Phys.* **2**, 67–68 (1957)].

192. A. G. Fischer and R. J. Poff, *J. Phys. Chem. Solids* **23**, 1479–1480 (1962).

193. M. R. Lorenz, in *Physics and Chemistry of II–VI Compounds*, Ed. by M. Aven and J. S. Prenner (North-Holland, Amsterdam, 1967), pp. 75–115.

194. M. Balkanski and R. D. Waldron, *Phys. Rev.* **112**, 123–135 (1958).

195. L. R. Shiozawa *et al.*, Final Report on Contract AF 33 (616-6865), Aerospace Research Laboratories, U.S.A.F. (June 1962), p. 139.

196. M. Grynberg, in *Proc. 7th Int. Conf. on Physics of Semiconductors*, Ed. by M. Hulin (Dunod, Paris, 1964), pp. 135–141.

197. A. D. Stuckes and G. Farrell, *J. Phys. Chem. Solids* **25**, 477–482 (1964).

198. S. Yamada, *J. Phys. Soc. Japan* **17**, 645–653 (1962).

199. B. J. Skinner and P. B. Barton, Jr., *Am. Mineralogist* **45**, 612–625 (1960).

200. W. W. Piper, *Phys. Rev.* **92**, 23–27 (1953).

201. D. T. F. Marple, G. Hite, and M. Aven, unpublished (1965).

202. H. E. Swanson, M. C. Morris, E. H. Evans, and L. Ulmer, Nat. Bur. Stand. Monograph 25 (September 1964).

203. S. Larach, R. E. Shrader, and C. F. Stocker, *Phys. Rev.* **108**, 587–589 (1957).

204. H. H. Woodbury, in *Physics and Chemistry of II–VI Compounds*, Ed. by M. Aven and J. S. Prenner (North-Holland, Amsterdam, 1967), pp. 223–264.

205. H. H. Woodbury, in *II–VI Semiconducting Compounds, 1967 Int. Conf.*, Ed. by D. G. Thomas (Benjamin, New York, 1967), pp. 244–276.

206. D. A. Stevenson, in *Atomic Diffusion in Semiconductors*, Ed. by D. Shaw (Plenum, New York, 1973).

207. A. S. Jordan, *Met. Trans.* **1**, 239–249 (1970).

208. A. S. Jordan and R. R. Zupp, *J. Electro. Chem. Soc.* **116**, 1264–1268 (1969).

209. B. M. Kulwicki, Ph.D. Thesis, University of Michigan, School of Engineering (July, 1963).

210. J. Carides and A. G. Fischer, *Solid State Comm.* **2**, 217–218 (1964).

211. P. Goldfinger and M. Jeunehomme, *Trans. Faraday Soc.* **59**, 2851–2867 (1963).

212. D. G. Thomas and E. A. Sadowski, *J. Phys. Chem. Solids* **25**, 395–400 (1964).

213. F. T. J. Smith, *J. Phys. Chem. Solids* **32**, 2201–2209 (1971).

214. D. Shaw and R. C. Whelan, *Phys. Stat. Sol.* **36**, 705–716 (1969).

215. G. H. Hershman and F. A. Kröger, *J. Solid State Chem.* **2**, 483–490 (1970).

216. V. Kumar and F. A. Kröger, *J. Solid State Chem.* **3**, 387–400 (1971).

217. F. T. J. Smith, *Solid State Comm.* **8**, 263–266 (1970).

218. W. D. Callister, Jr., C. F. Varotto, and D. A. Stevenson, *Phys. Stat. Sol.* **38**, K45–K50 (1970).

219. R. C. Whelan and D. Shaw, *Phys. Stat. Sol.* **29**, 145–152 (1968).

220. F. T. J. Smith, *Met. Trans.* **1**, 617–621 (1970).

221. E. F. Apple and J. S. Prener, *J. Phys. Chem. Solids* **13**, 81–87 (1960).
222. F. T. J. Smith, *Solid State Comm.* **7**, 1757–1761 (1969).
223. H. H. Woodbury, *Phys. Rev.* **134**, A492–A498 (1964).
224. L. R. Shiozawa and J. M. Jost, Aerospace Res. Lab. Rep. 69-0107, Office of Aerospace Research, United States Air Force, Wright-Patterson Air Force Base, Ohio (July 1969).
225. G. Brouwer, *Philips Res. Rep.* **9**, 366–376 (1954).
226. G. H. Hershman, V. P. Zlomanov, and F. A. Kröger, *J. Solid State Chem.* **3**, 401–405 (1971).
227. S. S. Devlin, in *Physics and Chemistry of II–VI Compounds* (North-Holland, Amsterdam, 1967), p. 603.
228. P. M. Borsenberger, D. A. Stevenson, and R. A. Burmeister, in *II–VI Semiconducting Compounds*, 1967 *Int. Conf.*, Ed. by D. G. Thomas (Benjamin, New York, 1967), pp. 439–461.
229. R. A. Reynolds and D. A. Stevenson, *J. Phys. Chem. Solids* **30**, 139–147 (1969).
230. P. M. Borsenberger and D. A. Stevenson, *J. Phys. Chem. Solids* **29**, 1277–1286 (1968).
231. H. H. Woodbury, *Phys. Rev.* **134**, A492–A498 (1964).
232. H. H. Woodbury and R. B. Hall, *Phys. Rev.* **157**, 641–655 (1967).
233. T. Soma and A. Morita, *J. Phys. Soc. Japan* **32**, 357–364 (1972).

Chapter 3

EFFECTS OF POINT DEFECTS
ON ELECTRICAL AND OPTICAL PROPERTIES
OF SEMICONDUCTORS

O. L. Curtis, Jr.

Northrop Corporate Laboratories
Hawthorne, California

1. INTRODUCTION

This chapter relates ways in which the characteristics of point defects in semiconductors may affect the electrical and optical properties of semi-conducting material. Convenience and other practical considerations have given impetus to radiation effect studies of semiconductors to the extent that most of the available information about point defects in semiconductors is acquired from these studies. This emphasis is apparent in the material discussed. Silicon and germanium will be stressed because they have been studied in far more detail than other semiconductors.

It is worthwhile to consider how restrictively we will define the term "point defects." Certainly the term should not be confined simply to vacancies and interstitials. These constitute a tiny fraction of the inventory of defects that have been studied, and furthermore, such defects are not stable at room temperature, where a major part of our interest lies. Until late in the 1950's it was customary to explain observations of radiation effects in semiconductors on the basis of simple Frenkel defects. Energy levels observed in the forbidden gap were compared with calculations of defect levels expected for vacancies and interstitials, as proposed by James and

Lark-Horovitz.[1] The point of turning away from this philosophy occurred in about 1959 when Brown, Augustyniak, and Waite[2] demonstrated the sensitivity of defect annealing to dopant impurities; while Watkins, Corbett, and Walker,[3] together with Bemski,[4] showed the effects of oxygen[3,4] and phosphorus[3] on electron spin resonance in silicon. The sensitivity of radiation behavior to dopant impurity concentrations of $\sim 10^{13}$ cm^{-3} (~ 20 parts per billion) was particularly surprising. As a result of these and related discoveries, investigators have since recognized the existence of many different defect–impurity complexes.

We include, then, such entities as vacancy–impurity complexes in our definition of point defects and will consider the properties of those that are more commonly observed. There is one very important area of radiation effect studies that, for the most part, will not be included in our discussions. This area includes defects in which the primary displaced atom acquires so much energy from the irradiating particle that multiple defects are produced. The prime example of such a situation is energetic neutron bombardment. In this case, the nature of semiconductors leads to perturbations in potential around the regions of disorder which have their own special affects on electrical and optical properties.

The qualitative nature of these disordered regions was first recognized by Gossick and Crawford,[5,6] and has been substantiated and made more quantitative by recent experiments. These disordered regions are not generally affected by lattice impurities[7,8] and have rather unique effects on electrical properties, particularly carrier recombination.[9] Such regions of disorder will not be considered further, since they apparently do not qualify for the term "point defects."

Since very small amounts of impurity can affect the response of semiconductor material to irradiation, an indefinite variety of defects can occur. It is often possible only to conjecture about the microscopic nature of these defects, but electron spin resonance and infrared absorption studies have helped identify a few of them. Although these techniques have only been useful for a portion of the types of existing defects, many investigators tend to explain all their experimental results in terms of these identified defects, probably because of the satisfaction of being able to express their results in terms of a specific model. Thus, caution should be exercised in accepting the attributes ascribed to specific microscopic configurations.

Effects on optical properties can be observed in convenient wavelength regions, but usually require large amounts of damage, while electrical properties are usually affected by small quantities of defects which introduce levels in the forbidden gap. For example, recombination of minority carriers in germanium and silicon can be substantially affected by 10^{10} defects/cm^3. Such sensitivity of a variety of properties to lattice defects is a mixed blessing. The reconciliation of data obtained from a wide variety of measurements over many decades in defect density (because even some electrical measurements are so much more sensitive than others) at times seems frustrating. We will try to assemble as many pieces of the puzzle as possible, but the reader will find that there are many areas that require considerable study before the nature of point defects in semiconductors is satisfactorily understood.

Since it is difficult to consider such a broad subject in a limited space, we will limit ourselves to a brief treatment of six major topics. Electrical properties considered will include carrier concentration, mobility, and carrier lifetime. Optical properties discussed will be optical absorption, photoconductivity, and luminescence. We will not examine device effects, even though semiconductor devices have informative responses to defect introduction and investigations that are difficult or impossible with bulk material can often be performed with devices.

To conserve space and make the material readily digestible, we have tried to explain important qualitative features of the analysis in descriptive terms, rather than in complex mathematical formulas. In each section we have tried to provide as many experimental examples as possible, yet there are so many studies from which to draw data that many have been neglected. For example, although several studies of annealing behavior of various properties are mentioned, we do not begin to summarize completely all the work in this area.

An overview such as this will, of course, reflect the author's personal interests. The reader will soon deduce what these are. But hopefully, a broad enough perspective will be provided to give the reader some insight into the subject and a foundation for further study.

2. CARRIER CONCENTRATION

Carrier concentration is normally determined using Hall effect measurements. In this experiment a magnetic field B is applied perpendicularly to an electric current of density J flowing through the sample. An electric field E is developed in the sample along the remaining axis. The Hall coefficient R is defined as E/JB, a quantity that can be easily determined experimentally. It is found that

$$R = \frac{p r_h \mu_h^2 - n r_e \mu_e^2}{e(p\mu_h + n\mu_e)^2} \qquad (1)$$

where p and n and μ_h and μ_p are hole and electron concentrations and mobilities, respectively, and r_h and r_e are statistical factors of the order of unity (usually between 1 and 2). The relation is greatly simplified for extrinsic material; for example, for $n \gg p$

$$R \approx -r_e/en \tag{2}$$

The factor r_e can be estimated theoretically, but is best determined experimentally from measurements on n- and p-type material with a similar concentration of scattering centers. The conductivity σ is measured in an n-type specimen and related to the carrier concentration through the expression

$$\sigma = \mu_{ed}en \tag{3}$$

Drift mobility measurements are performed on p-type material to determine μ_{ed}; then

$$r_e = -R\sigma/\mu_{ed} \tag{4}$$

Since carrier mobility tends to be less sensitive to lattice defects than carrier concentration, early studies of carrier concentration changes induced by lattice defects used conductivity measurements and disregarded the variation of mobility in (3). Among these were studies of the effects of alpha particles by Brattain and Pearson,[10] electrons by Brown, Fletcher, and Wright,[11] and quenched-in defects by Mayburg.[12] However, Hall effect studies, which enable observation of carrier concentration changes independently of mobility, were utilized early, particularly by the groups at Purdue[13] and Oak Ridge National Laboratory.[14] More recent investigations of carrier concentration behavior have relied primarily on this technique.

Measurement of carrier concentration as a function of temperature is probably the most direct way to determine the position of energy levels introduced by defects into the forbidden gap. The slope of a logarithmic plot of carrier concentration versus reciprocal temperature has been used for energy level determination, but such a technique is susceptible to large error, and it is better to deduce from such a curve the point at which the defect levels in question have a specific occupancy factor and determine the energy level position from the relationship

$$f = \{1 + \beta \exp[(E_l - \zeta)/kT]\}^{-1} \tag{5}$$

where f is the fraction of levels at E_l filled by electrons, β is a degeneracy factor,[15] and ζ is the position of the Fermi level. A major difficulty with this technique is the usual lack of knowledge of the factor β. In principle, it might appear that one could determine both β and energy level position by determining the occupancy factor f as a function of temperature. Sonder and Templeton[16] made such an attempt, but as they point out, the approach is invalid unless the energy level position is temperature invariant with respect to the band edge. Unfortunately, this is unlikely, and in fact, it appears in at least one instance[17] that the energy level position with respect to one band edge varies as rapidly as the product of the band-gap temperature dependence and the fraction of the energy gap between the level and the band edge. Because of the difficulty in separating degeneracy effects from temperature effects, the normal procedure is to calculate energy level position on the basis of a unity degeneracy factor. This caused an error of $kT \ln \beta$ in energy level position (± 0.018 eV for $\beta = 2$ or $1/2$ at room temperature). It turns out that this "effective" energy level position designation (assuming $\beta = 1$) is useful, since degeneracy affects in the same way measurements other than those of carrier concentration.

Another property of defects that can be deduced from carrier concentration studies is the charge state; i.e., whether the defect level is a donor or acceptor. When only one level is introduced, such a determination is made simply by observing the direction in which carrier concentration changes. However, since such a simple situation is not generally realized, it is usually necessary to make additional investigations. One technique that is useful for radiation effect studies is to observe the temperature dependence of carrier concentration for a given defect level population during anneal while noting the direction of change in carrier concentration. This eliminates the effect of defects other than the annealing ones, provided they do not both have the same temperature dependence of anneal.

2.1. Germanium

2.1.1. *Energy Levels*

Various investigators have used carrier concentration measurements to determine defect energy levels in germanium. Early measurements were performed at Purdue[18] on material irradiated with 10-MeV deuterons and 4.5-MeV electrons, while measurements on gamma-irradiated germanium were made at Oak Ridge.[19]

Low-energy processes dominate the interactions of charged particles with a solid, so the nature of the defects produced by 10-MeV deuterons and 4.5-MeV electrons should be reasonably similar to those produced by the electrons generated with ^{60}Co gamma rays through the Compton and photoelectronic processes. The 4.5-MeV electrons were reported[18] to create energy levels 0.02, 0.1, and 0.24 eV below the conduction band, and 0.01, 0.1, and 0.22 eV above the valence band. The presence of another level near the center of the gap was also inferred. Results for 10-MeV deuterons were almost identical. The Oak Ridge group observed[19] that irradiation by ^{60}Co gamma rays caused a decrease in electron concentration in n-type material of $\sim 1.4 \times 10^{-3}$ electron/photon, provided the Fermi level was well above the level ~ 0.2 eV below the conduction band. This level was also seen by Vitovskii and co-workers,[20,21] who additionally observed levels agreeing with the Purdue data at 0.02, 0.11, and 0.26 eV above the valence band. These studies and additional investigations including annealing studies by Ishino et al.[22] and Pigg and Crawford[23] have established the level 0.23 eV below the conduction band to be an acceptor, with the presence of additional deeper acceptors and a shallow donor near the valence band. While the $E_c - 0.23$ eV level always seems to be present, there is an indication that the lower-lying acceptors are sample dependent. Specifically, Cleland, Bass, and Crawford[24] observed that the dislocation density is very important, and, in fact, they conclude that "no low-lying acceptor-states are introduced in material of low dislocation density by Co^{60} photons."

2.1.2. Thermal Annealing

Annealing studies have been used extensively in the investigation of radiation-induced defects. The details of the annealing process are, of course, closely related to the microscopic nature of the defects that are annealing. Brown, Augustyniak, and Waite[2] provided the first strong indication of the importance of dopant impurities in the formation and annihilation of defects. Annealing "stages" are observed at many temperatures in germanium[25] for carrier concentration changes, as well as other properties. Unfortunately, the situation is so complicated that many different models, often inconsistent with each other, are used to explain the data. Space does not allow discussion of all the annealing studies that have been performed. Instead, we will emphasize some extensively studied, yet controversial, annealing processes.

There are prominent annealing stages observed in n-type germanium at ~ 35 and $\sim 65°$K. When germanium is irradiated at liquid helium temperature in the dark, and annealed in the dark, these are the first annealing stages observed, and thus should represent annihilation of the simplest types of defects. Naturally, these annealing data have no direct relationship to the energy levels discussed in the above section, which are based on defects that are stable at room temperature. Of course, the lower-temperature annealing processes may be responsible for the creation of more stable defects. The annealing stages at 35 and 65°K have been studied in detail by MacKay and Klontz and some of their students at Purdue University.[26–30] Figure 1 shows data of MacKay and Klontz[26] on heavily doped n-type germanium. High-carrier-concentration samples are required for such measurements because carrier freezeout will cause lightly doped samples to become insulating.

Observations of annealing kinetics, defect creation rates,[27] and stored energy released upon annealing[28] resulted in the 65°K annealing stage being interpreted as coupled vacancy–interstitial pair annihilation. However, complications in the data caused MacKay and Klontz[30] to change their earlier position and hypothesize a model in which the interstitial is trapped at dopant impurities and is then released by anneal to recombine with less mobile vacancies.

The 35°K anneal shows similarities to the 65°K anneal, and Klontz and MacKay[27] have suggested that it may be some type of modification (probably radiation-induced) of the 65°K defect. This viewpoint was extended by Penczer and DeAngelis.[31] On the other hand, Bourgoin and Mollet[32] correlated the degree of annealing occurring at 35°K with antimony

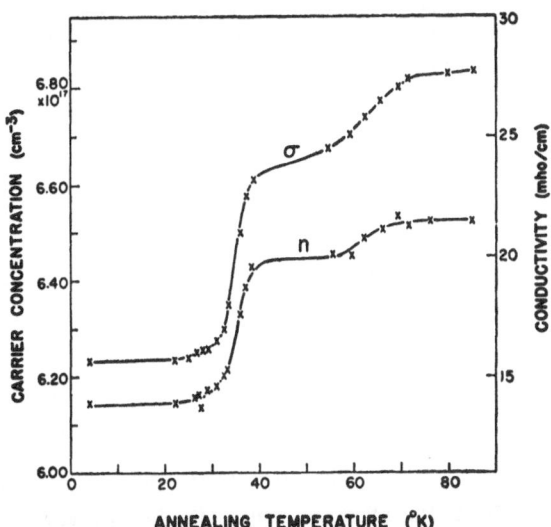

Fig. 1. Isochronal annealing of n-type Ge after 1-MeV electron irradiation. (After MacKay and Klontz.[26])

and arsenic concentrations and interpreted the results as corresponding to vacancy annihilation with an interstitial impurity. Qualitatively, such a dependence was observed much earlier by MacKay and Klontz. However, they seemed to suspect that the difference in electron concentration, rather than impurity concentration, was the important factor, while Burgoin and Mollet (apparently) showed that this was not the case.

Even though an impurity seems to be involved in the 35°K anneal, it may be incorrect to conclude that motion of a defect or impurity occurs, since the annealing temperature apparently does not shift with defect or impurity concentration. Perhaps vacancies coupled with impurity interstitials are left as a result of the nonequilibrium situation that exists during irradiation. This would be consistent with the conclusion[27,31] that the 35°K defects can be formed from 65°K defects by irradiation.

2.1.3. Radiation-Induced Annealing

An intriguing but controversial aspect of various annealing processes, particularly those just discussed for germanium at 35 and 65°K, is "radiation-induced annealing." Interest in this subject was particularly strong a few years ago.[29–37] It is illustrated in its simplest form by the data of Fig. 2, taken from MacKay and Klontz.[26] In experiments such as illustrated by

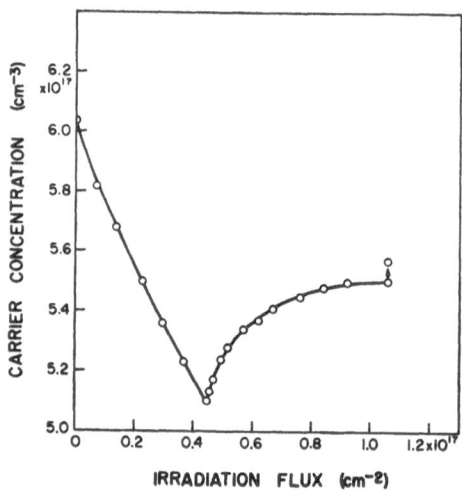

Fig. 2. Radiation annealing. (After MacKay and Klontz.[26]) Shown is removal of carriers by 1-MeV electrons, followed by recovery under 0.3-MeV irradiation for n-type Ge.

the figure, some semiconductor property such as conductivity, which has undergone a change due to irradiation, is observed to recover a substantial part of that change when irradiated at a temperature equal to or lower than the irradiation temperature or at least well below the temperature at which substantial annealing otherwise occurs. The recovery can also be induced by shining light on the sample. There have been objections* to the failure of considering adequately the effect of metastable charge states on the electrical properties being considered. We will attempt to place these objections in a more concrete framework and substantiate them with experimental observations, confining our discussion to experiments where conductivity is the observed quantity. The points considered will have relevance to experiments involving other parameters such as carrier lifetime, photoconductivity, and electron spin resonance, but the analysis in these cases is more subtle.

For illustration, let us consider conductivity changes that occur when traps, but no unstable defects, are present. Because of the wider band gap, deeper traps are possible in silicon than in germanium, and processes occur near room temperature in silicon which would occur only at low temperatures in germanium. Thus it is convenient to use silicon as an example. We employ for our example n-type, 10 Ω-cm, Czochralski-grown (high-oxygen content) silicon, having a "natural" (non-radiation-induced) trap concentration of $\sim 2 \times 10^{13}$ cm^{-3} (determined from trapping kinetics[39]). Since this represents only a small percentage of the majority carrier concentration (5×10^{14} cm^{-3}), conductivity changes due to trapping are substantially smaller than could be expected in a heavily irradiated sample. The source of carrier excitation was 150-kV X-rays filtered through ~ 3.5 mm of aluminum.

Figure 3 is a plot of fractional change in conductivity as a function of absorbed dose. The dose was administered 10 rad at a time at intervals of 1/2 min, and the temperature of measurement was the same as the temperature of irradiation. In two cases, saturation values of conductivity change are shown, measured after 450 rad. Figure 4 shows a similar plot; here the temperature was increased 10 deg between each irradiation. The points were taken approximately 3 min apart, the first 2 min being used to achieve temperature equilibrium, at which time the irradiations were performed. The conductivity was measured 1 min later.

The X-rays produced no defects, and the conductivity changes were caused entirely by trapping processes. While these measurements were

---

* See, for example, the discussion following Ref. 37.

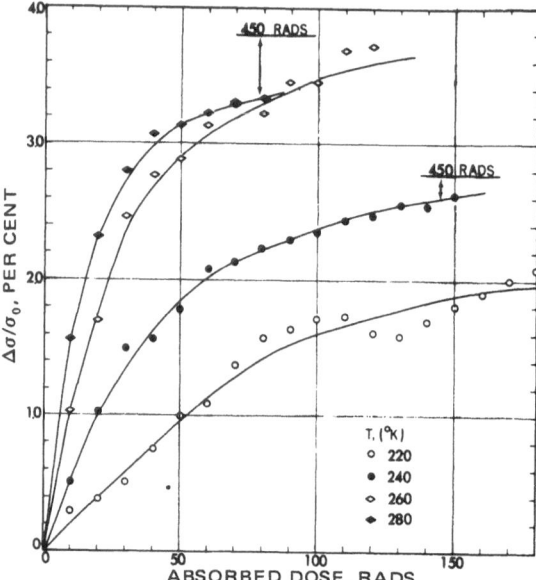

Fig. 3. Conductivity changes in silicon produced by X-ray irradiation at fixed temperatures.

performed on silicon for convenience in choosing a temperature range, similar data can be obtained for germanium, but at substantially lower temperatures. In these discussions we are not concerned with the specific nature of the traps, but merely use them to illustrate the type of behavior that can normally be expected under similar circumstances.

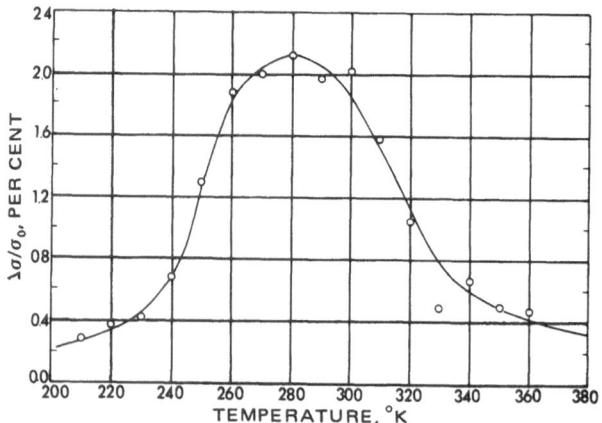

Fig. 4. Conductivity changes in silicon produced by irradiation at successively higher temperatures with 150-keV X-rays.

The behavior shown by this silicon sample, which has received no radiation damage, is fairly common, and it is well known that such effects do occur.[39] One surprising fact, however, is that although trapping is normally more effective at lower temperatures, for this particular example conductivity changes induced by trapping occur at a slower rate at lower temperatures. This is presumably due to enhanced recombination rates through competing centers. Thus it is not possible to discern the difference between trapping and annealing simply on the basis of the shape of an annealing curve where irradiation is held constant while the temperature is increased. The left-hand side of Fig. 4 is remarkably similar to many isochronal annealing curves.

Of course, at a high enough temperature, the traps begin to empty and equilibrium is restored. This is the process that is occurring in the right-hand side of the curve of Fig. 4. This reversible nature of traps is used as an argument against the existence of trapping in annealing experiments; i.e., if an observed process is not reversible, it is assumed not to be trapping. On the other hand, suppose that radiation-induced carrier removal centers are also minority carrier traps. In equilibrium at low temperatures, these centers remove majority carriers, but ionizing radiation causes minority carrier capture which restores the condition which existed before the defect was created. If such centers were unstable and annealed below the temperature at which untrapping occurred, they would never display release of trapped minority carriers and would then resemble the centers involved in "radiation-induced annealing," which are normally unstable at rather low temperatures. In view of the fact that traps may require fairly high temperatures to empty (Fig. 4), it would seem possible, even probable, that for radiation-induced defects, trap emptying would occur above the normal anneal temperature. If so, when trapping occurred, the carrier removed by the defect would be restored, and before it could be again removed by releasing the minority carrier, the defect would be annealed. This, of course, represents a nonreversible process.

It is important to realize that, in principle, it may be impossible to differentiate between trapping and radiation-induced annealing on the basis of conductivity measurements alone. The basic hypothesis of radiation-induced annealing is that the defect in question captures a minority carrier and thus becomes unstable. However, if the defect originally removed a majority carrier from its band, the instant it traps a minority carrier, the majority carrier is restored to its band and the effect of the defect (providing only one carrier is removed per defect) is nullified. An experiment based on conductivity measurements will register that defect as having annealed (except for the residual effects of a neutral center on mobility) at the instant trapping occurs, and conductivity will be insensitive to any subsequent migration and annihilation of that defect so long as it remains in the same charge state. Thus, even if it is true that trapping a minority carrier makes a defect more susceptible to anneal, it may not be possible to separate the kinetics of this annealing process from the process of trapping.

In spite of the apparent difficulty in differentiating between trapping and radiation-induced annealing, there are reasons to believe that the observations of radiation-induced annealing are actually related to trapping. It seems that changes in charge state caused by carrier excitation might sometimes make a defect more stable. Also, radiation annealing could change defect configurations to make them more effective in carrier removal. Neither type of behavior has been reported. Further, a defect in any charge state should be stable at an adequately low temperature, while trapping will probably occur to some extent at any low temperature. One of the most surprising aspects of "radiation-induced annealing" is the extremely low temperatures at which it occurs. The absence of definitive experimental observations which rule out a trapping process, and the fact that changes caused by "radiation-induced annealing" are always in the direction expected with trapping, support the trapping hypothesis and lead one to be critical of the conclusions that have been reached while disregarding trapping.

It should be noted that defects which are double acceptors are thought to exist[40] in many cases. In such a situation the partial annealing frequently observed may correspond to trapping at only one of these acceptor levels. On the other hand, double traps could also exist, particularly at low temperatures, provided the trapping process leads toward electrical neutrality, as in the case of a double acceptor. In fact, the most common situation may be one in which one level is filled during the irradiation and trapping is observable only at the second level.

There is an important point regarding the data of Klontz and Sivo for silicon[37]: In comparing data for similar samples measured at 35 and 101°K, one can deduce an extremely low activation energy ($\sim$0.001 eV) for the annealing process. It seems unlikely that the silicon lattice would offer such a small potential variation to any kind of defect. However, such a low activation energy is completely plausible if the observations are caused by trapping, and in fact, it is even possible to get negative activation energies with trapping.

Another observation can be made. If trapping tends to cancel the effects of carrier removal by defect centers, and since ionization always accompanies radiation, a number of defects would always be created that are ineffective for carrier removal. MacKay and Klontz's data[26] support this, since they observed factor-of-two changes in mobility for ~10% changes in carrier concentration (see Fig. 1). Evidently many more centers are being introduced than simply those responsible for carrier removal. If this effect were due to compensating defects, reverse conductivity anneal would be expected at the temperature at which those defects disappear.

If one considers in detail the experimental data,[26-37] he will observe the similarity between "radiation-induced annealing" and carrier trapping. We propose that it is possible that some, perhaps most, so-called radiation-induced annealing is simply the effect of trapping, and that actual annealing proceeds later or at a higher temperature. For this to be true, in view of the experimental data, a higher temperature is required to restore charge equilibrium than is necessary for the physical annealing of the defect. Such a possibility seems completely plausible.

This model is no panacea to explain all the unusual behavior observed, but it appears to be more consistent with the experimental observations than the explanations that disregard trapping. We do not contend that the stability of defects cannot be modified by changing charge state. Some data may require this and it seems to be a useful concept.[41] On the other hand, it appears that this idea has been overused while the process of trapping, which is known to occur, has generally been ignored. Even if the annealing models that disregard trapping are basically correct, a serious error has been made in ignoring the fact that observed changes in conductivity occur simultaneously with the trapping which the proponents of radiation-induced annealing require as a step for annealing. At low temperatures, it is to be expected that a metastable charge condition will result from this trapping with an inevitable change in conductivity. The annealing models generated by the observers of radiation-induced annealing are complicated, contra-dictory, and leave many questions unanswered. On the other hand, when the role of trapping is considered, the results seem simple and reasonable.

2.2. Silicon

2.2.1. Energy Levels

A number of investigations of energy levels[16,42-50] in electron- and gamma-irradiated silicon have been made using Hall effect measurements. This technique has established levels located 0.03, 0.14, 0.17, 0.21, 0.39, and 0.43 eV below the conduction band, and 0.05, 0.21, 0.27, and 0.31 eV above the valence band. More recent investigations[51] can be interpreted to indicate an additional level 0.54 eV below the conduction band, i.e., very near the center of the gap.

There is much better information regarding the microscopic nature of the defects that produce these energy levels for silicon than for germanium. We will consider some of the optical absorption and photoconductivity data later in this regard. Electron spin resonance data, which will be discussed in another chapter, have been particularly useful for silicon.[3,4,52-54] These studies have shown that a vacancy–oxygen complex ("*A* center") is an acceptor with an energy level $\sim$0.17 eV below the conduction band. A vacancy–phosphorus complex ("*E* center") is an acceptor with an energy level $\sim$0.4 eV below the conduction band. The divacancy appears to be a multilevel center,[51,53,55] with levels located 0.27 eV above the valence band, and 0.21, 0.39, and 0.54 eV below the conduction band. In a later section on photoconductivity we will discuss some of the evidence for these designations.

Let us emphasize that identification of energy levels associated with specific microscopic defects using electron spin resonance are inclusive rather than exclusive. For example, saying that the *A* center has an energy level 0.17 eV below the conduction band does not establish that a level located at that position belongs to an *A* center. Sonder and Templeton[46] showed that the introduction rate of $E_c - 0.17$ eV levels is approximately independent of oxygen concentration. Other evidence from carrier lifetime and optical absorption data, to be presented later, indicates that defects other than *A* centers lie near this position. Furthermore, levels other than those produced by *E* centers are located $\sim$0.4 eV below the conduction band (specifically, the divacancy has levels near here.) Thus, great care must be exercised in using the results of the ESR studies to establish the microscopic nature of particular defects studied by other experimental techniques. Further consideration of microscopic identification of defects will follow in later sections.

2.2.2. Temperature Dependence of Defect Introduction Rates

At first glance, one might not expect much unique information from an investigation of the temperature dependence of defect introduction rates. Since a decrease in temperature generally causes stability of a larger fraction of the radiation-induced defects, higher damage rates at lower temperatures might be expected. On the other hand, many different defect configurations are known to occur, so at least some of these reordering

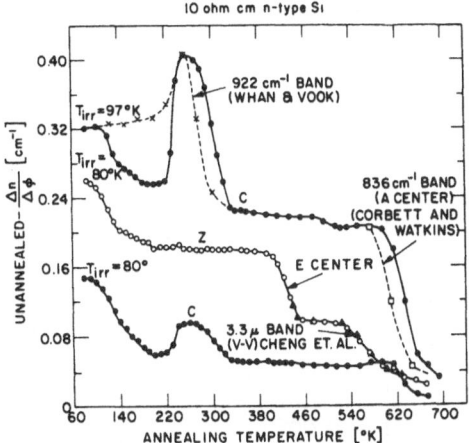

Fig. 5. Unannealed carrier-removal rate versus annealing temperature for crucible-grown silicon after irradiation with 1.7-MeV electrons at 80 and 97°K. (After Stein and Cook.[56])

processes should result in defects that are more effective for carrier removal. In this case, lower apparent defect introduction rates may occur at lower temperatures.

These effects, which correspond simply to normal and "reverse" annealing, are illustrated by the data of Stein and Vook[56] in Fig. 5. However, the most interesting feature of this figure is the dramatic increase, with increasing irradiation temperature, in apparent damage introduction rate between 80 and 97°K, which seemingly is not associated with a reverse annealing process. Further details of this temperature dependence are shown in Fig. 6. These and related data have been discussed[56] in terms of "ITI" (irradiation temperature independent), and "ITD" (irradiation temperature dependent) defects. The low-temperature annealing stage is apparently associated with the ITI defects, while the reverse annealing peak may be caused by creation of new defect configurations associated with disappearance of the ITI defects. There is some evidence that the reverse annealing peak has an energy level $E_c - 0.13$ eV associated with it. Annealing of the ITD defects seems to occur predominantly at temperatures $\gtrsim 600°$K.

Stein and Vook have analyzed[56,57] the data of Fig. 6 in terms of the creation of coupled vacancy–interstitial pairs. These metastable pairs exist in two charge states governed by the occupancy of an energy level ~ 0.07 eV below the conduction band. This model makes the hypothesis that such

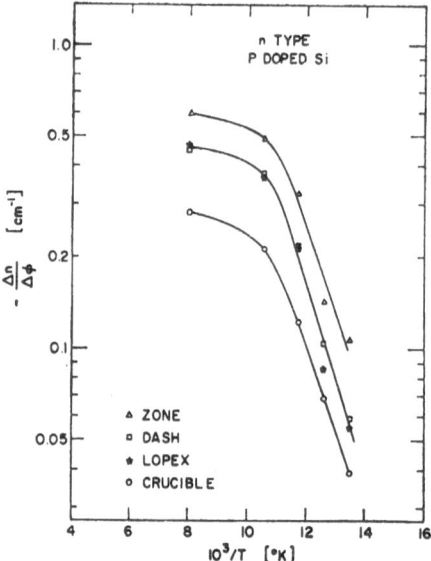

Fig. 6. Carrier removal rates versus reciprocal irradiation temperature for float-zone, quartz-crucible, Dash, and Lopex silicon. (After Stein and Vook.[56]) Measurements at 80°K after annealing to 200°K. Same samples used throughout by annealing at 400°C after each irradiation with 6.7×10^{14} electrons/cm².

defects are much less stable when the specified energy level is occupied, accounting for the lower introduction rate at lower temperatures. This model is related to an earlier one of Wertheim[58] in which the difference in barriers for recombination or separation would account for the temperature dependence. Yet another possible explanation (which seems less satisfactory) is given by considering the fact that these defects are produced by interactions very near threshold displacement energies. In this energy region the probability of atomic displacement is a very strong function of the energy of the displaced atom. Superimposing a distribution of thermal energies upon the distribution in energy imparted to the atoms by bombarding particles might produce the observed dependence of defect introduction rate upon temperature.

The idea that the ITD defects are associated with metastable pairs is supported by the fact that the threshold energy for introduction is apparently lower than for the ITI defects.[57] On the other hand, we cannot be dealing simply with vacancy–interstitial pairs, since the ITD defects show a strong dependence on impurity concentration. Evidently, after the temperature-dependent production of metastable pairs, the vacancies are released to produce defect complexes. This would seem to explain the data of Fig. 5 in which the annealing of A centers is compared to annealing of the ITD defects. Stein and Vook point to the lack of a dependence on impurities of the production rate of ITI defects, their production threshold energy, and their annealing temperature range to surmise that they are "nonre-orientable" divacancies, identified by Watkins from EPR measurements.

2.3. Compound Semiconductors

After observing the complexity of defect introduction and annihilation in the comparatively simple materials germanium and silicon, it would be surprising if an adequate understanding of the nature of defects in compound semiconductors existed. Nonetheless, a number of important studies have been made, and reasonable (but often controversial) models formulated to explain defect production and annealing in these materials. In particular, Eisen[59] has studied radiation damage in III–V compounds with emphasis on annealing kinetics,[60] threshold displacement energies,[61] and orientational effects.[62] Primarily because of disparities in masses, different threshold energies exist for the two constituents of the lattice, and it is often possible to create lattice defects associated with only one element. Table I, taken from Eisen,[59] lists measured threshold energies, including values for silicon and germanium for comparison.

Table I. Threshold Energies in Various Semiconductors (after Eisen[59])

Material	Atom displaced	Threshold displacement energy, eV
Si[63]	Si	15.8
Ge[64,65]	Ge	14.5
GaAs[66,67]	Ga	8.8
	As	10.1
InP[66,67]	In	6.6
	P	8.8
InAs[66,67]	In	6.7
	As	8.5
InSb[61]	Ib	5.8
	Sb	6.8

Among the compound semiconductors, threshold effects in indium antimonide have probably been studied most extensively. The threshold energy values given in Table I are comparatively low, and represent minimum bombarding electron energies of $\sim$170 and $\sim$400 keV for displacement. Still, damage appears to occur when even less displacement energy is available. Arnold and Vook[68] observed creation of damage by 100-keV (peak) X-rays. This indicates[69] that some process of defect creation other than simple displacement, such as that proposed by Varley,[70] is operative in this material.

Brehm and Pearson[71,72] have studied gamma-irradiated n- and p-type gallium arsenide extensively. Hall measurements have yielded energy levels for defects at $E_c - 0.13$, 0.16, and 0.30 eV and at $E_v + 0.059$ and 0.10 eV. It is possible that the latter two levels were present before irradiation and merely became observable because of the decrease in hole concentration caused by irradiation. Willardson[73] has reviewed the nature of lattice defects in gallium arsenide which has undergone increases in defect concentration through radiation and various types of heat treatment.

Compounds containing cadmium are particularly susceptible to thermal neutron damage because of cadmium's large cross section for neutron capture. Subsequent lattice damage occurs because of recoil by the cadmium atom when gamma rays are emitted. These defects are simple in nature and are not to be confused with the complex damage produced by energetic neutron irradiation. A study of n-type cadmium telluride irradiated with both thermal neutrons and 1-MeV electrons has been performed by Abramov and colleagues,[74] while Barnes and Kikuchi[75] studied thermal neutron recoil in p-type material.

Another appropriate material for study is silicon carbide.[76] Like most semiconductors, the carrier concentration of silicon carbide decreases with electron irradiation. This general tendency for a reduction in the net carrier concentration is usually observed, except in fairly high-resistivity material, since the limiting Fermi level position for large amounts of radiation is usually in the central portion of the band gap. However, for indium arsenide there appears to be an increase in electron concentration even for degenerate n-type samples.[77]

3. CARRIER MOBILITY

As already mentioned, carrier mobility is a quantity usually less sensitive to lattice defects than carrier concentration. This is particularly true in high-purity semiconductors near room temperature where lattice scattering dominates. Furthermore, deductions relating mobility changes to the number of lattice defects are indirect. For these and other reasons, studies relating changes in mobility to lattice defects are less common than those concerned with carrier concentration. Nonetheless, observation of mobility changes can be a powerful tool for understanding the nature of defects. Furthermore, such studies have strong practical implications due to the importance of mobility in many semiconductor device applications.

Most mobility studies utilize simultaneous measurements of Hall effect and conductivity to obtain Hall mobility. Referring to Section 2, Hall mobility is usually defined as $\mu_{\mathrm{H}} = |\,R\sigma\,|$, differing from the drift mobility by the factor r_e or r_h. Unfortunately, even beyond the two factors r_e or r_h there are often large discrepancies between Hall mobility and drift or "true" mobility. Nonetheless, Hall measurements can be quite useful, provided the sample under investigation is very homogeneous. Carrier concentration variations in a sample can create large errors and the meaning of the measurements then becomes obscure. Under conditions such as those produced by neutron irradiation, the damage tends to be localized and interpretation of Hall effect data is difficult. Also, Hall mobility measurements are not very useful for near-intrinsic samples, i.e., when the hole and electron concentrations do not differ by at least an order of magnitude. On the other hand, drift mobility measurements, generally obtained by some variation of the Haynes–Shockley experiment,[78] are tedious and feasible only for materials with long minority carrier diffusion length. Other methods such as the flying spot technique[79] show promise of being superior for these measurements, particularly when the resolution available with a scanning electron beam is used.[80]

To a good approximation, the mobility associated with various scattering processes can be determined by adding reciprocally the mobilities that would exist if each process were the only one active. Thus the change in reciprocal mobility is normally the quantity considered in defect studies. Calculations of the mobility associated with scattering at charged defects are fairly crude. However, the Conwell–Weisskopf[81] mobility relationship, based on simple Rutherford scattering, appears to be a reasonable representation for most situations, at least for qualitative purposes:

$$\mu_I = \frac{2^{7/2}\varkappa^2(kT)^{3/2}}{\pi^{3/2}(m^{*1/2}q^3Z^2N_I)}\ \frac{1}{\ln[1 + (3\varkappa kTrq^2N_I^{1/3})^2]} \tag{6}$$

where μ_I is the mobility associated with charged-defect scattering, $\varkappa$ is the dielectric constant, qZ is the charge associated with the defects whose concentration is N_I, and m^* is the density-of-states effective mass. This equation has been modified[82,83] to treat the scattering more realistically. In practice it is necessary to consider such aspects as the variation of the Hall factor r with impurity concentration and temperature to determine net and total charged defect concentration from Hall measurements.[84] This matter has been treated in some detail by Long and Myers.[85]

3.1. Germanium

An excellent example of the way that mobility measurements can be used in defect studies is illustrated in Fig. 7, taken from the work of Brown, Augustyniak, and Waite.[2] They use the fact that carrier concentration and mobility simultaneously decrease during 1-MeV electron irradiation. Because of the Z^{-2} term in Eq. (6), the relative mobility change will be much larger if the predominant defects are double acceptors than if they are singly charged, as is illustrated in the figure. The change in carrier concentration due to a given defect is proportional to Z, so that the ratio $\Delta(1/\mu)/\Delta n$ varies as Z. This is clearly an oversimplified viewpoint, since we know that various kinds of defects are introduced. Nonetheless, the

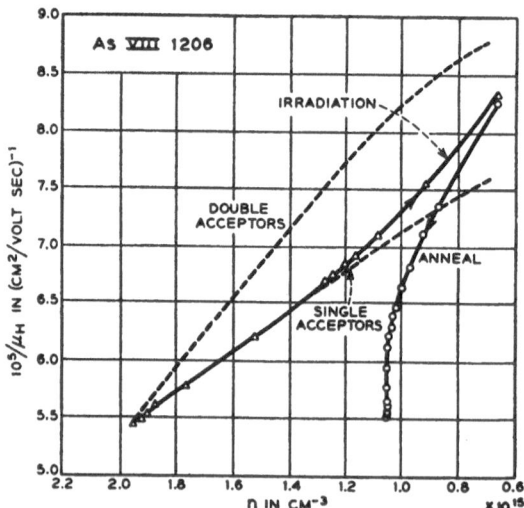

Fig. 7. Changes in $1/\mu_H$ during bombardment and anneal as a function of the change in carrier concentration. (After Brown, Augustyniak, and Waite.[2]) The data are for an As- doped sample. Note the difference between irradiation and anneal.

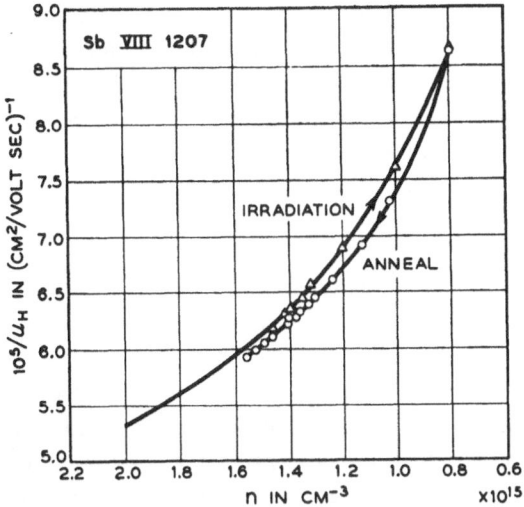

Fig. 8. Changes in $1/\mu_H$ versus n for an antimony-doped sample. (After Brown, Augustyniak, and Waite.[2]) Note the similarity between irradiation and anneal.

data of Fig. 7 indicate that during the early stages of irradiation, mobility changes appear to follow closely the single acceptor behavior, deviating only at higher defect concentrations.

Although this information is interesting, because of the multiplicity of defects likely to be present it is hardly definitive. However, the same type of information obtained during anneal gives remarkable insight into the nature of the annealing process. It is seen that virtually complete recovery of the mobility occurs during a comparatively small recovery of carrier concentration. As pointed out by Brown, Augustyniak, and Waite, there is one very logical explanation for this behavior. Single-acceptor-type defects are migrating to donor-impurity atoms to form complexes. The net carrier concentration in this process remains unchanged, but the resulting neutral center is ineffective for carrier scattering. (The original defect and donor atom may maintain their initial charge, producing a relatively ineffective dipole, or the new complex formed may be uncharged.) The small conductivity anneal which does occur may correspond to defects which find alternate sinks, or other types of defects that are not attracted by the impurity atoms.

Figure 8 shows similar data for antimony-doped material, where it is apparent that the associative process described for arsenic-doped material does not occur. The differences between the data of Figs. 7 and 8 provided much of the convincing evidence of defect–impurity interactions which have so influenced the course of radiation effect studies since their publication. The failure of the curves for antimony-doped material to follow the behavior either of single or double acceptors emphasizes further the complexity of this type of analysis.

3.2. Silicon

Sonder and Templeton have made extensive Hall mobility measurements on gamma-irradiated n- and p-type silicon.[16,46,49] Very large mobility changes were observed for $T \lesssim 100°$K. However, no fit to these data was possible with simple theory because of the anomalous shape of the mobility versus temperature curves. At low temperature the Hall mobility was found to decrease more rapidly with decreasing temperature than predicted by theory, and the mobility changes were so large for large doses that even using $Z = 5$ in Eq. (6) was insufficient to account for them. These problems led Sonder and Templeton to conclude that the effects were not real, but brought about by sample inhomogeneities, rather dramatically pointing out one of the major pitfalls of mobility measurements based upon Hall effect data. Similar observations were made earlier by Hill for electron irradiation.[45] However, for smaller amounts of radiation, reasonable behavior was observed by Wertheim, who deduced from the rate of mobility change that the defects were predominantly singly charged.[43] Wertheim's measurements were made on 0.4 Ω-cm, n-type material.

Clark, Fernandez, and Thompson observed some provocative behavior following 90°K irradiation with 2-MeV electrons.[86] Their n-type silicon ($n = 6.2 \times 10^{17}$ cm^{-3}) showed Hall mobility and carrier concentration following very similar annealing curves, although at intermediate temperatures ($\sim 170°$K) both carrier concentration and mobility attained values higher than before irradiation. It was the p-type material ($p = 10^{17}$ cm^{-3}), though, which behaved in a particularly remarkable fashion. Mobility was observed to decrease drastically during the annealing cycle, simultaneously with an increase in the carrier concentration. Thus a process opposite to that discussed above for germanium appeared to occur—either neutral complexes disassociated to produce defects or multiply charged complexes were formed from previously singly charged ones. The concomitant increase in hole concentration indicated an increase in acceptor concentration, i.e., negatively charged centers.

Studies of lithium-doped silicon produced results similar to those shown in Fig. 7 for germanium.[87] That is, mobility recovered more completely during anneal than did carrier concentration. In some cases, it appeared that the mobility recovery was virtually independent of carrier concentration change, indicating a highly efficient pairing mechanism, while at higher temperatures the opposite process occurred, appearing to demonstrate a break-up of such pairs. In fact, it seems that mobility recovery occurs during pairing of mobile lithium atoms with vacancy-containing complexes. In this case we may be observing formation and dissociation of Li–O–V complexes.

In their investigation of "irradiation-temperature-independent" defects, Stein and Vook deduced a doubly charged nature for these defects from the slope of a $1/\mu$ versus n plot.[88]

3.3. Other Materials

The high carrier mobility of gallium arsenide is important in certain device applications. Willardson has considered the effect of both chemical impurities and radiation-induced damage upon carrier mobility.[73] Grimshaw considered this problem[89] for electron irradiation in more detail, and by plotting electron mobility as a function of carrier removal in a manner similar to Figs. 7 and 8, has deduced that the primary defects introduced are multiply charged. On the other hand, at large doses, the mobility decreases at an anomalously high rate, similar to that observed at low temperatures in silicon by Sonder and Templeton.[16] Grimshaw also concludes that inhomogeneities are probably responsible for this effect. Brehm[72] has determined mobility changes for gamma-irradiated GaAs and concludes that the mobility changes and carrier removal rates are consistent with the presence of single acceptors in n-type material. In p-type GaAs the mobility changes are much larger that expected from carrier removal data, indicating severe compensation, or again possibly problems of inhomogeneity.

It seems evident, because of the problems associated with Hall mobility measurements, that more significant efforts in the area of drift mobility measurements are in order. Extensive measurements have been performed[90] in germanium for neutron irradiation where Hall measurements are especially ineffective, and it is hoped that such investigations will be performed for samples containing isolated defects. However, such experiments are difficult in germanium and silicon, and in other materials should be quite challenging.

4. MINORITY CARRIER LIFETIME

In materials such as germanium and silicon which have indirect band gaps that are fairly wide, very long minority carrier lifetimes would exist in the absence of lattice defects. The lifetimes associated with band-to-band transitions are ~ 0.3 and $\sim 10^4$ sec in germanium and silicon, respectively. Since the highest attainable lifetime in either of these materials is typically 10^{-4}–10^{-3} sec, it is clear that chemical or physical defects have a profound effect. The striking sensitivity of lifetime in silicon to defects is illustrated by the fact that, if one could neglect surface recombination, the effects of 10^4 defect centers/cm³ could be observed, provided they introduced energy levels in the central portion of the band gap and had favorable capture probabilities. Silicon or germanium crystals grown for high lifetime contain about 10^{11} recombination centers/cm³. Thus, even in the normal situation, small concentrations of defects are easily observed.

A number of investigations have shown that the recombination properties of defects are well described by the Hall[91]–Shockley–Read[92] recombination equation, which for the electron lifetime τ_n (n- or p-type material) in the presence of an arbitrary defect density N is

$$\tau_n = \frac{[(p + p_1 + \Delta p)/c_n N] + [(n + n_1 + \Delta n)/c_p N]}{p + n(\Delta p/\Delta n) + \Delta p} \tag{7}$$

In this equation n and p are the equilibrium electron and hole concentrations, Δn and Δp are the excess carrier concentrations, c_n and c_p are electron and hole capture probabilities, and n_1 and p_1 are the electron and hole concentrations that would exist if the equilibrium Fermi level coincided with the recombination energy levels, or

$$n_1 = N_c \exp[-(E_c - E_r)/kT] \tag{8a}$$

and

$$p_1 = N_v \exp[-(E_r - E_v)/kT] \tag{8b}$$

The expressions for electron lifetime and hole lifetime are identical when $\Delta n = \Delta p$. This situation exists for low defect concentrations in the absence of trapping. (We use the term "trapping" here specifically to refer to the process of temporary removal of a minority carrier with subsequent reemission into the minority carrier band and negligible recombination with majority carriers.) Evaluation of the ratio $\Delta n/\Delta p$ can be difficult for high concentrations of defects or in the presence of trapping, but there are many cases when this is not necessary, allowing us to use the simple form of Eq. (7) for $\tau = \tau_n = \tau_p$,

$$\tau = \frac{[(p + p_1 + \Delta n)/c_n N] + [(n + n_1 + \Delta n)/c_p N]}{n + p + \Delta n} \tag{9}$$

We will refer to this as the HSR (Hall–Shockley–Read) equation.

Recombination rates add, and since the recombination rate is inversely proportional to lifetime, the lifetimes due to various centers add reciprocally. The energy position of a recombination level is important, and a qualitative investigation of the above equation indicates a strong tendency for centers with energy levels near the middle of the band gap to dominate recombination. For a given situation, the capture probability ratio c_p/c_n, is also important. This can be shown quantitatively from the HSR equation, but is intuitively obvious for low excess densities upon realizing that, for a very large majority-to-minority-carrier ratio, the minority carrier capture process is likely to be the limiting one. Thus, centers with large minority carrier capture probabilities, generally corresponding to defects having the same charge as the majority carrier, are especially effective.

It is seen, then, that recombination studies tend to select defects with certain properties. This is an advantage in one sense, since it is easier to observe one defect at a time, whereas such properties as carrier concentration and mobility tend to reflect various defects simultaneously. This selectivity is also a disadvantage, since it is impossible to study the recombination properties of those defects that do not compete well in the recombination process.

Again, electron and gamma irradiation are the most convenient ways of introducing simple lattice defects into semiconductors. A number of studies have been performed on the recombination behavior of material so irradiated. However, most of these studies have been primarily concerned with temperature variations of lifetime at low excess-carrier densities, and have yielded inaccurate or ambiguous results because of the failure to take into account the temperature variation of all the terms in the HSR equation. Most investigators have simply assumed that the capture probabilities are temperature independent, and deduced energy level positions from the apparent variation of the terms p_1 or n_1. Others have attributed the temperature variation completely to capture probability, assuming the terms n_1 and p_1 were negligible.[93]

It is impossible to adequately determine recombination parameters by investigations of temperature dependence of low-level lifetime alone. However, the problem can be solved by performing complementary measurements of lifetime versus excess carrier density. The HSR equation can be written in the form (assuming n-type material)

$$\tau = \tau_0 \frac{1 + \Delta n/b}{1 + \Delta n/n} \tag{10}$$

where

$$b = \frac{c_p(p + p_1) + c_n(n + n_1)}{c_p + c_n} \tag{11}$$

The lifetime at low excess density is τ_0 and b can be determined from an experimental determination of τ versus Δn.

In specific instances, determining b at a single temperature can give enough additional information to determine energy level position. This is the case for n-type germanium[94] where the temperature dependence strongly indicates an energy level near the center of the band gap with a very large value for the ratio c_p/c_n. Assuming these facts yields a simplified form for Eq. (11), which gives a position for the recombination level consistent with the assumptions. The same energy level is obtained for various resistivities, provided the type of dopant is the same. This demonstrates the consistency of the analysis, the actual results of which will be discussed later. However, as will also be shown later, this relatively simple approach is not always adequate, and it may be necessary to determine not only lifetime at low excess densities as a function of temperature, and b at one temperature, but also the quantity b as a function of temperature.

Values of b and τ are, in general, dependent both on energy level position and the capture parameters c_n and c_p. However, Eqs. (9) and (11) can be manipulated to yield for n-type material

$$\frac{c_n c_p}{c_n + c_p} = \frac{b}{\tau_0 N n} = k \frac{b}{\tau_0 \phi n} \tag{12}$$

where ϕ is the irradiation dose and k is a damage factor. Thus the ratio $b/\tau_0 \phi n$ is a measure of certain recombination center characteristics, independent of energy level position. The ratio c_p/c_n will usually be large or small compared to unity. If, for instance, $c_p/c_n \gg 1$,

$$\frac{c_n c_p}{c_n + c_p} \approx c_n \approx k \frac{b}{\tau_0 \phi n} \tag{13}$$

and the temperature variation of electron capture probability is unambiguously given, provided both $b(T)$ and $\tau_0(T)$ are known. It is often possible to prove that the condition $c_p/c_n \gg 1$ is met, and this equation is a valuable starting point for analysis of complex recombination behavior.[95]

4.1. Two-Level Model for Recombination[96]

Any type of irradiation creates a variety of defect energy levels in the band gap of semiconductors, at least at room temperature, where various types of complexes have been allowed to form. In agreement with the preceding discussion, the usual observation at low excess carrier densities is that a particular level dominates the recombination behavior. This level lies deep within the forbidden gap, and has a large ratio of minority to majority capture probability. However, for high-level excitation, the relative importance of this type of level diminishes. This can be readily seen by inspecting Eq. (9).

For intermediate excitation, $\Delta n \sim n$, the effectiveness of all the deep levels tends to become comparable, and for high levels, $\Delta n \gg n$, provided also that $\Delta n \gg n_1, p_1$, the HSR equation reduces to

$$\tau \approx \frac{1}{c_n N} + \frac{1}{c_p N} \tag{14}$$

Now, the defects responsible for recombination will be those that minimize Eq. (14). At such high excitation levels, all energy level positions between the electron and hole quasi-Fermi levels ($\Delta n > p_1, n_1$) are equally effective. Thus only the defect concentration and the magnitude of the smaller capture probability are important, and there should be a tendency to favor those defects with near-unity capture probability ratios. Since the dependence on energy level has been removed, there is less selectivity in the recombination process, and a number of levels can contribute to recombination at high excitation. However, experimentally, one level often appears to dominate.

Thus it seems that if an assortment of levels are available for recombination, two levels will tend to determine the recombination behavior, one of which will dominate, depending on the range of excitation investigated. This is illustrated in Fig. 9, where the variation of lifetime with excess density for two types of levels discussed is shown, together with some actual data for neutral-irradiated (3×10^{12} neutrons/cm²) germanium, which in this case behaves in a manner very similar to that observed for gamma and electron irradiation. Since lifetimes add reciprocally, the lifetime versus excess density curve behaves as shown. This model[96] is in excellent agreement with experimental data for a variety of irradiated

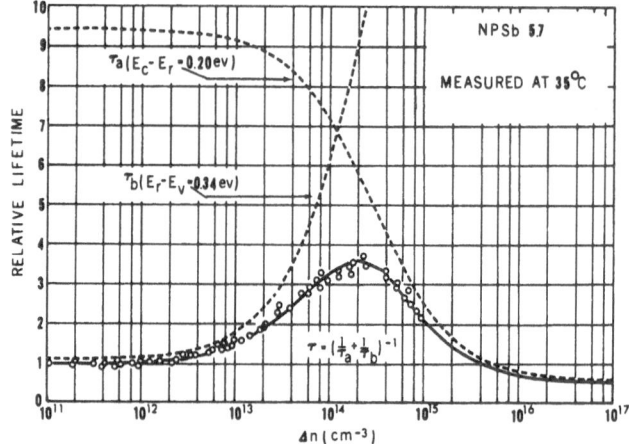

Fig. 9. Analysis of recombination data for 5.7 Ω-cm, Sb-doped Ge.

materials as seen from the figure and from data presented in the following
sections.

4.2. Recombination in Irradiated Germanium

Recombination behavior has been studied in detail in n-type germanium. The general characteristics of the recombination centers in n-type
material are quite similar for a wide range of resistivities and for various
types of dopants. Energy levels are near the center of the band gap, and
c_p/c_n ratios are very large, usually several hundred. There are clear differences in detail between samples with different types of dopants and those
of moderate ($1 \lesssim \varrho \lesssim 20\ \Omega$-cm) and low ($\varrho \lesssim 0.3\ \Omega$-cm) resistivity material. We will discuss these cases and the behavior in p-type material which,
though not investigated in detail, seem consistent with data for n-type
germanium.

4.2.1. Moderate-Resistivity n-Type Material

A number of measurements of low-excess-density lifetime as a function of temperature have been performed in n-type germanium.[97-99] Figure
10 illustrates the simplest behavior observed.[97] Often the behavior is more
complex because of trapping. The behavior shown in the figure is explained
simply if (1) the hole capture probability is temperature independent,
(2) the term containing p_1 in the HSR equation is dominant at higher
temperatures, (3) $c_p/c_n \gg 1$, and (4) $E_r - E_v \sim 0.36\ \text{eV}$, neglecting any
temperature dependence of c_n. (The indicated slope must be corrected for

the effects of initial lifetime and the temperature dependence of the density-of-states function.) Figures 11 and 12 illustrate the effect of excess carrier density on lifetime in ^{60}Co-gamma-irradiated n-type germanium.[94]

A summary of the parameters obtained[94] from analyzing the data on the basis of the two-level recombination model is presented in Table II. Determined positions of the level dominant at low excess densities are indicated with respect to the valence band, while the position of the second level is shown with respect to the conduction band. The soundness of the analysis is indicated by the agreement of the results for various resistivities, being almost exact between 1 and 7 Ω-cm material for each doping. The slight deviation for higher resistivities was probably due to neglect of the hole concentration in the HSR equation. The position of the deep level is difficult to confirm with carrier concentration measurements, but the position of the shallow level appears to agree well with one at $E_c - 0.23$ eV commonly observed and discussed earlier. Differences between samples containing different dopants are confirmed by annealing studies.[100]

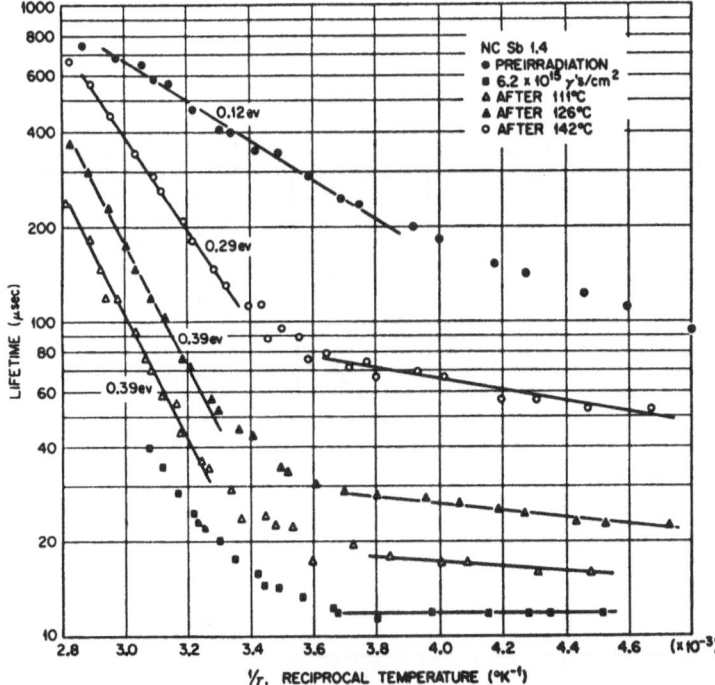

Fig. 10. The recombination behavior of 1.4 Ω-cm, antimony-doped germanium following irradiation by ^{60}Co gamma rays and successive 1-hr anneals.[82]

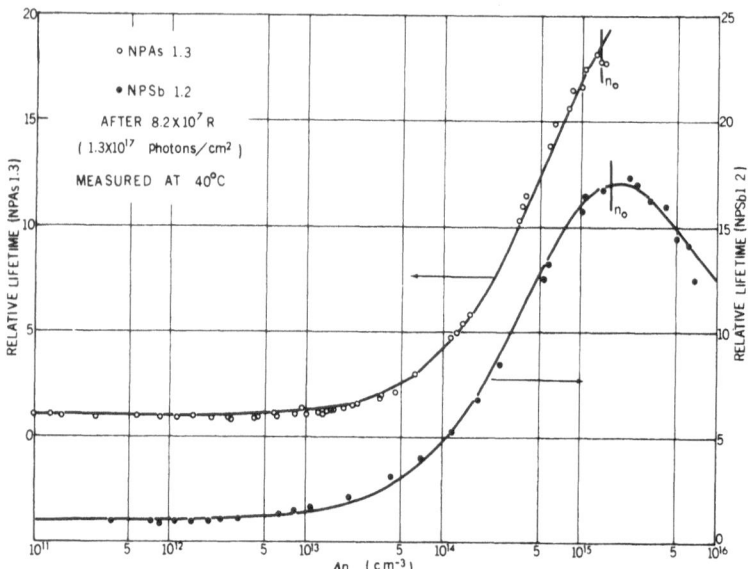

Fig. 11. The dependence of carrier lifetime on excess density for 1 Ω-cm, As-doped and 1 Ω-cm, Sb-doped germanium following ^{60}Co gamma irradiation.[100]

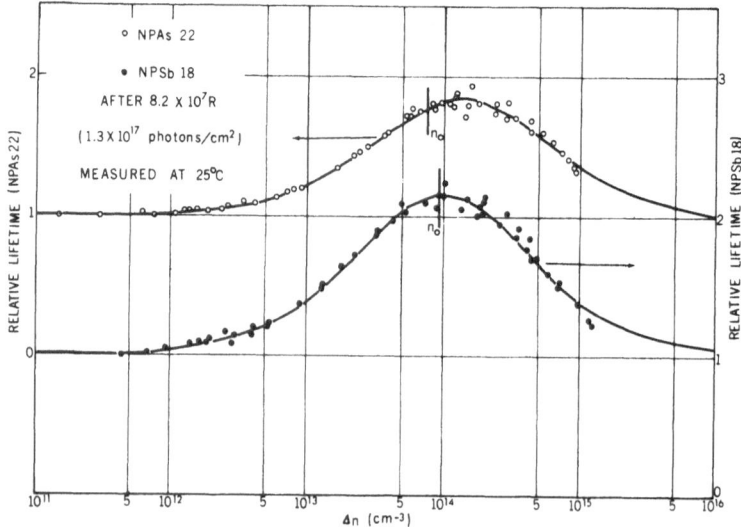

Fig. 12. The dependence of carrier lifetime on excess density for 22 Ω-cm, As-doped and 18 Ω-cm, Sb-doped germanium following ^{60}Co gamma irradiation.[100]

Table II. Recombination Level Parameters Obtained from Dependence of Lifetime on Excess Density in Gamma-Irradiated Germanium[100] [a]

Sample	$E_r - E_v$ (deep level), eV	$E_c - E_r$ (shallow level), eV
NPAs 1.0	0.332	0.23
NPSb 1.0	0.338	0.19
NPAs 7.1	0.331	0.20
NPSb 7.0	0.338	0.23
NPAs 22	0.327	0.23
NPSb 18	0.333	0.25

[a] Unity statistical weight factors have been assumed.

4.2.2. Low-Resistivity n-Type Material

Even more detailed studies have been performed on low-resistivity ($\sim$0.2 Ω-cm) arsenic- and antimony-doped germanium.[17] In this material, almost exact single-level behavior is observed for low excess densities, and data for a second level were not obtainable at the levels of excitation available. The dominant recombination center in this material appears to be different from that of the higher resistivity materials, probably because a center effective as a trap[99] in higher resistivity material is transformed into a recombination center at higher carrier concentrations. Low-excess-density lifetime and the parameter b were determined as a function of temperature for both arsenic- and antimony-doped material. The magnitude of b showed conclusively that $c_p/c_n \gg 1$. This allowed the determination of the temperature variation of c_n independent of assumptions concerning the other recombination parameters using Eq. (13).

The results are shown in Fig. 13. Irradiation with ^{60}Co gamma rays was performed at dry-ice temperature and the "unannealed" samples were measured following 1 hr at room temperature. The "annealed" specimens underwent an additional $\frac{1}{2}$-hr treatment at 72°C. A marked change in the behavior of the arsenic-doped samples was observed after this anneal, but no changes occurred for the antimony-doped specimens. It was possible to show that for both cases the hole capture probability was very nearly temperature independent, i.e., in terms of an apparent activation energy, the value would be less than 0.01 eV. (The use of activation energies to describe the temperature dependence of capture probabilities is not meant to imply the physical origin of that dependence. A power law dependence could also be used to fit the data.)

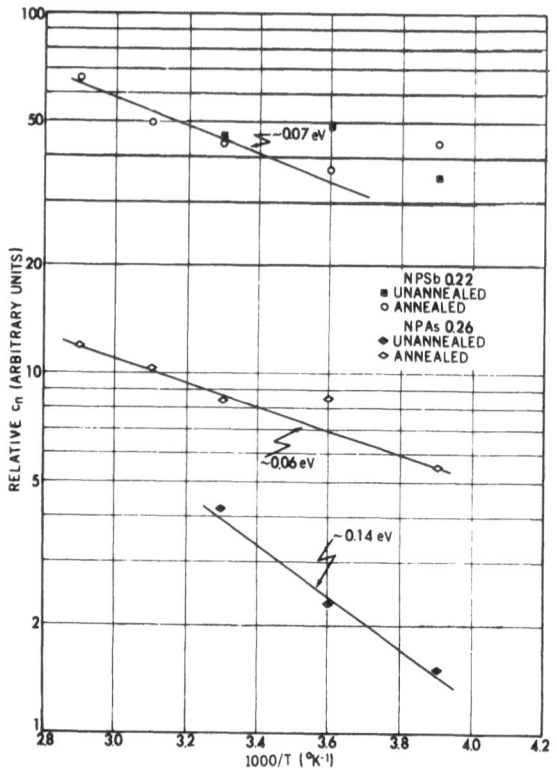

Fig. 13. Relative electron capture probability c_n versus reciprocal temperature for low-resistivity Sb- and As-doped germanium. No comparison should be made between any of the curves with regard to relative magnitude. The curves are plotted on the same figure for convenience only.[17]

The capture probability ratio and energy level position for antimony-doped germanium are shown in Fig. 14. Note the large magnitude of the capture probability ratio and the apparent temperature dependence of the energy level position. While it may not be physically significant, the sensitivity of the measurements is indicated by the line drawn through the data points for energy level positions that corresponds to one-half the variation of band gap with temperature. Figure 15 is a corresponding plot for arsenic-doped material which has undergone 72°C anneal (necessary to achieve a suitable range of measurement temperature). The capture probability ratio and the temperature dependence are similar to those of the antimony-doped case. The energy level is somewhat shallower, but apparently displays the same temperature dependence.

This set of measurements on low-resistivity n-type germanium illustrates the thorough understanding of recombination parameters that can be obtained with adequate measurements. The temperature dependences of both hole and electron capture probabilities and the magnitude of capture probability ratio were obtained, and the energy level position was determined within a few millielectron volts.

4.2.3. p-Type Material

While measurements for p-type material are not nearly as complete as in the n-type case, Fig. 16 illustrates[97] the dependence of lifetime on temperature at low excess densities. It seems quite likely, particularly in view of the preceding results, that the temperature variation observed is simply that of the electron capture probability and that the lifetime is given by $1/c_n N$ [obtained from Eq. (9) for low excess densities when terms containing n_1 and p_1 are negligible]. This does not adequately explain the variation of lifetime with resistivity,[97] but the nature of the recombination

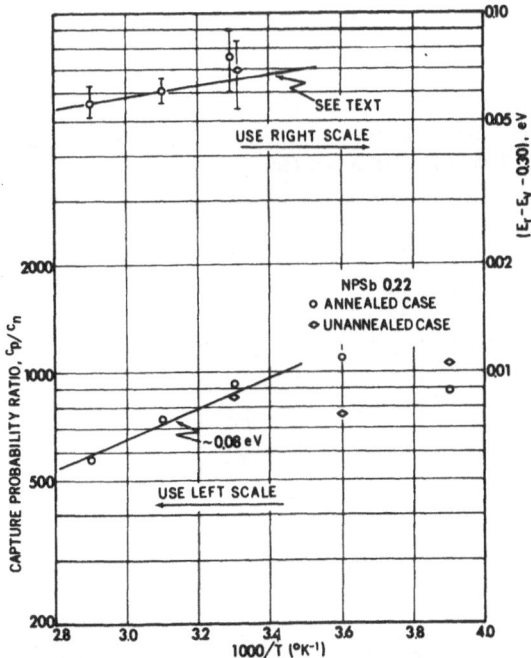

Fig. 14. Capture probability ratio and recombination-center energy level position versus reciprocal temperature for 0.22 Ω-cm, Sb-doped Ge.[17]

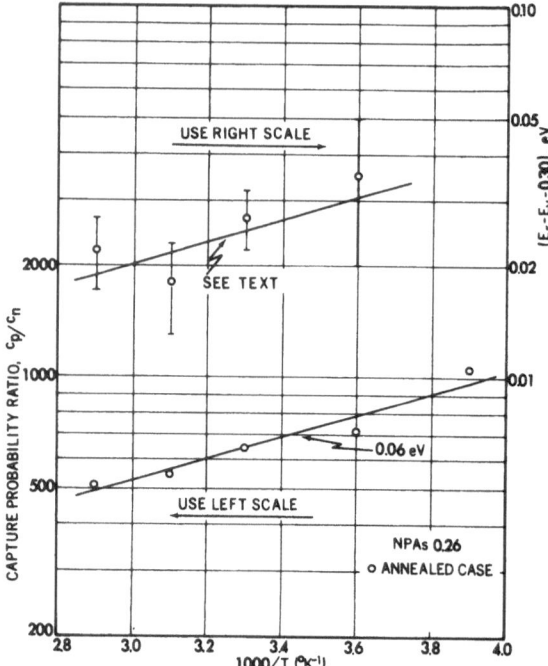

Fig. 15. Capture probability ratio and recombination-center energy level position versus reciprocal temperature for 0.26 Ω-cm, As-doped Ge.[17]

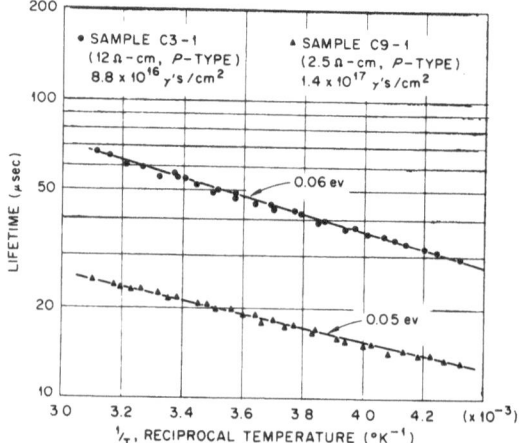

Fig. 16. Recombination behavior of Ga-doped Ge following irradiation by ^{60}Co gamma rays.[97]

centers is not known, and it may be that the dopant atoms play an important role in their creation. The agreement between the temperature dependence shown in Fig. 16 and that of Figs. 13–15 for capture probabilities appears too good to be fortuitous.

4.3. Silicon

A number of studies have been performed on recombination in electron- and gamma-irradiated silicon.[43,101–108] However, these studies have been concerned primarily with analysis of low-excess-density lifetime to determine recombination parameters. As has already been discussed, such an analysis is at best inconclusive and may be considerably in error. Because the lifetime temperature variation observed is often reasonably consistent with an energy level ~ 0.17 eV from a band edge, there has been a tendency to conclude that the observed behavior is associated with the A (oxygen-vacancy) center. However, studies we have performed on 10-MeV electron-irradiated material have shown that if such a level is dominant, it is introduced at a lower rate in material which contains a higher concentration of oxygen, and so must not be associated with the A center. Similar results were obtained for ^{60}Co gamma-irradiated silicon.[108]

4.3.1. *Analysis for n-Type Material with Two Levels Effective at Low Excess Density*

There has been considerable evidence from several observers that two levels are effective for recombination in *n*-type silicon even at low excess densities. This, of course, makes the analysis substantially more difficult, and severely limits the usefulness of low-excess-density temperature-dependence measurements used alone. Again, studies of lifetime as a function of excess density serve to unravel the problem. Figure 17 is an illustration of such measurements.[109] These data are well represented by the two-level model, but the parameter values obtained by curve fitting confirm that both levels are effective at low excess densities. If one defines the quantity k as τ_{02}/τ_{01}, the ratio of the low-excess-density lifetimes corresponding to the two levels, the resultant lifetime can be written for the entire range of excess densities used as in the following expression[96]:

$$\tau = \left(\frac{1}{\tau_1} + \frac{1}{\tau_2}\right)^{-1} = \tau_0 \left[\frac{kb}{(k+1)n}\frac{n+\Delta n}{b+\Delta n} + \frac{c}{(k+1)n}\frac{n+\Delta n}{c+\Delta n}\right]^{-1}$$

(15)

where τ_0 is the measured low-level lifetime.

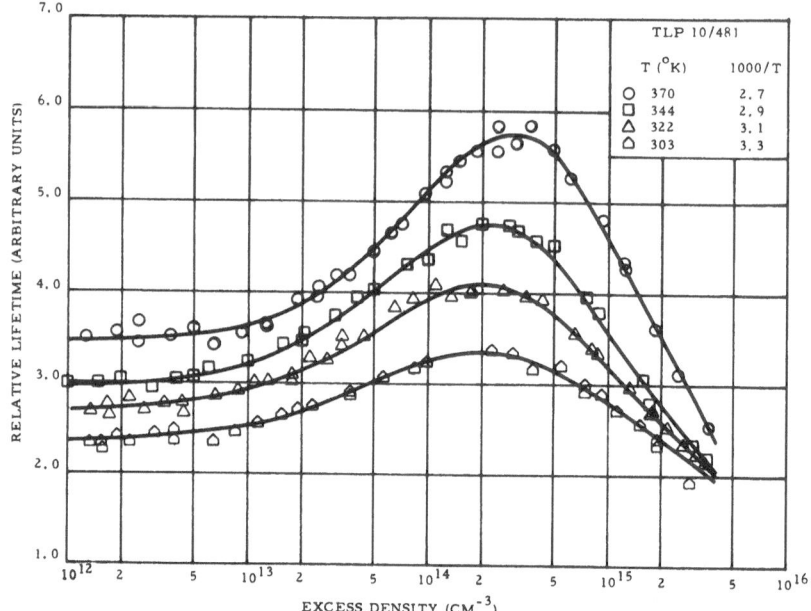

Fig. 17. Relative lifetime versus excess density at four temperatures for gamma-irradiated (6.9×10^5 R), 10 Ω-cm, n-type silicon after a 1/2-hr anneal at 417°K. The solid lines are least-squares computer fits to the data using a two-level Hall–Shockley–Read model.[109] No comparison should be made of the relative magnitudes of the four curves shown. They are on the same plot for convenience only.

The quantity b has the definition given earlier, and c is the corresponding parameter for the second level. Using the definition of k and the fact that the low-excess-density lifetimes add reciprocally to give the observed lifetime τ_0, the following expressions hold:

$$\begin{aligned} \tau_{01} &= \tau_0(1 + 1/k) \qquad (\text{``}b\text{ level''}) \\ \tau_{02} &= \tau_0(1 + k) \qquad (\text{``}c\text{ level''}) \end{aligned} \qquad (16)$$

Although laborious, this indicates a way to determine the individual temperature dependences of the two recombination levels. For each temperature at which a complete lifetime versus excess density curve is obtained, k can be determined, yielding the two lifetime components from the above expressions. Figure 18 shows just such an analysis, where the individual temperature dependences have been determined. This information and the temperature dependences of b and c allow a complete analysis of the recombination parameters associated with both levels in a manner similar to that described above for n-type germanium.

The simultaneous importance of the two levels decreases the precision with which the parameters can be determined, but nonetheless, performance of extensive measurements has removed practically all the ambiguity. That is, fitting the temperature dependence at low excess density and the temperature dependence of the parameters b and c determines whether the temperature dependence of low-excess-density lifetime arises partially or completely from capture probability.

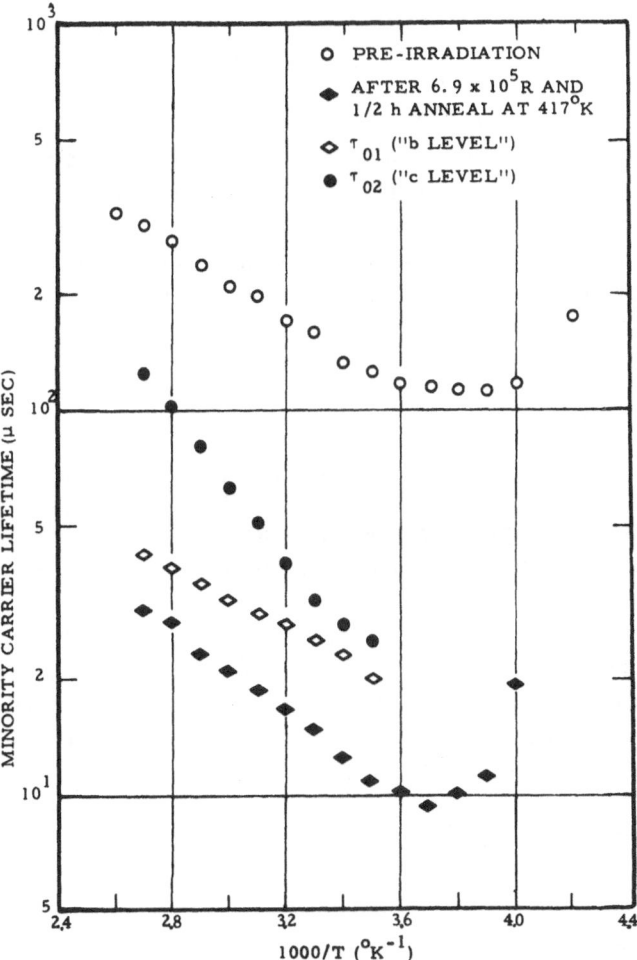

Fig. 18. Minority carrier lifetime at low excess density versus reciprocal temperature for gamma-irradiated, 10 Ω-cm, n-type silicon. The separation of measured post-irradiation lifetime into two components (associated with two recombination levels) is shown.[109]

In this analysis,[109] for data fitting purposes, exponential temperature dependences were assumed for the capture probabilities. As previously stated, a power law dependence could have been used, but the results would be essentially the same, except for the manner of expression of resultant capture probability temperature variations.

The data for Figs. 17 and 18 were obtained from 10 Ω-cm, low-oxygen silicon. The recombination parameters for the "b" level (the center of greatest importance at low excess densities) are as follows: The energy level position is at $E_c - 0.39$ eV, the electron capture probability varies as $\exp(0.06 \text{ eV}/kT)$ with the ratio c_p/c_n ranging from $\sim$9 at 303°K to $\sim$10 at 370°K. For the "c" level, the recombination energy level is at $\sim E_c - 0.21$ eV, $c_n \propto \exp(0.01 \text{ eV}/kT)$, and c_p/c_n ranges from $\sim$5 at 303°K to $\sim$8 at 370°K. It is interesting to note that energy levels associated with the divacancy have been reported near both of these positions. However, the behavior is inconsistent with coupled energy levels.

Although perhaps to some extent fortuitous, these data are in reasonable agreement with those of Glaenzer and Wolf,[105] who, neglecting any temperature dependence of capture probability, deduced levels at $E_c - 0.17$ eV and $E_c = 0.40$ eV. They ascribed the shallow level to the A center, evidently because of its position.

4.3.2. p-Type Material

Hewes and Compton[107] performed an analysis on p-type silicon similar to that performed by Glaenzer and Wolf for n-type material, again neglecting any variation of capture probabilities with temperature. In the absence of detailed lifetime versus excess density measurements, the parameters obtained give an indication of some of the recombination properties, and if, as in the case of n-type material, the capture probabilities are only weakly temperature dependent, the values obtained for energy level position may be reasonably good. In Hewes and Compton's investigations, ^{60}Co-gamma-irradiated, p-type silicon with resistivities from 25 to 1150 Ω-cm was employed, and they also deduced two energy levels, one located 0.3 eV from the conduction band, and the other 0.17 eV from one of the bands. Although it was not possible to tell from lifetime measurements alone to which band the second level was nearer, other considerations led the authors to conclude that it was more likely nearer the valence band and thus not the A center.

Even though we tend to discount the A center as responsible for recombination in the examples given, Corbett et al.[110] indicate that the annealing data of Bemski and Augustyniak[111] for lifetime in 700-keV electron-irradiated, *n*- and *p*-type material correspond closely to those for the *A* center.

In *p*-type silicon the recombination behavior depends strongly upon the oxygen concentration, but in both Czochralski and vacuum-float-zone material it is apparently more nearly dominated by a single level than has been observed for *n*-type material. Typical data[112] for 10-MeV electron irradiation are shown in Figs. 19 and 20. For the reasons discussed earlier, we do not feel that it is advisable to interpret these data in terms of specific energy levels without support from measurements as a function of excess density. (The indicated slopes have not been corrected for the temperature dependence of the density-of-states function.)

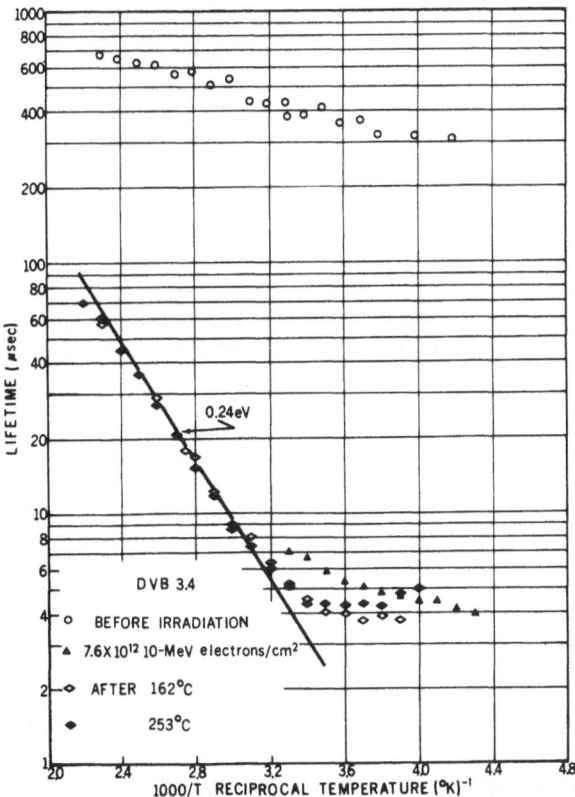

Fig. 19. Temperature dependence of lifetime before and after electron irradiation and following anneal (3.4 Ω-cm, B-doped, vacuum-float-zone Si).[112]

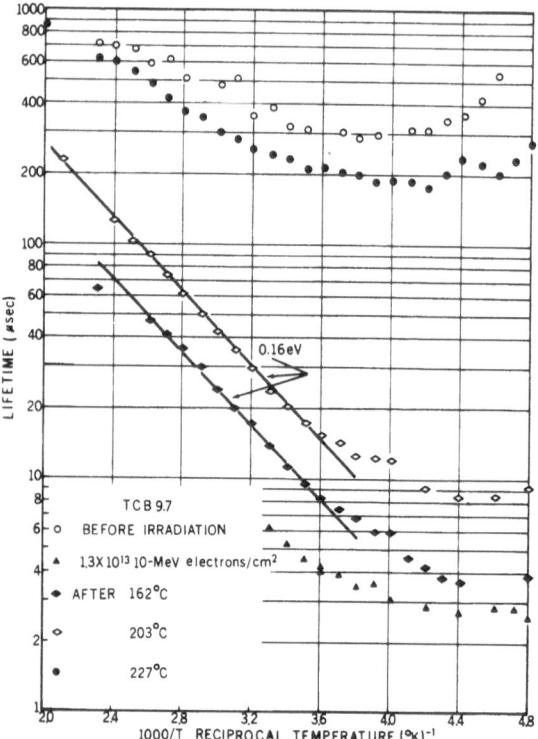

Fig. 20. Temperature dependence of lifetime before and after electron irradiation and following anneal (9.7 Ω-cm, B-doped, Czochralski-grown Si).[112]

4.3.3. Annealing Studies

It should be emphasized that the parameters for recombination given here or in any other work pertain to very specific situations. It is observed that as annealing progresses, the relative importance of a given recombination level varies. The source of the material is important, particularly in regard to its oxygen content. The nature of the irradiating particle is also important. The differences between starting material appear to be less with samples irradiated by 10-MeV electrons than those receiving low-energy electron or gamma irradiation. Figure 21 shows annealing curves for Czochralski (high-oxygen) and vacuum-float-zone (low-oxygen) material.[112] The sample designation in the figure indicates the material source (K for Knapic, and T for Texas Instruments); C and V stand for Czochralski and vacuum-float-zone, respectively; the chemical symbol of the dopant is given next, followed by the numerical value of the resistivity in Ω-cm. While the annealing behavior of these various samples is quite similar, the greatest difference is exhibited by two quite similar samples. The source of this difference will be clear shortly.

These curves, in addition to indicating similar recombination behavior for the various samples, show the presence of a very well-defined annealing stage near 227°C. This same stage is indicated in Czochralski p-type material, but not that which is produced by the vacuum-float-zone technique. Figure 22 displays this difference dramatically.[112] The vacuum-float-zone samples (obtained from Dow Corning) showed strong reverse annealing and did not recover within the annealing range. This difference is consistent with the recombination data of Figs. 19 and 20.

Continued investigation has shown that the annealing stage near 227°C is very similar in ^{60}Co-gamma-irradiated material and is strongly dose dependent, in both n- and p-type material. (This accounts for the differences between the TVP samples of Fig. 21.) A number of identical samples

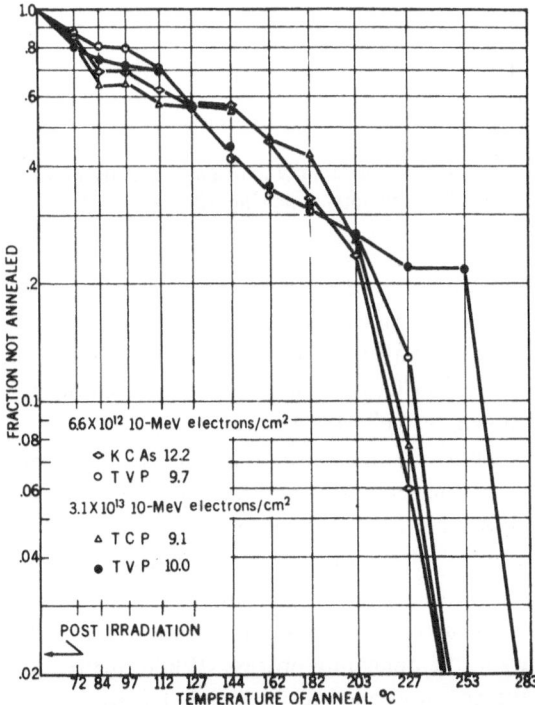

Fig. 21. Fraction of lifetime damage remaining in 10 Ω-cm, n-type samples following successive 30-min anneals.[112]

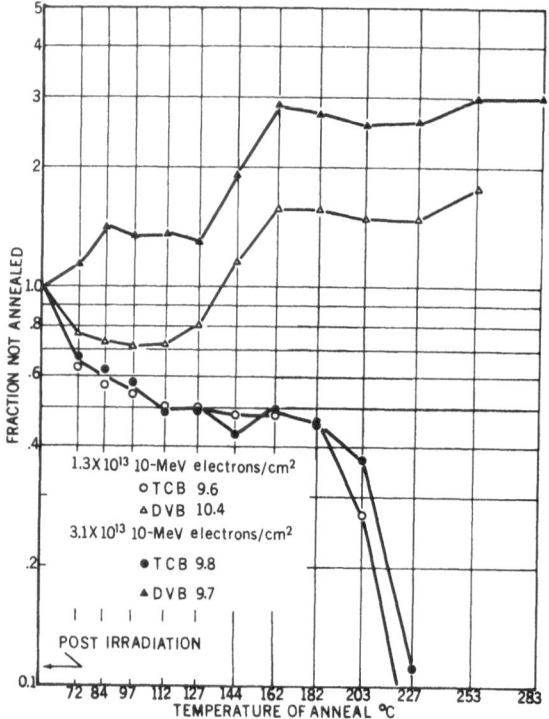

Fig. 22. Fraction of lifetime damage remaining in 10 Ω-cm, p-type samples following 30-min anneals.[112]

(TCB10 in the notation just described) were prepared and irradiated to various levels in a ^{60}Co source. The dose ranged from 8×10^6 to 2×10^8 R, and while the annealing stage remained well defined, it shifted in temperature from $\sim$210 to $\sim$280°C over the range of irradiation described. In all cases, complete anneal was effected with less than 1% of the initial damage remaining. The same five samples were reirradiated to a level of 1.4×10^7 R and subsequently reannealed. Although the initial damage was very nearly the same in these samples, the difference in annealing behavior persisted, indicating a "memory" of the preceding irradiations.

Because of the abruptness of this annealing step, it appears that a simple process is occurring. Evidently it involves a population of sinks which are used up in the annealing process. While this is a very interesting effect, studies have not proceeded to the point that the nature of the interactions responsible for the annealing stage being discussed can be determined.

While it is difficult to draw conclusions from existing recombination studies about the microscopic nature of recombination centers, one experiment performed by Srour[113] does appear to give a strong indication of the nature of one type of recombination center in p-type silicon. These are centers not involved in any of the behavior discussed thus far because they are unstable at room temperature. The method of observation involves the phenomenon known as "short-term annealing." In these experiments a short burst of radiation is given, and the property of interest (in this case carrier lifetime, deduced from steady-state photoconductivity) is observed at a fixed temperature as a function of time.

Such studies using electron irradiation have been performed in both n- and p-type silicon,[114-116] but Srour's data for p-type material are particularly informative. In the room-temperature range, typically about two-thirds of

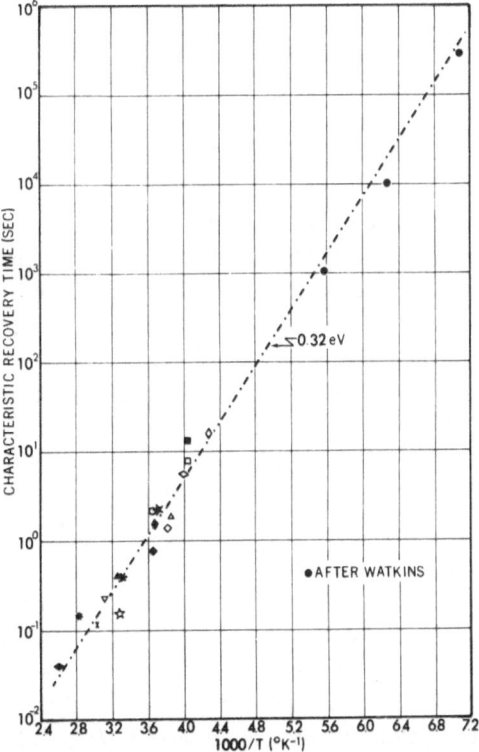

Fig. 23. Characteristic short-term annealing recovery time versus reciprocal temperature for a number of samples. (After Srour.[113])

the damage introduced anneals with a characteristic recovery in a convenient time range. This recovery is first order, i.e., it is dose independent and follows an exponential law. The time constant of this recovery was determined for a number of samples over a temperature range above and below room temperature, and the results are shown in Fig. 23.

A line drawn through the data, based on a least-squares fit, is extrapolated to a region where Watkins performed EPR measurements on electron-irradiated p-type material.[117] The activation energy obtained by Watkins was 0.33 ± 0.01 eV. The agreement of Srour's and Watkins' data in terms of activation energy and absolute magnitude strongly indicates that the defect observed is the same in both instances. Watkins states[118] that this defect is a neutral vacancy. It seems unlikely that a center which is neutral in equilibrium would be effective for recombination, for reasons discussed earlier. However, it is possible that the neutral charge state pertains to the vacancy after its capture of an electron, so that it is neutral to hole capture but positively charged for electron capture. This would produce the large c_n/c_p ratio desirable for an effective recombination center.

5. OPTICAL ABSORPTION

To this point we have been concerned with the effects of defects through the energy states produced in the forbidden gap of semiconductors. These energy levels could also be expected to affect optical absorption because of optically induced carrier transitions between defect energy levels and the conduction or valence band. However, because it is a more sensitive technique, photoconductivity is usually employed to investigate electronic transitions. Still, there are other processes which produce absorption without producing photoconductivity. In some cases these other processes can be used to obtain microscopic information about the nature of the individual defect.

For photon energies greater than the band gap, absorption is determined by electron–hole pair generation, and in wavelength regions corresponding to a significant fraction of band-gap energy, electronic transitions between localized energy levels and the band edges tend to dominate defect effects. In the longer wavelength portion of the spectrum, however, nonelectronic absorption bands are observed. Even electrically inactive defects produce infrared absorption. A notable example of this is the 9.1 μm band in silicon, which is directly related to the oxygen concentration and has been used as a monitor of this impurity.[119] In this section we will deal exclusively with infrared absorption for photon energies that are small compared to the band gap for germanium, but will also discuss some electronic transitions for silicon. Newman has prepared a useful review of infrared vibrational bands involving impurity complexes.[120]

5.1. Germanium

Historically, studies of the electronic properties of defects in germanium have tended to lead those in silicon, but the opposite is true of optical absorption studies. Much more optical work has been performed in silicon, evidently due to the influence of electron paramagnetic resonance studies in silicon, which led to microscopic identification of certain defects. Optical absorption measurements complement EPR studies. However, EPR measurements are much more difficult in germanium than in silicon.

For germanium, absorption studies tend to fall into two regions of the spectrum; the first ranging from the band gap ($\sim$1.9 μm) to $\sim$9 μm, and the second from $\sim$9 to $\sim$17 μm. Absorption in the first range tends to reflect electronic transitions, but has not been studied extensively, since photoconductivity measurements usually serve this purpose better. However, a number of interesting bands appear in irradiated material in the range 9–17 μm (1100–600 cm^{-1}); these will be the object of our discussion here. These bands only appear in material containing oxygen and thus evidently are associated with oxygen-defect complexes. That this is indeed the case was demonstrated conclusively by Whan,[121] who in line with previous studies of silicon,[110] used material which had been doped with ^{18}O-enriched oxygen. She showed that these bands were not present immediately following 2-MeV electron irradiation at 25°K, but appeared upon annealing or after higher temperature irradiation as shown in Figure 24. Corresponding roughly to the $\sim$65°K annealing stage discussed in Section 2.1, two absorption lines appear at 719 and 736 cm^{-1}. These subsequently begin to disappear above $\sim$100°K, and a new line emerges at 620 cm^{-1}. Following subsequent annealing, this line disappears and others appear.

This figure well illustrates the usefulness of infrared absorption studies for investigating defect structures. As one type of structure disappears and others are formed, sometimes no difference in electrical properties will be observed. The defects may be neutral and not observed electrically or they may remain in the same charge state as they change from one configuration to another and not change electrical behavior. However, any such change will be reflected in absorption measurements, since, if there is any change in defect structure, there must be a corresponding shift in the infrared bands. On the other hand, it must be noted that infrared measurements, made in a restricted energy range, will not reflect the presence of every defect.

The growth and disappearance of the various bands were followed by Becker and Correlli[122] (who used 40- to 60-MeV electrons), and by Whan.[121] These data help to clarify the relationships that exist between the various bands. Whan concluded that the 620 cm^{-1} band was associated with the germanium A center. EPR studies have indicated the presence of a germanium A center which anneals in the same temperature range as the 620 cm^{-1} band.[123] Furthermore, the position of the band for the germanium A center can be predicted from that observed in silicon, resulting[121] in a value of 630 cm^{-1}, in good agreement with experiment.

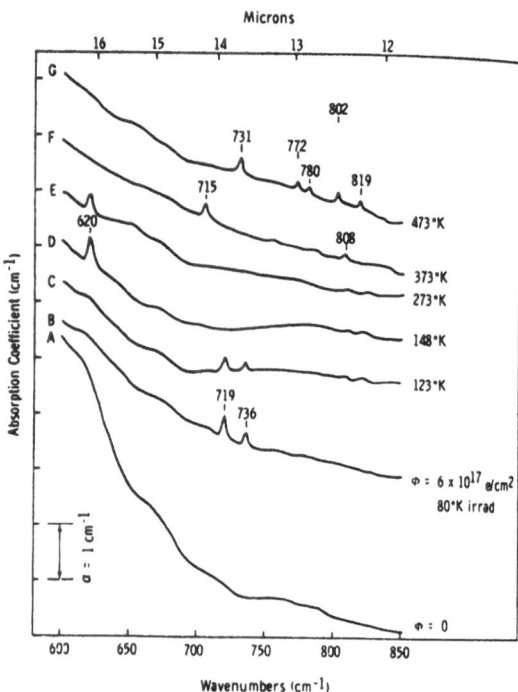

Fig. 24. Infrared spectra of oxygen-doped germanium. Spectra have been vertically displaced for clarity. All spectra recorded at 80°K. A. Preirradiation. B. After 80°K irradiation to $\Phi = 6 \times 10^{11}$ electrons/cm^2. C. After 20-min anneal at 123°K. D. After 20-min anneal at 148°K. E. After 20-min anneal at 273°K. F. After 20-min anneal at 373°K. G. After 20-min anneal at 473°K. (After Whan.[121])

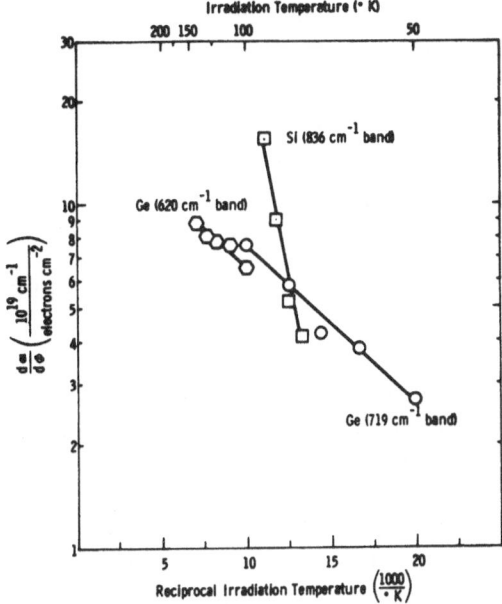

Fig. 25. Dependence of the production rates of the
719 and 620 cm^{-1} bands in Ge on the reciprocal
irradiation temperature. (After Whan.[124])

In a later study,[124] Whan studied some of the bands in greater detail.
Certain bands were dependent upon irradiation temperature, and thus
related to the ITD defects discussed in Section 2.2.2. Temperature dependence for the production rate of two of the bands is shown in Fig. 25 and
is comparatively slight, corresponding to an activation energy of only
~0.01 eV. The 836 cm^{-1} band in silicon, shown for comparison, displays
a much stronger temperature dependence. Whan explains these data on a
basis similar to that discussed in Section 2.2.2.

5.2. Silicon

Many studies have been concerned with the infrared absorption spectra
produced in silicon by radiation, and the early work in this area has been
summarized by Fan and Ramdas.[125] Evidently inspired by the success of
EPR measurements in identifying the oxygen–vacancy complex (A center)
in silicon, Corbett and co-workers[110] correlated the same defect with infrared absorption. They introduced techniques that have influenced many
subsequent studies, including those presented in the preceding section for

germanium. Among the techniques used were the application of uniaxial stress (also to be considered in the subsequent section on photoconductivity) and the use of isotopically enriched oxygen (already discussed in connection with Whan's study of germanium[121]).

The band that Corbett *et al.* concluded to be caused by the *A* center is located at 830 cm^{-1} ($\sim$12 μm). The use of isotopically enriched oxygen showed that it was definitely an oxygen-related center, and that it was specifically the *A* center was supported by the following arguments: (1) It annealed the same way as did the *A* center, based on EPR measurements; (2) it exhibited similar dichroism (dependence on polarization) under stress. Under uniaxial stress, certain types of centers have a preferred orientation, significantly modifying the absorption of polarized light. In these experiments, the absorption of polarized light was monitored at several temperatures as a function of time following removal of the stress. The time constants for return to equilibrium was the same for the 630 cm^{-1} absorption as for the *A* center as observed from EPR measurements. Furthermore, the amount of defect alignment for a given stress applied along particular directions was found to be the same for the 630 cm^{-1} defect as for the *A* center.

In a later paper,[126] Corbett, Watkins, and McDonald considered additional bands in electron-irradiated silicon that had been annealed above 300°C, where the 830 cm^{-1} band (*A* center) begins to disappear. The major bands observed were at 887, 904, 968, and 1000 cm^{-1}. Evidently, these bands grow at the expense of the 830 cm^{-1} band, and thus are due to new complexes formed following disassociation of the *A* center. The authors proposed defect configurations for the origin of these bands, but in view of the limited information available, these models were considered highly tentative.

Ramdas and Rao[127] investigated yet other bands in the infrared region associated with oxygen–defect complexes. Possibly the most interesting aspect of their work was an attempt to determine quantitatively the relative concentrations of defects associated with various bands. They concluded, first of all, that *the total number of defects associated with infrared-active bands was small compared to the number of defects determined on the basis of carrier removal.* Specifically, and quite surprisingly, they concluded that even in oxygen-containing material, the concentration of *A* centers was only $\sim$1% of the total concentration of electrically active defects.

If correct, this conclusion has serious implications in view of the attempt by many to "blame" the *A* center for many changes in electrical properties. In fact, it even raises a question as to whether the level at $E_c - 0.17$ eV belongs exclusively to the *A* center. It is quite possible, for instance, that this level is essentially that which would occur for a free vacancy. Interaction with either a neutral oxygen atom or other neutral impurity might immobilize the vacancy without materially affecting its electronic structure. Even if this is not the case, there may be other vacancy configurations that have the same energy level as the A center. This might, for instance, explain why the A center is often regarded as a recombination center, and yet, the increase in recombination rate in silicon due to irradiation is significantly lower for material containing oxygen.[108]

A surprising point made by Ramdas and Rao is that the A-center absorption band (they give the value as 835 cm^{-1}) does not change position with the charge state of the A center. While this observation may not hold for other bands, whenever it does, it yields a distinct advantage of infrared absorption measurements compared to EPR measurements, because in the latter case, specific charge states are required for defect observation.

Whan also explored the infrared region to 1000 cm^{-1} and observed additional bands,[128] some of which were present following irradiation at 100°K. One particular observation was made which may be relevant to the question of "radiation-induced annealing" (Section 2.1.3), at least to the extent of demonstrating the existence of metastable charge configuration. During thermal anneal, a band at 945 cm^{-1} disappears with a corresponding growth of a band at 956 cm^{-1}. Regeneration of the 945 cm^{-1} band at the expense of the 956 cm^{-1} band occurs upon exposing the sample to light or additional radiation in a manner very similar to that observed in radiation-induced annealing. Whan concludes, reasonably, that the two bands correspond to two charge states of the same defect and that during thermal anneal, the defect does not, in fact, disappear, but merely changes to a charge state which is reversed by light or ionizing radiation. These data seem to give support to some of the arguments presented in Section 2.1.3, that processes that appear to be annealing may simply be changes in charge configuration. These experiments happen to be reversible, so the electronic nature is clear, but this is not necessarily always the case.

Whan and Vook[129] extensively studied the temperature dependence of the creation of certain absorption bands. Again, these bands appear to be associated with the ITD defects discussed in Section 2.2.2. The more intense group located at 836 (A center), 865, 922, and 932 cm^{-1} evidently has its origin in the same primary defect, and was found to display an activation energy for production of 0.05 eV. A less intense set of lines at 936, 945, and 956 cm^{-1} yielded an activation energy of 0.10 eV. These results were explained on the basis of a 0.05-eV difference in the barriers to liberation and recombination of vacancy–interstitial pairs. The resultant temperature

dependence of free vacancy production in turn affects the rate of creation
of vacancy–oxygen complexes. Whan and Vook also formulated a model
where the 0.10-eV activation energy is explained in terms of the creation
of defects containing two initially correlated vacancies.

Oxygen is not the only neutral impurity to be associated with radiation-
induced absorption bands in silicon. Recent studies have considered the
role of carbon. Table III, taken from Bean, Newman, and Smith,[130] is a
good summary of experimental observations and illustrates the complex
nature of the interactions which occur. Their data indicate that complexes
involving both oxygen and carbon are formed, as well as those involving
either carbon or oxygen alone. By considering those bands that correspond
to interstitial and substitutional sites for oxygen and carbon, respectively,
the degree of participation of these impurities in annealing processes can
be observed. Bean, Newman, and Smith observed that the rates of loss of
carbon and oxygen bands were interrelated. They further surmised that
interstitial silicon atoms interact with substitutional carbon atoms to produce interstitial carbon. These atoms are then thought to be mobile at
room temperature and move about to form carbon–oxygen complexes.
Somewhat more recent work of Bean and Newman[132] concentrated on the
922 and 932 cm^{-1} bands in carbon-doped silicon irradiated by 2-MeV
electrons at 105°K. They concluded that these bands do not involve oxygen
as was previously supposed.[130] The involvement of carbon was confirmed
by doping with carbon isotopically enriched with ^{13}C. These bands, which
are not stable at room temperature, were hypothesized by Bean and Newman
to arise from an interstitial carbon atom in a particular site.

Throughout these discussions, attention has been focused on the silicon
vacancy and it appears, to a large extent, that the observations do relate
primarily to one kind of primary defect, since creation of new bands is
usually accompanied by annihilation of others. The question naturally
arises as to the role of the interstitial. Watkins gave a clue to this,[54] at least
in aluminum-doped material, using EPR measurements complemented with
observations of infrared absorption. From analysis of infrared and EPR
spectra, a good argument can be made that the interstitial is mobile, even
at 4.2°K. The interstitials migrate to aluminum substitutional sites where
they interchange positions, creating interstitial aluminum atoms. (See the
conclusion for further discussion.) Devine and Newman[133] found that the
removal rate of substitutional aluminum atoms in heavily doped material
was ∼2 atoms/electron-cm. This is close to the anticipated rate of introduction of silicon interstitials. It is possible that similar processes take place
in materials doped with other impurities.

Table III. Vibrational Bands Observed in Electron-Irradiated Silicon Containing Carbon and Oxygen[130]

Wavelength, μm	Wave number (vacuum), cm^{-1}	Anneal temperature, °C	Maximum peak intensity, cm^{-1}	Inter-pretation	Ref.
13.6–14.2	735–704	(^{a})	0.5	C	—
18.12	551.7	unchanged	0.15	C	—
18.46	541.5	240	0.4	C	—
18.60	537.6	unchanged	0.1	C	—
18.92	528.5	(^{a})	0.4	C	—
8.962	1115.5	340	0.7	O, C?	—
9.260	1079.6	unchanged	0.3	O, C?	—
9.271	1078.3	340	0.8	O, C?	—
9.415	1061.9	360	0.2	O, C?	—
9.631	1038.0	250	0.1	O, C?	—
9.696	1031.1	250	0.06	O, C?	—
9.773	1023.2	180	1.0	O	127
9.956	1004.1	450	0.1	O	126, 127
10.031	996.6	250	0.4	O	—
10.097	990.1	450	0.2	O	126
10.256	974.7	450	0.3	O	126, 127
10.351	965.9	400	0.2	O, C?	127, 131
10.426	958.9	250	0.1	O, C?	127, 131
10.476	954.4	350	0.1	O, C?	127
10.577	945.3	300	0.2	O, C?	—
10.651	938.7	200	0.6	O	—
10.686	935.6	250	0.2	O, C	129
10.744	930.6	400	0.2	O, C?	—
10.989	909.8	450	0.2	O	126, 127
11.179	894.3	330, 480	0.8	O(O_2V)	126, 127
11.476	871.2	350	0.2	O, C?	—
11.556	865.2	350	1.8	O, C?	127
11.876	841.8	480	1.1	O, C?	127, 131
11.975	834.9	350	7.7	O(O–V)	12, 110
12.006	832.7	280, 400	2.0	O	127, 131
12.063	828.8	280, 400	1.4	O	127, 131
12.486	800.7	350	0.15	O	—
12.801	781.0	350	0.15	O	—
13.48	741.5	340	0.6	C, O	—
18.20	549.2	350	0.6	C, O	—

^{a} Apparent superposition of bands with more than one annealing stage observed in the range 250–350°C.

Absorption bands at shorter wavelengths are not directly related to oxygen content, and have also yielded information concerning microscopic defect structure. Those bands of specific interest are a set at 1.8, 3.3, and 3.9 μm. The absorption in this region is illustrated[134] in Figure 26. The 1.8 and 3.3 μm bands were discussed by Fan and Ramdas,[125] who suggested that both belonged to the same defect. However, they were not seen at the same time. Remember that Fig. 26 is a schematic, and it does not necessarily indicate that Cheng saw both levels simultaneously. Evidently the two bands correspond to electronic transitions to a level from the valence and conduction bands, respectively. Preliminary considerations of stress-induced dichroism indicated that electronic reorientation was occurring, and that a certain defect symmetry existed. Cheng and co-workers[134] studied these bands in greater detail, and concluded that they belong to the divacancy. These observations were made on material irradiated with 45-MeV electrons, while those of Fan and Ramdas were primarily with deuterons and neutrons. In these cases the production of divacancies relative to vacancies would be expected to be comparatively high, but these bands evidently are also observed with lower energy electron irradiation.

Cheng's technique for investigating dichroism and some of the results

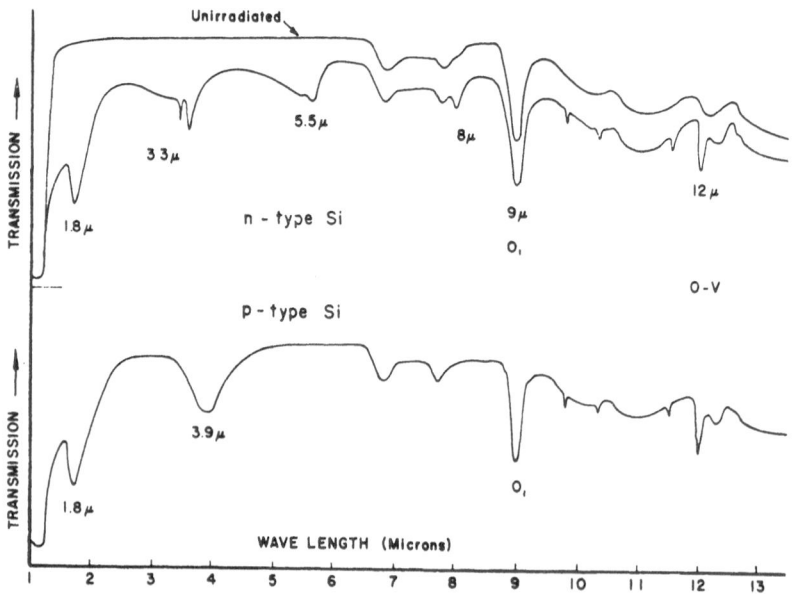

Fig. 26. Schematic diagram of radiation-induced defect infrared absorption bands in silicon (both n and p type) in the wavelength region 1–13 μm. (After Cheng et al.[134])

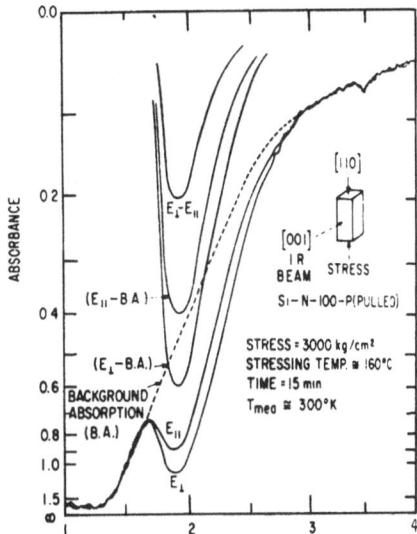

Fig. 27. The quenched-in dichroism exhibited by the 1.8 μm band at room temperature as a result of 15-min stress at ∼160°C. (After Cheng *et al.*[134])

are illustrated in Fig. 27. The nature of the symmetries discovered and the similarities of these observations to EPR measurements were primarily responsible for the authors' conclusion that the defect under observation was the divacancy. Table IV gives experimental values of the dichroic ratios observed for various stress and beam directions, with calculated values based upon the symmetry expected for the divacancy. The authors conclude that these data themselves are a convincing argument that the defect observed is a divacancy, but this is further supported by comparisons with the EPR data, which were also interpreted in terms of a divacancy.[53] The isochronal annealing behavior of dichroism is quite similar for all three bands (1.8, 3.3, 3.9 μm) and agrees well with the EPR data. Furthermore, thermal annealing of the defects as measured by the disappearance of the absorption bands shows consistency with the EPR data.

Thus, Cheng and co-workers give quite convincing evidence that the three bands belong to the divacancy and provide a very useful tool for investigating divacancy behavior. For this purpose, a technique based on observation of these bands is preferable to electronic measurements, because no electronic property is uniquely dependent upon the divacancy, as are these absorption bands.

Table IV. Dichroic Ratios of the 1.8, 3.3, and 3.9 μm Bands after Stressing at 160° for 15 min[a]

Stress direction	Magnitude of stress, kg/cm²	IR beam direction	$\alpha(E_\perp)/\alpha(E_\parallel)$					
			Experimental			Calculated		
			1.8 μm	3.3 μm	3.9 μm	$\theta = -5°$	$-10°$	$-15°$
[110]	2000	[001]	1.29	1.24	—	1.29	1.35	1.40
[110]	3000	[001]	1.44	1.46	—	1.51	1.62	1.73
[110]	3000	[1$\bar{1}$0]	1.27	1.24	—	1.27	1.32	1.47
[110]	1580	[001]	1.26	—	1.25	1.21	1.25	1.29
[111]	2100	[1$\bar{1}$0]	1.33	—	—	1.20	1.24	1.28
[112]	2000	[111]	1.26	—	1.22	1.15	1.19	1.22
[001]	2100	[1$\bar{1}$0]	1.0	1.0	—	1.00	1.00	1.00
[001]	1500	[$\bar{1}$10]	1.0	—	1.04	1.00	1.00	1.00

[a] The angle θ refers to dipole orientation. After Cheng et al.[134]

5.3. III–V Compound Semiconductors

As would be expected, absorption spectra in compound semiconductors are rather complicated. Nonetheless, a number of studies have been performed, particularly with respect to compounds containing various chemical impurities. Newman[120] has summarized the results of various studies on gallium antimonide, gallium arsenide, and indium antimonide. More recently, Spitzer, Kahan, and Bouthillette[135] have investigated electron-irradiated gallium arsenide which contains silicon as a major impurity. Samples with an initial carrier concentration of $\sim 10^{18}$ cm^{-3} were irradiated until a high degree of compensation had been achieved. (Some type of compensation is required to eliminate free carrier absorption. Earlier investigations utilized lithium or copper for this purpose.) Most of the structure observed in the absorption spectra was not associated with radiation-induced defects, since they have been observed in unirradiated material. However, one new band at 369 cm^{-1} appears to be directly related to electron irradiation. The origin of this band has not been established.

6. PHOTOCONDUCTIVITY

Because of their comparative sensitivity, photoconductivity measurements tend to be more useful than absorption measurements for studying processes involving excitations of electrons between one of the bands and a defect level. This very sensitivity does, however, have its drawbacks. In particular, surface-related photoconductivity is readily observable,[136] and care must be taken not to confuse surface and bulk effects. Like other optical measurements, photoconductivity is easier to utilize over a wide temperature range than many electrical measurements. As discussed before for absorption, the combination of uniaxial stress and polarized light is useful in deducing defect structures from photoconductivity behavior. Various aspects of photoconductivity are studied, including the kinetics of filling and emptying defect levels. However, we shall be concerned primarily with steady-state observations of light-induced conductivity measured as a function of photon energy.

In deducing energy level positions from photoconductivity measurements, it is generally assumed that only direct optical transitions are involved. There seems to be no evidence for phonon involvement in the longer wavelength regions. However, near the band edge, oscillations in the photoconductivity of irradiated material have been observed, and these oscillations have been attributed to the emission of transverse acoustical phonons associated with electron capture at an A center.[137]

6.1. Germanium

Much of the work on photoconductivity in irradiated germanium has been performed by a group of Russian authors. Akimchenko, Ginzberg, and Plotnikov[138] utilized very effectively the possibility of performing photoconductivity measurements at very low temperatures. For this study the very high sample resistance that occurs for all but degenerate samples is actually an advantage. The authors described measurements at irradiation temperatures of 5.2 and 100°K, following irradiation with 1-MeV electrons. At these temperatures the Fermi level was very near the majority carrier band so that in p-type material the observed transitions could all be attributed to excitation of electrons from the valence band into impurity levels, whereas in n-type material the excitations must correspond to electrons being excited from the defect levels to the conduction band.

Following irradiation at 5.2°K, a surprising assortment of energy levels was observed. These were given by the authors as 0.22, 0.36, 0.42, and 0.62 eV above the valence band. If all these levels are actually present, they must be associated with simple defects unless some motion of the interstitial and/or vacancy to form complexes has occurred at these low temperatures. The results of Whan[121] discussed earlier for infrared absorption indicate that vacancy–oxygen complexes have not formed, and so the vacancy presumably is not mobile. However, the possibility of interstitial complexes existing at this temperature cannot be ruled out. Even if there were not enough thermal energy for interstitial motion, energy might be derived from the radiation.[139]

Irradiation and measurement at 100°K indicated the same levels, except for the one 0.22 eV above the valence band, which was absent. There was evidence in some samples of at least a few defects with levels at 0.28 and 0.33 eV above the valence band. (All the levels quoted are based on p-type germanium, but results from n- and p-type samples were reasonably consistent.) Measurements were repeated at low temperature following irradiation at room temperature. The levels at 0.36 and 0.42 eV above the valence band were still present, but in addition to the 0.22-eV level, the one at 0.62 eV above the valence band had also disappeared. Two new levels were present at 0.26 and 0.33 eV above the valence band. The kinetics of the photoconductivity process following irradiation at room temperature were different, indicating the presence of many more traps after room-temperature irradiation. It is interesting to note that the level near the center of the gap corresponds closely to that observed from recombination studies (Section 4.2), even though in the latter case the defect appears related to the dopant atoms. It also seems significant that the level ~0.2 eV below the conduction band, which is the dominant level observed from the Hall effect measurements in n-type material irradiated at room temperature, was not observed.

In a later paper Akimchenko, Vavilov, and Plotnikov[140] discussed a level at $E_c - 0.09$ eV, observed from photoconductivity measurements. This level is presumably different from the one 0.62 eV above the valence band, since it is produced only at higher temperatures where the latter is unstable. The apparent dependence upon oxygen concentration led the authors to conclude that this level was due to a vacancy–oxygen complex. However, in view of the small amount of evidence, such a conclusion should be considered quite tentative.

Konopleva and co-workers[141] investigated photoconductivity and Hall effect in material which had been irradiated with high-energy (660-MeV) protons, and found a large number of defect levels. Unfortunately, the damage produced by such high-energy protons is expected to have the same complication that we discussed earlier for neutron irradiation, and thus great care must be exercised in assigning energy level positions from these data to simple defects. Vasil'ev and Smirnov[142] have investigated the photoconductivity of n-type germanium irradiated at room temperature by ^{60}Co gamma rays. Ten different levels are reported, some of which also were observed in infrared quenching experiments. Unfortunately, the brevity of the presentation of these results makes it difficult to assess their significance.

Fischer, Zizine, and Cherki[143] made both photoconductivity and infrared quenching measurements on n-type germanium following electron bombardment near 4.2 and 20°K. Their primary observations led to the conclusion that both photoconductivity and quenching could be attributed to a level 0.20 eV from a band edge, probably nearer the valence band. This presumably would correspond to the level at $E_v + 0.22$ eV observed by Akimchenko, Ginzberg, and Plotnikov.[138] Since the measurements of Fisher, Zizine, and Cherki were made near 4°K, there is additional evidence that this level should belong to one of the simplest types of defects.

6.2. Silicon

The early work on photoconductivity in silicon has been summarized by Fan and Ramdas.[125] Among their conclusions was that the 1.8 and

3.3 μm absorption bands are produced by electronic transitions from the valence and conduction bands, respectively, to an energy level 0.21 eV from the conduction band. Since later work,[134] previously described, shows that these bands belong to the divacancy, the contention that the 0.21-eV level belongs to the divacancy appears to be strengthened. Fan and Ramdas also concluded that the band at 5.5 μm is associated with energy levels in the region $\sim E_c - 0.17$ eV. Thus this band might be associated with the A center. However, as emphasized earlier, it is quite possible, even likely, that defects other than the A center have energy levels near this position.

Vavilov and co-workers[144] investigated the photoconductivity behavior of special high-purity silicon. Following irradiation by 1-MeV electrons at liquid nitrogen temperature, they observed a photoconductivity response apparently caused by a wide distribution of energy levels in the forbidden gap. However, upon heating to room temperature, a few distinct levels emerged. One, 0.45 eV above the valence band, was observed in both oxygen-containing and oxygen-free materials, whereas additional levels 0.16 eV below the conduction band and 0.30 eV above the valence band were observed only if oxygen was present.

Matsui and Baruch[145] reported measurements on n- and p-type silicon irradiated at 20°K by 1.5-MeV electrons. The p-type material indicated energy levels at 0.29, 0.445, 0.525, and 1.0 eV above the valence band. In n-type material, levels were seen at 0.16 eV below the conduction band, and, depending upon whether the material was phosphorus- or arsenic-doped, 0.43 or 0.48 eV below the conduction band. The authors propose tentative models for the microscopic nature of the defects associated with each of these levels.

Kalma and Correlli[146] used a technique reported by Cheng[147] to study the defects giving rise to photoconductivity in silicon. In particular, they studied the photoconductivity associated with absorption bands already shown to be associated with the divacancy. The photoconductivity spectra observed depended upon whether the Fermi level was above or below $E_c - 0.22$ eV, again lending support to the idea that a divacancy level might exist here. With the Fermi level below $E_c - 0.22$ eV, the only level observed was that at $E_c - 0.54$ eV. For higher Fermi level positions, defect levels at 0.36 and 0.39 eV below the conduction band were observed. The dichroism of the latter level was investigated, and is shown in Fig. 28. Curves were obtained by subtracting the effects of the other levels so that the photoconductivity due to the $E_c - 0.39$ eV level was observed alone. The data are in excellent agreement with the results from EPR and IR absorption measurements.[3,134] The dichroism associated with the

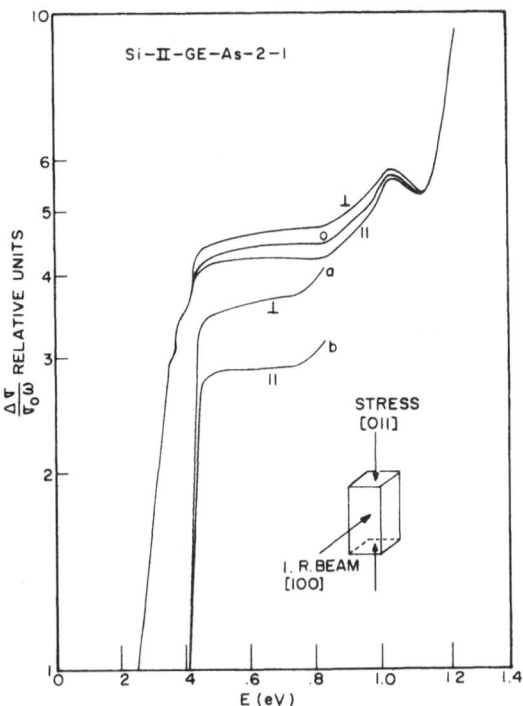

Fig. 28. Dichroism of the $E_c - 0.39$ eV level after high-temperature stress. (After Kalma and Corelli.[146]) As-doped, 1 Ω-cm Si irradiated with 2×10^{16} electrons cm^{-2} at 1.5 MeV. 1540 kg/cm^2 stress applied at 157°C for 15 mm.

$E_c - 0.54$ eV level was very similar, as was the annealing of the stress-induced alignment. Thus it was concluded that the two defect levels belong to the same defect, and are associated with the divacancy.

Cherki and Kalma[148] reported studies on the "D_i" defect in silicon. This defect is slightly different in boron- and aluminum-doped silicon, having an energy level position 0.430 and 0.395 eV above the valence band for boron- and aluminum-doped silicon, respectively. Irradiations and measurements were performed at 4.2, 20, 77, and 300°K, and the defects were produced at all temperatures. Since the nature of the defect was affected by the nature of the dopant atoms, it seems reasonable to conclude that some impurity complex is involved. Furthermore, since the absorption measurements discussed in Section 5.2 strongly indicate that the vacancy does not move below ∼60°K, it appears that an interstitial is somehow involved. The authors investigated the dichroism of this level and concluded that it was consistent with an interstitial impurity atom, evidently in agreement with a model proposed earlier by Watkins.[118] According to this model, silicon interstitials are mobile at very low temperatures and migrate to dopant atoms where an exchange process takes place, leaving behind an interstitial dopant atom. Thus, the D_i defects are dopant atoms in interstitial positions.

Photoconductivity measurements have been performed by Swanson[149] on silicon which had been quenched from high temperatures ($\sim$1000°C). Swanson interpreted his results to indicate an absence of single vacancies, divacancies, or A centers in quenched samples, even though approximately 10^{15} donors/cm³ were introduced into 10 Ω-cm, boron-doped silicon. A prominent photoconductivity step corresponding to an energy level at 0.4 eV above the valence band was attributed to a defect which anneals at about 300°K. The nature of this defect was not identified.

6.3. Other Materials

Extensive use has been made of photoconductivity measurements to investigate defect structure in high-resistivity, wide-band-gap materials. Since these materials are typically much less pure than available silicon and germanium, ample defects for study are found in undamaged material. However, studies also have been performed on various materials following irradiation. For example, Schweinberger and Wruck[150] studied defects introduced by 13.5-MeV deuterons into cadmium sulfide single crystals. Energy levels 0.10, 0.42, and 0.65 eV below the conduction band were indicated. Tentative suggestions as to the nature of these defects were made by the authors. For a discussion of the many ways in which photoconductivity measurements can be used to study defects in solids, the reader is referred to Bube's book[151] on this subject.

7. LUMINESCENCE

When hole–electron pairs are created in semiconducting material, carrier recombination processes often produce light emission. This phenomenon has been widely studied, and indeed has important applications. The light-emitting diode or, in its more sophisticated form, the semiconducting laser, depends on this process for its operation. Scintillation counters and phosphorescent screens are applications of this behavior for wider band-gap materials.

In the materials with which we have primarily dealt, silicon and germanium, only a very small amount of recombination produces luminescence. However, those recombining pairs that do yield light emission provide useful information about the levels through which their recombination occurs, although it is almost certainly true that the centers being investigated are not the dominant recombination centers. Thus the study of defects in germanium and silicon using recombination luminescence has at once an advantage and a disadvantage. The advantage comes about because of the potential capability of studying a level otherwise difficult to investigate, while the disadvantage is that the levels that are important in other processes may not be observable with this technique.

Investigations of luminescence are numerous. However, they primarily involve materials not discussed in this chapter, and the defect centers studied are probably mainly of chemical origin. In this section we will confine ourselves to two of the more interesting and most studied topics in this area—recombination luminescence in irradiated silicon and effects of irradiation on luminescence in gallium arsenide.

7.1. Recombination Luminescence in Irradiated Silicon

Although recombination luminescence in silicon has been studied for some time to obtain information containing chemical impurities,[152–155] radiation effect studies are comparatively new. Yukhnevich and Ivanov studied the emission of light from forward-biased $p-n$ junctions at liquid nitrogen temperature.[156,157] They observed emission lines corresponding to the transition that would occur between the valence band and a level 0.18 eV below the conduction band. On the basis of this observation, they deduced that the silicon A center was involved. Consideration of the line structure led to the hypothesis that light emission was sometimes accompanied by the simultaneous emission of a transverse acoustical phonon. The Russian authors[158–160] extended these studies to observe additional bands associated with deeper levels, and made the observation[159] that certain emission lines were narrow, with widths considerably less than kT. They termed this an optical analog of the Mössbauer effect because of the reasonable argument that such narrow lines could only arise through a Mössbauer-type mechanism.

A comprehensive study of luminescence in irradiated silicon was conducted by Spry and Compton.[161,162] In their experiments, the samples were excited by light whose wavelength was shorter than 7000 Å, and a grating monochromator was used to observe the emission lines. Typical data for

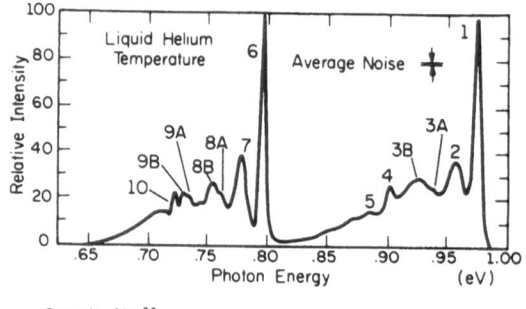

Sample No. 11
Silicon N-Type Pulled
100 ohm cm

Co^{60} Gamma Ray Flux: 10^8 R
14 Days After Irradiation

Band Number N_i	Peak (eV)	Band N_i - Band N_1 (eV)	Band N_i - Band N_6 (eV)	Corresp Bands	Separation Between Corresponding Bands (eV)	Phonon Emitted
1	0.971			6	0.179	zero
2	0.955	0.016		7	0.180	TA
3-A	0.936	0.035		8-A	0.179	2TA
3-B	0.925	0.046				L
4	0.900	0.071		10	0.181	TO+TA
5	0.883	0.088				TO+2TA
6	0.792			1	0.179	zero
7	0.775		0.017	2	0.180	TA
8-A	0.757		0.035	3-A	0.179	2TA
8-B	0.751		0.041			L
9-A	0.730		0.062			TO
9-B	0.727		0.065			O
10	0.719		0.073	4	0.181	TO+TA

Fig. 29. Luminescent spectrum for a Czochralski-grown silicon sample following gamma-ray irradiation. (After Spry and Compton.[161]) Data presented in terms of a constant number of emitted quanta in a constant wavelength interval. Table indicates the energy location of the numbered bands.

oxygen-rich material are shown in Fig. 29. These measurements were made at 4°K. One striking feature of these data is the similarity of the two sets of lines in the higher and lower energy portions of the spectrum. Although there are some differences between the two sets of bands, the fact that they are near replicas of each other is emphasized by the separation between corresponding lines, which is almost identical in every case. The narrow, intense lines must be zero-phonon bands, and the origins of the remaining lines in terms of phonon emission have been analyzed and are shown with the figure.

In material containing less oxygen, only the higher energy group of lines appears. Thus, in spite of their similarity, the two groups must be independent, and the lower energy group must be related to an oxygen–defect complex. By implication, then, the defects that produce the higher energy group do not involve oxygen.

In order to analyze such data, a model must be formulated to account for the transitions giving rise to the observed luminescence. The transitions may be: (1) from a bound state of one defect to that of another, (2) between two bound states of the same defect, and (3) between a bound state of the defect and a free state in one of the bands. For the first process to occur, the participating sites must lie within a certain critical distance of each other, so that the trapped-carrier wave functions overlap. For the donor and acceptor concentrations present in the luminescence experiments, it seems unlikely that appreciable donor–acceptor pair recombination could be occurring. Furthermore, there seems to be no dependence of luminescence on type and amount of III–V dopant atoms.[162] A choice between possibilities (2) and (3) in principle could be based on the linewidth since the analogy to the Mössbauer effect would seem to depend upon bound-to-bound transitions. (Mössbauer-type luminescence spectra have also been reported in gamma-irradiated germanium.[163]) There is some disagreement among the authors[159,162] about the width of the zero-phonon lines, but it does appear that while there is a dependence of linewidth on temperature, the width is always less than kT, and it is difficult to rule out interlevel transitions. It appears that this question is yet to be resolved.

If one assumes that the phononless transition corresponds to a band-to-level process, then the energy level of the defect producing the higher energy spectrum is 0.194 eV from the band edge. This level belongs to a defect which is stable at room temperature, and is apparently independent of oxygen concentration. Continued investigations using uniaxial stress and variation of intensity with temperature indicate it to be a divacancy level.[164] In oxygen-rich material the additional set of lines which appears apparently corresponds to an energy level 0.374 eV from a band edge. Although these experiments are elegant and have energy resolution better than any other technique, there is nonetheless considerable difficulty in analyzing the data. Additional studies in this area should be useful, particularly a detailed study of linewidth as a function of temperature. The ratio of intensities for the sharp line and broad band can also indicate whether Mössbauer-type spectra are being observed.[165]

7.2. Effects of Irradiation on Luminescence in Gallium Arsenide

Gallium arsenide light-emitting diodes are normally fabricated by diffusing zinc or cadmium into an *n*-type substrate. Most of the emitted light comes from the recombination of carriers injected into this *p*-type region. The effect of electron irradiation on the luminescence process in zinc- and cadmium-doped material has been investigated by Arnold, Brice, and Gobeli.[166-168] An example of the emission intensity for zinc-doped gallium arsenide before and after irradiation plus heat treatment is shown in Fig. 30. The irradiations were performed near 80°K and, following irradiation, the broad band was also reduced drastically in intensity. However, annealing near 200°C restored that intensity, and added the structure shown in the figure. The narrow band, near 1.49 eV, is presumably due to the transition associated with electron capture by a neutral acceptor.

The broad band present before irradiation and shown in Fig. 30 is present also in cadmium-doped material, although its peak appears to occur at a slightly lower energy. Several explanations can be made for the

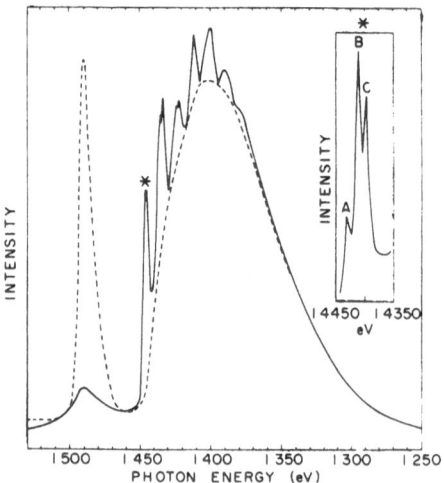

Fig. 30. Emission intensity at 4.2°K for Zn-doped GaAs $(3.8 \times 10^{17}$ holes/cm^3) versus photon energy. (After Arnold and Gobeli.[166]) Dashed line represents intensity before irradiation; solid line represents intensity after 0.6-MeV electron irradiation followed by 15-min anneals at 190, 200, and 210°C. Details of the fine-line structure near 1.44 eV are shown in the insert. (After Arnold and Gobeli.[166])

origin of this band, and Arnold and Brice used the results of radiation effects studies to support one of these. The sharp line structure shown in Fig. 30 appears to be related to the wide band initially present because of the wavelength region over which it occurs. This is substantiated by the temperature dependence of the intensity of the broad band. An Arrhenius plot of the broad-band intensity yields an activation energy of 0.082 eV. On the other hand, the highest energy line of the sharp set that occurs after irradiation and anneal is at 1.44 eV, just 0.08 eV less than E_g. Since this difference is quite close to the above activation energy, there appears to be a relationship between the broad band and the superimposed lines. The structure of the fine lines is evidently caused by phonon replication of the 1.44-eV line, a premise in good agreement with the spacing between those lines. It appears, then, that the original broad band is also due to transitions at 1.44 eV, but broadened and its peak shifted to a lower energy due to coupling to the lattice. Evidently, the appearance of the structure following irradiation indicates that the irradiation has partially decoupled the center from the lattice, but the manner in which this occurs is not clear.

The sharp line at 1.44 eV has fine structure as shown in the inset to the figure. This is thought to arise from the annihilation of an exciton bound to an ionized acceptor. This was confirmed by investigating the line shape as a function of uniaxial stress. The fine structure occurs because of the j–j coupling of the $s_{1/2}$ electron and the $p_{3/2}$ hole. The temperature dependence of the line structure is consistent with this model.

Even though the initial broad-line spectra are very similar in cadmium- and zinc-doped material, the fine structure which develops after irradiation for zinc-doped material just discussed does not occur in cadmium-doped material.[168] There is some line structure, but it is much simpler. Perhaps the lattice decoupling thought to be responsible for the structure in irradiated zinc-doped material does not occur with cadmium.

In cadmium-doped material, the relative intensity of both the narrow- and broad-band emissions were observed to decrease exponentially with electron fluence. This was explained on the basis of a model in which each radiation-induced defect captures all of the injected electrons within a certain radius, thereby quenching emission from that volume. The exponential dependence arises only when these volumes begin to overlap. The exponential intensity decrease and the energy shifts of the narrow and wide bands are interpreted by Arnold to imply that the emissions are all produced by transition between donor–acceptor pairs. The acceptor is neutral cadmium in two different charge states for the narrow- and broad-band emissions, while the donor is unknown, but may be silicon.

The two analyses[167,168] for electron-irradiated zinc- and cadmium-doped material appear inconsistent. In the first case, exciton recombination, and in the second case, donor–acceptor pair recombination, is said to account for the emission bands which nonetheless appear very similar in unirradiated material. Further clarification of this point may be useful. In this connection it is interesting to note the work of Hwang[169,170] on the effect of copper on luminescence in zinc-doped gallium arsenide. After annealing at 550°C in the presence of copper, Hwang observed sharp lines very similar to those observed by Arnold and Brice after electron irradiation and annealing. The fine structure of the highest energy sharp line again shows three sharp peaks, and Hwang also concluded that it is due to bound-exciton recombination. Hwang further agreed that the lower energy lines arise from phonon replication of the one at highest energy. He believed the exciton recombination to be occurring at a defect center, probably an arsenic vacancy copper complex.

There might be some concern that the difference between the work of Arnold and Brice on zinc-doped material[167] and the later work of Arnold on cadmium-doped material[168] might arise because of diffusion of copper into zinc-doped material during anneal. Such a possibility is of particular concern, since the maximum annealing temperature used by Arnold and Brice, 210°C, is above the temperature where significant copper diffusion into gallium arsenide occurs.[171] Furthermore, Hwang observed the same structure in cadmium-doped material into which copper had been diffused.[172] It is, of course, quite possible that, for some reason, copper impurities and radiation-induced defects operate in a similar way in zinc-doped material, but not in material which has been doped with cadmium. This is a point that should be resolved. It might be noted that Hwang interpreted the broad-band emission in unirradiated material differently than did Arnold, presenting a model which attributes it to self-activated luminescence involving arsenic vacancy–acceptor centers. Structure on the broad band has been observed in tin-doped material also.[173]

Brehm[72] studied photoluminescence of gamma-irradiated GaAs, with irradiation performed at room temperature and measurements at 77°K. Measurements were performed on several n- and p-type specimens and in all cases the narrow- and broad-band emissions decreased in intensity, with attendant growth of longer wavelength bands. Barnes[174] considered the degradation of intensity from a laser diode and found, among other things, that intensity degradation was a linear function of dose over two orders of magnitude decrease in intensity. Spectral details were not considered.

8. CONCLUSION

The most convenient way to catalog electrically active defects is to list them according to the position of the energy levels which they introduce into the forbidden gap. Tables V and VI do this for germanium and silicon, respectively. Since the nature of the defects produced by gamma rays and by electrons with energies $\lesssim 2$ MeV are expected to be quite similar, no distinction is made in these tables with respect to irradiating particles except

Table V. Energy Levels in Electron- and Gamma-Irradiated Germanium

Level	Position	Method of observation	Note
_____	$E_c - 0.02$	Carrier concentration	(a)
_____	$E_c - 0.10$	Carrier concentration, photoconductivity	(b)
_____	$E_c - 0.23$	Carrier concentration, lifetime	(c)
_____	$E_v + 0.370$	Lifetime, photoconductivity	(d)
_____	$E_v + 0.338$	Lifetime, photoconductivity	(e)
_____	$E_v + 0.326$	Lifetime, photoconductivity	(f)
_____	$E_v + 0.26$	Carrier concentration, photoconductivity	(g)
_____	$E_v + 0.21$	Photoconductivity	(h)
_____	$E_v + 0.11$	Carrier concentration	(i)
_____	$E_v + 0.02$	Carrier concentration	(j)

a Observed with 4.5-MeV electrons.[18]

b Observed with 4.5-MeV electrons,[18] but apparently absent in γ-irradiated germanium.[19]

c A level near here is frequently observed[18,19,22,23] with specified values varying slightly among observers. It is an acceptor[22,23] and is an important recombination center in n-type material at high excess carried densities.[94] It apparently is not observable with photoconductivity.

d Dominant recombination center in low-resistivity, antimony-doped germanium.[17] Level at $E_v + 0.36$ eV obtained from photoconductivity measurements.[138] Double acceptor in n-type material.[17]

e Dominant recombination center in moderate-resistivity ($1 \lesssim \varrho \lesssim 20\ \Omega$-cm) antimony-doped material appears real, but this level may be due to mixture of 0.326 and 0.370 levels. Double acceptor in n-type germanium.[17]

f Dominant recombination center in arsenic-doped germanium.[17,94] Photoconductivity measurements indicate[138] a level at $E_v + 0.33$ eV. Double acceptor in n-type material.[17]

g Sample-dependent acceptor level.[24] Photoconductivity[138] and carrier concentration data[20,21] in good agreement.

h Value is average of 0.20 and 0.22 from two observers.[138,143] Is produced at 4.2°K and thus probably is a simple defect. Is unstable at 100°K. Perhaps best studied defect in germanium from photoconductivity standpoint. Probably associated with the 35 or 65°K annealing stage.[26]

i Sample-dependent acceptor level.[20,21,24]

j Donor level.[20,21,23]

Table VI. Energy Levels in Electron- and Gamma-Irradiated Silicon

Level	Position, eV	Method of observation	Note
————	$E_c - 0.03$	Carrier concentration	(a)
————	$E_c - 0.14$	Carrier concentration	(b)
————	$E_c - 0.17$	Carrier concentration, photoconductivity	(c)
————	$E_c - 0.21$	Carrier concentration, lifetime, luminescence? photoconductivity	(d)
————	$E_c - 0.36$	Photoconductivity	(e)
————	$E_c - 0.39$	Carrier concentration, lifetime, photoconductivity	(f)
————	$E_c - 0.43$		(g)
————	$E_c - 0.48$	Photoconductivity	(h)
————	$E_c - 0.54$	Carrier concentration, photoconductivity	(i)
————	$E_v + 0.45$	Photoconductivity	(j)
————	$E_v + 0.430$	Photoconductivity	(k)
————	$E_v + 0.395$		
————	$E_v + 0.31$	Carrier concentration, photoconductivity	(l)
————	$E_v + 0.27$		(m)
————	$E_v + 0.21$		(n)
————	$E_v + 0.05$	Carrier concentration	(o)

a Produced by 4.5-MeV electrons.[45]

b Very sample dependent.[44,46] Another level, possibly unrelated, was reported[36] at $E_c - 0.13$ eV, and is observed only above ~220°K and below ~300°K. It appears to arise from disappearance of ITI defects.

c An acceptor, the most commonly observed[42-47] level position in silicon. Position of energy level for A center,[52-54] but strong evidence that other centers, probably other vacancy-neutral impurity complexes, have levels here.

d Evidently a level of the divacancy.[51,138,143,146] Luminescence measurements[162,163] place a divacancy level 0.194 from a band edge at low temperature.

e Present following 1.5-MeV electron irradiation, but not after 45–50-MeV irradiation.[146]

f Shown from photoconductivity studies to be a divacancy.[146]

g EPR investigations[52-54] and photoconductivity data show this level to be that of an "E center" (vacancy–phosphorus pair) and it is an acceptor. The corresponding antimony center appears to be somewhat shallower.[46]

h Observed only in As-doped material,[145] and may be arsenic–vacancy pair pair, analagous to E center.

i Shown to be a divacancy level from photoconductivity studies.[146]

j Observed in high-purity material independent of oxygen concentration.[144]

k This pair of levels is evidently produced by centers arising from interstitial boron or aluminum atoms resulting from exchange with silicon interstitial.[148] ($E_c + 0.430$ level belongs to boron.)

l Only observed in oxygen-containing material.[144,46] Level reported by other observers at $E_v + 0.29$ eV[145] and 0.35 eV.[46]

m Another divacancy level.[53] This makes four, the other three at $E_c - 0.21$, 0.39, and 0.54 eV. The divacancy is evidently neutral when only the $E_v + 0.27$ eV level contains an electron.

n Observed only in vacuum-float-zone silicon.[46,48]

o Produced by 4.5-MeV electrons.[45]

in a few cases where the level is observed following irradiation by 4.5-MeV electrons. It is anticipated, however, that experiments using lower energy electrons or gamma rays will show these levels under the proper conditions. Information concerning the identification of the defects associated with these levels is given where possible. However, because of the limited data for microscopic defect identification in germanium, no attempt has been made to give specific models for this material.

These are not the only defect energy-level positions which can be found in the literature. However, we have used discretion in listing those whose existence seems questionable, or which seem to occur only for very special circumstances, e.g., with special dopants such as tellurium[175] or lithium.[87, 176,177] In several cases where various observers have reported levels close together and which appear to be the same, we have selected a representative value. This could, of course, produce errors. For instance, in silicon, levels 0.39 and 0.43 eV below the conduction band might be confused, but strong evidence indicates that both the divacancy and the E center (vacancy-phosphorus pair) have levels in this region. Again, levels 0.3 and 0.27 eV above the valence band might be thought to belong to the same defect, but the deeper level is only observed in oxygen-containing material, while the other is apparently due to a divacancy. Thus those levels that we have lumped together may also belong to more than one defect, but this is certainly less confusing than listing every observed value.

Statistical weight has been neglected in all cases. Furthermore, no effort has been made to account for the variation of energy level position with temperature. The energy levels correspond to their positions at the temperature of measurement. Of course, these levels do not represent all defects in gamma- or electron-irradiated material. Others which are present may not be observable by electrical measurement, but can be seen through their influence on optical absorption. Other materials are not nearly so well catalogued as are germanium or silicon, and we will make no attempt here to list the defects or the energy levels seen in those materials.

In this chapter on the effect of point defects on the electrical and optical properties of semiconductors, we have, for the most part, provided only a brief introduction to each topic along with a summary of various observations of the particular property being considered. The investigation of a specific electrical or optical property involves its own peculiar experimental techniques and analytical methods. In no case have we attempted to consider in detail the experimental problems associated with the measurements discussed. Neither have we, in general, considered in adequate detail the analysis that is required to determine defect properties from the experimental

observations. However, in the case of recombination studies, where observations of minority carrier lifetime are performed to investigate defect
properties, a fairly complete outline of the analysis with specific examples
has been given. From this outline the reader should be able to understand
just what measurements are required to perform an adequate analysis.

We have tried to point out two general areas in which we feel errors
often occur. First, it seems that "radiation-induced annealing" is often
confused with metastable changes in electronic states of defects. This is a
controversial point. The evidence seems quite strong that a change in the
charge state of a defect can cause that defect to anneal more or less rapidly.
However, in this author's opinion, this is a badly overworked explanation
for observed phenomena. Another frequent error is caused by the desire
of many to associate their observations with a specific microscopic defect.
Since several defects have been identified, the tendency is to choose some
property, usually an energy level position, which is similar to that identified
defect, and thus conclude that the same defect has been observed. For
instance, from Table VI it is easy to see how, for silicon, confusion could
occur between A centers and divacancies, E centers and divacancies, and
the oxygen-related center which has a level 0.3 eV above the valence band
and the divacancy. These are just a few of the more obvious possibilities
for error.

Perhaps the next most common way of identifying defects, which is
probably as dangerous, is to observe annealing behavior and assume that
the defects are the same if they anneal in the same temperature range.
Annealing can be observed at almost any temperature, and often it is defect-
concentration dependent. The optical and EPR measurements which have
been most successful in identifying microscopic defects usually require quite
heavy irradiations, whereas many other experimental techniques use radiations that are relatively light. Defect inventories in the two cases may be
drastically different, as may be the annealing behavior of the same defect.

While we have referred many times to the divacancy, the existence of
which seems to be well established by various techniques, an alternate
model for the experimental observations has been proposed by Masters.[178]
This interesting model postulates that "semivacancy pairs" are responsible
for the observations that have been attributed to divacancies. A semivacancy
pair occurs when two adjacent lattice sites are unoccupied, but a silicon
interstitial exists midway between. In several ways this model is appealing,
particularly with regard to the apparently low activation energies for
vacancy motion, which Masters interprets as being the energy associated
with creation of semivacancy pairs from vacancies rather than actual

vacancy motion. There are difficulties with the model, however, and it will be interesting to see how the questions it raises are resolved in the future. It may be possible, of course, that only part of the observations attributed to divacancies are due to semivacancy pairs. In this case one or more of the[124] "divacancy" levels may actually arise from the semivacancy pair.

Another point seems worthy of special consideration. Defects involving impurities are produced at very low temperatures, even as low as 4.2°K. An example of this is seen in the data of Cherki and Kalma,[148] from which they were able to identify interstitial impurity atoms which had been displaced somehow through the process of irradiation. Related data were obtained earlier by Watkins.[118] In heavily doped material Devine and Newman[133] observed loss of substitutional aluminum atoms at a rate close to the expected rate of silicon interstitial production. Generally, the data have been explained on the basis of silicon interstitial motion and replacement of the impurity by the silicon atom. On an intuitive basis, it has been more appealing to us to suppose that the impurity atoms are displaced during the actual irradiation process. The mechanism of this would be the propagation of energy along an atomic chain with displacement occurring at weak points caused by impurities. Such energy propagation processes occur in close-packed structures[38,179] but it is not clear how effective they might be in a diamond lattice. In any case, it appears dangerous to deduce extremely low activation energies from results of these low-temperature experiments. Koehler and McKeighen have proposed[139] that energy for motion may come from the irradiating particle rather than from heat. The model just outlined would be a special case of their more general statement, but other types of processes might be conceived which would produce the same results.

Consideration of optical absorption data[121–129] leads to the proposal of an experiment which may shed light on the topic of radiation-induced annealing and defect mobility at very low temperatures. Proponents of radiation-induced annealing invoke[34] the concept that defects in the proper charge states are highly mobile, even near 4°K. In the absorption spectra of both irradiated germanium and silicon there are a number of bands that require oxygen for their formation. They are not present following irradiation at low temperatures,[121,128,129] but appear upon thermal annealing. One should be able to irradiate a silicon or germanium sample below the temperature at which these bands are formed and measure the optical absorption. The temperature could then be held constant or decreased to a lower value, even 4.2°K, and the conditions of radiation anneal supplied. Upon returning the sample to the initial irradiation and measurement temperature a reexamination of the data could be made to test for the appearancet o absorption lines. The results should indicate whether defect motion of oxygen atoms has occurred. Appearance of bands could possibly arise from changes in defect charge state,[128] but Ramdas and Rao[127] showed that the bands corresponding to A centers do not depend upon charge state, so if such bands are observed, the existence of true radiation-induced annealing would be strongly indicated.

ACKNOWLEDGMENTS

The author is pleased to acknowledge many helpful discussions with J. R. Srour, A. Bahraman, and S. Othmer.

REFERENCES

1. H. M. James and K. Lark-Horovitz, *Z. Phys. Chem.* **198**, 107 (1951).
2. W. L. Brown, W. M. Augustnyiak, and T. R. Waite, *J. Appl. Phys.* **30**, 1258 (1959).
3. C. D. Watkins, J. W. Corbett, and R. M. Walker, *J. Appl. Phys.* **30**, 1198 (1959).
4. G. Bemski, *J. Appl. Phys.* **30**, 1995 (1959).
5. B. R. Gossick and J. H. Crawford, Jr., *Bull. Am. Phys. Soc.* **3**, 400 (1958).
6. B. R. Gossick, *J. Appl. Phys.* **30**, 1214 (1959).
7. O. L. Curtis, Jr., *IEEE Trans. Nucl. Sci.* **13**, 6, 33 (1966).
8. O. L. Curtis, Jr., in *Lattice Defects in Semiconductors*, Ed. by R. R. Hasiguti (Univ. of Tokyo Press, Tokyo, 1968), p. 333.
9. O. L. Curtis, Jr., *J. Appl. Phys.* **39**, 3109 (1968).
10. W. H. Brattain and G. L. Pearson, *Phys. Rev.* **80**, 846 (1950).
11. W. L. Brown, R. C. Fletcher, and K. A. Wright, *Phys. Rev.* **92**, 591 (1953).
12. S. Mayburg, *Phys. Rev.* **95**, 38 (1955).
13. E. E. Klontz and K. Lark-Horovitz, *Phys. Rev.* **86**, 643 (1952).
14. J. W. Cleland, J. H. Crawford, Jr., K. Lark-Horovitz, J. C. Pigg, and F. W. Young, *Phys. Rev.* **83**, 312 (1951).
15. J. S. Blakemore, *Semiconductor Statistics* (Pergamon, New York, 1962), p. 118.
16. E. Sonder and L. C. Templeton, *J. Appl. Phys.* **31**, 1279 (1960).
17. J. R. Srour and O. L. Curtis, Jr., *Phys. Rev.* **2B**, 4977 (1970).
18. H. Y. Fan and K. Lark-Horovitz, "Irradiation Effects in Semiconductors," Spec. Rept., Purdue University (June 1957).
19. J. W. Cleland, J. H. Crawford, Jr., and D. K. Holmes, *Phys. Rev.* **102**, 722 (1956).
20. N. A. Vitovskii, T. V. Mashovets, and S. M. Ryvkin, *Soviet Phys.—Solid State* **1**, 1266 (1959).
21. N. A. Vitovskii *et al.*, *Soviet Phys.—Solid State* **3**, 727 (1961).
22. S. Ishino, F. Nakazawa, and R. R. Hasiguti, *J. Phys. Chem. Solids* **24**, 1033 (1963).
23. J. C. Pigg and J. H. Crawford, Jr., *Phys. Rev.* **135**, A1141 (1964).
24. J. W. Cleland, R. F. Bass, and J. H. Crawford, Jr., *Appl. Phys. Lett.* **2**, 113 (1963).
25. R. R. Hasiguti and S. Ishino, in *Radiation Damage in Semiconductors* (Dunod, Paris, 1965), p. 259.

26. J. W. MacKay and E. E. Klontz, *J. Appl. Phys.* **30**, 1269 (1959).
27. E. E. Klontz and J. W. MacKay, *J. Phys. Soc. Japan* **18**(Suppl. III), 216 (1963).
28. M. P. Singh and J. W. MacKay, *Phys. Rev.* **175**, 985 (1968).
29. T. A. Calcott and J. W. MacKay, *Phys. Rev.* **161**, 698 (1967).
30. J. W. MacKay and E. E. Klontz, in *Radiation Effects in Semiconductors* (Plenum, New York, 1968), pp. 175–185.
31. R. E. Penczer and H. M. DeAngelis, *Phys. Rev.* **171**, 862 (1968).
32. J. Bourgoin and F. Mollet, *Phys. Lett.* **30A**, 264 (1969).
33. J. Zizine, in *Radiation Effects in Semiconductors* (Plenum, New York, 1968), pp. 186–194.
34. I. Arimura and J. W. MacKay, in *Radiation Effects in Superconductors* (Plenum, New York, 1968), pp. 204–209.
35. H. M. DeAngelis and R. E. Penczer, *J. Appl. Phys.* **39**, 5842 (1968).
36. H. Hiraki, J. W. Cleland, and J. H. Crawford, Jr., *Phys. Rev.* **177**, 1203 (1969).
37. E. E. Klontz and L. L. Sivo, in *Radiation Effects in Semiconductors* (Plenum, New York, 1968), pp. 136–141.
38. J. B. Gibson *et al.*, *J. Appl. Phys.* **30**, 1322 (1959).
39. O. L. Curtis, Jr., *J. Appl. Phys.* **41**, 5297 (1970).
40. T. A. Calcott and J. W. MacKay, in *Radiation Damage in Semiconductors* (Dunod, Paris, 1965), p. 27.
41. B. L. Gregory, *J. Appl. Phys.* **36**, 3765 (1965).
42. T. A. Longo, Ph.D. thesis, Purdue University (1957).
43. G. K. Wertheim, *Phys. Rev.* **110**, 1272 (1958).
44. J. Stannard, personal communication.
45. D. E. Hill, *Phys. Rev.* **114**, 1414 (1959).
46. E. Sonder and L. C. Templeton, *J. Appl. Phys.* **34**, 3295 (1963).
47. C. A. Klein, *J. Appl. Phys.* **30**, 1222 (1959).
48. V. S. Vavilov *et al.*, *Soviet Phys.—Solid State* **4**, 1442 (1963).
49. J. R. Carter, Jr., *J. Phys. Chem. Solids* **27**, 913 (1966).
50. H. J. Stein and R. Gereth, *J. Appl. Phys.* **39**, 2890 (1968).
51. I. E. Konozenko, A. K. Semenyuk, and V. I. Khivrich, *Phys. Stat. Sol.* **35**, 1043 (1969).
52. G. D. Watkins and J. W. Corbett, *Phys. Rev.* **134**, 1359 (1964).
53. J. W. Corbett and G. D. Watkins, *Phys. Rev.* **138A**, A555 (1965).
54. G. D. Watkins, *Phys. Rev.* **155**, 802 (1967).
55. J. H. Crawford, Jr., *IEEE Trans. Nucl. Sci.* **10**, 6, 1 (1963).
56. H. J. Stein and F. L. Vook, *Phys. Rev.* **163**, 790 (1967).
57. F. L. Vook and H. J. Stein, in *Radiation Effects in Semiconductors* (Plenum, New York, 1968), pp. 99–123.
58. G. K. Wertheim, *Phys. Rev.* **115**, 568 (1959).
59. F. H. Eisen, in *Radiation Damage in Semiconductors* (Dunod, Paris, 1965), pp. 163–171.
60. F. H. Eisen, *Phys. Rev.* **123**, 736–44 (1961).
61. F. H. Eisen and P. W. Bickel, *Phys. Rev.* **115**, 345–6 (1959).
62. F. H. Eisen, *Phys. Rev.* **135**(5A), 1394–9 (1964).
63. H. Flicker, J. J. Hoferski, and J. Scott-Monck, *Phys. Rev.* **128**, 2557 (1962).
64. J. J. Loferski and P. Rappaport, *Phys. Rev.* **111**, 432 (1958).
65. W. L. Brown and W. M. Augustyniak, *J. Appl. Phys.* **30**, 1300 (1959).
66. R. Bauerlein, *Z. Naturforsch.* **14a**, 1069 (1959).

67. R. Bauerlein, *Z. Physik* **176**, 498 (1963).
68. G. W. Arnold and F. Vook, *Phys. Rev.* **137**, A1839 (1965).
69. R. J. Khansevarov *et al.*, in *Lattice Defects in Semiconductors* (Univ. of Tokyo Press, Tokyo, 1968), p. 381.
70. J. H. Varley, *Phys. Chem. Solids* **23**, 985 (1963).
71. G. E. Brehm and G. L. Pearson, *IEEE Trans. Electron Devices* **17**, 475 (1970).
72. G. E. Brehm, Ph.D. thesis, Stanford Univ. (1970).
73. R. K. Willardson, in *Lattice Defects in Semiconductors* (Univ. of Tokyo Press, Tokyo, 1968), p. 221.
74. A. A. Abramov, V. S. Vavilov, and L. K. Vodopyanov, *Soviet Phys.—Semiconductors* **4**, 219 (1970).
75. C. E. Barnes and C. Kikuchi, *Rad. Eff.* **2**, 243 (1970).
76. E. W. J. Mitchell and M. J. Moore, *Radiation Damage in Semiconductors* (Dunod, Paris, 1964), p. 235.
77. J. W. Cleland and J. H. Crawford, *Bull. Am. Phys. Soc.* **3**, 142 (1958).
78. J. R. Haynes and W. Shockley, *Phys. Rev.* **81**, 835 (1951).
79. I. A. Baev, *Soviet Phys.—Solid State* **6**, 217 (1964).
80. C. Munakata, *Microelectronics and Reliability* (*Great Britain*) **5**, 267 (1966).
81. E. M. Conwell, and V. F. Weisskopf, *Phys. Rev.* **77**, 388 (1950).
82. P. P. Debye and E. M. Conwell, *Phys. Rev.* **93**, 693 (1954).
83. H. Brooks, in *Advances in Electronics and Electron Physics*, Ed. by L. Morton (Academic, New York, 1955), Vol. 7, pp. 85–152.
84. I. V. Dakhouskii, D. I. Levinzon, and V. A. Shershel, *Soviet Phys.—Solid State* **3**, 896 (1970).
85. D. Long and J. Myers, *Phys. Rev.* **115**, 1107 (1959).
86. C. D. Clark, A. Fernandez, and D. A. Thompson, *Phil. Mag.* **20**, 951 (1969).
87. G. J. Brucker, *Phys. Rev.* **183**, 712 (1969).
88. H. J. Stein and F. L. Vook, in *Radiation Effects in Semiconductors* (Plenum, New York, 1968), pp. 115–122.
89. J. A. Grimshaw, in *Radiation Damage in Semiconductors* (Dunod, Paris, 1964), pp. 378–384.
90. B. R. Gossick and T. H. Yeh, personal communication.
91. R. N. Hall, *Phys. Rev.* **87**, 387 (1951).
92. W. Shockley and W. T. Read, Jr., *Phys. Rev.* **87**, 835 (1952).
93. R. Leadon and J. A. Naber, *J. Appl. Phys.* **40**, 2633 (1969).
94. O. L. Curtis, Jr. and C. A. Germano, in *Radiation Effects in Semiconductors* (Plenum, New York, 1968), pp. 331–338.
95. J. R. Srour and O. L. Curtis, Jr., *J. Appl. Phys.* **41**, 4200 (1970).
96. O. L. Curtis, Jr., *J. Appl. Phys.* **36**, 2094 (1965).
97. O. L. Curtis, Jr. and J. H. Crawford, Jr., *Phys. Rev.* **124**, 1731 (1961).
98. O. L. Curtis, Jr. and J. H. Crawford, Jr., in *Radiation Damage in Semiconductors* (Dunod, Paris, 1964), pp. 143–149.
99. B. G. Streetman, *J. Appl. Phys.* **37**, 3145 (1966).
100. O. L. Curtis, Jr. and J. H. Crawford, Jr., *Phys. Rev.* **126**, 1342 (1962).
101. G. N. Galkin, N. S. Rytova, and V. S. Vavilov, *Soviet Phys.—Solid State* **2**, 1819 (1960).
102. J. A. Baicker, *Phys. Rev.* **129**, 1, 1174 (1963).
103. K. Nakashima and Y. Inuishi, in *Radiation Effects in Semiconductors* (Plenum, New York, 1968), p. 162.

104. T. Nakano and Y. Inuishi, *J. Phys. Soc. Japan* **19**, 851 (1964).

105. R. H. Glaenzer and C. V. Wolf, *J. Appl. Phys.* **36**, 2197 (1965).

106. M. Hirata, M. Hirata, and H. Saito, *J. Appl. Phys.* **37**, 1867 (1966).

107. R. A. Hewes and W. D. Compton, in *Lattice Defects in Semiconductors* (Univ. of Tokyo Press, Tokyo, 1968), p. 291.

108. Y. Inuishi and K. Matasuura, *J. Phys. Soc. Japan* **18**, Suppl. III (1963).

109. O. L. Curtis, Jr. J. R. Srour, and R. B. Rauch, (being prepared for publication).

110. J. W. Corbett, G. D. Watkins, R. M. Chrenko, and R. S. McDonald, *Phys. Rev.* **121**, 4, 1015 (1961).

111. G. Bemski and W. M. Augustyniak, *Phys. Rev.* **108**, 3, 645 (1957).

112. R. F. Bass and O. L. Curtis, Jr., *IEEE Trans. Nucl. Sci.* **15**, 6, 47 (1958).

113. J. R. Srour, *IEEE Trans. Nucl. Sci.* **17**, 6, 118 (1970).

114. I. Arimura and R. R. Freeman, *J. Appl. Phys.* **40**, 2570 (1970).

115. B. L. Gregory and H. H. Sander, *Proc. IEEE* **58**, 1328 (1970).

116. C. E. Mallon and J. A. Naber, *IEEE Trans. Nucl. Sci.* **17**, 6, 123 (1970).

117. G. D. Watkins, *J. Phys. Soc. Japan* **18**(Suppl. II), 22 (1963).

118. G. D. Watkins, in *Radiation Damage in Semiconductors* (Dunod, Paris, 1965), p. 97.

119. W. Kaiser and P. H. Keck, *J. Appl. Phys.* **28**, 882 (1957).

120. R. C. Newman, *Adv. Phys.* **18**, 545 (1969).

121. R. E. Whan, *Phys. Rev.* **140**, A690 (1965).

122. J. F. Becker and J. C. Corelli, *J. Appl. Phys.* **36**, 3606 (1965).

123. J. A. Baldwin, *J. Appl. Phys.* **36**, 793 (1965).

124. R. E. Whan, in *Radiation Effects in Semiconductors* (Plenum, New York, 1968), pp. 195–203.

125. H. Y. Fan and A. K. Ramdas, *J. Appl. Phys.* **30**, 1127 (1959).

126. J. W. Corbett, G. D. Watkins, and R. S. McDonald, *Phys. Rev.* **135**, A1381 (1964).

127. A. K. Ramdas and M. G. Rao, *Phys. Rev.* **142**, 451 (1966).

128. R. E. Whan, *Appl. Phys. Lett.* **8**, 131 (1966).

129. R. E. Whan and F. L. Vook, *Phys. Rev.* **153**, 814 (1967).

130. A. R. Bean, R. C. Newman, and R. S. Smith, *J. Phys. Chem. Solids* **31**, 739 (1970).

131. A. K. Ramdas and H. Y. Fan, *J. Phys. Soc. Japan* **18**(Suppl. II), 33 (1963).

132. A. R. Bean and R. C. Newman, *Solid State Comm.* **8**, 175 (1970).

133. S. D. Devine and R. C. Newman, *J. Phys. Chem. Solids* **31**, 685 (1970).

134. L. J. Cheng, J. C. Corelli, J. W. Corbett, and G. D. Watkins, *Phys. Rev.* **152**, 761 (1966).

135. W. G. Spitzer, A. Kahan, and L. Bouthillette, *J. Appl. Phys.* **40**, 3398 (1969).

136. V. G. Litouchenko, V. P. Kovbasyuk, and P. T. Sviridenko, *Soviet Phys.—Solid State* **8**, 915 (1966).

137. I. P. Akimchenko, V. S. Vavilov, V. A. Vdovenkov, and A. F. Plotnikov, *Soviet Phys.—Solid State* **10**, 2917 (1969).

138. I. P. Akimchenko, M. I. Ginzberg, and A. F. Plotnikov, *Soviet Phys.—Solid State* **8**, 932 (1966).

139. J. S. Koehler and R. E. McKeighen, *Bull. Am. Phys. Soc.* **16**, 396 (1971).

140. I. P. Akimchenko, V. S. Vavilov, and A. F. Plotnikov, *Soviet Phys.—Solid State* **8**, 1577 (1967).

141. R. F. Konopleva, S. R. Novikov, E. E. Rubinova, Yu. A. Zaporoschenko, V. N. Pokrovskii, and L. N. Nikityuk, *Soviet Phys.—Semiconductors* **3**, 948 (1970).

142. A. V. Vasil'ev and L. S. Smirnov, *Soviet Phys.—Solid State* **8**, 484 (1966).

143. J. E. Fischer, J. L. Zizine, and M. Cherki, in *Radiation Effects in Semiconductors* (Plenum, New York, 1968), pp. 219–223.
144. V. S. Vavilov, S. I. Vintovkin, A. S. Lyutovich, A. F. Plotnikov, and A. A. Sokolova, *Soviet Phys.—Solid State* **7**, 399 (1965).
145. K. Matsui and P. Baruch, in *Lattice Defects in Semiconductors* (Univ. of Tokyo Press, Tokyo, 1968), p. 282.
146. A. H. Kalma and J. C. Corelli, in *Radiation Effects in Semiconductors* (Plenum, New York, 1968), p. 153.
147. L. J. Cheng, *Physics Lett.* **24A**, 729 (1967).
148. M. Cherki and A. H. Kalma, *IEEE Trans. Nucl. Sci.* **16**, 6, 24 (1969).
149. M. L. Swanson, *Phys. Stat. Sol.* **33**, 721 (1969).
150. W. Schweinberger and D. Wruck, *Phys. Stat. Sol.* **15**, 355 (1966).
151. R. H. Bube, *Photoconductivity of Solids* (Wiley, New York, 1960).
152. J. R. Haynes, *Phys. Rev. Lett.* **4**, 361 (1960).
153. Ya. E. Pokrovskii and K. I. Svistunova, *Soviet Phys.—Solid State* **6**, 13 (1964).
154. A. Honig and R. Enck, in *Radiative Recombination in Semiconductors*, Ed. by C. B. Guillaume (Dunod, Paris, 1964).
155. Ya. E. Pokrovskii and K. I. Svistunova, *Soviet Phys.—Solid State* **7**, 1478 (1965).
156. Yu. L. Ivanov and A. V. Yukhnevich, *Soviet Phys.—Solid State* **6**, 2965 (1965).
157. A. V. Yukhnevich, *Soviet Phys.—Solid State* **7**, 259 (1965).
158. A. V. Yukhnevich and V. D. Tkachev, *Soviet Phys.—Solid State* **7**, 2746 (1966).
159. A. V. Yukhnevich and V. D. Tkachev, *Soviet Phys.—Solid State* **8**, 1004 (1966).
160. M. V. Bortnik, V. D. Tkachev, and A. V. Yukhnevich, *Soviet Phys.—Semiconductors* **1**, 290 (1967).
161. R. J. Spry and W. D. Compton, in *Radiation Effects in Semiconductors* (Plenum, New York, 1968), p. 421.
162. R. J. Spry and W. D. Compton, *Phys. Rev.* **175**, 1010 (1968).
163. R. J. Spry and J. D. Henes, in *Proc. 1970 Int. Conf. on Radiation Effects in Semiconductors* (to be published).
164. C. E. Jones and W. D. Compton, in *Proc. 1970 Int. Conf. on Radiation Effects in Semiconductors* (to be published).
165. D. B. Fitchen, R. N. Silsbee, T. A. Fulton, and E. L. Wolf, *Phys. Rev. Lett.* **11**, 275 (1963).
166. G. W. Arnold and G. W. Gobeli, in *Radiation Effects in Semiconductors* (Plenum, New York, 1968), p. 435.
167. G. W. Arnold and D. K. Brice, *Phys. Rev.* **178**, 1399 (1969).
168. G. W. Arnold, *Phys. Rev.* **183**, 777 (1969).
169. C. J. Hwang, *J. Appl. Phys.* **39**, 4307 (1968).
170. C. J. Hwang, *J. Appl. Phys.* **39**, 4313 (1968).
171. R. N. Hall and J. H. Racette, *J. Appl. Phys.* **35**, 379 (1964).
172. C. J. Hwang, *Phys. Rev.* **180**, 837 (1969).
173. H. Kressel, H. Nelson, and F. Z. Hawryio, *J. Appl. Phys.* **39**, 5647 (1968).
174. C. E. Barnes, *Phys. Rev. B* **1**, 4735 (1970).
175. J. W. Cleland and J. H. Crawford, Jr., *Phys. Rev. B* **1**, 713 (1970).
176. V. S. Vavilov, in *Radiation Damage in Semiconductors* (Dunod, Paris, 1965), p. 115.
177. J. J. Wysocki, *IEEE Trans. Nucl. Sci.* **13**, 6, 168 (1966).
178. B. J. Masters, *Solid State Comm.* **9**, 283 (1971).
179. R. H. Silsbee, *J. Appl. Phys.* **28**, 1246 (1957).

Chapter 4

ELECTRON PARAMAGNETIC RESONANCE OF POINT DEFECTS IN SOLIDS, WITH EMPHASIS ON SEMICONDUCTORS

George D. Watkins

Solid State and Electronics Laboratory
General Electric Corporate Research and Development
Schenectady, New York

1. INTRODUCTION

Electron paramagnetic resonance (EPR) has proven to be an extremely powerful tool for the study of point defects in semiconductors and insulators. As the various chapters in this and the preceding volume of this series attest, there is hardly an area of point defect study today that does not rely heavily either directly or indirectly upon EPR results for interpretation.

The reason for this is that the EPR spectrum of a defect, properly interpreted, contains highly detailed microscopic information about the structure of the defect which often cannot be obtained in any other way. In the field of color centers and radiation effects in solids, for instance, the ability of EPR simply to sort out and identify the defects has proven in most cases to be the essential first step before any meaningful understanding could be obtained. In addition to identification, EPR experiments can reveal much about the electronic and mechanical properties of the defects, shedding valuable insight into their role in altering the macroscopic electrical, optical, and mechanical properties of the solids in which they are incorporated.

Much of this information can be obtained directly from an analysis of the EPR spectrum alone. The main body of this chapter will be devoted to building up an elementary understanding of what this information is, and how it can be extracted from a spectrum. The three principal interaction terms which are revealed in an EPR spectrum will be described: (1) the hyperfine interaction; (2) the g-tensor; and (3) the fine structure terms. Several examples will be cited to illustrate how each of these may show up in a spectrum and how, in turn, they can be used to reveal the microscopic properties of a point defect.

A rich variety of additional information is often available about point defects when auxiliary techniques are used in conjunction with EPR. Techniques such as applied uniaxial stress, electric fields, optical illumination *in situ*, etc., offer different ways to perturb the defect—to push at it, to probe it, to see how it responds—which reveal important dynamic and energetic features of the defect not available from the static spectrum alone. A special section is included to enumerate some of the more important types of experiments that can be performed using these techniques.

No attempt will be made to catalog all of the point defects that have been studied by EPR. Several reviews are already available for this purpose,* in addition to the various chapters in this series which deal with specific point defect systems.† Instead we will present only a few selected examples for illustrative purposes in the treatment. Consistent with the inclusion of this chapter in the present volume of this series, we have taken most of the examples from defect studies in semiconductors. Equally familiar examples could have been taken from studies in ionic crystals. For many of these, the reader can refer to the excellent reviews in Chapters 4–7 of Vol. 1.

The treatment in this chapter is elementary. It is not aimed at the EPR expert. Rather it is directed at the solid-state physicist and student interested in point defects. The intent is to introduce him sufficiently to the basic concepts involved in EPR and its analysis that he can read an EPR paper and extract intelligently the information relevant to his problem. Perhaps it also may serve to convince him that he should include EPR in his own experimental bag of tools.

---

* Recent reviews of EPR point defect studies include those of Seidel and Wolf[1] (alkali halides), Henderson and Wertz[2] (alkaline earth oxides), Title[3] and Schneider[4] (II–VI compounds), and Feher,[5] Ludwig and Woodbury,[6] and Lancaster[7] (semiconductors).

† See particularly Chapters 6 and 7, Vol. 1.

2. BASIC CONCEPTS

2.1. Origin of Paramagnetism

A fundamental property of an electron is that it has an intrinsic *spin* angular momentum S of magnitude $\frac{1}{2}(h/2\pi)$. Here h is Planck's constant and $h/2\pi$ is the basic quantum unit of angular momentum. In addition, an electron may have *orbital* angular momentum L associated with its planetary circulation around an attractive force center (such as, for instance, in an atom). Orbital angular momentum always has integral values of $h/2\pi$ (0 for an atomic s function, 1 for a p function, 2 for a d function, etc.). Because the electron has an electrical charge, this angular momentum produces circulating electrical currents and therefore a magnetic dipole moment.

We can estimate the magnitude of the dipole moment as follows: Consider an electron of mass m and charge $-e$ in a circular orbit of radius r and period T. The angular momentum L, in units of $h/2\pi$, is

$$Lh/2\pi = mvr = m(2\pi r/T)r$$

and the magnetic moment (in cgs units) is

$$\mu = iA = (-e/cT)(\pi r^2)$$

where c is the velocity of light. This gives

$$\mu = -(eh/4\pi mc)L$$

In conventional vector notation, this becomes

$$\mu_L = -\beta \mathbf{L} \tag{1}$$

where $\beta = eh/4\pi mc$ is the Bohr magneton, the atomic unit for magnetic dipole moments.

This simple classical derivation gives the correct result for the orbital contribution to the magnetic dipole moment. There is no simple classical analog for spin. For this we must rely on the Dirac theory of the electron as refined by a small quantum electrodynamics correction. This theory gives

$$\mu_s = -2.0023\beta \mathbf{S} \tag{2}$$

2.2. Diamagnetic Solids

Electrons in completely filled inner shells of atoms always have their orbital and spin angular momenta each exactly canceling. Also, in insulators and semiconductors, the electrons in the outer valence shells tend to have their orbital and spin angular momenta canceling due to the strong electrical forces associated with the chemical bonds between the atoms. As a result, most such solids in their pure state are diamagnetic.*

In these materials, the paramagnetism of interest arises from point defects in the otherwise perfect diamagnetic solid. A point defect may have a local magnetic moment for any of several reasons: It may be a transition element impurity with a partially filled inner d shell. It may be paramagnetic because it has trapped an extra electron or "hole." It may have unpaired electrons in dangling ruptured chemical bonds as, for instance, near a lattice vacancy. In each case, the magnetic dipole arises from the spin and orbital moments of the "left over" electrons.

2.3. Concept of Resonance

Consider an isolated defect which has a total angular momentum **J** which could arise from either or both spin and orbital sources,

$$\mathbf{J} + \mathbf{L} + \mathbf{S}$$

As before, the magnetic moment can be written

$$\boldsymbol{\mu} = -g\beta\mathbf{J}$$

where g is a constant whose magnitude depends upon the relative contributions of **L** and **S**.

In a magnetic field **H** applied along the z-axis direction, the energy of interaction is given by

$$\mathscr{H} = -\boldsymbol{\mu} \cdot \mathbf{H} = g\beta H J_z$$

It is a familiar result of simple quantum theory that J_z, the z component of the angular momentum J, can assume only $2J + 1$ discrete values, given by $M = -J, -J + 1, \ldots, J$. This is turn leads to $2J + 1$ allowed energy

---

* Notable exceptions to this are the magnetic materials made from transition and **rare** earth elements. For these the magnetism per atom arises from partially filled inner d and f shells. We will not be concerned with these materials in this chapter.

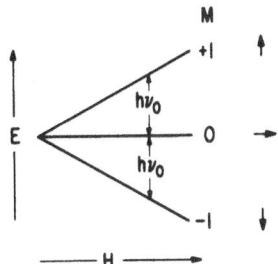

Fig. 1. Quantized energy levels vs. magnetic
field H for $J = 1$.

states

$$E(M) = g\beta HM$$

This is shown in Fig. 1 for the case $J = 1$.

The selection rule for magnetic dipole transitions is $\Delta M = \pm 1$, so that an oscillating magnetic field perpendicular to the z axis at the resonance frequency ν_0, given by the quantum condition

$$h\nu_0 = E(M) - E(M - 1) = g\beta H \tag{3}$$

can cause transitions between the adjacent levels. Transitions of $\Delta M = +1$ (absorption of a quantum $h\nu_0$) and $\Delta M = -1$ (emission of quantum $h\nu_0$) are equally probable. However, for an assembly of such defects in thermal equilibrium, there will be slightly more defects in the lower energy states (reflecting a Boltzmann distribution between the $2J + 1$ states) and a net absorption of energy will result.

This can conveniently be expressed in terms of a complex magnetic susceptibility for the assembly of defects

$$\chi_{x,y} = \chi' - i\chi''$$

as shown in Fig. 2. Here χ'' reflects the absorption due to resonance and χ',

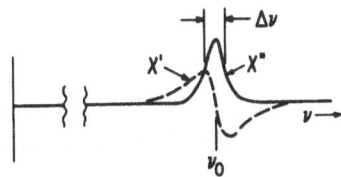

Fig. 2. Complex magnetic susceptibility near
resonance.

the accompanying dispersion. The energy levels of Fig. 1 will always have a finite width and this will show up as a width $\Delta\nu$ for the resonance peak in the susceptibility. It can be shown that near resonance,

$$\chi'_{\text{peak}} \sim \chi''_{\text{peak}} \sim (\nu_0/\Delta\nu)\chi_0$$

where

$$\chi_0 = \tfrac{1}{3}Ng^2J(J+1)/kT \tag{4}$$

is the dc static susceptibility associated with the defects.* Here N is the density of the defects, k is the Boltzmann constant, and T is the absolute temperature. Herein lies one of the big advantages of resonance: its sensitivity. The quantity $\nu_0/\Delta\nu$ defines a "Q," or *quality factor* for the resonance, in direct analogy to the Q of an electrical tuned circuit. For the sharp lines often encountered, it can be $\sim 10^3$–10^4. Resonance methods are therefore proportionately more sensitive than static susceptibility measurements. More important, the resonance condition allows the resolution of different defects present at the same time (different g-values), impossible in static measurements.

It is instructive to have in mind the classical description of magnetic resonance. The classical equation of motion for a gyroscope with angular momentum $\mathbf{J}(h/2\pi)$ is

$$(d\mathbf{J}/dt)(h/2\pi) = \mathscr{E}$$

The torque $\mathscr{E}$ is provided by the interaction of the magnetic moment $\boldsymbol{\mu}$ with $\mathbf{H}$

$$\mathscr{E} = \boldsymbol{\mu} \times \mathbf{H}$$

With $\boldsymbol{\mu} = -g\beta\mathbf{J}$, we obtain

$$d\mathbf{J}/dt = -(2\pi/h)g\beta\mathbf{J} \times \mathbf{H}$$

The solution is well known and corresponds to a simple precessional motion around $\mathbf{H}$ with J_z constant and a precessional frequency

$$\nu_p = g\beta H/h$$

The resonance condition of (3) therefore corresponds classically to the precessional frequency of the electronic top under the forces of the magnetic field trying to align it along the z direction. From Fig. 3 it is also

---

* See Ref. 8. This article, concerned with nuclear magnetic resonance, applies equally well for EPR.

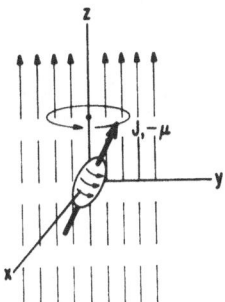

Fig. 3. Precessional motion of an electronic
magnetic moment in a magnetic field.

apparent why only the microwave magnetic field perpendicular to the z
axis is effective in causing transitions. In the precession it is only the x
and y components of μ that oscillate and can therefore interact with an
external oscillating field.

2.4. Experimental Apparatus

With $g \sim 2$ in Eq. (3), a magnetic field of 3500 G corresponds to a
resonance frequency of 10 GHz (X band). At 8500 G, the frequency is
24 GHz (K band). Homogeneous magnetic fields in this range are readily
available in commercial electromagnets, as are the microwave components,
so that most spectrometers tend to be in these bands.

It is general practice to keep the frequency fixed and sweep the magnetic field for resonance. This allows the sample to be placed in a fixed,
tuned resonant cavity where the microwave magnetic field can be highly
concentrated. Resonance can then be detected either as a loss in Q of the
cavity (absorption of energy via χ'') or as a change in its resonance frequency (dispersion via χ'). A simplified block diagram of a spectrometer
used by the author is shown in Fig. 4. Here resonance is monitored as a
change in the power reflected from the cavity. Either dispersion or absorption can be studied by adjusting the phase of the reference signal derived
directly from the klystron source. As shown, provision for cooling the
sample is provided. Operation at cryogenic temperatures can often greatly
increase the sensitivity through the temperature dependence of χ_0, Eq. (4).
Further gains also often occur through the decrease in the linewidth $\Delta \nu$
at low temperature.

To further increase the sensitivity, low-frequency magnetic field modulation is used. This serves to bring the sample in and out of resonance

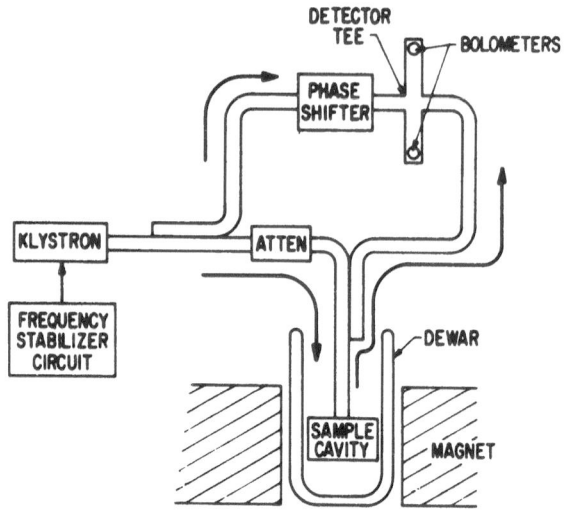

Fig. 4. Block diagram of an EPR spectrometer.

at the modulation frequency and the corresponding ac signal is amplified, phase-sensitive-detected, and fed to a recorder as the dc magnetic field is slowly swept through resonance. For study of the angular dependence of an EPR spectrum, provision for rotating the magnet (or rotating the crystal in the cavity) is required.

Several additional refinements can greatly increase the power of EPR in defect studies. These include provision for applying *in situ* to the sample: (1) uniaxial stress, (2) dc electric fields, (3) monochromatic light (preferably polarized), (4) X-rays, γ-rays, or high-energy particle irradiation, and (5) an additional rf magnetic field for ENDOR studies (see Section 3.2.4). The ability to change and control the temperature allows the study of annealing processes, and complex motional processes associated with the defect. These refinements will be discussed in detail in Section 5.

3. THEORY OF EPR FOR DEFECTS IN SOLIDS

3.1. Quenching of Orbital Angular Momentum

For many of the defects encountered in solids, the orbital angular momentum is strongly quenched. Viewed physically, electrical charges of nearby atoms cause hills and valleys in the potential for the electrons, hindering their free circulation.

Quantum mechanically, the quenching of orbital angular momentum means that the expectation value for all components of **L** is zero (or very close to zero) in the ground state. To see how this comes about, consider an electron in a p function of an atom. For the free atom, the p state is threefold degenerate with orbital angular momentum $L = 1$. In terms of the three functions directed along the perpendicular axes x, y, z [i.e., $p_x = xf(r)$, $p_y = yf(r)$, etc.], the three eigenstates of L_z are

$$(p_x + ip_y)/\sqrt{2}, \qquad \langle L_z \rangle = M_L = +1$$

$$p_z, \qquad M_L = 0$$

$$(p_x - ip_y)/\sqrt{2}, \qquad M_L = -1$$

corresponding to whether **L** is pointing parallel, perpendicular, or antiparallel to the z direction. These derive directly from the matrix elements for L_z

$$\langle p_y | L_z | p_x \rangle = i = -\langle p_x | L_z | p_y \rangle \qquad (5)$$

all others being zero. [Those for L_x and L_y also follow from (5) by cyclic permutation of the indices x, y, z.]

Placed in a crystal field of low symmetry, this degeneracy can be removed. An example is shown in Fig. 5. Here the p_x orbital is lowered in energy because the lobe of the wave function concentrates the electronic charge density along the x axis where it interacts with the positive charges. In the same way, the p_z orbital is raised in energy. Now the correct eigenstates are p_x, p_y, and p_z and the expectation value of **L** is zero in each of

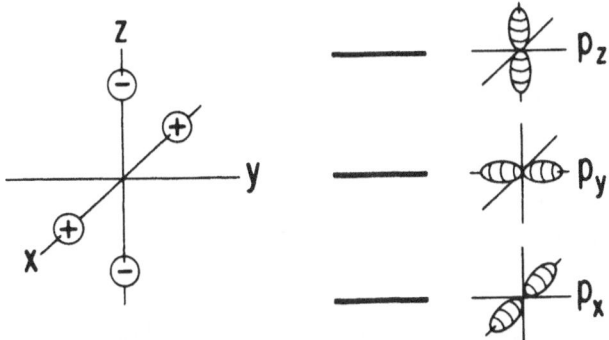

Fig. 5. Electrical charges of nearby ions produce an electrostatic field that lifts the degeneracy of a p function on an atom at the origin.

the states. This follows directly from (5), i.e.,

$$\langle i \mid L_j \mid i \rangle = 0$$

The quenching of orbital angular momentum is therefore directly related to the lifting of the orbital degeneracy by the crystalline electric fields.

When this quenching occurs, the magnetism arises primarily from the spin,* and

$$\mu \simeq -g\beta S \tag{6}$$

with g very close to 2.0023.

Placed in a magnetic field H,

$$\mathscr{H} = -\mu \cdot H \simeq g\beta S \cdot H \tag{7}$$

and the resonance condition is simply

$$h\nu_0 = g\beta H \tag{8}$$

3.2. Hyperfine Interactions

3.2.1. Changes in the Spectrum

Strictly speaking, the magnetic field in Eqs. (7) and (8) should have been the *total* field seen by the electron. This is the externally applied field H plus any additional fields H_{loc} arising from local sources near the defect. In particular, neighboring atomic nuclei also can have spin angular momentum I and small associated magnetic dipole moments

$$\mu_n = g_n \beta_n I \tag{9}$$

Here β_n is the nuclear magneton $eh/4\pi Mc$, where M is the proton mass ($\sim 1840m$). Nuclear moments are therefore characteristically a factor ~ 2000 smaller than electron moments. The weak stray fields from neighboring atomic nuclei may add to the total field seen by the electron and produce an apparent shift in the resonance frequency.

We can write this as

$$\mathscr{H} = -\mu \cdot H_{tot} = g\beta S \cdot (H + H_{loc}) \tag{10}$$

---

* Because this is often the case, EPR is often referred to as ESR (electron *spin* resonance).

Consider a single nearby nucleus as shown in Fig. 6. The magnitude and direction of the field seen by the electron clearly depend upon the orientation of the nuclear spin $\mathbf{I}$, and the direction and magnitude of the separation vector $\mathbf{r}$. In a most general way we can write it as

$$\mathbf{H}_{\text{loc}} = \mathbf{A} \cdot \mathbf{I}/g\beta \tag{11}$$

where the tensor $\mathbf{A}$ contains the magnitude of the nuclear dipole and all of the position- and angle-dependent parts. Substituting (11) into (10), we obtain the compact notation generally found in the literature

$$\mathscr{H} = g\beta\mathbf{S} \cdot \mathbf{H} + \mathbf{S} \cdot \mathbf{A} \cdot \mathbf{I} \tag{12}$$

The quantum mechanical solution of (12) gives for the allowed energy states (first-order perturbation theory)

$$E(M, m) \simeq g\beta HM + AMm \tag{13}$$

with the resonance condition now ($\Delta M = \pm 1$, $\Delta m = 0$)

$$h\nu_0 \simeq g\beta H + Am = g\beta(H + Am/g\beta) \tag{14}$$

Here m is the nuclear azimuthal quantum number which, in direct analogy to M for the electronic states, takes on the $2I + 1$ values $-I$, $-I + 1, \ldots, +I$. A is given by

$$A = (A_1{}^2 n_1{}^2 + A_2{}^2 n_2{}^2 + A_3{}^2 n_3{}^2)^{1/2} \tag{15}$$

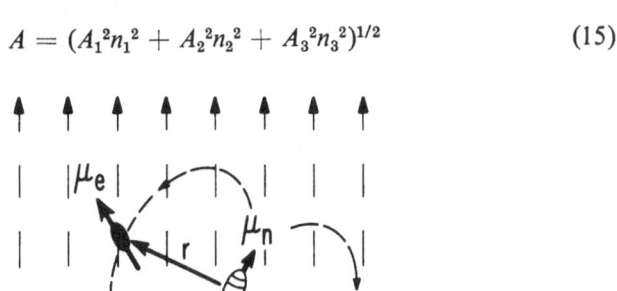

Fig. 6. The magnetic interaction between an electronic dipole μ_e and that of a nearby nucleus μ_n.

where A_1, A_2, A_3 are the principal (extremum) values of the hyperfine tensor **A** and n_1, n_2, n_3 are the direction cosines of the magnetic field direction with respect to the 1, 2, 3 principal axes of the A-tensor.*

A simple physical derivation of (13)–(15) is as follows. The electronic spin quantizes along (precesses about) the direction of **H** (unit vector direction $\hat{n}$). The nucleus **I** cannot follow the rapid precessional motion of the electron, so that it sees only its average value

$$\langle \mathbf{S} \rangle = M\hat{n}$$

The Hamiltonian seen by the nucleus is then, from (12),

$$\langle \mathcal{H} \rangle \simeq g\beta HM + M\hat{n} \cdot \mathbf{A} \cdot \mathbf{I}$$

(we ignore here the small direct interaction of **I** with **H**). Thus **I** quantizes along $\hat{n} \cdot \mathbf{A}$, leading directly to (13), with A given by (15) just the magnitude of $\hat{n} \cdot \mathbf{A}$.

At any one site, the nucleus will have a specific single value for m, and, according to (14), resonance will occur at a magnetic field value shifted by the local field $Am/g\beta$. Since in an EPR experiment we are observing many similar defect centers at the same time, statistically there will be equal numbers of the centers with each of the $2I + 1$ values for m. The result is the observation of $2I + 1$ equally intense, equally spaced lines with separation $A/g\beta$. The magnitude of the separation will depend upon the orientation of the magnetic field. According to (15) when the magnetic field is pointing along the 1 axis ($n_1 = 1$, $n_2 = n_3 = 0$), A is given by A_1. Along the 2 axis, $A = A_2$, etc. At intermediate orientations, the value of A varies smoothly between the three principal values, as given analytically by (15). Throughout, the overall pattern remains centered around $h\nu_0/g\beta$.

3.2.2. Examples

An example of hyperfine interaction with a single nucleus is given in Fig. 7. This is a spectrum produced by high-energy electron irradiation of

---

* An orthogonal set of axes can always be found which allows the hyperfine constant A to be expressed in this form. We define these axes (1, 2, 3) as the *principal axes* and the values A_1, A_2, A_3 as the *principal values* of the hyperfine tensor. If **A** is a symmetric tensor, as is often assumed in EPR analysis, these axes correspond to the coordinate axes that diagonalize the tensor ($A_{ij} = 0$ if $i \neq j$) and the principal values are the diagonal components A_{ii}. It is not necessary to assume **A** to be symmetric, however, to derive (15).

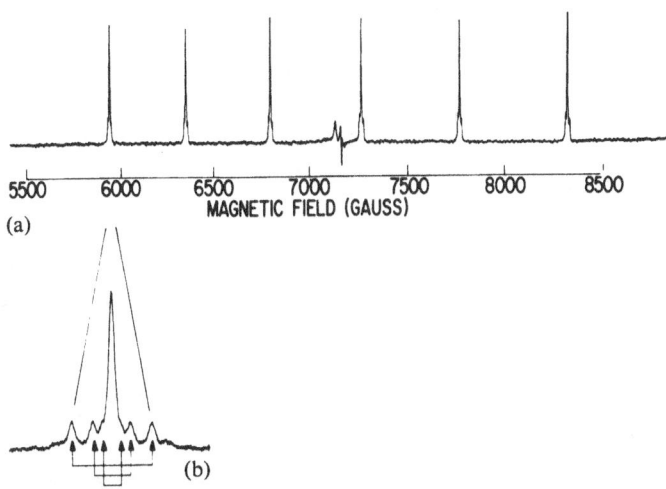

Fig. 7. (a) EPR spectrum of interstitial Al^{2+} in irradiated silicon, with $\nu_0 \sim 20$ GHz. (b) Expanded magnetic field scale reveals satellites arising from hyperfine interaction with nearby ^{29}Si nuclei.

aluminum-doped silicon.[9] The $2I + 1 = 6$ hyperfine components immediately identify it as arising from aluminum (^{27}Al, $I = 5/2$, 100% abundant). Note also that each hyperfine line displays weak satellite structure when the magnetic field is expanded. This can be explained by an additional weaker hyperfine interaction with neighboring silicon atoms, the interaction occurring with 4.7% abundant ^{29}Si, $I = 1/2$. Satellites can arise each time one of the near-neighbor silicon atoms happens to contain a ^{29}Si nucleus. The additional hyperfine interaction with it causes a further splitting into $2I + 1 = 2$ lines symmetrically disposed around each of the unsplit lines. In the figure, several pairs of satellites can be seen corresponding to interaction with a corresponding number of nonequivalent silicon neighbor sites. The satellite pairs are weak, reflecting the low (4.7%) abundance of ^{29}Si.

In cases like this where interaction with more than one nucleus is involved, (12) should be written

$$\mathscr{H} = g\beta \mathbf{S} \cdot \mathbf{H} + \sum_j \mathbf{S} \cdot \mathbf{A}_j \cdot \mathbf{I}_j \qquad (16)$$

where the sum is over each of the relevant nuclei. The resonance condition now becomes

$$h\nu_0 = g\beta H + \sum_j A_j m_j \qquad (17)$$

By studying the angular dependence of the satellites in Fig. 7, the hyperfine

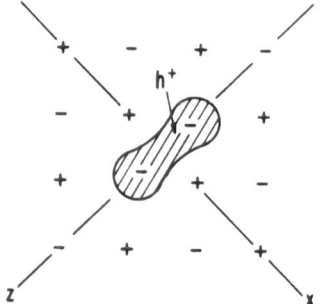

Fig. 8. Model of the V_k center in KCl. A hole is trapped on two Cl⁻ ions to form a localized Cl_2^- molecule.

tensors A_j have been determined[9,10] and shown by their symmetry to arise from ^{29}Si in the nearest several silicon site shells surrounding the *interstitial* site in the lattice. In this way it could be concluded that the aluminum ion must be in the interstitial position. Originally substitutional, irradiation apparently caused the ions to be ejected into their interstitial position. We will not describe this analysis in detail. Rather let us consider a different center, which involves the same kind of analysis.

A particularly dramatic example of the use of hyperfine structure as a diagnostic tool is in the study of the self-trapped hole in KCl. Originally discovered by Kanzig and Castner,[11] this "color center" is known as the V_k center. The model deduced by these workers is shown in Fig. 8. Here a "hole" is shared equally between the two halogen ions which have pulled together to form a Cl_2^- ion oriented along a $\langle 110 \rangle$ direction in the crystal.

The spectrum is shown in Fig. 9. All of the prominent lines follow immediately from the model when it is recognized that there are two isotopes of chlorine, ^{35}Cl and ^{37}Cl, both of spin 3/2. Chlorine-35 is 75% abundant, chlorine-37 is 25% abundant, and $\mu_{37}/\mu_{35} = 0.83$. The prominent equally spaced lines of $1:2:3:4:3:2:1$ intensity arise from ^{35}Cl–^{35}Cl molecules, which can be seen as follows: From Eq. (17) with A the same for both nuclei, the lowest field line arises from $(m_1 = 3/2, m_2 = 3/2)$. The next line arises from both $(3/2, 1/2)$ and $(1/2, 3/2)$ and hence has twice the intensity. The next arises from $(3/2, -1/2)$, $(1/2, 1/2)$, and $(-1/2, 3/2)$, of three times the intensity, etc., the separation between two adjacent lines being $A/g\beta$ throughout. This is illustrated in Fig. 10. Shown also are the predicted spectra for the ^{37}Cl–^{35}Cl and ^{37}Cl–^{37}Cl molecules where the percentage abundance and relative magnetic moments are scaled proportionately. The $1:1$ correspondence between Figs. 9 and 10 serves as a striking

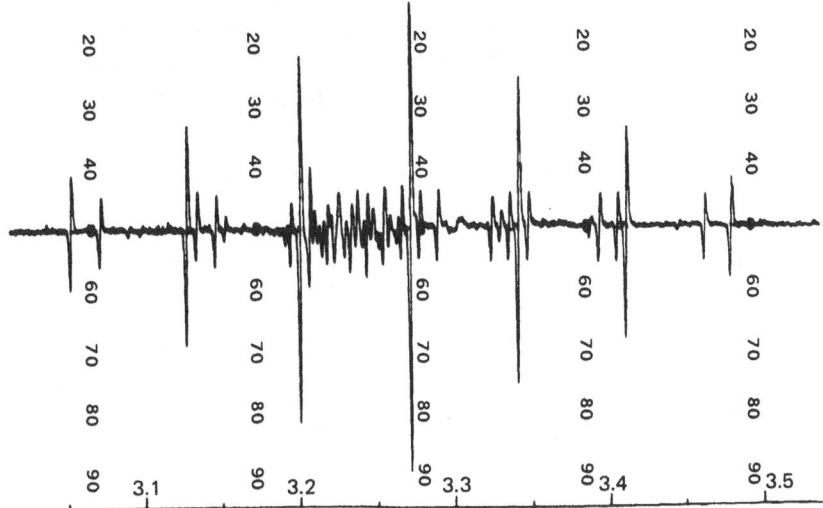

Fig. 9. EPR of the V_k center in KCl, $\nu_0 \sim 9$ GHz, with the magnetic field oriented along a $\langle 100 \rangle$ axis of the crystal. (From Castner and Kanzig.[11])

proof of the model. The extra lines near the center of the spectrum arise from V_k centers which are oriented along other $\langle 110 \rangle$ directions (perpendicular to the applied **H** direction) and can also be accounted for in a complete analysis.

The hyperfine tensors **A** were found to have axial symmetry around the $\langle 110 \rangle$ molecular axis consistent with the model. They are strongly aniso-

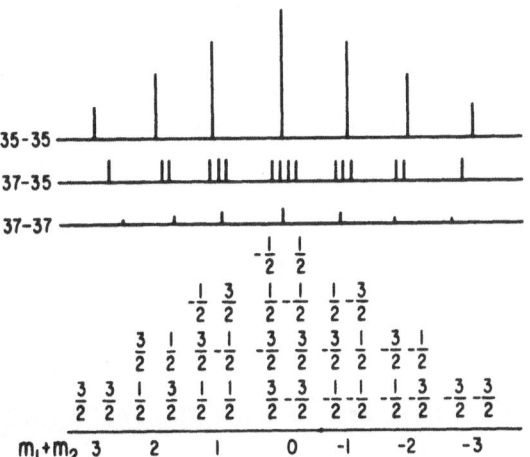

Fig. 10. Predicted spectrum for the V_k center in KCl, showing isotope effects. (After Castner and Kanzig.[11])

tropic, the splitting $A_{35}/g\beta$ ranging from 101 G with **H** along the molecular axis to 9 G in the perpendicular direction. This tells us additional information about the defects which we explore in the next section.

3.2.3. *Quantitative Aspects of the hf Interaction*

The classical dipole–dipole interaction between two magnetic dipoles $\boldsymbol{\mu}_e$ and $\boldsymbol{\mu}_n$ separated by **r** (Fig. 6) is

$$\mathscr{H}_{dd} = \left[\frac{\boldsymbol{\mu}_e \cdot \boldsymbol{\mu}_n}{r^3} - 3 \frac{(\boldsymbol{\mu}_e \cdot \mathbf{r})(\mathbf{r} \cdot \boldsymbol{\mu}_n)}{r^5} \right] \tag{18}$$

To this must be added the "Fermi contact term"*

$$\mathscr{H}_c = -(8\pi/3)\boldsymbol{\mu}_e \cdot \boldsymbol{\mu}_n \, \delta(\mathbf{r}) \tag{19}$$

which is applicable when there is no separation ($\mathbf{r} = 0$).

Substituting $\boldsymbol{\mu}_e = -g\beta\mathbf{S}$ and $\boldsymbol{\mu}_n = +g_n\beta_n\mathbf{I}$, these can be written in the form of (12)

$$\mathbf{S} \cdot \mathbf{A} \cdot \mathbf{I}$$

where the tensor components of **A** are given by

$$A_{ij} = gg_n\beta\beta_n \left[\left\langle \frac{3x_i x_j}{r^5} - \delta_{ij} \frac{1}{r^3} \right\rangle + \frac{8\pi}{3} |\Psi(0)|^2 \, \delta_{ij} \right] \tag{20}$$

By the angular brackets, we mean an average taken over all of the electron positions in its orbit. In quantum-mechanical language, this is the expectation value, or matrix element, of the enclosed function in the ground electronic state. The term with $|\Psi(0)|^2$, the amplitude of the wave function at the nucleus, comes directly from the matrix element of the Dirac delta function $\delta(\mathbf{r})$.

Equation (20) gives a direct recipe for calculating the hyperfine interaction tensor once the electronic wave function is known. For an atomic s function, the interaction with the central nucleus comes only from the $|\Psi(0)|^2$ term, the remaining terms averaging to zero over the spherically symmetric s state. Therefore for an s function, **A** is isotropic with

$$A_{xx}^s = A_{yy}^s = A_{zz}^s = a = (8\pi/3)gg_n\beta\beta_n |\Psi_s(0)|^2 \tag{21}$$

---

* A simple classical derivation is given for this term in Ref. 12, pp. 84ff, where a rigorous quantum-mechanical derivation is also outlined.

For an atomic p function, $|\Psi(0)|^2$ is zero. For the function p_z with its lobe directed along the z axis, the hyperfine tensor is axially symmetric around the z axis with

$$A_{zz}^p = -2A_{xx}^p = -2A_{yy}^p = 2b = \tfrac{4}{5}gg_n\beta\beta_n\langle r^{-3}\rangle_p \qquad (22)$$

Tabulated Hartree–Fock wave functions for most atoms are available from which $|\Psi(0)|^2$ and $\langle r^{-3}\rangle$ can be estimated for the orbital of interest. In addition, accurate experimental estimates of $\langle r^{-3}\rangle$ can be obtained from atomic fine structure splittings. Table I shows estimated values for these quantities for the outer $3s$, $3p$ valence states of neutral ^{27}Al, ^{29}Si, ^{31}P, and ^{35}Cl along with the corresponding estimates for a and b.

Returning to the interstitial aluminum spectrum, Fig. 7, the ^{27}Al hyperfine interaction is isotropic and we can conclude therefore that it arises from an s function. The splitting $A/g\beta$ of 471 G corresponds to a/h of 1321 MHz, which is $\sim 1/3$ that expected for a free atom $3s$ function. From this it has been concluded that the charge state is Al^{2+} (free ion configuration $1s^22s^22p^63s^1$), the reduction of a from the free ion value indicating a spreading of the wave function into the surrounding lattice. Consistent with this is the sizable overlap with the neighboring silicon atoms as indicated by the ^{29}Si hyperfine satellite structure in Fig. 7.

In analyzing such problems, where the electronic wave function overlaps many atom sites, it is often convenient to construct a wave function for the unpaired electron as a linear combination of atomic orbitals centered on each of the sites involved

$$\Psi = \sum_j \eta_j\psi_j \qquad (23)$$

Table I. Hyperfine Parameters Estimated^a for the $3s$ and $3p$ Wave Functions of Neutral ^{27}Al, ^{29}Si, ^{31}P, and ^{35}Cl Atoms

Atom	$\|\psi_{3s}(0)\|^2$, 10^{24} cm^{-3}	$\langle r_{3p}^{-3}\rangle$, 10^{24} cm^{-3}	a/h, MHz	b/h, MHz
^{27}Al	20.4	8.95	3520	73.8
^{29}Si	31.5	16.1	4150	101
^{31}P	41.6	24.2	11150	310
^{35}Cl	80.4	49.5	5220	153

^{a} Watkins and Corbett.[25]

Here $\eta_j{}^2$ represents the fraction of the total wave function that is localized in the atomic orbital ψ_j centered on each atomic site j. At site j, ψ_j is further broken down into a linear combination of the various s, p, etc. *atomic orbitals on that site*

$$\psi_j = \alpha_j(\psi_{ns})_j + \beta_j(\psi_{np})_j + \cdots \tag{24}$$

Since inner atomic shells are filled, the principal quantum number n refers to the first outer unfilled (valence) shell for the atom (3 for Al, Si, P, Cl of Table I).

In Eqs. (21) and (22), $|\Psi_s(0)|^2$ and $\langle r^{-3} \rangle_p$ strongly weight only those parts of the wave function very close to the nucleus. Therefore the hyperfine interaction at site j results primarily from ψ_j, the part of the total wave function which is atomic-like around site j. The hyperfine interaction at site j therefore allows a direct estimate of $\eta_j{}^2$, the fraction of the total wave function at site j, as well as $\alpha_j{}^2$ from the isotropic part of the interaction, and $\beta_j{}^2$ from the anisotropic part, etc. Such an analysis on the ^{29}Si hyperfine satellites of the Al^{2+} spectrum reveals that an additional $\sim$45% of the wave function can be accounted for on the four nearest-neighboring silicon sites (2.35 Å from the Al^{2+} interstitial), the six next-nearest ones (2.72 Å), and four others—probably the fourth-nearest ones (4.70 Å). In this way most of the wave function ($\sim$75%) has been accounted for. At the same time a rather detailed picture of its distribution among the neighboring sites is obtained.

The V_k center can also be analyzed in this manner. Equation (23) now refers primarily to the two chlorine atom sites with $\eta_j{}^2$ the same for both. The isotropic part of the hyperfine interaction indicates only a very small 3s contribution ($\sim$2–3%). The wave function is therefore primarily made up from a 3p wave function on each chlorine site, the p-lobe pointing in each case along the chlorine–chlorine axis (Fig. 8).

3.2.4. ENDOR

For the Al^{2+} or V_k center, hyperfine interactions also exist for more distant neighbors but they are too weak to be resolved. They contribute only to the width of the lines. A powerful technique that often allows these weaker interactions to be measured is ENDOR. First introduced by Feher,[13] ENDOR stands for electron nuclear double resonance. In the ENDOR experiment, nuclear resonance [$\Delta m = \pm 1$, Eq. (13)] is performed on the distant nucleus by applying an additional radiofrequency magnetic field while simultaneously observing the EPR. When a nuclear resonance transition takes place, it is detected as a change in the intensity of the EPR line.

Because of the increased resolution of ENDOR (and the difference in selection rules), the approximations leading to Eq. (13) are no longer accurate enough. The direct nuclear interaction with the external field **H** should be included as well as nuclear quadrupole interactions if they exist. This can be done in a straightforward way and will not be included here. The important point is that with the increased resolution of ENDOR the hyperfine interactions can be probed out farther from the center of the defect. The additional information concerning the nuclear quadrupole interaction (which probes the electric charge distribution around the site) is also a bonus of these studies.

An ENDOR study has been performed on the V_k center in LiF[14] and NaF[15] (directly analogous to the one discussed here in KCl). Through these studies, hyperfine interactions have been measured for three additional sets of equivalent lithium (or sodium) neighbors and four sets of fluorine neighbors, for a total of 28 atom sites surrounding the F_2^- center. From the analysis of the symmetry and disposition of these sites, the precise location of the F_2^- center in the lattice as shown in Fig. 8 has been confirmed.

This was a confirmation. Sometimes, ENDOR reveals surprises. In LiF an EPR center originally identified as an isolated F_2^- molecular ion on a single fluoride ion site[16] was later found by ENDOR to have a Na impurity as a near neighbor.[17] The real isolated center has been subsequently found in EPR[18] and verified by ENDOR.[19]

3.3. The g-Tensor

3.3.1. *Changes in the Spectrum*

So far, we have considered the orbital angular momentum as completely quenched. For Al^{2+}, the measured g-value is 2.0019 ± 0.0003 and is isotropic. Comparing this to 2.0023 for the spin-only g-value, we see that this assumption was indeed very good. For the Cl_2^- center in KCl, the g-values are also close to the spin-only value. However, there are small shifts ($\sim 2\%$) which are easily measurable. In particular, the g-value is found to be anisotropic with the principal values

$$g_x = 2.0428, \quad g_y = 2.0447, \quad g_z = 2.0010 \tag{25}$$

where the x, y, z axes are those of the Cl_2^- molecule in Fig. 8.

Classically, the origin of g-shifts can be visualized as follows. The magnetic dipole $g_0\beta S$ will induce a small dipole moment μ_{ind} into its surroundings—in much the same way that an electric dipole will induce a dipole moment when placed into a dielectric. The interaction with the applied magnetic field must then be written

$$-\mu_{tot} \cdot H = (g_0\beta S - \mu_{ind}) \cdot H$$

The induced dipole is proportional to $g_0\beta S$. However, due to the anisotropic nature of the defect and its surroundings, the dipole may be easier to induce in some directions than others. This means that, in general, the constant of proportionality is a *tensor*. Writing it as

$$\mu_{ind} = -\beta S \cdot \Delta g \tag{26}$$

the Hamiltonian then becomes

$$\mathscr{H} = \beta S \cdot g \cdot H + \sum_j S \cdot A_j \cdot I_j \tag{27}$$

where g is a tensor given by

$$g = g_0 1 + \Delta g$$

The solution to (27) gives, as before, (17)

$$h\nu_0 = g\beta H + \sum_j A_j m_j$$

but with g now given by

$$g = (g_1^2 n_1^2 + g_2^2 n_2^2 + g_3^2 n_3^2)^{1/2} \tag{28}$$

Here g_1, g_2, and g_3 are the principal values of g and n_1, n_2, n_3 are the direction cosines of the magnetic field H with respect to the 1, 2, 3 principal axes of the g-tensor. Here, as in the arguments leading to (15) for the hyperfine interaction, the spin quantizes along $g \cdot H$, with gH the magnitude of the vector $g \cdot H$. The principal axes of the g-tensor need not be the same as the hyperfine tensors.

Because the anisotropy in g results from the induced dipole, it necessarily reflects the symmetry of the defect and its environment. Often this of itself is an important clue in piecing together a model for a defect observed in EPR. For the V_k center, for instance, the x, y, z axes of (25) reflect the symmetry axes of the defect (Fig. 8).

3.3.2. *Quantitative Treatment of the g-Tensor*

Consider the spin and orbital magnetic interactions for an electron in an atomic orbital on a single atom,

$$\mathcal{H} = g_0 \beta \mathbf{S} \cdot \mathbf{H} + \beta \mathbf{L} \cdot \mathbf{H} + \lambda \mathbf{L} \cdot \mathbf{S} \tag{29}$$

The second term is the previously considered orbital dipole interaction with **H**, Eq. (1). The last term, not previously considered, is the familiar magnetic interaction between the spin and orbital dipoles where λ is the spin–orbit constant (directly obtainable from optical spectra). In a crystal field which quenches **L**, first-order perturbation theory gives only the first term, the expectation value of **L** being zero in the ground state.

The terms involving **L** can contribute, however, in second-order perturbation theory

$$\mathcal{H}' = g_0 \beta \mathbf{S} \cdot \mathbf{H} - \sum_n \frac{|\langle 0 | \beta \mathbf{L} \cdot \mathbf{H} + \lambda \mathbf{L} \cdot \mathbf{S} | n \rangle|^2}{E_n - E_0} \tag{30}$$

Here 0 denotes the ground state and the sum is over all excited states n. Expanding the terms in (30) and retaining only the one containing both **S** and **H**, we obtain

$$\mathcal{H} = \beta \mathbf{S} \cdot \mathbf{g} \cdot \mathbf{H}$$

where

$$\mathbf{g} = g_0 \mathbf{1} - 2\lambda \mathbf{\Lambda} \tag{31}$$

and the components of the $\mathbf{\Lambda}$ tensor are

$$\Lambda_{ij} = \sum_n \frac{\langle 0 | L_i | n \rangle \langle n | L_j | 0 \rangle}{E_n - E_0} \tag{32}$$

Equations (31) and (32) give a direct quantitative expression for the g-shift. Physically they express the fact that a small amount of orbital angular momentum does in fact exist in the ground state,

$$\langle \mathbf{L} \rangle = -2\lambda \mathbf{\Lambda} \cdot \mathbf{S} \tag{33}$$

It exists because the spin–orbit interaction $\lambda \mathbf{L} \cdot \mathbf{S}$ is inducing it. It is small because the crystal field is trying to quench it as measured by the energy denominator. $E_n - E_0$. The amount induced is thus of the order of $[\lambda/(E_n - E_0)]\mathbf{S}$. Equation (33) multiplied by β is therefore $\boldsymbol{\mu}_{ind}$ in our previous classical argument.

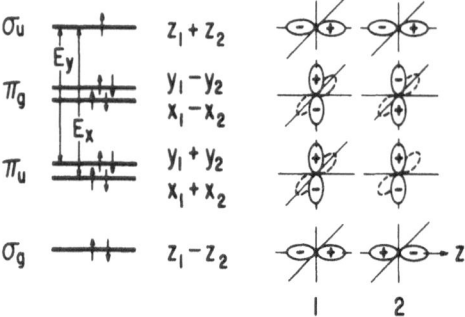

Fig. 11. Simple one-electron molecular orbital treatment of the V_k center. The wave functions are linear combinations of p functions on the two chlorine atoms (labeled 1 and 2) as shown. All orbitals are doubly occupied except the highest energy one, and EPR results from the unpaired spin in this orbital.

The derivation given was for an electron on a single atom. With care the treatment can also be extended to defect states which spread over many atoms, by considering the spin–orbit contribution at each atom site separately.*

3.3.3. The g-Shift of the V_k Center

Figure 11 gives a simple one-electron molecular orbital treatment[11] for the V_k center. Shown are the six molecular states that can be made from the three p functions on each atom. The state labeled σ_g is lowest because the z lobes on each atom point toward each other constructively building up charge density between the two positively charged nuclei. Conversely, the σ_u is highest because the two z lobes destructively interfere in this region. The same arguments apply for the π_u and π_g orbitals except that their overlap is weaker and the energy splittings therefore are also correspondingly less. For Cl_2^-, we lack one electron from completely filling all states, as shown in the figure. The EPR arises therefore from the unpaired electron in the σ_u orbital. (Note the similarity of the molecular V_k problem to the crystal field effect previously considered for a single p function, Fig. 5.)

By symmetry, orbital angular momentum matrix elements exist only to the π_u states (the two atom core contributions cancel for the π_g states).

---

* There are pitfalls, however. See Ref. 12, pp. 201ff, where this is discussed in considerable detail.

Reducing the matrix elements of (32) to those at each atom core, we obtain, with (5),

$$g_x \simeq g_0 - 2\lambda/E_y, \qquad g_y \simeq g_0 - 2\lambda/E_x, \qquad g_z \simeq g_0 \qquad (34)$$

We see that the general features of the observed g-values (25) are explained quite satisfactorily. In particular, the g_z value has only a small departure from g_0. The positive shifts for g_x and g_y result from the negative spin–orbit interaction at the chlorine site (-587 cm^{-1} for the free atom) indicating values for E_x, E_y of $\sim$2–3 eV.

3.4. Fine Structure Terms for $S > 1/2$

So far we have considered only terms linear in the spin S, Eq. (27). This is usually all that is required when the magnetism arises from a single unpaired electron with $S = 1/2$. Most defect problems fall into this class, and for them, Eq. (27) is sufficient to completely describe the EPR spectrum. Occasionally, however, a defect has two or more paramagnetic electrons coupled together to give a resultant spin $S > 1/2$. For such a defect it is necessary to include higher order terms in S. These can cause additional structure in the spectrum and are called *fine structure* terms.

3.4.1. *Changes in Spectrum*

The most important of these terms is usually the quadratic one, which in its most general form can be written in tensor form

$$\mathbf{S} \cdot \mathbf{D} \cdot \mathbf{S} \qquad (35)$$

Its effect can be seen with the simple Hamiltonian

$$g\beta \mathbf{S} \cdot \mathbf{H} + \mathbf{S} \cdot \mathbf{D} \cdot \mathbf{S} \qquad (36)$$

If $D \ll g\beta H$, first-order perturbation theory gives for the energy levels

$$E(M) \simeq g\beta H M + \tfrac{1}{2}[D_1(3n_1{}^2 - 1) + D_2(3n_2{}^2 - 1) + D_3(3n_3{}^2 - 1)]M^2 \qquad (37)$$

The EPR transitions are then

$$h\nu(M \to M - 1) \simeq$$
$$g\beta H + [D_1(3n_1{}^2 - 1) + D_2(3n_2{}^2 - 1) + D_3(3n_3{}^2 - 1)](M - \tfrac{1}{2}) \qquad (38)$$

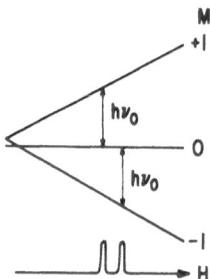

Fig. 12. The presence of a fine structure term
produces a splitting of the EPR spectrum
for $S = 1$.

Here D_1, D_2, D_3 are the principal values of the **D** tensor and n_1, n_2, n_3 are the direction cosines of **H** with respect to the 1, 2, 3 principal axes. Again, in situations where **g** and **A** also exist, the axes of **D** need not be the same as either of the others—although they often are.

The effect of the fine structure term is to destroy the equal energy spacings produced by $g\beta\mathbf{S} \cdot \mathbf{H}$ alone (Fig. 1). There are now $2S$ equally spaced lines* (one for each $M \to M - 1$ transition), the spacing given by the bracketed term in (38). This is illustrated in Fig. 12 for $S = 1$. Angular dependence studies of the splittings give the principal values and axes of **D** directly.

One of the immediate bonuses of the fine structure interaction is that we can count the number of spectral lines and determine S. In particular, it is the absence of fine structure splittings that really allows us to conclude that $S = \frac{1}{2}$ for the V_k center. In addition, the principal values and axes of **D** can sometimes be related to a detailed model of the center when their origin is considered.

3.4.2. Origin of D

Expansion of (30) gives directly a term of the form of Eq. (35) from the matrix element of $\lambda\mathbf{L} \cdot \mathbf{S}$ taken twice. The corresponding components of **D** are given by

$$D_{ij} = -\lambda^2\Lambda_{ij} \tag{39}$$

where Λ_{ij} is given by (32).

---

* There is little danger of confusing fine structure splittings with hyperfine splittings. In the angular dependence of (38), the fine structure splittings go through zero. For the hyperfine interaction (15), they do not. Also, for $S > 1$, the intensities for the fine structure lines are unequal, as opposed to equal intensities for the hyperfine lines.

Classically, this simply expresses the lowering of the total energy as a result of the magnetic interaction between the induced dipole (26) and the inducing dipole $g\beta S$. This can be seen simply as follows. From Eq. (29), this magnetic interaction energy is

$$\lambda \langle L \rangle \cdot S$$

Half of this, however, is stored in the crystal field interaction as a result of inducing $\langle L \rangle$, so that the total energy is

$$\tfrac{1}{2}\lambda \langle L \rangle \cdot S$$

This, with Eq. (33), leads directly to Eqs. (35) and (39).

Equation (39) implies a relation between the g-shift (Δg) and D:

$$D_{ij} = \tfrac{1}{2}\lambda \Delta g_{ij} \tag{40}$$

which is often useful and serves as a test of this origin for D.

Equations (39) and (40) often give a good description, for instance, for transition-element ion impurities in crystals. However, extreme care must be used in applying them to defects involving lattice vacancies, host atom interstitials, etc. The reason is that Eq. (29) is only strictly valid for electrons so tightly coupled together that those components of the spin-orbit coupling that tend to decouple the electrons need not be considered. This, in effect, requires that the exchange interaction J which couples the electrons together

$$J S_1 \cdot S_2$$

be large with respect to the crystal field energies $E_n - E_0$ in Eq. (30). This is apt to be true for transition element ions where the paramagnetic electrons couple together strongly in the inner d shells but interact less so with the surrounding atoms in the crystal. However, for intrinsic lattice defects, the paramagnetic electrons are often formed from outer shell "valence" electrons and, because of their strong overlap with neighboring atoms, the crystal field energies become comparable and often exceed the exchange energies. Specifically, for valence electrons on a single atom or coupled between adjacent atoms in a lattice, the interactions tend to be comparable. As a result, a proper treatment for such compact defects is extremely difficult and has not been given in the literature in any general form.

However, more extended defects again can be treated simply. This is because exchange involves the overlap of the wave functions between the electrons and therefore tends to decrease extremely rapidly (exponentially) vs. separation. Already at second nearest neighbor positions in a lattice, for instance, the exchange will be very much smaller than the crystal field energies. In this limit, the spin–orbit contribution to D is reduced in magnitude by $\sim J/\langle E_n - E_0\rangle_{av}$, giving

$$D_{ij} \sim (J/\langle E_n - E_0\rangle_{av})\lambda \, \Delta g_{ij} \sim J(\Delta g_{ij})^2 \qquad (41)$$

Another contribution to $\mathbf{D}$ originates from the simple magnetic dipole-dipole interactions between the electrons. In analogy to (18), this has the form

$$\mathscr{H}_{dd} = (g\beta)^2\left[\frac{\mathbf{S}_1 \cdot \mathbf{S}_2}{r^3} - \frac{3(\mathbf{S}_1 \cdot \mathbf{r})(\mathbf{r} \cdot \mathbf{S}_2)}{r^5}\right] \qquad (42)$$

where $\mathbf{r}$ is the vector separating the two spins. For two electrons coupled together to form $S = 1$, it is straightforward to show that this leads directly to[20]

$$D_{ij} = \tfrac{1}{2}g^2\beta^2\langle(r^2\delta_{ij} - 3x_ix_j)/r^5\rangle \qquad (43)$$

where the brackets denote the average over the electronic wave function for the two electrons. Because this varies as $1/r^3$, it is a longer range interaction than (41) and tends to be the dominant origin of $\mathbf{D}$ for defects whose wave function is spread over several atom sites.

3.4.3. *Higher Order Terms for $S > 3/2$*

In cubic symmetry, $D_1 = D_2 = D_3$ and the fine structure splitting (38) is zero. The D-tensor therefore shows up in the spectrum only for symmetry lower than cubic. Higher order perturbation theory introduces weaker terms that are quartic in S and reflect the cubic symmetry. Since they are proportional to

$$S_1{}^4 + S_2{}^4 + S_3{}^4$$

it is easy to show that these cause splittings only for $S > 3/2$.

3.5. Orbital Angular Momentum Not Quenched

In some cases, the symmetry of the crystalline field is such that orbital angular momentum would not normally be expected to be quenched. For instance, a pure cubic crystalline field will not quench $\mathbf{L}$ for a p state. (The existence of three equivalent perpendicular directions means that the degeneracy of p_x, p_y, and p_z is not lifted.) Often in these circumstances, however, the orbital angular momentum does turn out to be quenched after all, because of a phenomenon known as the Jahn–Teller effect.* In the Jahn–Teller effect, a spontaneous local distortion occurs around the defect, lowering the symmetry and removing the degeneracy. The V_k center can actually be considered as a Jahn–Teller defect. Consider the hole as first formed on a single chlorine atom—a site of cubic symmetry. The threefold degeneracy of the p function is lifted because the chlorine distorts toward one of its 12 nearest chlorine neighbors to form the Cl_2^- center.

There still remain cases where the Jahn–Teller effect does not occur and orbital angular momentum remains. The rare earth elements are good examples. This is sometimes true for the $3d$ transition element ions. Defects with very diffuse wave functions also occasionally fall in this category (see Section 5.1.5). These cases are in the minority, however, and we will not consider them in detail in this chapter.

3.6. Summary

For most defects, orbital angular momentum $\mathbf{L}$ is quenched in first order by the crystalline electric fields. For these the EPR spectrum can be described by the spin Hamiltonian

$$\mathcal{H} = \beta \mathbf{S} \cdot \mathbf{g} \cdot \mathbf{H} + \mathbf{S} \cdot \mathbf{D} \cdot \mathbf{S} + O(S^4, S^6, \ldots) + \sum_j \mathbf{I}_j \cdot \mathbf{A}_j \cdot \mathbf{S}$$

1. The first term is the spin Zeeman term. The departure of the g-tensor from 2.0023 arises from the small orbital angular magnetic moment induced into the surroundings via spin–orbit interaction at the atomic cores.

2. The next two terms are fine structure terms. They arise via spin–orbit interaction as well as spin–spin magnetic interactions averaged over the wave function.

3. The remaining term is the magnetic hyperfine interaction with neighboring nuclei.

The selection rule for EPR transitions is $\Delta M = \pm 1$, $\Delta m_j = 0$. From a study of the angular dependence of the EPR spectrum in a single crystal, $\mathbf{g}$, $\mathbf{D}$, and $\mathbf{A}_j$ can be determined. $\mathbf{g}$ and $\mathbf{D}$ reflect the overall character and symmetry of the defect, and from the number of fine structure lines, the

* A recent review of the Jahn–Teller effect has been given by Sturge.[21]

spin S can be determined. The $\mathbf{A}_j$ reflect the symmetry and character of the electronic wave function at each of the j atom sites in the immediate vicinity of the defect. The intensity of the hyperfine satellites determines the isotopic abundance of the j nucleus, and the number of hyperfine components determines its nuclear spin. This is usually sufficient to identify the nucleus involved. With this microscopic information, a detailed model for the defect can often be constructed.

ENDOR involves nuclear transitions $\Delta m_j = \pm 1$, $\Delta M = 0$. Often hyperfine interactions with distant nuclei, not resolved by EPR, can be resolved and studied in ENDOR. The nuclear Zeeman term and quadrupole interaction, normally not seen in EPR studies, can also be measured.

4. ADDITIONAL EXAMPLES

4.1. Defects in Irradiated Silicon

4.1.1. Annealing of Interstitial Aluminum

We described earlier the interstitial aluminum spectrum produced by radiation damage in silicon (Fig. 8). Annealing produces interesting changes in the spectrum[9] which gives further insight into the complex processes involved in the radiation damage behavior of this material. A new spectrum emerges (Fig. 13) which still displays a large six-line hyperfine splitting characteristic of Al^{2+}, but each "line" is further split into six closely spaced ones indicating the presence of a nearby *second* aluminum atom. This weaker "superhyperfine" interaction is anisotropic, reflecting a $\langle 111 \rangle$ axis of the crystal. From this it can be concluded that in this annealing stage, interstitial $(Al)_i^{2+}$ has diffused through the lattice until trapped by substitutional $(Al)_s$ [not yet coverted to $(Al)_i^{2+}$ by the irradiation] to form $(Al)_i^{2+} \cdot (Al)_s^-$ pairs. The $\langle 111 \rangle$ axis reflects the direction from the substitutional to nearby interstitial site in silicon.

4.1.2. The Oxygen–Vacancy Pair

The dominant effect of high-energy particle or γ-irradiation on the electrical properties of n-type pulled silicon is the formation of a defect which has an acceptor level 0.17 eV below the conduction band. The EPR spectrum of a center that has been identified with this defect[22,23] is shown in Fig. 14(a).

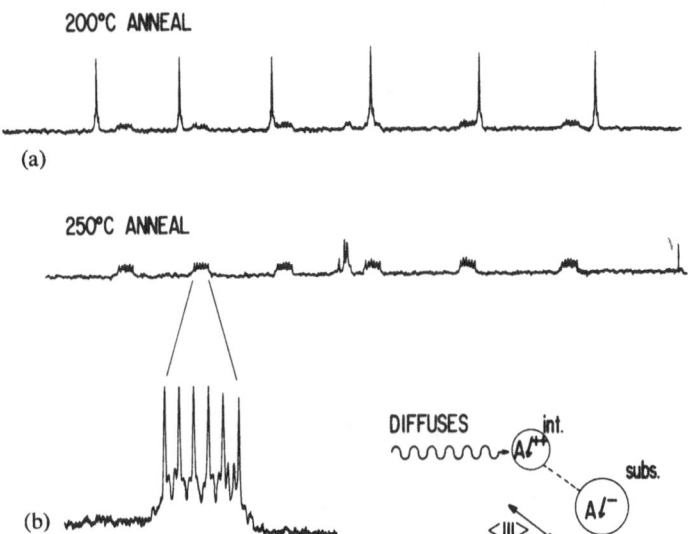

Fig. 13. Loss of interstitial Al_i^{2+} spectrum (Fig. 7) and emergence of that for $Al_i^{2+} \cdot Al_s^-$ pairs. This is interpreted as resulting from diffusion of Al_i^{2+} to form the pairs as shown.

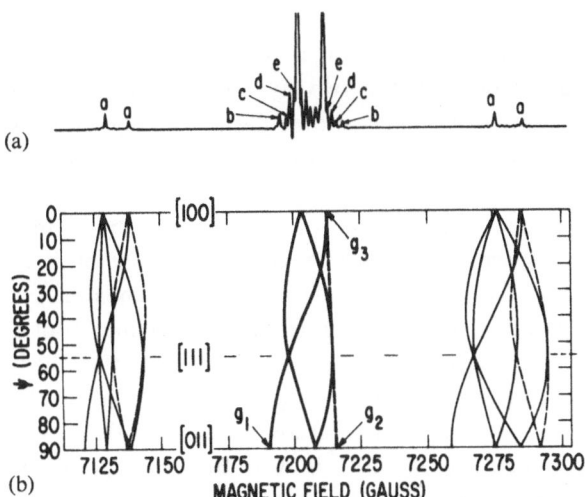

Fig. 14. (a) EPR spectrum of the oxygen–vacancy pair in irradiated silicon with the magnetic field oriented along a ⟨100⟩ crystallographic direction. (b) Angular dependence of the spectrum with **H** in the (0$\bar{1}$1) plane. Shown dashed are the spectral components associated with the defect orientation of Eq. (44) and Fig. 15.

The angular dependence of the strong central group of lines and the satellites labeled a is shown in Fig. 14(b). The central group reflect the angular dependence of the g-tensor for an anisotropic $S = 1/2$ defect of orthorhombic symmetry and with no resolved hyperfine splittings. For a defect of this symmetry, there must be six equivalent orientations in the cubic lattice and the central lines are a superposition of the spectra from all six. The principal values of the g-tensor are

$$g_1[0\bar{1}1] = 2.0093, \qquad g_2[011] = 2.0025, \qquad g_3[\bar{1}00] = 2.0031 \quad (44)$$

which can be determined directly from Fig. 14(b), as shown. The principal axes indicated in (44) are those for one of the six equivalent defect orientations. The spectrum associated with this specific orientation is shown dashed in Fig. 14(b).

The satellites are due to hyperfine interaction with ^{29}Si. Analysis shows that for the satellites labeled a, *two* pairs can be associated with each central line, indicating that the electronic wave function is highly localized on *two* silicon sites. The hyperfine interaction is the same for both sites with

$$A_1/h = 458.7 \text{ MHz}, \qquad A_2/h = A_3/h = 386.1 \text{ MHz}$$

but with the principal axis (1) being a different $\langle 111 \rangle$ axis for each. For the defect orientation of (44), the two hyperfine axes are $[111]$ and $[1\bar{1}\bar{1}]$. (Note the two sets of satellites shown dashed in Fig. 14b.) The satellites labeled b–e also reflect the same set of axes and arise from ^{29}Si hyperfine interactions at more distant positions.

The model deduced from the EPR analysis is shown in Fig. 15. It is a lattice vacancy which contains an oxygen atom impurity. The oxygen bonds to two of the four silicon neighbors. The unpaired electron is accommodated in an orbital made up from the dangling "broken" bonds of the other two silicon atom neighbors. Consistent with this, the ^{29}Si hyperfine interaction, analyzed according to Eqs. (21)–(24) and Table I, indicates that 68% of the wave function is accounted for on these two atoms. The orbital on each is 30% $3s$ and 70% $3p$, with the p lobe pointing in the direction from the atom toward the vacancy. (This is very close to the 25% $3s$ and 75% $3p$ orbital that a physical chemist would assign to a tetrahedral covalent bond,[24] confirming the "broken" bond character.) Most of the remaining wave function can be accounted for by the resolved hyperfine interactions b–e which, from their relative intensities, account for $\sim$12–16 neighboring silicon sites.

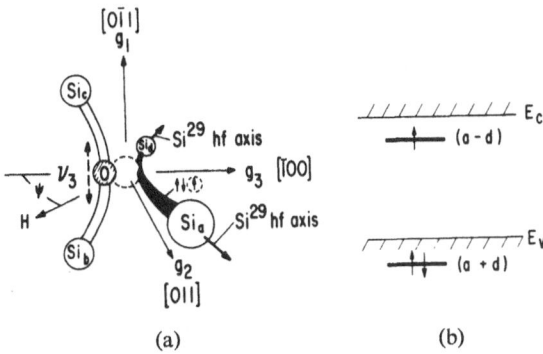

Fig. 15. (a) Model of the oxygen–vacancy pair as a lattice vacancy containing an oxygen atom impurity. The unpaired electron is shared between two of the silicon atom neighbors, giving rise to the g-tensor and hyperfine axes as shown. (b) Simple molecular orbital model for the center accounting for the $E_c - 0.17$ eV electrical level associated with the defect. The oxygen atom has an infrared-active vibrational band (ν_3) that is observed at 828 cm^{-1}.

A simple one-electron molecular orbital model for the center is shown in the figure. Here the two "broken" bonds recombine to form simple bonding $(a + d)$ and antibonding $(a - d)$ orbitals. The bonding orbital is filled with two electrons and is probably located electrically within the valence band as shown. The antibonding orbital has the unpaired resonance electron and is located in the forbidden gap of the material, 0.17 eV below the conduction band edge (the acceptor level position determined from electrical measurements). With such a model, the g-value can be accounted for roughly in terms of Eqs. (31) and (32) where the excited states (n) are identified with the localized filled bonding and empty antibonding orbitals between the two silicon atoms and their other three immediate neighbors.

The presence of oxygen in the defect was initially inferred from the fact that a trace amount of oxygen impurity in the silicon was found to be necessary for the formation of the center. Its presence could not be determined directly from the EPR because no hyperfine interaction exists with the majority isotope ^{16}O, $I = 0$. This therefore is as far as one can go using only the static EPR spectrum analysis. Actually, however, the existence of the oxygen atom and its exact position in the defect as shown in Fig. 15 have been confirmed in considerable detail. This has been achieved primarily by studying the effect of uniaxial stress on the EPR and correlating this with optical studies. We will describe these experiments further in Section 5, where these techniques are elaborated.

4.1.3. *The Phosphorus–Vacancy Pair*

The dominant effect of irradiation damage on the electrical properties of *n*-type silicon especially prepared to have low oxygen content is the formation of a deep acceptor level at $\sim$0.4 eV below the conduction band. EPR studies have identified the defect responsible for this level as a lattice vacancy adjacent to one of the original group V chemical donors.[25,26] In phosphorus-doped material, for instance, this center gives the spectrum shown in Fig. 16(a). The angular dependence of the spectrum is shown in Fig. 16(b).

As found for the oxygen–vacancy pair, there is a strong central group of anisotropic lines plus ^{29}Si hyperfine satellites. There are differences, however. Here there is a *single* pair of satellites for each central line, reveal-

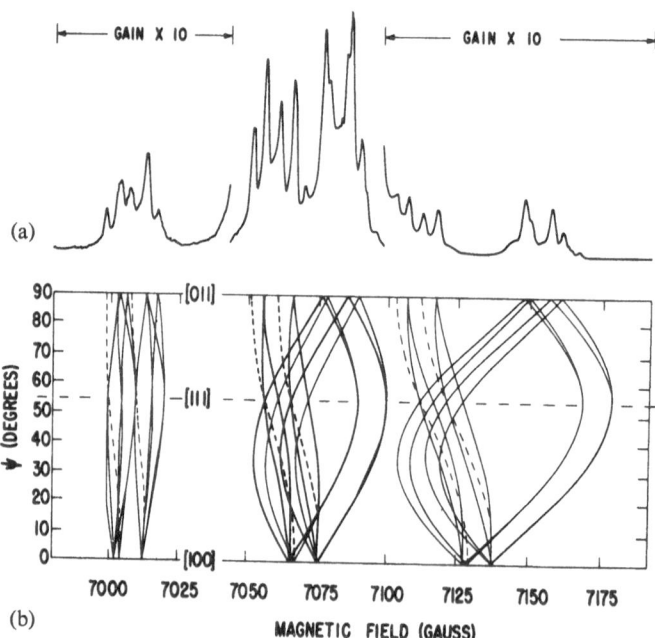

Fig. 16. (a) EPR spectrum of the phosphorus–vacancy pair in irradiated silicon, $\nu_0 \sim 20$ GHz, with the magnetic field oriented along a $\langle 110 \rangle$ crystallographic direction. (b) Angular dependence of the spectrum with **H** in the $(0\bar{1}1)$ plane. Shown dashed are the spectral components associated with the defect orientation in Fig. 17.

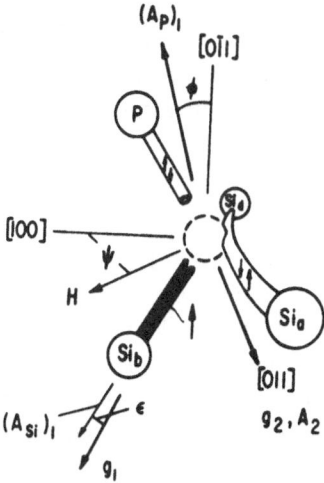

Fig. 17. Model of the phosphorus–vacancy pair
showing axes of the g and hyperfine tensors.

ing that the wave function is primarily located on a *single* silicon atom site. In addition, the spectrum is made up of doublets revealing an additional hyperfine interaction with a single 100% abundant ^{31}P atom, $I = 1/2$.

Unraveling the spectrum is straightforward. The center of gravity of each doublet in the central group gives the g-value for each differently oriented defect. The splitting of each central group doublet gives the corresponding phosphorus hyperfine interaction. The ^{29}Si doublets in turn give the silicon hyperfine interaction for each orientation. The angular dependence of each therefore leads directly to the principal values and axes for the g and hyperfine tensors associated with each defect. With these, the model of Fig. 17 has emerged.

Here the unpaired electron is primarily located in the broken bond of a single silicon atom neighboring a vacancy. Consistent with this, the g-tensor

$$g_1 = 2.0005, \qquad g_2 = 2.0112, \qquad g_3 = 2.0096$$

reveals only small departures from axial symmetry around the $\langle 111 \rangle$ broken bond direction. Using Eqs. (31) and (32), these values can be shown to be generally consistent with those expected for a single dangling silicon bond. The ^{29}Si hyperfine interaction is axially symmetric along this $\langle 111 \rangle$ direction with

$$A_1 = 450 \text{ MHz}, \qquad A_2 = A_3 = 295.5 \text{ MHz}$$

With Eqs. (21)–(24), this indicates 14% $3s$ and 86% $3p$ character on the atom with 59% of the total wave function accounted for on this atom alone. The phosphorus hyperfine tensor reflects the symmetry of the defect as seen from the phosphorus nucleus, which leads to the assignment of its position as shown. The values

$$A_1 = 31.7 \text{ MHz}, \qquad A_2 \simeq A_3 \simeq 26.0 \text{ MHz}$$

are small, however, and from Table I, only ~1% of the wave function is required on the phosphorus atom to account for them.

A simple molecular orbital model for the defect is shown in Fig. 18. In the undistorted lattice, the phosphorus–vacancy pair would have axial symmetry along the $\langle 111 \rangle$ vacancy–phosphorus axis. (C_{3v} symmetry). However, in the neutral state, this symmetry would leave electronic degeneracy associated with the unpaired electron in the doubly degenerate (e) orbital. The reduced symmetry (C_{1h}) observed by EPR is interpreted as the result of a Jahn–Teller distortion, as shown, which lifts this degeneracy.

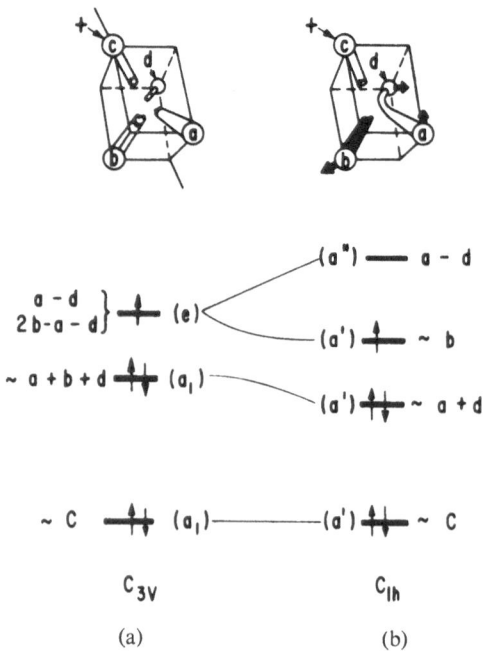

Fig. 18. Simple molecular orbital model for the phosphorus–vacancy pair (a) before and (b) after Jahn–Teller distortion.

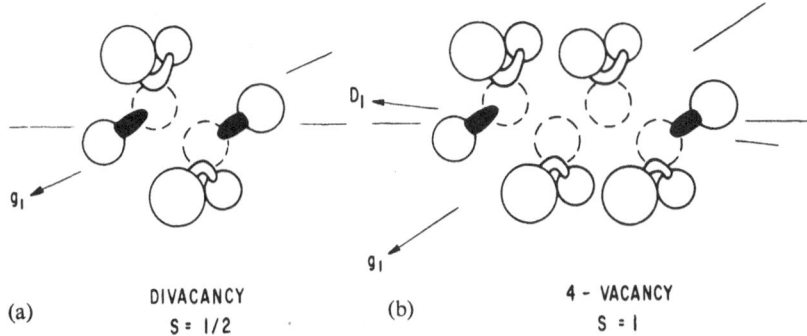

Fig. 19. (a) The divacancy in silicon. (b) A model for a four-vacancy center in silicon. (After Corbett.[29])

Similar EPR centers have been observed in the II–VI compounds ZnS and ZnSe.* These have been identified as analogous zinc vacancy–donor impurity pairs. Labeled A centers, they have been identified as the important defects responsible for "self-activated" luminescence in these materials.

4.1.4. Multiple-Vacancy Defects

The divacancy has been studied by EPR in both its singly positive and negative $S = 1/2$ charge states.[27] The model, shown in Fig. 19(a), is that of two adjacent vacancies, the C_{3v} symmetry again being lowered by a Jahn–Teller distortion. As for the phosphorus–vacancy pair, this distortion takes the form of a pulling together of the neighboring atoms by pairs, the single unpaired electron in this case being divided equally between the dangling bonds on the two atoms at opposite extremes of the center.

The EPR spectrum for the divacancy is therefore again given by an anisotropic set of central lines plus [29]Si hyperfine satellites. The symmetry and principal values of the g-tensor are similar to those for the phosphorus–vacancy pair since in both cases they reflect primarily the properties of a dangling bond (one for the phosphorus pair, two equivalent ones for the divacancy) pointing in a single $\langle 111 \rangle$ direction. The [29]Si hyperfine interactions are now divided between two silicon sites. Because they are fully equivalent, no splitting of the satellites occurs, however. Instead, the fact that *two* sites are involved is revealed solely by the relative intensity of the satellites, which is twice the isotopic abundance (4.7%) of [29]Si.

---

* See the reviews by Schneider[4] and Title.[3]

The structure of the divacancy deduced from these EPR studies has served as a model for understanding more complex many-vacancy aggregates. In particular, several $S = 1$ EPR centers have been observed in neutron-irradiated silicon[28] which have been postulated to arise from

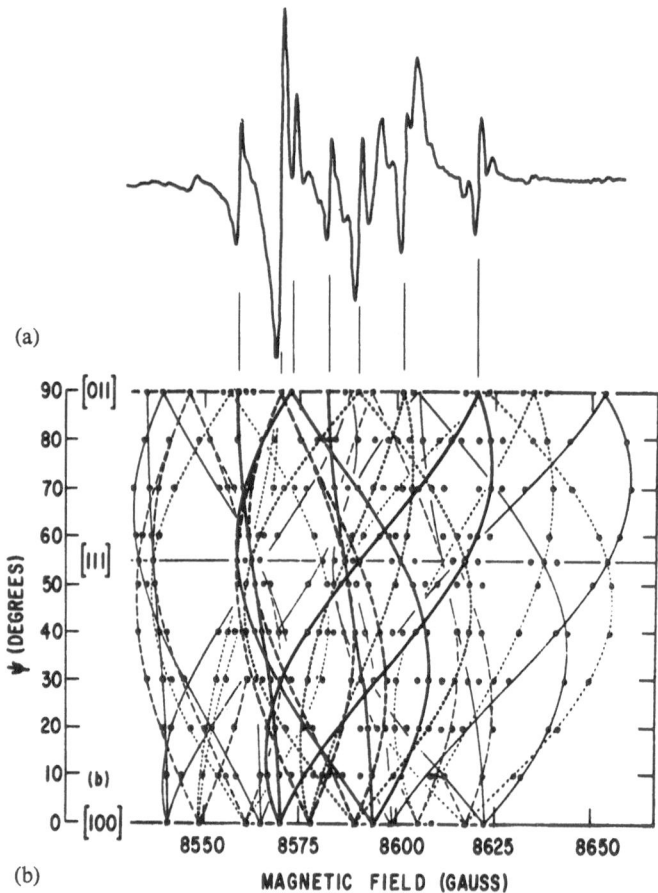

Fig. 20. (a) EPR spectrum of an $S = 1$ center observed in neutron-irradiated silicon, $\nu_0 \sim 24$ GHz, with the magnetic field oriented along a $\langle 110 \rangle$ axis of the crystal. (b) Angular variation of this spectrum with H in the $(0\bar{1}1)$ plane. The heavy lines represent the central group of lines and the light lines the ^{29}Si hyperfine satellites. The branches have been coded by solid, dashed, and dotted lines to assist in identifying fine structure pairs associated with specific defect orientations (after Jung and Newell[28]). This center, labeled (II, III)[28] or Si–P3[9] in the literature, has been identified with the four-vacancy complex of Fig. 19.[29,30]

multiple-vacancy aggregates.[29,30] The symmetries of the EPR spectra suggest chainlike arrays of vacancies in a {110} plane as if simply building upon and continuing the divacancy, the unpaired electrons again always ending up on the two extremes of the center. A possible four-vacancy center[29] is shown in Fig. 19(b).

The angular dependence of one of the $S = 1$ spectra observed in neutron-irradiated silicon is given in Fig. 20. Although the spectrum appears extremely complex, the analysis is actually again fairly straightforward, once the overall pattern is visualized: The strong central group (thick lines) can be decomposed into sets of equal-intensity doublets, their center of gravity giving the g-values and their separation giving, in this case, the strongly anisotropic fine structure splittings, Eq. (38), for each defect orientation. The ^{29}Si hyperfine satellites are also observed (fine lines) and reflect, as for the divacancy, an equal distribution of the wave function between the dangling bonds of two fully equivalent sites.

The spectrum has been identified with the four-vacancy center[29,30] (Fig. 19b). The symmetry and principal values for the **g** and **A** tensors are consistent with this identification, being very similar to those for the divacancy. The symmetry and principal values of the **D** tensor provide an important additional clue. For such an extended defect, spin–orbit contributions to **D** should be small and the principal contribution should come from dipole–dipole interactions between the spins. Estimates of the dipole–dipole contribution using Eq. (43) averaged over the two-electron wave function, as deduced from the ^{29}Si hyperfine interactions, give remarkable agreement with the four-vacancy model.[30] In fact, ordering the many $S = 1$ centers in descending magnitude of D has provided a systematic way of sorting out candidates for one-, two-, three-, four-, etc. vacancy centers.[30] Similar analysis has been applied to $S = 1$ centers in irradiated diamond.[31]

4.2. Transition Elements in Silicon

A substantial fraction of all EPR studies that have been performed over the past 25 years has centered on the properties of transition element ions in solids.* Much of this vast literature is of only marginal interest, however, in the traditional field of "point defects." One reason for this is that these studies have been for the most part in ionic crystals where these ions often have little direct effect on the properties of interest in the ma-

---

* See, for instance, the comprehensive text on the subject by Abragam and Bleaney.[32]

terial. The thrust of these studies has been more on understanding the properties of the ion as perturbed by the crystal, than the reverse.

This cannot be said for semiconductors, however. The transition element ions introduce multiple donor and acceptor states in silicon, for instance, strongly affecting its electrical properties. In compound semiconductors, they play a vital role in luminescent properties. They represent, therefore, bona fide point defects in every sense of the word and considerable interest centers on their role in these materials.

In this section we will summarize, therefore, what has been learned about the electronic structure and configuration of $3d$ transition element ions when incorporated into the silicon lattice. This will be an abbreviated account of the work of Ludwig and Woodbury,[6] who systematically unraveled a large variety of EPR spectra in transition element-doped silicon to answer this question.

They found that transition element ions may enter either substitutionally or interstitially and that, for each, several different charge states may be stable. In spite of this complexity, a fairly simple picture of the electronic structure of these defects has emerged.

To understand what has been learned, let us first consider the effect on a single d electron of the crystalline electric fields associated with the atoms surrounding an interstitial or substitutional site in silicon. Superficially the substitutional and interstitial sites appear to have almost identical environments. For both there are four equidistant silicon neighbors (2.35 Å away) on the corners of a regular tetrahedron. There are differences in the next-nearest shell (six for the interstitial site, 12 for the substitutional), but let us ignore that for the moment. We will assume that the effective crystalline field seen by the d electrons is determined primarily by the nearest neighbors.

In an ionic crystal it is rather straightforward to visualize a crystalline electric field because the surrounding ions have net charges. In a covalent crystal, however, it is not so obvious how to visualize the origin of such a field nor to know in advance even what its sign is. Therefore we assume only that there *is* an effective crystalline field—of tetrahedral symmetry—and that its sign is something that will have to be deduced by experiment.

The question of sign can be reduced simply to the question of whether the four neighbors appear to the central ion as being effectively positively or negatively charged. These two possibilities are illustrated in Fig. 21. The effect of the field will be to split the fivefold degeneracy of a single d orbital into a doublet (e) and triplet (t_2). In analogy to our crystal field arguments for a p function (Section 3.1), the lobes of the three t_2 states

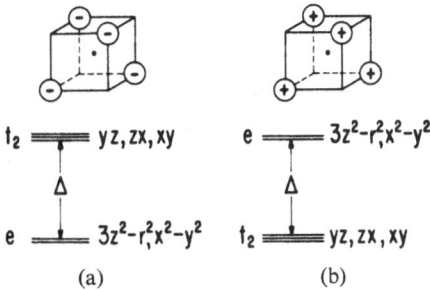

t_2 ══ yz, zx, xy   e ── $3z^2-r^2, x^2-y^2$

e ─⏐─ $3z^2-r^2, x^2-y^2$ t_2 ══ yz, zx, xy

(a)   (b)

Fig. 21. The energy states of a single d electron in a tetrahedral crystalline field. (a) The four nearest neighbors appear negatively charged. (b) They appear positively charged. The angular dependences of the d functions are indicated.

(angular dependence xy, yz, zx) overlap more strongly with the four neighbors than the e states (dependence $3z^2 - r^2$, $x^2 - y^2$), determining which states are lowest in energy. It is straightforward to show that the orbital angular momentum is quenched in the e states. However, in analogy to the p function in a cubic field, the t_2 states still have unquenched orbital angular momentum.

A simple model that satisfactorily accounts for all of the EPR results is summarized in Fig. 22. Here the many-electron state is constructed simply by populating one at a time the one-electron d states as split by the crystal field. In populating these states, Hund's rules are found to

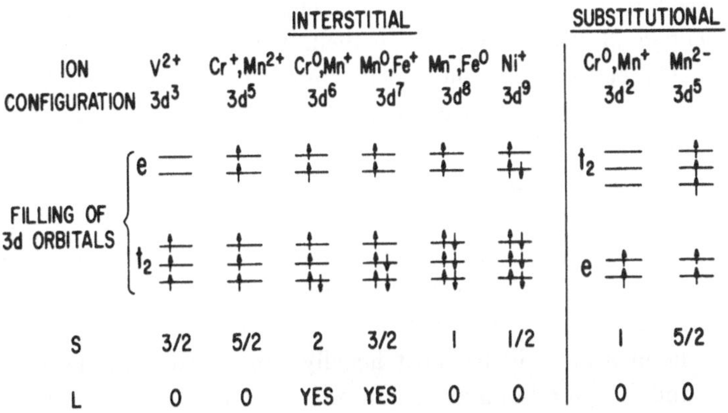

		INTERSTITIAL						SUBSTITUTIONAL	
ION	V^{2+}	Cr^+, Mn^{2+}	Cr^0, Mn^+	Mn^0, Fe^+	Mn^-, Fe^0	Ni^+	Cr^0, Mn^+	Mn^{2-}
CONFIGURATION	$3d^3$	$3d^5$	$3d^6$	$3d^7$	$3d^8$	$3d^9$	$3d^2$	$3d^5$
FILLING OF 3d ORBITALS							t_2	
S	3/2	5/2	2	3/2	1	1/2	1	5/2
L	0	0	YES	YES	0	0	0	0

Fig. 22. Summary of the electronic structure of transition elements in silicon. (After Ludwig and Woodbury.[6])

apply, meaning that, as in the free ion, the levels are filled in such a way as to yield maximum total spin, consistent with the Pauli exclusion principle.

Experimentally, let us consider how such a table can be constructed. First, from the hyperfine interaction, the identification of the specific ion can be made. Second, the total spin S can be deduced from the fine structure splittings. This in turn gives direct information concerning how many d electrons are involved and whether Hund's rule is being followed or not. Third, the sign of the cubic field splitting can be deduced from the degree to which orbital angular momentum is quenched. In particular, all of the configurations in the table except the $3d^6$ and $3d^7$ interstitials have g-values very close to two, indicating quenched orbital angular momentum. These ions must therefore have either empty, half, or completely filled t_2 states. Otherwise the unquenched orbital angular momentum of a single t_2 electron would contribute. The $3d^6$ and $3d^7$ interstitial ions, on the other hand, have complex resonances and large departures from the free-electron g-values, indicating the unquenched orbital angular momentum contribution of a partially filled t_2 state. The signs of the cubic field splitting indicated in Fig. 22 represent the only possible consistent choice that satisfactorily works for all of the ions.

Figure 22 reveals a number of interesting differences between the substitutional and interstitial ions. In the first place, the sign of the crystal field is reversed between the two. Second, even the ion configuration is different. For instance, Mn$^+$ is $3d^2$ as a substitutional ion. As an interstitial, it is $3d^6$. These observations lead to a simple physical picture for the incorporation of the ions into the lattice as follows:

1. In the *substitutional* position, four electrons first go into the $4s4p$ shell of the ion and form the normal tetrahedral bonds to its four neighbors. The *remaining* electrons are in the d shell. This bonding concentrates electronic charge between the ion and its four neighbors and the sign of the crystalline field can be visualized as resulting from the excess negative charge density in these directions.

2. The interstitial ion cannot bond with its neighbors because they are already completely bonded to their respective four nearest neighbors. All of the electrons of the ion are therefore squeezed into the d shell. The sign of the crystal field is now reversed, reflecting the *lack* of electronic charge density in the direction of the neighbors. (The electronic density on the nearest silicon atom will be most heavily concentrated in the region between it and its four bonding neighbors. As a result, the positive silicon atom core can be considered to be partially exposed to the transition element ion.)

5. AUXILIARY TECHNIQUES

So far we have considered the information available from direct analysis of the EPR (and ENDOR) spectrum itself. The examples we have chosen clearly demonstrate that this often represents a wealth of information indeed. It is often sufficient to *identify* the defect and, at the same time, it can reveal considerable microscopic detail concerning its electronic structure.

However, the power of EPR as a tool in the study of point defects can often be greatly expanded by the use of *auxiliary techniques* in conjunction with the EPR study. In this section we will discuss some of these techniques. To illustrate the types of information that these techniques can uncover, we will describe their use on the defects we have discussed in the previous sections.

5.1. Applied Uniaxial Stress

In the study of point defects, perhaps the most powerful single addition to an EPR spectrometer is the facility to apply uniaxial stress to the sample *in situ*. Some of the properties that can be studied using stress are as follows.

5.1.1. *Jahn–Teller Alignment*

When a defect undergoes a Jahn–Teller distortion it is, in effect, trading *electronic* degeneracy for *orientational* degeneracy: A local distortion lowers the symmetry of the defect environment and thereby removes its electronic degeneracy. However, there are always several equivalent such distortions, each of which produces a different equivalent *orientation* of the distorted defect in the lattice.

Consider the phosphorus–vacancy pair in silicon[25] as an example. The distortion illustrated in Fig. 18(b) is just one of three possible distortions which could result from the pair of Fig. 18(a). Any two of the three silicon atoms neighboring the vacancy could have pulled together, leaving the unpaired electron on the third. The three distortions are equally probable, equally energetic, and as a result, the EPR spectrum reveals all three, equally intense.*

---

* Each particular defect in the lattice has only one vacancy–phosphorus orientation and for it only one distortion direction at a time. However, the EPR spectrum results from interaction with many ($\sim 10^{16}/cm^3$) such defects and therefore reflects a statistical average over all possible orientations.

   The application of uniaxial stress can destroy the equivalence of the different Jahn–Teller orientations. For some orientations the stress will aid the distortion, further lowering the energy of the defect. For others, the stress will oppose the distortion, raising the energy. If the defects can reorient at the temperature of the experiment, they will tend to do so and this can be observed directly by a change in the relative intensities of the corresponding lines in the EPR spectrum. This is illustrated for the phosphorus–vacancy pair in Fig. 23. Since each line in the spectrum is identified with a specific defect orientation, as shown, a highly specific, unambiguous characterization of the elastic coupling of the defect to the applied strain field emerges.

   Such an experiment reveals immediately several important bits of information about the defect. First, and foremost, it demonstrates convincingly that the lowered symmetry (C_{1h}) of the defect is indeed the result of a Jahn–Teller distortion: The fact that this reorientation can occur at

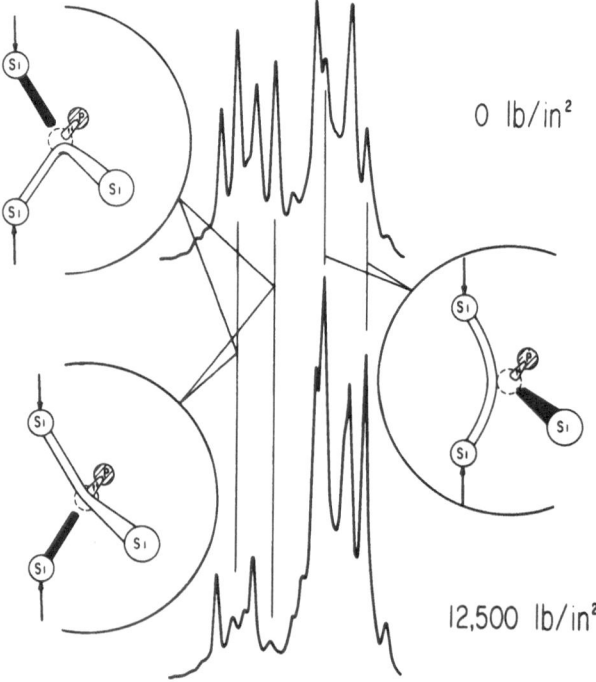

Fig. 23. Changes in the phosphorus–vacancy pair spectrum, H ∥ [011], revealing preferential alignment of the Jahn–Teller distortion under [0$\bar{1}$1] compressional stress at 20.4°K. The insets show a typical defect orientation for each multiplet.

the low temperature of 20.4°K strongly argues against the alternate possibility that the reduced symmetry could have been due to a nearby defect locked in the lattice. This could not have been ruled out from the static spectrum alone. Second, it confirms the *sign* of the distortion, i.e., that the two silicon atoms *pull together* as shown in Fig. 18. Third, it allows the study of the kinetics of the reorientation process. For the phosphorus–vacancy pair, the alignment achieved in Fig. 23 does not occur instantaneously but rather builds up over a period of several minutes at 20.4°K. Monitoring these rates vs. temperature allows a direct measurement of the kinetics of the reorientation process.

Such an experiment therefore is the direct analog of an internal friction experiment, familiar in the more traditional point defect studies. It is, however, the *ideal* internal friction experiment. Rather than detect the stress-induced alignment indirectly as a macroscopic reaction to the stress apparatus, a *direct* observation of the alignment is observed.

The degree of alignment per unit stress, measured accurately from the amplitudes of the spectral components, also gives a direct *quantitative* measure of the coupling of the defect to the stress field. For this, the difference in energy $E_1 - E_2$ for two defect orientations is reflected in their relative amplitudes n_1 and n_2 via a Boltzmann distribution function

$$n_1/n_2 = \exp[-(E_1 - E_2)/kT] \qquad (45)$$

By applying stress in different crystallographic directions, a complete characterization of the coupling between the defect and applied stress can be achieved in this manner. Such studies, in turn, allow a direct estimate of the magnitude of the Jahn–Teller distortion, a quantity not easily available by any other technique. Briefly, this analysis proceeds as follows.

The change in energy of a defect undergoing a specific local Jahn–Teller distortion mode Q can be written

$$E = -VQ + \tfrac{1}{2}kQ^2 + \cdots \qquad (46)$$

Here V is the "Jahn–Teller coupling coefficient," which expresses the linear lowering of the energy arising from the lifting of the electronic degeneracy vs. Q. The second term represents the elastic restoring forces of the system, which oppose the distortion. Allowing the system to relax to lowest energy gives, from (46), the Jahn–Teller energy

$$E_{\mathrm{JT}} = -V^2/2k \qquad (47)$$

The distortion under externally applied stress is always very small. Therefore its effect can be written, from (46), as

$$\Delta E = -V \Delta Q \qquad (48)$$

where ΔQ is the additional component of mode Q induced by the applied stress. Using conventional elastic theory, ΔQ can be estimated from the applied stress and with (48) and (45), V can be determined directly. For the phosphorus–vacancy pair, its value turns out to be[26] 4.4 eV/Å. The force constant has been estimated indirectly from the bulk compressibility of silicon to be[26] $k \sim 7.25$ eV/Å². This gives, with (47), an estimate of the Jahn–Teller energy of $\sim$1.4 eV.

Similar studies have also been performed on divacancies, and single vacancies in silicon as well as the group V–vacancy pairs.[26,33] The Jahn–Teller energy estimates range from $\sim$0.4 to $\lesssim$2.0 eV for these vacancy-associated defects. Such large Jahn–Teller energies, originally unanticipated, have serious consequences for the theoretical treatment of such defects, a point of considerable current concern. These stress studies with EPR represent a unique way of estimating these quantities directly.

5.1.2. Defect Alignment

At higher temperatures, stress may also induce alignment which involves atomic rearrangements in the lattice. For the group V atom–vacancy pairs, for instance, a preferential alignment of the impurity–vacancy axis has been observed directly in the EPR spectrum by stressing at elevated temperature and then cooling down to cryogenic temperatures for the EPR measurement.[25,26] Similarly, preferential orientation of the vacancy–vacancy axis of the divacancy has been achieved and studied by EPR.[27] Figure 24 illustrates the corresponding experiment for the oxygen–vacancy pair.[22] Here $\langle 110 \rangle$ stress was applied at 125°K and then cooled to 77°K, with stress on. The stress was then removed and the EPR spectrum taken. The large changes in the intensities of the lines reveal a substantial alignment of the defect frozen in by this process.

In such experiments, monitoring the EPR alignment vs. subsequent annealing allows a direct study of the kinetics of the reorientation process. In Table II, we summarize the results of such a study for several defects in silicon, where the characteristic reorientation time τ is given by

$$\tau^{-1} = \tau_0^{-1} \exp(-E/kT) \qquad (49)$$

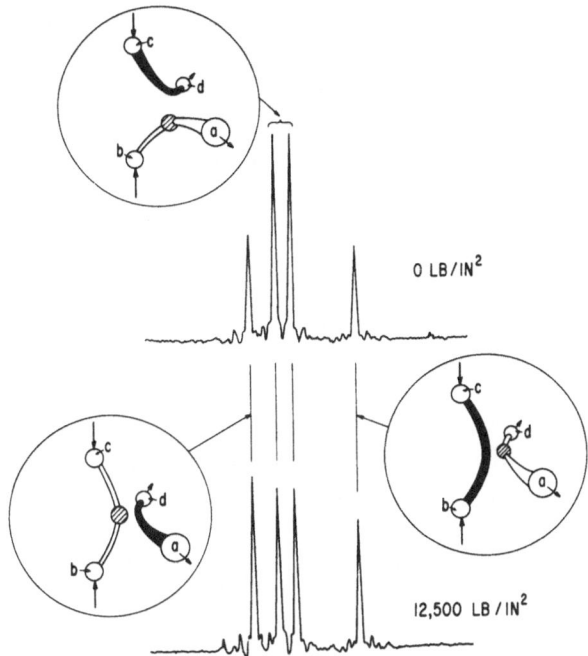

Fig. 24. Changes in the oxygen–vacancy pair spectrum resulting from alignment under [0$\bar{1}$1] stress. (The stress was applied
at 125°K and the resulting alignment was frozen in by quenching to 77°K, where the spectra were taken.) The insets show
the defect orientation corresponding to each multiplet.

Table II. Reorientation Kinetics Determined by EPR for Defect Alignment in Silicon

Defect	Ref.	τ_0^{-1}, sec^{-1}	E, eV
P·V	25	$1.6(10^{13})$	0.93 ± 0.05
As·V	26	$1.0(10^{13})$	1.07 ± 0.08
Sb·V	26	$9.4(10^{14})$	1.29 ± 0.10
V·V	27	$(10^{13})^a$	~ 1.3
V·O	22	$3.0(10^{12})$	0.38 ± 0.04

^{a} Assumed to derive E. Detailed kinetic studies were not performed.

These results, in turn, reveal important information about the defects: For the divacancy, for instance, a glance at the silicon lattice (Fig. 19) reveals that reorientation automatically involves a displacement of the defect from one pair of atom sites to another. For it, therefore, the 1.3-eV activation energy must also be the activation energy for *migration*, an important property measured unambiguously and for the first time by this simple procedure. For the group V atom–vacancy pairs, on the other hand, migration through the lattice requires both reorientation and impurity atom–vacancy interchange. Here help in interpretation comes from electrical measurement studies.[34] In each of the correspondingly doped materials, an annealing stage in the loss of the $E_c - 0.4$ eV level identified with the defect has been found which agrees in activation energy, within experimental error, with the corresponding energy given in Table II. This has been interpreted to indicate that impurity–vacancy interchange occurs readily with the reorientation being the limiting step in the diffusion.

5.1.3. *Correlation with Optical Dichroism*

Optical transitions associated with an anisotropic point defect often have well-defined polarization properties with respect to the defect axes. As a result, stress-induced alignment of the defect can produce dichroism* in the optical absorption bands. Therefore, by combining such studies with stress-induced alignment studies observed directly in EPR, a direct identification of optical transitions associated with a specific defect is possible.

An interesting example of this is afforded by the oxygen–vacancy pair in silicon. A sharp optical absorption band at 828 cm^{-1} (near infrared) was found to be produced by irradiation of pulled silicon.[35] In a sample intentionally doped with oxygen enriched with ^{18}O the band was found to be shifted to 791 cm^{-1}. This demonstrated that the band is vibrational in origin, involving primarily the motion of a single oxygen atom. The application of uniaxial stress was found to produce a substantial dichroism in the band. This dichroism could be frozen in if the temperature was lowered below $\sim$100°K with the stress on, and then removed.

The kinetics of this alignment process was measured directly by observing the recovery of the infrared dichroism as the temperature was raised. The recovery was found to be identical to that observed by EPR for the oxygen–vacancy pair (Table II), proving that the infrared absorp-

---

* A difference in the optical absorption with the **E** vector of the light parallel and perpendicular to the stress direction.

tion arises from the same defect as the EPR. A detailed study of the magnitude and sense of the dichroism vs. alignment seen in the EPR was made for stress along several crystallographic directions. This led in turn to an unambiguous assignment for the direction of the vibration with respect to the EPR axis as shown in Fig. 15. The frequency of vibration (828 cm^{-1}) is consistent with that usually observed for Si–O covalent bond stretching vibrations, confirming the bonding role of the oxygen between the two silicon atoms.

This example is an instructive one because the complete understanding of the defect required both the EPR and the optical studies. EPR looks primarily at one side of the defect. The presence of oxygen could only be inferred indirectly from it alone. The infrared study looks directly at the oxygen on the other side. Together, a rather complete picture of the defect emerges.

Electronic transitions, as observed by optical absorption and/or photoconductivity, have also been correlated with EPR studies via this technique. For the divacancy, for instance, comprehensive studies have been performed, involving both Jahn–Teller and vacancy–vacancy axis alignment, to identify several electronic transitions with different charge states of the divacancy.[36–38a]

5.1.4. *Electrical Level Determination*

In some cases stress can be used to locate the electrical level position associated with a defect. An example is the oxygen–vacancy pair.[22] For this center, it was found that in heavily irradiated material, applied stress also induces changes in the amplitudes of the EPR lines at temperatures well below those at which atomic reorientation occurs. A study of the kinetics of this process (rate vs. temperature) revealed an activation energy of 0.20 ± 0.03 eV.

This has been interpreted as reflecting the activation barrier for thermally exciting the trapped electron into the conduction band. Under stress, the electrical level position of each defect is altered, being raised or lowered depending upon the orientation of the stress with respect to the defect axes As a result, in heavily irradiated material where there are many more oxygen–vacancy pair defects than electrons to occupy them, the electrons can flow via the conduction band to seek out the lower energy traps. The measured barrier 0.20 ± 0.03 eV agrees well with the electrical level position $E_c - 0.17$ eV determined via bulk electrical measurements for the dominant defect produced by irradiation in such material.

In addition, the changes in the amplitudes of the lines in this type of experiment can again, via (45), be used to give a detailed quantitative description of the interaction of the electrical level position and the applied stress tensor. This gives different information from that of the stress alignment studies at higher temperatures, which reflect the *total energy* of the defect. For instance, the sign of the amplitude changes observed in the EPR confirms immediately that the electron is in an *antibonding* orbital (see Fig. 15) which is *raised* in energy when the two silicon atoms are pushed together.

5.1.5. *Removing Orbital Degeneracy*

As indicated in Section 3, for most defects, orbital degeneracy is quenched either by the crystalline electric fields of the undistorted crystal, or via a spontaneous Jahn–Teller distortion. This is not always true. Interesting examples where this does not occur are shallow acceptors in silicon,[39] group IV shallow donors in GaP,[40] and the shallow lithium donor in silicon.[41] In these particular cases a Jahn–Teller distortion does not occur, primarily because the wave function is too diffuse. The elastic restoring force in (46) is too large because the elastic energy has to be stored over the large volume of the wave function.

In these cases, applied uniaxial stress serves to supply this elastic energy and lift the degeneracy. By this method, EPR has been observed and studied for these systems,[39–41] where without stress they are broadened beyond detectibility.

5.1.6. *Distortion of the Wave Function*

We mentioned in discussing the transition elements in silicon that the spin S could be deduced from the fine structure terms. Actually, an isolated ion in an undistorted substitutional or interstitial position in silicon is in tetrahedral (cubic) symmetry. Therefore only quartic terms exist in S and fine structure splittings occur only for $S > 3/2$. Applied stress can reduce the symmetry and may introduce quadratic terms (D-tensor). This is illustrated in Fig. 25(a) for interstitial Fe^0 in silicon.[42] The splitting into two lines reveals that $S = 1$. Such experiments again are potentially quantitative ones. The magnitude, sign, and symmetry of the induced D-tensor reflect important microscopic information about the electronic structure of the defect.

Substantial distortion of the wave function is sometimes also evidenced by changes in the **g** and hyperfine tensors for an $S = 1/2$ defect. This is apt to occur when excited orbital states of the defect are not too far removed

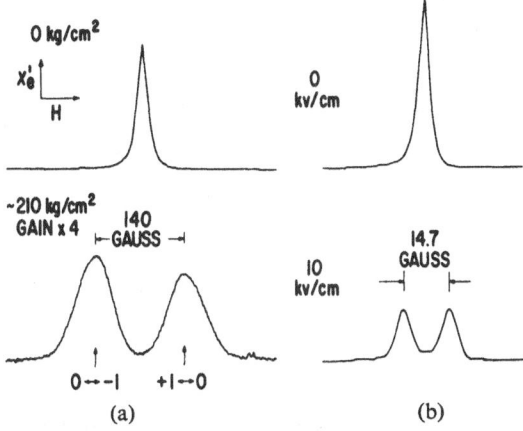

Fig. 25. (a) Applied compressional stress induces a fine structure splitting in the EPR spectrum of interstital Fe^0 in silicon. (From Woodbury and Ludwig.[42]) (b) A similar splitting produced by an applied electric field. (From Ludwig and Woodbury.[44])

in energy. A good example here is the shallow donor in silicon. By analyzing stress-induced changes in the spin Hamiltonian parameters, it has been possible to locate the energy position of an excited state for these defects as well as unravel a number of other finer features of these centers.[43]

5.2. Applied Electric Fields

Potentially, externally applied electric fields should be able to produce many of the same types of effects as applied uniaxial stress. An example is shown in Fig. 25(b), where an electric field applied across the sample also serves to produce a fine structure splitting for interstitial Fe^0 in silicon.[44] Other examples include electric field-induced shifts of hyperfine interactions in F centers (observed by ENDOR)[45] and an attempt to estimate the singlet–triplet energy separation for the neutral divacancy in silicon from hyperfine studies in EPR under electric fields.[27]

Electric fields have not been widely used, however. Difficulties often exist in sustaining large enough electric fields in the crystal without breakdown and in the lack of a linear effect for defects with inversion symmetry.[46] This technique may find increasing use, particularly in III–V and II–VI materials where sizable linear effects may be anticipated. Because an electric field couples to the defect by a different mechanism than stress, different, complementary information can be obtained from such studies.

5.3. Optical Illumination *in situ*

The ability to introduce light into the microwave cavity *in situ* opens up many experimental opportunities. Excited states of defects may be generated for study. New EPR centers can be produced, others removed, as charge states are changed by photoconductive processes. Defect alignment can be produced by polarized light.

5.3.1. *Excited States*

The ground state of the neutral oxygen–vacancy pair in silicon (no trapped electron in the $E_c - 0.17$ eV level) is diamagnetic, $S = 0$, and gives no EPR. However, illumination with band gap light ($\gtrsim 1.2$ eV) produces a new $S = 1$ EPR center that has been identified as an excited triplet state of this defect.[47] (Convincing proof of this identification was provided again by the stress alignment properties of the defect. A one-to-one correlation of the alignment monitored both by the excited $S = 1$ center and some remaining $S = 1/2$ oxygen–vacancy centers in the same crystal was used to establish this identity.) For this center, hyperfine interactions have been determined again for the two principal silicon atoms plus, in a specially enriched ^{17}O crystal, those for the oxygen atom as well.[48] Adding this to the rich store of information gathered from the $S = 1/2$ state, the oxygen–vacancy pair emerges as one of the best-documented point defects in solids.

Other optically generated excited states studied in EPR and ENDOR include M[49] and R[50] centers in alkali halides, F-center pairs in CaO,[51] and an aluminum–vacancy pair in silicon.[52]

5.3.2. *Metastable Charge States*

At cryogenic temperatures where EPR is often studied, electronic equilibration times in insulators and even narrow-band-gap semiconductors can be extremely long. As a result, metastable charge states of defects can often be produced and studied by selective generation and bleaching with monochromatic light. From the wavelength dependence of these processes, the electrical level positions of defects can be deduced and their role in complex photoconductive processes unraveled.

Many examples exist in the literature. Here we consider the divacancy in silicon.[27] In highly irradiated material, most divacancies are neutral and hence diamagnetic, $S = 0$. Illumination with band gap light, $h\nu \gtrsim 1.2$ eV, produces free electrons and holes, some of which are subsequently trapped at divacancies to produce both the single negative and single positive $S = 1/2$ charged states which are then observed in EPR. Subsequent illumination with lower energy monochromatic light removes the positively charged state. A study of this process vs. energy of the light reveals the onset of this bleaching at 0.25 eV. Identifying this as the optical excitation of a hole from the divacancy to the valence band edge serves to locate the corresponding donor state at $\sim E_v + 0.25$ eV.

5.3.3. *Optical Alignment*

Illumination with *polarized light* into an absorption band characteristic of a defect can often produce alignment of the defect which can be directly monitored in the EPR. This can result, for instance, if easy reorientation occurs in the excited state to which the defect is carried in the optical transition. The net effect in this case is to bleach orientations that couple most strongly to the polarized light, with a corresponding increase in the population of the others. Such an experiment unambiguously identifies the optical bands associated with a specific defect because alignment can only result if the transition is a local one occurring at the defect site. From the sense of the alignment, the selection rules for the transition can also be determined directly.

The power of this technique was first demonstrated by Delbecq, Smaller, and Yuster[53] for the V_k center in KCl. They located a band at 3.40 eV which produced the dramatic effects shown in Fig. 26. Here, with the magnetic field along the [011] direction in the crystal, there are three sets of spectra: one with the Cl_2^- axis along this [011] direction, labeled (0°), one made up of the four that have their axes at 60°, and a third (not indicated) for the one at 90° to the magnetic field direction. Illumination with 3.40-eV light polarized along the [0$\bar{1}$1] direction tends to bleach all of the centers except the 0° one, which increases in intensity as seen in Fig. 26(b). This immediately establishes the dipole moment for the transition as parallel to the Cl_2^- axis. A second transition, at 1.65 eV, was also established for the center from its ability to destroy the alignment.

These two transitions at 1.65 and 3.40 eV in turn could be identified, respectively, with the two allowed optical transitions $\sigma_u \rightarrow \pi_g$ and $\sigma_u \rightarrow \sigma_g$ predicted by the simple molecular orbital model for the center in Fig. 11. These values are consistent with the $\sigma_u \rightarrow \pi_u$ energy separation of 2–3 eV deduced from the g-shifts (Section 3.3.3), confirming the general correctness of the molecular orbital model for the center.

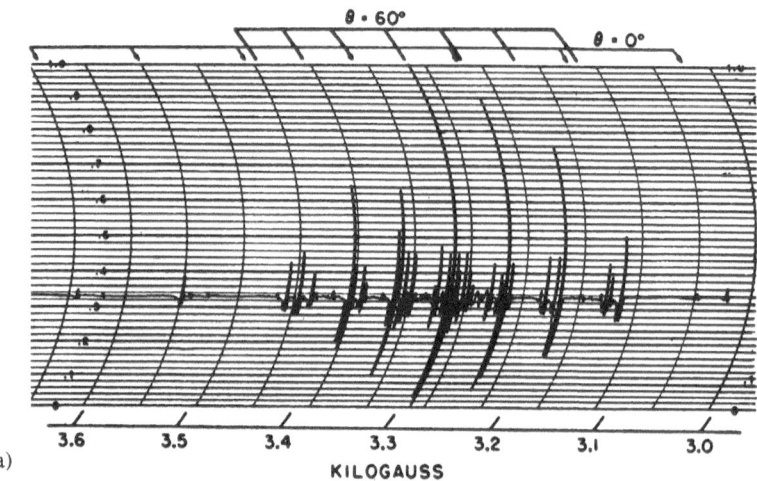

(a)

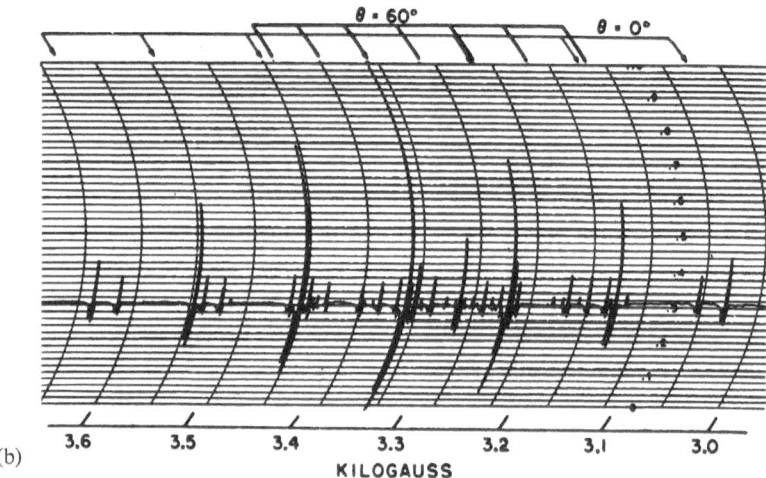

(b)

Fig. 26. EPR spectrum at 77°K of the V_k center in KCl, **H** ∥ [011]: (a) before and (b) after bleaching with [0$\bar{1}$1] polarized light at 3.40 eV. (After Delbecq, Smaller, and Yuster.[53])

5.4. Studies of the Effect of Temperature

5.4.1. Annealing

We have already discussed determining the kinetics of Jahn–Teller and atomic reorientations vs. temperature. Complex annealing processes in solids involving long-range diffusion of defects can also often be unraveled unambiguously by EPR studies because the defects involved can often be followed throughout the whole process both in the initial and final configurations. An example of this has been given for the conversion of interstitial Al_i^{2+} to Al_i^{2+}–Al_s^- pairs (Fig. 13). Another important example is the isolated lattice vacancy in silicon, observed by EPR after irradiation at low temperature[33,54] with its subsequent migration to form divacancies, phosphorus–vacancy pairs, and oxygen–vacancy pairs. In these studies accurate kinetic studies have allowed the determination of the activation energy for vacancy migration in silicon and its dependence upon charge state, important quantities unavailable by other techniques.

For annealing studies, it is not necessary to observe EPR at the temperature of anneal. Isochronal or isothermal annealing sequences with intermittent EPR observation at cryogenic temperatures are often sufficient. There are, however, other interesting effects which must be observed at temperature. These involve changes in the character of the spectrum itself.

5.4.2. Linewidth

A particularly striking example of linewidth changes vs. temperature is illustrated in Fig. 27, for the phosphorus–vacancy pair in the temperature region 60–150°K. Some of the lines broaden and disappear as the temperature is raised, and at higher temperatures, a simplified spectrum emerges.

What is being observed is the effect of thermally activated reorientation from one Jahn–Teller distortion direction to another. At these temperatures, the rate becomes so fast that it begins to exceed the natural width of the lines themselves (measured in frequency) and the lines become broadened directly by the hopping process. At higher temperatures, a phenomenon known as "motional narrowing" sets in. Here the hopping rate is occurring so rapidly that the resonance begins to take on properties associated with the average over each of the Jahn–Teller distortions. In effect, the hopping is too fast for the experiment to distinguish whether the defect is actually reorienting or simply in a single configuration which is spread equally between the various distortions.

A quantitative theory has been given for this effect which permits direct extraction of the lifetime of the reorientation process from the measured linewidth in this temperature region.[55] The results of such an analysis are given for the phosphorus–vacancy pair[25] in Fig. 28. Included also is the point at 20.4°K, determined from the stress alignment recovery studies discussed earlier. We note that by combining these two studies we have points spanning over 13 decades for a single process, allowing unprecedented accuracy in the determination of the kinetics!

The stress alignment and linewidth studies provide essentially the same information about the reorientation kinetics. Each, however, also gives somewhat unique information as well, making it worthwhile to study a defect by both techniques when possible. In the stress alignment studies, the sign and magnitude of the Jahn–Teller coupling coefficients were also determined. Studies in the high-temperature motionally averaged state also give an important piece of information not available by stress studies: ^{29}Si hyperfine satellites also exist for the motionally averaged spectrum. A study of their angular dependence reveals that the "averaged" center is now a *three*-silicon center, the wave function uniformly distributed over three silicon atoms. This confirms unambiguously that the hopping is "electronic" between the three equivalent silicon sites neighboring the vacancy, as predicted by the model of Fig. 17. (From the stress alignment studies alone, one could not rule out the possibility that a different kind of motion might

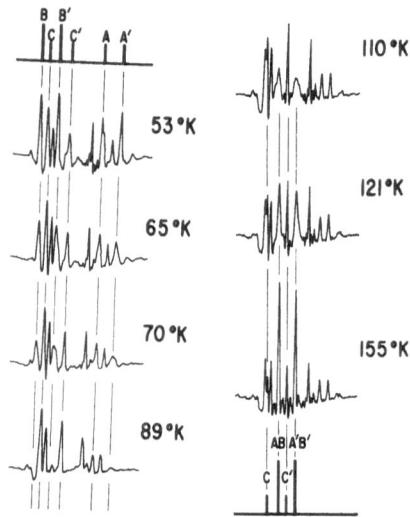

Fig. 27. Spectrum of the phosphorus–vacancy pair in silicon vs. temperature, showing the effect of thermally activated reorientation from one Jahn–Teller distortion to another. The lines labeled A, A', B, B' broaden and disappear as the temperature is raised, eventually re-emerging as a single pair AB, $A'B'$. C and C' remain unaffected throughout.

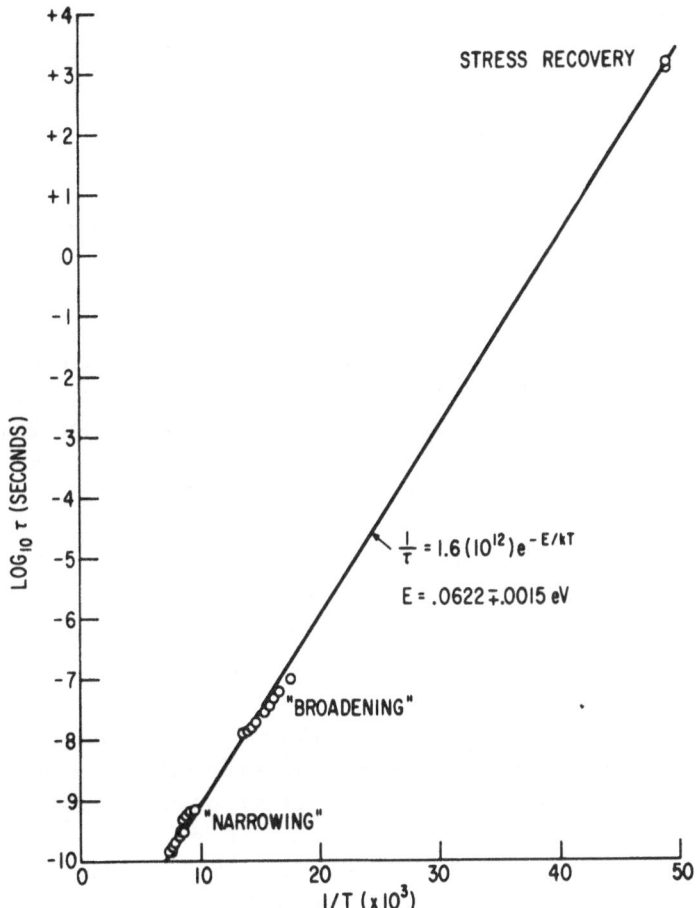

Fig. 28. Kinetics of the lifetime τ for Jahn–Teller reorientation of the phosphorus–vacancy pair in silicon as determined from linewidth and stress alignment studies.

be involved, requiring a completely different model for the defect, in which "bond switching" on a single atom were occurring.)

5.4.3. Relaxation Times

The transverse (T_2) and longitudinal (T_1) relaxation times represent the characteristic times for transfer of energy within the spin system (i.e., between defects) and between the spins and the heat reservoir of the lattice, respectively.

As such, they reflect important static and dynamic aspects of an EPR center. In the linewidth studies of the previous section, for instance, T_2 was actually the quantity being measured, being directly related in that case to the hopping time for reorientation. There it was so short ($\gtrsim 10^{-7}$ sec) that it could be studied directly from the linewidth. These quantities can still be studied, however, at lower temperatures when they are too long to have any observable effect on the linewidth. In these cases they reveal themselves in more subtle ways in the resonance experiment.

A great deal of the general EPR literature has been devoted to the study and understanding of relaxation times. However, because these experiments are more difficult to perform and more complex to interpret, they have tended not to play a leading role in studies aimed specifically at point defect characterization. Instead, the scientist interested in point defects has tended to settle for the already rich store of information available from straight EPR spectroscopy experiments alone. As a result, we will not discuss this complex subject in this chapter. We recognize, however, that in the future such studies may play an increasingly important role because of the potential information they contain about a defect.*

The EPR spectroscopist still is often unavoidably dealing with these parameters because they affect the character of the resonance signal. For instance, if the relaxation time T_1 is too long, the signal observed in absorption (χ'') can become very weak. Under these conditions, strong signals can still often be seen with the spectrometer tuned to dispersion (χ'). In addition, the exact shape of the resonance seen in dispersion depends strongly upon the relative magnitudes of T_1, T_2, the magnetic field modulation frequency, the microwave field in the cavity, etc. Signals can also appear both in and out of phase with respect to the modulation. This is a complicated subject, again, and the interested reader should refer to the literature where these effects are discussed in detail.[57,58] We mention them here primarily because the EPR spectroscopist often can use them to advantage, without necessarily studying them in detail: Consider a complex spectrum hopelessly confused by the superposition of overlapping spectra from several defects present at the same time. This happens often in point defect studies. Each defect will in general have a different relaxation time. Therefore it is often possible to

---

* An interesting example of such a study is that of Wolf[56] on the EPR of F centers in KCl at temperatures $> 400°C$. By measurements of T_2 and T_1, he was able to extract the kinetics of F-center diffusion as an entity (carrying its electron with it). Since macroscopic F-center diffusion measurements tend to be dominated by long-range diffusion of electrons between anion vacancies, this fundamental information was previously unavailable.

separate out the different spectra simply via their different resonance response characteristics. Because of the large temperature dependence of relaxation times, this analysis is greatly facilitated by studies of behavior vs. temperature.*

5.5. Defect Production

A simple and often unambiguous way of introducing point defects for study is via irradiation. For defects in highly ionic crystals, ionizing radiation (UV, X-rays, etc.) is often sufficient. For more covalent crystals, the required displacement of atoms tends to occur only by direct knock-on collisions with energetic particles (electrons, neutrons, etc.). For many materials it has been found that the primary defects (single vacancies, interstitials, close pairs, etc.) are unstable at room temperature. The facility to irradiate the sample *in situ* in an EPR apparatus therefore greatly enhances the power of EPR in these studies.

For covalent materials, where damage is produced only by atomic collisions with the irradiating particles, high-energy electrons have a distinct advantage. In the first place, relatively simple damage (isolated simple vacancies, interstitials, etc.) tends to be produced because the light electron does not transfer energy efficiently to the atoms. In addition, because the electrons are characterized by a well-defined energy and momentum direction, several unique experiments can be done with EPR.

An example is afforded by the divacancy in silicon[59]:

1. By comparing the intensity of the divacancy spectrum and that for the oxygen–vacancy pair (a monitor of single-vacancy formation) vs. electron beam energy, the expected higher threshold energy for formation of the divacancy was confirmed.

2. A study of the effect of incident beam direction revealed the interesting result shown in Fig. 29. Here the intensity of the line on the right of each recording measures the number of divacancies with their vacancy–vacancy axis along the magnetic field direction. The two recordings demonstrate clearly that a substantial preferential alignment of the divacancies

---

* The spectra of Figs. 9, 20, and 26 were taken with the spectrometer tuned to absorption, the magnetic field modulation and subsequent phase-sensitive detection giving a derivative of absorption curve (second derivative in Fig. 26). The remainder of the spectra in this chapter were taken with the spectrometer tuned to dispersion. The difference in the response of Fig. 27 (derivative of dispersion) and Figs. 16 and 23 (absorption-like) for the same EPR center results from the difference in the relaxation times at the low and higher temperatures.

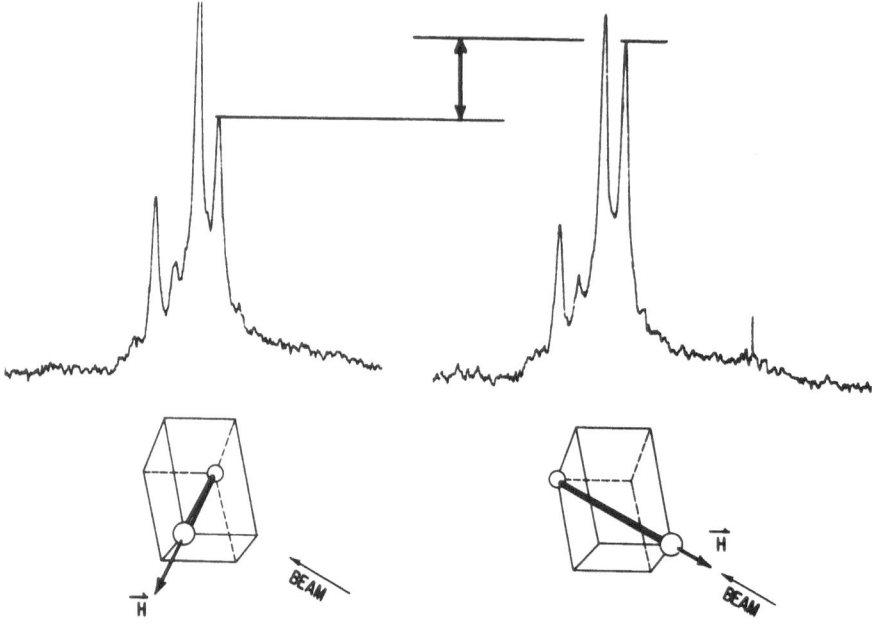

Fig. 29. Spectra of the silicon divacancy ($V \cdot V^+$) produced by a 1.0-MeV electron beam in the $\langle 111 \rangle$ direction. The high field multiplet in each spectrum (to the right) measures the number of divacancies with their vacancy–vacancy axis along the magnetic field **H**. A substantial alignment of the divacancies along the incident beam direction has been frozen in.

along the incident beam direction has been frozen in. This alignment is a strong function of bombarding energy, being greatest near threshold. These studies have been used to recreate a detailed microscopic picture of the damage event.

Ion implantation also represents an interesting technique for producing defects for EPR study. Several such studies have been reported. One unique advantage is the ability to introduce selective impurities and isotopes. The studies discussed earlier on the ^{17}O hyperfine interaction for the excited $S = 1$ state of the oxygen–vacancy pair were actually achieved in this manner.[48] The ^{17}O was first implanted in silicon and annealed, and then the damage was produced by ion-implanted helium ions.

5.6. Optical Detection of EPR

Just as in ENDOR, where nuclear resonance is detected by its effect on EPR, techniques have been developed over the past several years which under favorable circumstances permit EPR to be detected by its effect on optical absorption or fluorescence. Such a double resonance experiment serves of course to unambiguously identify optical bands associated with a defect. More important, however, it serves to convert optical sensitivity to the EPR experiments, making it possible to study EPR in sparsely populated excited states. For instance, a recent experiment of this type has allowed for the first time the detection of EPR and ENDOR in the relaxed excited state of the F center.[60] A description of these techniques is outside the scope of this chapter. However, they represent an exciting new direction for point defect study and the interested reader is referred to a recent review by Geschwind.[61]

REFERENCES

1. H. Seidel and H. C. Wolf, in *Physics of Color Centers*, Ed. by W. B. Fowler (Academic, New York, 1968), Chapter 8.
2. B. Henderson and J. E. Wertz, *Advan. Phys.* **17**, 749 (1968).
3. R. S. Title, in *Physics and Chemistry of II–VI Compounds*, Ed. by M. Aven and J. S. Prener (North-Holland, Amsterdam, 1967), Chapter 6.
4. J. Schneider, in *II–VI Semiconducting Compounds*, Ed. by D. G. Thomas (Benjamin, New York, 1967), pp. 40–67.
5. G. Feher, in *Paramagnetic Resonance*, Vol. II, Ed. by W. Low (Academic, New York, 1963), pp. 715–748.
6. G. W. Ludwig and H. H. Woodbury, in *Solid State Physics*, Vol. 13, Ed. by F. Seitz and D. Turnbull (Academic, New York, 1962), pp. 223–304.
7. G. Lancaster, *Electron Spin Resonance in Semiconductors* (Plenum, New York, 1968).
8. G. E. Pake and E. M. Purcell, *Phys. Rev.* **74**, 1184 (1948).
9. G. D. Watkins, in *Radiation Damage in Semiconductors* (Dunod, Paris, 1965), p. 97.
10. K. L. Brower, *Phys. Rev. B* **1**, 1908 (1970).
11. T. G. Castner and W. Kanzig, *J. Phys. Chem. Solids* **3**, 178 (1957).
12. C. P. Slichter, *Principles of Magnetic Resonance* (Harper and Row, New York, 1963).
13. G. Feher, *Phys. Rev.* **114**, 1219 (1959).
14. R. G. Gazzinelli and R. L. Mieher, *Phys. Rev.* **175**, 395 (1968).
15. D. F. Daly and R. L. Mieher, *Phys. Rev.* **175**, 412 (1968).
16. W. Kanzig and T. O. Woodruff, *J. Phys. Chem. Solids* **9**, 70 (1958).
17. M. L. Dakss, and R. L. Mieher, *Phys. Rev. Lett.* **18**, 1056 (1967).
18. Y. H. Chu and R. L. Mieher, *Phys. Rev. Lett.* **20**, 1289 (1968).
19. Y. H. Chu and R. L. Mieher, *Phys. Rev.* **188**, 1311 (1969).
20. A. Carrington and A. D. McLachlan, *Introduction to Magnetic Resonance* (Harper and Row, New York, 1967), p. 116.
21. M. D. Sturge, in *Solid State Physics*, Vol. 20, Ed. by F. Seitz, D. Turnbull, and H. Ehrenreich (Academic, New York, 1967), pp. 91–211.
22. G. D. Watkins and J. W. Corbett, *Phys. Rev.* **121**, 1001 (1961).
23. G. Bemski, *J. Appl. Phys.* **30**, 1195 (1959).

24. L. C. Pauling, *The Nature of the Chemical Bond*, 2nd ed. (Cornell Univ. Press, Ithaca, New York, 1948), Chapter 3.
25. G. D. Watkins and J. W. Corbett, *Phys. Rev.* **134**, A1359 (1964).
26. E. L. Elkin and G. D. Watkins, *Phys. Rev.* **174**, 881 (1968).
27. G. D. Watkins and J. W. Corbett, *Phys. Rev.* **138**, A543 (1965).
28. W. Jung and G. S. Newell, *Phys. Rev.* **132**, 648 (1963).
29. J. W. Corbett, in *Solid State Physics*, Suppl. 7, Ed. by F. Seitz and D. Turnbull (Academic, New York, 1966), p. 79.
30. K. L. Brower, in *Radiation Effects in Semiconductors*, Ed. by J. W. Corbett and G. D. Watkins (Gordon and Breach, New York, 1971), p. 189.
31. J. F. Lomer and A. M. A. Wild, *Rad. Eff.* **17**, 37 (1973).
32. A. Abragam and B. Bleaney, *Electron Paramagnetic Resonance of Transition Ions* (Oxford Univ. Press, London, 1970).
33. G. D. Watkins, in *Radiation Effects in Semiconductors*, Ed. by F. L. Vook (Plenum, New York, 1968), p. 67.
34. M. Hirata, M. Hirata, and H. Saito, *J. Phys. Soc. Japan* **27**, 405 (1969).
35. J. W. Corbett, G. D. Watkins, R. M. Chrenko, and R. S. McDonald, *Phys. Rev.* **121**, 1015 (1961).
36. L. J. Cheng, J. C. Corelli, J. W. Corbett, and G. D. Watkins, *Phys. Rev.* **152**, 761 (1966).
37. A. H. Kalma and J. C. Corelli, *Phys. Rev.* **173**, 734 (1968).
38. L. J. Cheng, in *Radiation Effects in Semiconductors*, Ed. by F. L. Vook (Plenum, New York, 1968), p. 143.
38a. C. A. J. Ammerlaan and G. D. Watkins, *Phys. Rev. B* **5**, 3988 (1972).
39. G. Feher, J. C. Hensel, and E. A. Gere, *Phys. Rev. Lett.* **5**, 309 (1960).
40. F. Mehran, T. N. Morgan, R. S. Title, and S. E. Blum, *Phys. Rev. B* **6**, 3917 (1972).
41. G. D. Watkins and F. S. Ham, *Phys. Rev. B* **1**, 4071 (1970).
42. H. H. Woodbury and G. W. Ludwig, *Phys. Rev.* **117**, 102 (1960).
43. D. K. Wilson and G. Feher, *Phys. Rev.* **124**, 1068 (1961).
44. G. W. Ludwig and H. H. Woodbury, *Phys. Rev. Lett.* **7**, 240 (1961).
45. Z. Usmani and J. F. Reichert, *Phys. Rev.* **180**, 482 (1969).
46. N. Bloembergen, *Science* **133**, 1363 (1961).
47. K. L. Brower, *Phys. Rev. B* **4**, 1968 (1971).
48. K. L. Brower, *Phys. Rev. B* **5**, 4274 (1972).
49. H. Seidel, *Phys. Lett.* **7**, 27 (1963).
50. H. Seidel, M. Schwoerer, and D. Schmid, *Z. Physik* **182**, 398 (1965).
51. D. H. Tanimoto, W. H. Ziniker, and J. C. Kemp, *Phys. Rev. Lett.* **14**, 645 (1965).
52. G. D. Watkins, *Phys. Rev.* **155**, 802 (1967).
53. C. J. Delbecq, B. Smaller, and P. H. Yuster, *Phys. Rev.* **111**, 1235 (1958).
54. G. D. Watkins, *J. Phys. Soc. Japan* **18**(Suppl. II), 22 (1963).
55. H. S. Gutowsky and A. Saika, *J. Chem. Phys.* **21**, 1688 (1953).
56. E. L. Wolf, *Phys. Rev.* **142**, 555 (1966).
57. A. M. Portis, *Phys. Rev.* **100**, 1219 (1955); Technical Note No. 1, Sarah Mellon Scientific Radiation Laboratory, Univ. of Pittsburgh, November 1955.
58. M. Weger, *Bell Syst. Tech. J.* **103**, 834 (1956).
59. J. W. Corbett and G. D. Watkins, *Phys. Rev.* **138**, A555 (1965).
60. L. F. Mollenauer and S. Pan, *Phys. Rev. B* **6**, 772 (1972).
61. S. Geschwind, in *Electron Paramagnetic Resonance*, Ed. by S. Geschwind (Plenum, New York, 1972), pp. 353–425.

Chapter 5

PHONON–DEFECT INTERACTION*

D. Walton

Physics Department
McMaster University
Hamilton, Ontario, Canada

1. INTRODUCTION

In this chapter we wish to discuss the interaction of phonons with defects. We will only consider the effect of this interaction on the phonons themselves. The study of the properties of defects in crystals is a vast area, and has been the subject of a great deal of attention in recent years. For a recent compilation of results which bear on the effect of phonons on defect properties we refer the reader to the proceedings of a recent conference, and the reviews contained therein.[1] Our interest lies in the modification of the lattice vibrations of a crystal caused by the presence of defects. The local modes associated with the defect will not be considered here. For a discussion of these phenomena the reader is referred to Ref. 1.

Two things happen to the phonons: First, additional dispersion is introduced and the relationship between the phonon frequency and wave vector is changed; second, the lifetime of the phonon is decreased. This can be conveniently described as follows: Let a phonon of frequency ω be a propagating wave in the perfect crystal, $e^{ik \cdot R}$. Then in the crystal containing defects this phonon whose frequency is ω will propagate as $e^{ink \cdot R}$, where n is a complex index of refraction. Because n is complex, it is immediately

* Research supported by the National Research Council of Canada.

obvious that the real part of the index of refraction will give the additional dispersion, and the imaginary part yields the damping of the wave.

For light the real and imaginary parts, as is well known, are connected by the Kramers–Kronig relations. Since we are considering propagating waves, the same relations will hold in our case. Therefore it is important that the frequency shift and the damping which are calculated satisfy these relations.

It is easy to appreciate that the phonon damping caused by the interaction is frequency dependent, generally increasing as some high power of the frequency (usually 4) until a resonance is reached. The resonance frequencies and most of the interesting information lie in the thermal frequency range above 10^{10} Hz.

Our aim will be to present a theory for the interaction which will yield the frequency shift and damping for a particular phonon, given the appropriate parameter which measures the strength of the interaction of that phonon with a particular defect. We will then compare theory and experiment. For the model we will use, one parameter is sufficient. But we must recognize that a great deal of information is included in that parameter. In principle, this detailed information is available from static stress measurements. While at the present time the stress coupling coefficients of some defects have been studied in some detail, we find that because of the limitations of present techniques the details of the interaction are lost in experiments with phonons of frequencies in the thermal range and above. In fact the experimental situation is the major stumbling block in understanding these effects.

There are three main experimental techniques which have been used: lattice thermal conductivity, neutron diffraction, and microwave ultrasonic techniques. Optical techniques such as infrared absorption and Raman scattering have been used up to the present mainly to study the defect local modes, and these lie outside the scope of our discussion.

The information available in the literature for neutron diffraction is limited to frequencies above 10^{11} Hz. Because of the difficulty of obtaining the very high resolution which is necessary, it is not a useful technique for studying phonon lifetimes. It does yield some information on phonon frequency shifts, but since these are at high frequencies, the interaction must be quite strong, and the defect concentration quite high, for measurement to be practical.

Conventional ultrasonic techniques are limited to frequencies below 10^{10} Hz, because of the necessity of polishing two opposing crystal surfaces flat to about 1/10 of a wavelength, and at the same time keep them parallel to a corresponding degree. However, they do measure both the frequency shift and the damping. Techniques are now becoming available which do not require coherent detection of the phonons generated, thus obviating the stringent polishing requirements.[2] However, there is one potential difficulty in all ultrasonic techniques which arises from the large number of excitations generated in a few modes. This can lead to saturation of internal levels for the defect if the rate at which the defect decays to the ground state from its excited state is too slow. At high power levels interaction between the phonons also can become important.[3]

The problems outlined above with the other two techniques have left the thermal conductivity, almost by default, as the only means of conveniently studying phonon–defect interactions. This technique depends on the fact that the phonons responsible for the transport are peaked about an energy of $\sim 4kT$. Thus a measurement of the thermal conductivity as a function of temperature can be translated into a measure of the damping as a function of frequency. However, the peak is a broad one: The phonons follow a Planck distribution peaked at $\sim 4kT$, which means that phonons whose energy is far from $4kT$ are important. The stronger the scattering at the peak, moreover, the more important the phonons in the "wings" of the line become. In fact, the thermal conductivity is such a broad-band spectrometer that it is no spectrometer at all, and an integral must be performed over all phonon states to compare theory with experiment. Nevertheless, this technique has yielded most of the information on defect interactions to date. This information is primarily about the phonon lifetime; the conductivity is not very sensitive to the phonon frequency shifts. In some cases, however, the frequency shift can have an important effect, as we shall show toward the end of this chapter.

Both neutron diffraction and ultrasonic techniques yield results which can be interpreted directly: Neutron inelastic scattering gives the frequency of a phonon of given k. Ultrasonic techniques usually yield the group velocity of a sound pulse of given ω, and the attenuation, which is equal to $k \cdot I_m(n)$.

The interpretation of the thermal conductivity data is less direct, however. It rests on the solution of a Boltzmann equation using a relaxation-time approximation. The derivation of this expression can be found in so many reviews we will not go through it here. The result is[4]

$$K = \frac{k_B}{4\pi^2} \sum_\lambda \int_0^{k_D} \left[\frac{\hbar\omega/k_B T}{\sinh(\hbar\omega/2k_B T)} \right]^2 \frac{k_\lambda^2(\omega)\,d\omega}{\sum_i l_i^{-1}(\omega)} \tag{1}$$

where k_B is the Boltzmann constant, k_λ is the phonon wave vector, λ is the polarization index, $l_i^{-1}(\omega)$ is the relaxation time due to scattering process i for phonon k_λ, whose energy is $\hbar\omega$.

For point defects $l^{-1}(\omega)$ is related to the total scattering cross section σ by

$$l^{-1}(\omega) = \varrho_D \sigma \tag{1a}$$

where ϱ_D is the number of defects per unit volume.

The situation, then, is that there is not a great deal of direct experimental data which bear on the phonon properties, and there is a great deal from thermal conductivity which is to a large extent indirect. Thus we present here the simplest theory we can formulate which is in agreement with the available data. The systems which have been studied using thermal conductivity will undoubtedly have to be reexamined if and when better techniques become available and at that time this theory may well prove deficient.

In the next section we develop the general theoretical background including a brief discussion of multiple scattering effects. We will find that, for the experimental situations of interest, multiple scattering can be largely ignored.

In the two subsequent sections specific interaction mechanisms are discussed. We consider four possible ways in which the defect can scatter phonons: through a change in the mass of the unit cell, a change in force constant, a change in volume of the unit cell, and by a resonant interaction with internal states of the defect. The theory is similar for the first three, so they are treated together. The theory of resonant scattering is treated separately using elementary time-dependent perturbation theory. The formalism is applicable to all scattering due to transitions between internal states, whether they be spin states or not. For the sake of classification, and no other reason, we will call the first three "external interactions," and "internal interactions" will constitute the fourth.

In each section we compare our results with appropriate experimental evidence. This is not intended to be a complete and exhaustive survey of the literature. We have simply selected examples which bear directly on the question being discussed.

Finally, we should explain that we do not discuss the density of states function at all. A change in the density of states of the host lattice is a consequence of the interaction we are concerned with, but these changes are subtle and are not measurable with present techniques such as infrared absorption or superconductive tunneling. This is not to say that changes do not occur upon introduction of defects. It is just that these result mainly from modes introduced by the defect itself, and this is outside the scope of this chapter.

2. THEORETICAL BACKGROUND

2.1. General

Consider a perfect harmonic crystal: The phonons are true eigenstates and can be represented as plane waves with infinite lifetimes. The problem of interest here is what happens to these eigenstates upon the introduction of point defects. In general terms it is expected that the energy of the states will change, i.e., the phonon frequencies will shift, and the lifetime of the states will become finite. The problem, then, is to calculate these two quantities.

The phonon can be considered to propagate as a plane wave, $e^{i(k \cdot x - \omega t)}$ in a perfect crystal, and $e^{i(nk \cdot x - \omega t)}$ in an imperfect crystal, where k is the wave vector of a phonon of frequency ω and n is an index of refraction. The index of refraction in general is a function of phonon frequency, and is a complex quantity, i.e., $n(\omega) = n_{re}(\omega) + i n_{im}(\omega)$. The phonon now propagates as a damped wave with a modified wave vector for a given frequency ω. In other words, the imaginary part expresses the decreased phonon lifetime due to the scattering, and the real part the change in the phonon self-energy. It has been recognized since the work of Kramers and Kronig that these two quantities are related:

$$n_{re}(\omega) = 1 + \frac{2}{\pi} P \int_0^\infty \frac{\omega' n_{im}(\omega')}{\omega'^2 - \omega^2} \, d\omega' \tag{2}$$

where ω is real and P denotes the Cauchy principal value.

We now associate a mean free path l with n_{im}, i.e.,

$$k \cdot n_{im} = 1/l \tag{2a}$$

and

$$l^{-1} = \varrho_D \sigma \tag{2b}$$

where σ is the total scattering cross section and ϱ_D is the number of defects per unit volume.

Equation (1) can be written in terms of k, where, using (2a) and (2b), we obtain

$$n_{re}(k) = 1 + \frac{\varrho_D}{\pi} P \int_0^\infty \frac{\sigma(k') \, dk'}{k'^2 - k^2} \tag{3}$$

The inverse relationship is, of course, readily obtained.

For a given phonon of frequency ω the wave vector is now $k \cdot n_{re}(k)$. Equation (3) then relates the "shift" in wave vector to the total scattering cross section σ. It is a trivial exercise, of course, given the index of refraction $n(k)$, to obtain the shift in ω at a particular wave vector k.

Dispersion relations, as equations such as (3) are called, have their usefulness limited, however, to those cases where $\sigma(k)$ is known for all k and to those cases where the integral can be performed.

Given σ, we can, in principle, calculate the energy shift ΔE or vice-versa. We now discuss how to obtain these quantities.

Let the Hamiltonian for the perfect crystal be H_0, and the unperturbed eigenfunction be ϕ. Let the interaction be V; then

$$H_0\phi = E_0\phi = (\hbar\omega_0)\phi$$
$$(H_0 + V)\psi = H\psi = E\psi = (\hbar\omega)\psi$$

where E_0 and E are the unperturbed and modified energy levels for the system, and the ψ are the new wave functions.

Now formally a transition operator can defined such that

$$\psi = \phi + G_0T\phi$$

where G_0 is the Green's function for the perfect crystal, and

$$T = V/(1 - G_0V) \tag{4}$$

The T operator is a useful mathematical device: The energy shift can now be obtained,[5] remembering that $\langle\phi \mid H_0 = \langle\phi \mid E_0$,

$$\langle\phi \mid H \mid \psi\rangle = E\langle\phi \mid \psi\rangle$$
$$\langle\phi \mid H_0 + V \mid (\phi + G_0T\phi)\rangle = E\langle\phi \mid (\phi + G_0T\phi)\rangle$$
$$E_0(1 + \langle\phi \mid G_0T\phi\rangle) + \langle\phi \mid V \mid \phi\rangle + \langle\phi \mid V \mid G_0T\phi\rangle = E(1 + \langle\phi \mid G_0T\phi\rangle)$$

Thus

$$(E - E_0) = \Delta E = \hbar(\omega - \omega_0) = \frac{\langle\phi \mid T \mid \phi\rangle}{1 + \langle\phi \mid G_0T\phi\rangle} \tag{5}$$

The denominator of Eq. (5) contains $\langle\phi \mid G_0T\phi\rangle$, which is due to interference between the incident and scattered waves. To the extent that this correction is negligible, the energy shift is $\langle\phi \mid T \mid \phi\rangle$, a standard result. The total scattering cross section is given by the standard relation

$$\sigma = (2/\hbar C_g) \int |\langle\phi \mid T \mid \phi\rangle|^2 \varrho(E) \, d\Omega$$

where $\varrho(E)$ is the density of states, C_g is the group velocity, and $d\Omega$ is the element of solid angle. The unperturbed Green's function is diagonal in ϕ, so that it involves no additional labor to keep the correction when calculating ΔE. Computation of $\langle \phi \mid T \mid \phi \rangle$ is the difficult task and now we discuss this further.

In general $V = \sum_i V_i$, where i labels a defect site. (It should be emphasized that the number of defects and their distribution have not been specified as yet). Thus

$$T = \frac{\sum_i V_i}{1 - G_0 \sum_i V_i} = \sum_i V_i + \left[\sum_i V_i \right] G_0 \left[\sum_j V_j \right] + \cdots \qquad (6)$$

Now the unperturbed Green function allows for the field at i due to a source at j.[6] Thus it connects the defect sites. That is, there will be terms in the T matrix like $V_i G_0 V_j$, $i \neq j$.[*] However, the predominant effect will still come from the terms for which $i = j$, those which refer to a single defect site. Their contribution is readily calculated to be

$$T_1 = N_d V_1 / (1 - G_0 V_1) \qquad (7)$$

where V_1 is the interaction potential for a single defect and N_d is the number of defects. In this approximation, then, it is only necessary to obtain the solution to the single-defect problem.

Some care should be exercised when performing the sum in Eq. (6). If multiple scattering is neglected, this can be rewritten as

$$T = \sum_i T_i$$

where T_i is the T matrix for an individual scatterer acting alone.

Now what is really required is a matrix element of T which will be of the form $\langle k', b \mid T \mid k, a \rangle$, where b and a are final and initial states of the scattering center and k' and k are final and initial wave vectors.

It is a well-known result that[7]

$$\langle k', b \mid T_i \mid k, a \rangle = \{ \exp[i(k - k') \cdot R_i] \} \langle k', b \mid T_0 \mid k, a \rangle$$

where T_0 is the matrix element for scattering center i if the origin were at i, and R_i is the vector connecting i and the origin.

However, we will find that we only need the matrix elements for elastic scattering in the forward direction: The real part gives the energy shift directly and the imaginary part yields the cross section via the optical theorem. Thus $k = k'$, and we obtain

$$\langle k, b \mid T \mid k, a \rangle = \sum_\alpha N_\alpha \langle k, b \mid T_\alpha \mid k, a \rangle$$

where α now labels the type of defect of which there are N_α in the crystal. Thus for elastic scattering both the real and imaginary parts of the scattered waves should be added.

Obviously it is far simpler to use Eq. (7) than to do the complete calculation indicated by Eq. (6). When is this a good approximation? A criterion may be obtained by examining the magnitude of the first neglected term, $V_i G_0 V_j$.

Mott and Massey[7] indicate how the scattering of massive particles should be handled for two scattering centers alone. A brief summary is presented below and we can use this to estimate, very crudely, when Eq. (7) would be expected to break down.

For two identical centers, one located at the origin and the other at a distance R from the origin, the scattered wave amplitude consists of a series of terms of which the first two are

$$f_2(\mathbf{k}_0, \mathbf{k}_1) = -(2\pi m/k^2)\{\langle k_1 \mid t_0 + t_0 G_0 t_R \mid k_0 \rangle$$
$$+ \langle k_1 \mid t_R + t_R G_0 t_0 \mid k_0 \rangle\}$$

where k_0 is the incident and k_1 the scattered wave vector, t_0 and t_R are the t matrix operators for the centers at 0 and R, respectively, and G_0 is the unperturbed Green's function.

The first term in each matrix element represents single scattering; e.g., $\langle k_1 \mid t_0 \mid k_0 \rangle$ gives single scattering for the center at 0. The second term represents the scattering from that center of a wave that has already been scattered from the other center.

For a random distribution of a large number of scatterers diffraction effects vanish and the first term leads to a cross section which is just N times the scattering cross section for a single center. Mott and Massey estimate the contribution of the second term to be

$$\{2R^{-1} \exp(ik \cdot R) \exp[i(\mathbf{k}_0 + \mathbf{k}_1) \cdot \mathbf{R}/2]\} f_0(k, k)^2$$

If $k \cdot R \ll 1$

$$f_2 \sim (2/R) f_0^2 \tag{8}$$

Thus the corrections are only important if $f_0/R \approx \sqrt{\sigma}/R \gtrsim 1$.

This result indicates that this correction becomes important when the amplitude of the scattered wave at a neighboring defect is large, an eminently reasonable criterion. If the average defect separation is much greater than one wavelength, a correction for double (or higher order) scattering is unnecessary: The largest possible scattering cross sections will be those for resonant scattering at resonance. There is a well-known theorem[8] which states that the cross section at resonance is $4\pi/k^2$. Thus the amplitude of the scattered wave can be no larger than $\sim\lambda$. For long-wavelength phonons, however, these effects can become important.

2.2. Multiple Scattering

Reference to Eq. (6) reveals that at high concentrations evaluation of the T matrix element is complicated by the necessity of summing over a random distribution of defects. Since we are primarily interested in evaluating the shift in self-energy, we now discuss briefly a treatment which provides a considerably improved approximation to this quantity.

2.2.1. The Coherent Potential Approximation (CPA)

This approach, an extension of Lax's[9] multiple scattering theory, was introduced by Taylor,[10] Soven,[11] Yonezawa,[12] and Onodera and Toyozawa.[13] While these authors were, respectively, dealing with phonons, electrons, and excitons, they all use essentially the same method. The objective of these approaches is to provide a reasonably self-consistent treatment across the whole concentration range. Our treatment here follows Taylor's work on phonons.

Following Taylor[10] and Velicky,[14] we seek the poles of the Green's function, which yield the phonon self-energies.

In the perfect lattice the Green's function is

$$G_0 = [\omega^2 - \omega^2(k)]^{-1}$$

where k should be taken to include the polarization of the mode. In the imperfect crystal this becomes

$$G = [\omega^2 - \omega^2(k) - \Sigma(k)]^{-1} \qquad (9)$$

The shift in the phonon self-energy is given by $\Sigma(k)$, which is the quantity we wish to calculate. Let us define the "coherent potential" E, which will be an approximation to Σ, as

$$H = (H_0 + E) + (V - E)$$

Now, we define a new Green's function

$$\bar{G} = [\omega^2 - \omega^2(k) - E(k)]^{-1}$$
$$G = \bar{G} + \bar{G}(V - E)G = \bar{G} + G(V - E)\bar{G}$$

which can be rewritten in terms of a new T matrix $\bar{T}$ as

$$G = \bar{G} + \bar{G}\bar{T}\bar{G} \tag{10}$$

We are free to choose the energy E, and we do this in such a way that if a configuration average is performed on Eq. (10), $G = \bar{G}$. This clearly requires

$$\langle \bar{T} \rangle = 0$$

If we examine the new perturbing potential $(V - E)$ we see that it is a sum of contributions from each impurity site since $V = \sum_i V_i$. We rewrite it as $V = \sum_n V_n$, where now the sum is over all the atoms in the crystal, and V_n has a value corresponding to the atom which occupies the site n. In other words, if there are two kinds of atoms in the crystal A and B, V_n is either V_n^A if the site is occupied by A, or V_n^B if it is occupied by B. Correspondingly, the single-atom T matrix $\bar{T}_n$ can be written

$$\bar{T}_n = (V_n - E_n)[1 + \bar{G}\bar{T}_n]$$

and the full $\bar{T}$ is then

$$\bar{T} = \sum_n \bar{T}_n + \sum_{n \neq m} \bar{T}_m \bar{G} \bar{T}_m + \sum_{n \neq m = l} \bar{T}_n \bar{G} \bar{T}_m \bar{G} \bar{T}_l + \cdots$$

The self-consistency condition that $\langle \bar{T} \rangle = 0$ now becomes

$$\langle \bar{T}_n \rangle = 0 \tag{11}$$

There are as many equations (11) as there are sites in the crystal, but they are all equivalent. If the defect concentration is χ, the Eq. (11) becomes

$$(1 - \chi)\bar{T}_n^A + \chi \bar{T}_n^B = 0$$

Hence

$$E_n = (1 - \chi)V_n^A + \chi V_n^B - (V_n^A - E_n)\bar{G}(V_n^B - E_n) \simeq \Sigma$$

or, if we take $V_n{}^A = 0$ and let the defect interaction energy be V,

$$E = \chi V - E\bar{G}(V - E) \tag{12}$$

For phonons we can evaluate the Green's function, replacing a sum over k by an integral, from[10]

$$\bar{G} = \int \frac{\nu(\omega')\,d\omega'}{[\omega^2 - E(\omega)] - \omega'^2} \tag{13}$$

where $\nu(\omega')$ is a phonon density of states normalized to unity.

In general $E(\omega)$ must be determined numerically. However, for the important case of the Debye approximation we can evaluate Eq. (13) and find

$$\bar{G} = \int_0^\omega \frac{\omega'^2\,d\omega'}{[\omega^2 - E(\omega)] - \omega'^2} = \frac{3}{2\omega_M{}^2} \left[\frac{\tilde{\omega}}{\omega_M} \ln \left| \frac{\tilde{\omega} + \omega_M}{\tilde{\omega} - \omega_M} \right| - 2 \right] \tag{14}$$

where $\tilde{\omega} = [\omega^2 - E(\omega)]^{1/2}$.

This is not a great deal of help since $E(\omega)$ can still only be obtained numerically, except at low frequencies such that $\tilde{\omega} \ll \omega_M$. Then the logarithm can be expanded and we obtain for $\bar{G}$

$$\bar{G} \simeq 3/\omega_M{}^2 \tag{15}$$

However, in this case the self-energy becomes

$$E(\omega) = \chi V - 3E(\omega)[V - E(\omega)]/\omega_M{}^2 \tag{16}$$

Since $E(\omega) \ll \omega^2$, $E(\omega) \ll \omega_M{}^2$, the correction represented by the second term on the right is quite negligible and can usually be ignored. However, close to a resonance it may be possible for V to become large enough for the correction to become significant.

2.2.2. Range of Validity of CPA

As we have already seen, it is difficult to be precise about the error involved in any multiple scattering approximation scheme. However, Soven[11] shows that CPA is exact to third order in the interaction energy V. On the other hand, it does not properly account for scattering from pairs of defects. Thus we expect that this treatment is limited to those cases where the defect separation is large enough for the amplitude of the scattered wave at a neighboring defect to be negligible, the criterion contained in Eq. (8). This is not to say that the scattered wave is ignored completely. On the contrary, its contribution and interference with the incident wave are included in a self-consistent fashion. However, the calculation is obviously insensitive to the distribution of defects, since we have made a mean field approximation.

Some progress in the direction of including scattering from pairs of defects has been made by Aiyer *et al.*[15] However, this is a difficult problem and it is still not clear that a satisfactory solution is at hand, and we will not include pair corrections here.

2.2.3. *Summary and Conclusion*

The CPA is certainly a useful approximation for phonon wavelengths less than the defect separation. For longer wavelengths it is probably appropriate provided the amplitude of the scattered wave at a neighboring defect is small. If this is not the case, scattering from pairs, triplets, etc. must be taken into account.

At the present time we conclude that the theory of multiple scattering is incomplete. While this is an active area and progress is being made, it is not completely clear, at the time of writing, what form the solution will take.

For the remainder of this chapter we will confine our attention to determining the energy shift and scattering cross section resulting from the interaction of the defect with the phonons. We will do this in the single-defect approximation, but although this will not be done here, our results can be employed in a CPA or other multiple scattering formalism. The reason we will not do this here is that the majority of the experimental information bearing on the subject of this chapter comes from the thermal conductivity at low temperatures, which is not sufficiently sensitive to reveal such effects.

3. EXTERNAL INTERACTIONS

3.1. Theory

We will now consider the effect on the phonons of the different mass, different force constants, and displacement field introduced by the defect.

3.1.1. *Mass-Defect and Force Constant Changes*

The first interaction we wish to discuss is the one that is best understood. Indeed the theory is exact for a single defect in a crystal which only interacts through a different mass and changed force constants. We discuss these together since they usually occur together. The experimental situation is excellent for mass defects; in fact they are really the "fruit flies" of this area of solid-state physics. Mass-defect scattering can also be observed in the absence of force constant changes in isotopically impure crystals.

Just the opposite is true of force constant changes: Not knowing precisely how a defect modifies the force constants in the crystal precludes a reliable comparison between theory and experiment in this case.

We now derive the expression for the interaction energy V, following Elliott and Taylor.[16]

In the harmonic approximation the Hamiltonian becomes

$$H = \tfrac{1}{2} \sum_{i,\alpha} m_i(\dot{u}_i^\alpha)^2 + \tfrac{1}{2} \sum_{\substack{i,j \\ \alpha,\beta}} u_i^\alpha \phi_{i,j}^{\alpha,\beta} u_j^\beta \tag{17}$$

where i, j refer to the lattice site, and α, β refer to the Cartesian coordinates.

If all the atoms are not identical, it is convenient to write

$$m_i = (1 - \varepsilon_i)m \tag{18}$$

and if the force constants are changed from those for the perfect lattice $\Phi_{i,j}^{\alpha,\beta}$, we can write

$$Y_{i,j}^{\alpha,\beta} = \Phi_{i,j}'^{\alpha,\beta} - \phi_{i,j}^{\alpha,\beta}$$

and Eq. (17) becomes

$$H = \tfrac{1}{2} \sum_{i,\alpha} m(\dot{u}_i^\alpha)^2 + \tfrac{1}{2} \sum_{\substack{i,j \\ \alpha,\beta}} u_i^\alpha \Phi_{i,j}^{\alpha,\beta} u_i^\beta - V_{i,j}^{\alpha,\beta} \tag{19}$$

where

$$V_{i,j}^{\alpha,\beta} = \tfrac{1}{2} \sum_{i,\alpha} \varepsilon_i m(\dot{u}_i^\alpha)^2 + \tfrac{1}{2} \sum_{\substack{i,j \\ \alpha,\beta}} u_i^\alpha Y_{i,j}^{\alpha,\beta} u_j^\beta \tag{20}$$

Using

$$u_i^\alpha = \sum_k \left(\frac{\hbar}{2M\omega_k}\right)^{1/2} (a_k e^{ik\cdot R_i} - a_k^+ e^{-ik\cdot R_i}) e_k^\alpha \tag{21}$$

where $M = Nm$, and specializing to a single defect at the origin, this becomes, keeping only those terms that conserve energy,

$$V_{0,j}^{\alpha,\beta} = \frac{1}{2} \sum_{k,k'} \left[\frac{\hbar k^2}{2M\omega_k} \varepsilon(a_k a_{k'}^+ + a_k a_{k'}^+) \right.$$
$$\left. + \sum_{\substack{\alpha \\ \alpha'}} \sum_j \frac{\hbar}{2M} \left(\frac{1}{\omega_k \omega_{k'}}\right)^{1/2} (a_k a_{k'}^+ + a_k^+ a_{k'}) e^{ik\cdot R_j} Y_{0,j}^{\alpha,\beta} e_k^\alpha e_{k'}^\alpha \right] \tag{22}$$

Bearing in mind that the force constant changes are phenomenological parameters to be determined by experiment, we let

$$Y_k = \sum_j e^{ik \cdot R_j} Y_{0,j}^{\alpha,\beta} \tag{23}$$

and since $\omega_k = \omega_{k'}$ by energy conservation,

$$V = \frac{\hbar}{4M} \sum_{k,k'} \frac{k^2 \varepsilon + Y_k}{\omega_k} (a_k a_{k'}^+ + a_k^+ a_{k'}) \tag{24}$$

We now wish to compute

$$\langle k \mid T \mid k \rangle = \langle k \mid V/(1 - G_0 V) \mid k \rangle$$

Now, $T - G_0 VT = V$; thus

$$\langle k \mid T \mid k \rangle - \sum_{k'} \langle k \mid G_0 \mid k \rangle \langle k \mid V \mid k' \rangle \langle k' \mid T \mid k \rangle = \langle k \mid V \mid k \rangle \tag{25}$$

The potential is usually short range, because the mass difference is a delta function, and the important force constant changes are confined to the nearest neighbors. If we confine our attention to values of k such that $k'a < 1$, the scattering is predominantly isotropic, i.e., S wave and $\langle k' \mid T \mid k \rangle = \langle k \mid T \mid k \rangle$, in which case Eq. (25) can be simplified to

$$\langle k \mid T \mid k \rangle \left\{ 1 - G_0^k \langle k \mid V \mid k \rangle \left[\frac{\Omega}{(2\pi)^3} \int \delta(k - k')\, d^3k' \right] \right\} = \langle k \mid V \mid k \rangle$$

where we have transformed the sum to an integral using $\sum_k = [\Omega/(2\pi)^3] \times d^3k$, where Ω is the volume of the crystal.

Finally,

$$\langle k \mid T \mid k \rangle = \frac{\langle k \mid V \mid k \rangle}{1 - (\Omega k^2/2\pi^2) G_0^k \langle k \mid V \mid k \rangle} \tag{26}$$

We observe that the denominator of Eq. (26) may become zero. The phonon would then be totally absorbed—its energy becoming a vibrational mode of the defect. In this case, of course, we have to be much more careful, if we are close to the resonance, and include the imaginary part of the Green's function to keep the matrix element finite.

At low frequencies the term $G_0 V$ in the denominator is usually negligible. We find in this case that the elements of T are

$$\langle k \mid T \mid k \rangle \simeq \langle k \mid V \mid k \rangle = \frac{\hbar}{4M} \frac{k^2 \varepsilon + Y_k}{\omega_k} [n_k(n_k + 1)]^{1/2} \tag{27}$$

3.1.2. *Strain-Field Scattering*

A defect in a crystal lattice in general occupies a different atomic volume than a host atom. This means that the atoms in the lattice are displaced from their equilibrium positions by an amount that depends inversely on the square of the distance from the defect. Such a displacement field can scatter phonons.

The theory of the interaction was given first by Klemens.[17] Here we will summarize a later treatment by Carruthers.[18]

A displacement field interacts with the phonons through the anharmonic part of the potential. In other words, the harmonic terms cannot possibly lead to any effect since a fixed displacement leads to no change in the Hamiltonian. The leading anharmonic term, the cubic term in the potential, can be expanded in powers of the displacements. This term is

$$V_3 = (1/3!) \sum_{\substack{i,j \\ \alpha,\beta,\gamma}} B_{i,j}^{\alpha,\beta,\gamma}(u_i{}^\alpha - u_j{}^\alpha)(u_i{}^\beta - u_j{}^\beta)(u_i{}^\gamma - u_j{}^\gamma) \qquad (28)$$

There are four contributions:

(a) All three displacements are constant and due to the stress fields, and we just get a constant shift in the energy of the crystal which does not interest us here.

(b) If two displacements are constant, we have a term linear in the displacements. Using the argument that the crystal must be in equilibrium, we can show that this term is identically zero.

(c) If one displacement is independent of time and two are not, we have a term that scatters phonons.

(d) Finally, the term in which all three displacements are time dependent leads to three-phonon anharmonic processes.

The term of interest here with the defect at the origin can be written

$$V = (1/3!) \sum_{\substack{j \\ \alpha,\beta,\gamma}} B_{0,j}^{\alpha,\beta,\gamma} V^\alpha(R_j)(u_0{}^\beta - u_j{}^\beta)(u_0{}^\gamma - u_j{}^\gamma) \qquad (29)$$

where $V^\alpha(R_j)$ is the α component of the static displacement at j.

We observe that this is quadratic in the phonon displacements, as were the interactions due to mass-defect and force constant changes. There is one important difference, however. Here the interaction is long range and cannot be assumed to be confined to the nearest neighbors of the defect.

Writing the displacements in terms of phonon operators leads to an expression which is identical in form to the second term in Eq. (22):

$$V = \sum_{k,k'} (\hbar/4M)(\omega_k \omega_{k'})^{-1/2}(a_k a_{k'}^+ + a_k^+ a_{k'}) Y_k^s \qquad (30)$$

$$Y_k^s = \tfrac{1}{3} \sum_{\substack{j \\ \alpha,\beta,\gamma}} B_{0,j}^{\alpha,\beta,\gamma} V^\alpha(R_j) e^{ik \cdot R_j} \qquad (31)$$

Energy conservation requires that $\omega_k = \omega_{k'}$. This expression is identical to Eq. (23), except that we must sum over all atoms rather than just the nearest neighbors.

We stop here with the comment that Y_k^s is an experimental parameter very difficult to calculate from first principles since $V(R_j)$ is not known, in general.

If assumptions are made as to the elastic properties of the crystal, it is possible to calculate the form of $V(R_j)$. The simplest of these, of course, is that it is an isotropic continuum. Results for this case can be found in Refs. 17 and 18.

We have already observed that Eqs. (30) and (31) have the same form as (22) and (23). This is no coincidence since the physics we have used has been to treat the effects of a strain field as a change in elastic constants. Thus we can include a contribution from the strain field and write for the potential

$$V_{\text{total}} = \frac{\hbar}{4M} \sum_{k,k'} \frac{k^2 \varepsilon + Y_k + y_k^s}{\omega_k} (a_k a_{k'}^+ + a_k^+ a_{k'}) \qquad (32)$$

Of the three quantities that measure the strength of the interaction, only ε can be calculated with confidence. At the time of writing the best that can be done with the other two is to treat them as parameters to be determined by experiment.

3.2. Experiment

There are two techniques which have yielded data that bear on this question: neutron diffraction and thermal conductivity. They complement each other in the sense that the former is sensitive to the energy shift whereas the latter is primarily sensitive to the lifetime. In practice, though, the two techniques are usually most useful in nonoverlapping frequency ranges.

3.2.1. Mass-Defect Scattering—Isotopes

The obvious system in which to study scattering due to mass defects is an isotopically impure crystal. Unfortunately, damping due to the scattering is so weak that it is not detectable by neutron scattering, and the only experimental technique sensitive enough is the thermal conductivity.

The first experiment of this nature was performed by Geballe and Hull on germanium.[19] The relaxation time deduced from their data by Callaway[20] is in good agreement with that calculated using the interaction in Eq. (24) with $Y_k = 0$ and ε given by the isotopic mass charge. Figures 1 and 2 show experimental results obtained by two different groups, Berman and Brock[21] and Thacher.[22] The remarkable sensitivity of the technique is clearly revealed by the fact that the natural isotopic content of LiF depresses the conductivity at the peak by about a factor of six.

An experimental system which should show even stronger effects is solid helium. Here the isotopic mass change is truly large. However, it develops that the differences in zero-point motion are such that a ³He atom

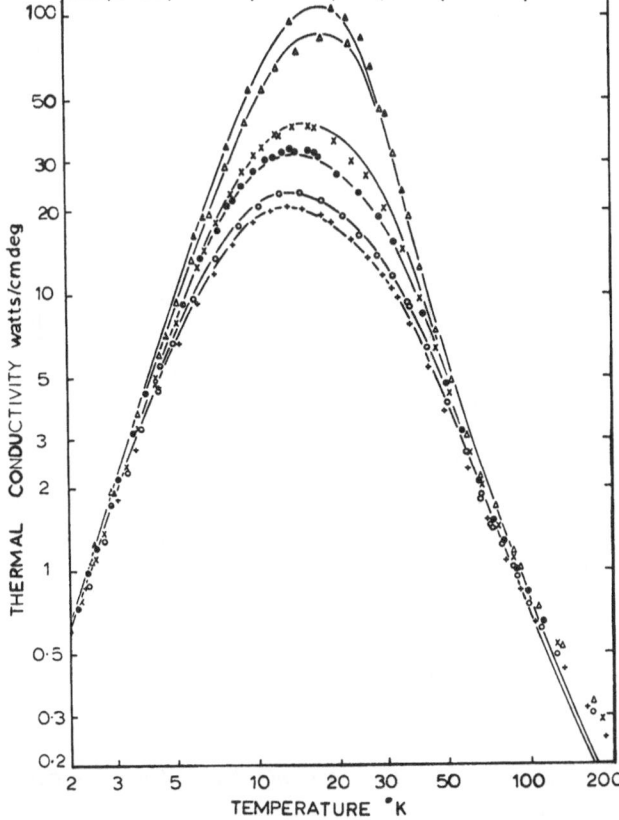

Fig. 1. Thermal conductivity data for LiF with different isotopic ratios obtained by Berman and Brock.[21] Solid curves are calculated.

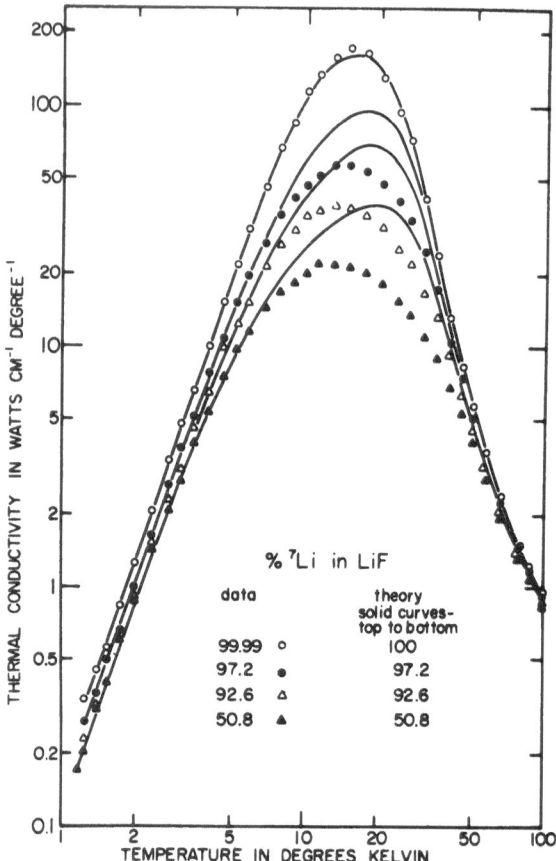

Fig. 2. Data of Thacher[22] on the thermal conductivity of
LiF with different isotopic ratios.

requires a volume different from that of a ^{4}He atom. The result is that most
of the scattering is due to the strain induced by the volume change and the
isotope scattering is swamped by the strain field scattering.

To extract the isotopic scattering cross section from the data, one must
assume a normal process relaxation time.[21,22] The latter, however, is not
well understood. Therefore one should approach the good agreement
between theory and experiment with some caution.

3.2.2. Scattering Due to Mass and Force Constant Changes

a. *Thermal Conductivity.* In general, in addition to changing the
mass of the unit cell in a crystal, a defect leads, as we have said, to changed

force constants and a possible strain field. It is difficult to untangle the two. Reference to Eq. (32) immediately reveals that the net result depends on the signs of Y_k and Y_k^s. In other words, the effect of strain fields and force constant changes can either add to or subtract from the mass difference effect alone. In general, though, the principal source of scattering is the mass difference, and many systems behave as if this alone were important.[23] There is a large body of thermal conductivity data which bears on this question. The thermal conductivity is extremely sensitive to the presence of defects. Since this has been the subject of a very recent and comprehensive review, we refer the reader to it.[23]

Using this technique, Walker and Pohl[24] were the first to identify the presence of resonance modes referred to in the theory outlined above [Eq. (26)]. Since that time numerous experiments have revealed the same resonance phenomenon. Figure 3 shows an example: The dip at $\sim 30°K$ in the doped samples is caused by the increased scattering of phonons at and near resonance. Unfortunately, however, attempts to compare theory and experiment in a quantitative fashion have not been entirely successful. Caldwell and Klein[25] have attempted to calculate thermal conductivities as a function of temperature, using parameters deduced from other experiments. Radosevich and Walker[26] have also performed a similar exercise,

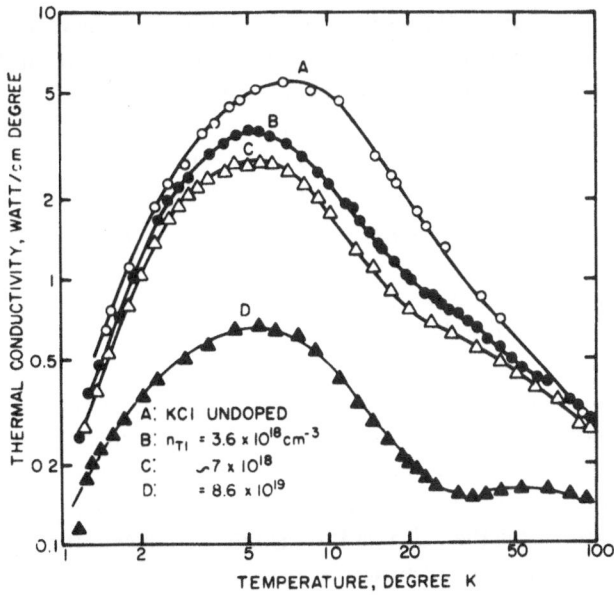

Fig. 3. Thermal conductivity of KCl crystals doped with various concentrations of TlCl. Data taken from Ref. 33.

but did not actually compute the conductivities. Elliott and Taylor[27] have also attempted to fit Walker and Pohl's results for KCl:KI with qualified success. Thus the agreement between the calculated and experimental curves near a resonance is not entirely satisfactory.

One cannot help but observe that so many other mechanisms contribute to the conductivity that it is difficult to be sure what has gone wrong: the theory for the defect or the theory for the "intrinsic" scattering. For this reason we will not discuss the thermal conductivity data any further in this connection. Instead we refer the interested reader to the several excellent reviews.[18,23]

b. *Neutron Diffraction.* We now turn our attention to the neutron diffraction experimental results. We point out that we are now considering the energy shift, whereas the thermal conductivity is sensitive to the lifetime. In other words we are now talking about an experiment which is concerned with the real part of the self-energy. In 1965 Elliott and Maradudin[28] calculated the effect on the phonon dispersion curves of a heavy substitutional atom in a crystal. They were particularly interested in the resonance which would result upon the vanishing of the denominator in Eq. (26). In this calculation they considered only the mass change.

Shortly after Elliott and Maradudin's work, Svensson, Brockhouse, and Rowe[29] determined the dispersion curve for the transverse branch of a copper crystal containing 9.3% gold. They found a distortion due to the presence of the gold which revealed the expected resonance. Figure 4 shows their results. However, the shape of the dispersion curves did not agree with the prediction of Elliott and Maradudin, which ignored force constant change. Because the concentration was so high, it was difficult to know if the difference was due to changed force constants to possible higher order multiple scattering effects.

Subsequently, Svensson and Brockhouse[30] and Svensson and Kamitakahara[31] determined the dispersion curves for a 3% alloy. At these concentrations the low-concentration theory should be applicable, and using this theory, Bruno and Taylor[32] have recently calculated dispersion curves in reasonably good agreement with the 3% results. In order to do this, they found they had to take into account the changes in force constants. However, they found that in addition to the change in force constants between the Cu and Au atoms they also had to include a change in the Cu–Cu force constant due to an increase in lattice parameters. The addition of 3% Au to Cu expands the lattice by 1.6%. It was found that this volume effect contributed roughly 75% of the observed frequency shift. The experimental

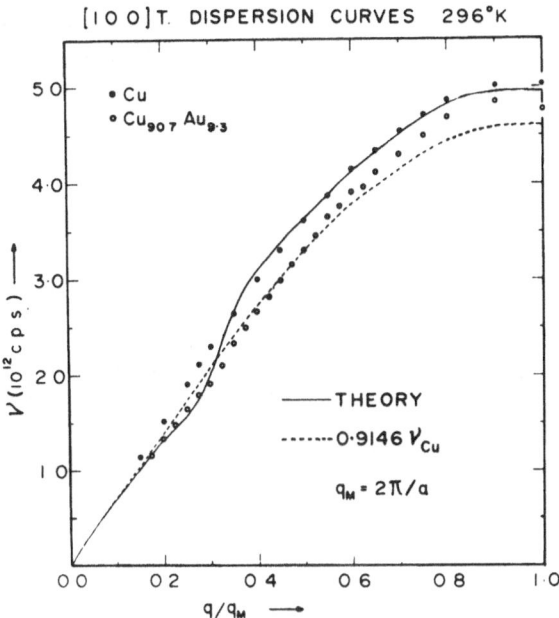

Fig. 4. Dispersion relation for a Cu:9.3% Au alloy showing the effect of the resonance on the phonon energy. Data from Ref. 29.

results with Bruno and Taylor's calculated curves are shown in Fig. 5. Nevertheless, as Bruno and Taylor conclude, the lack of a detailed knowledge of interatomic force constants makes it difficult to understand the effect of defects on phonon dispersion curves.

3.2.3. Strain Fields

There is, once again, no experimental data outside those from thermal conductivity studies. The results therefore bear on the scattering cross section. However, it appears likely that the scattering-induced dispersion in ^{3}He–^{4}He mixtures is large enough to be measured directly.

The strain field produced by most point or molecular defects appears to be so small that its effect is negligible. However, one should bear in mind that it is impossible to untangle the strain field effects from the force constant changes (this is not surprising since we have included the strain field through its effect on the force constants).

One system where strain field effects appear to predominate is in isotopic scattering in ^{3}He–^{4}He mixtures.[34,35] (Results obtained by Bertman *et al.* are shown in Fig. 6.) The minor constituent in the He crystal has, because of its different mass, a considerably different zero-point motion. This difference is large enough for the volume of the unit cell to be very different, and hence a large strain results. The resultant scattering is very strong. In fact it is strong enough for the changes in the phonon self-energy to lead to significant additional dispersion due to the scattering, i.e., a change in the velocity of sound. A rather puzzling fact, though, is that while the scattering is very strong, there does not appear to be any evidence of a resonance in the thermal conductivity results. The scattering cross section is two to three orders of magnitude larger than that for other defect–crystal systems which interact through changes in force constant, mass, or strain field, and do show resonance effects: compare Figs. 3 and 6. In fact the

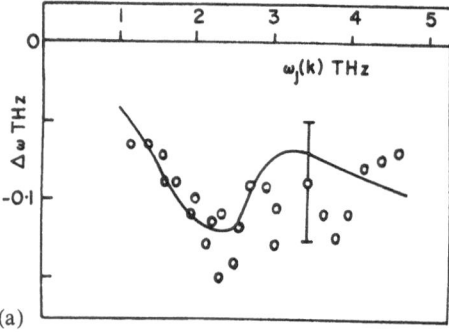

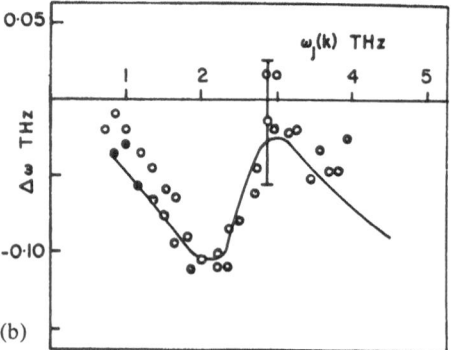

Fig. 5. Frequency shifts for the (a) transverse $(0, 0, \zeta)$ and (b) the transverse $(\zeta, \zeta, 0)$ branch of a Cu:3% alloy. Circles are experimental points from Ref. 31; the solid line is a calculated curve from Ref. 33.

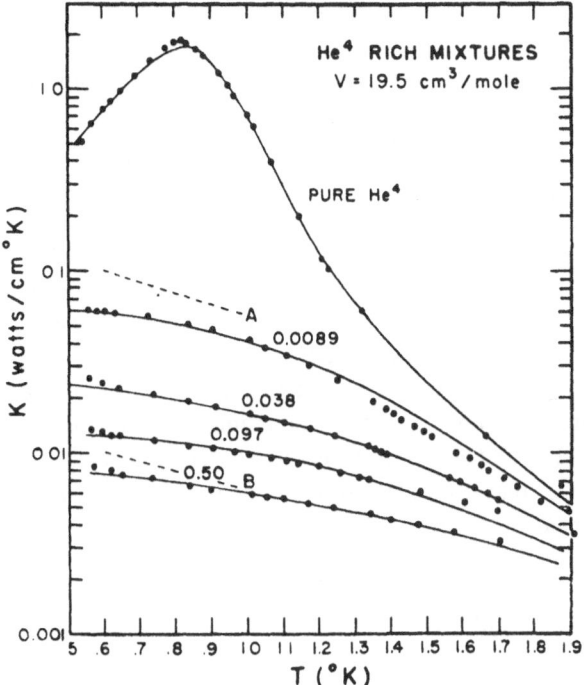

Fig. 6. Thermal conductivity of ^{3}He–^{4}He mixtures. Data taken from Ref. 35. The numbers on the curves designate the atom fractions of ^{3}He.

scattering is so strong in ^{3}He–^{4}He mixtures that the velocity of sound should be changed to a significant degree.[36]

4. INTERNAL INTERACTIONS

4.1. Introduction

In this section we consider the interaction of phonons with "internal states" of the defect. The distinction is obvious when one considers molecular defects, such as the NO_2^- ion in the alkali halides. Here the ion would be expected to have well-defined states when removed from the crystal.

The distinction is less clear for defects such as the Li$^+$ ion in KCl: Since the Li$^+$ ion is much smaller than the K$^+$, for which it is a substitute, the center of the unit cell does not constitute a stable location. The result is that the ion moves off center. After some confusion, theory and experiment managed to agree that the stable minima were at the $\langle 111 \rangle$ corners of the unit cell (for a review see Ref. 37). There are eight such locations, and in the absence of other considerations the Li^+ ground state would be eightfold degenerate. However, the Li^+ wave functions are sufficiently extensive for significant overlap to occur. Thus the Li^+ ion can tunnel between these minima, and the tunneling effectively removes the degeneracy. Needless to say, the fascination of such a system has captured many, and this problem has received a great deal of attention. The theory of the effect has been given by Gomez et al.[38] and the level spacing has been measured directly.[39] For a review of the literature on this and similar defects blessed with tunnel split ground states there is a recent review.[40]

Returning to the point we wish to make, the Li^+ ion does not have internal states in the phonon range when removed from the crystal. We could consider the Li^+ and its nearest neighbors: The molecule so formed could conceptually display the internal states of interest. But, then, how would this differ from a resonance due to a large mass difference which would make the denominator of Eq. (26) vanish? Nor would the distinction exist for nearest neighbor force constant changes, or those confined to the close vicinity of the defect. Here we can also form a molecule by extracting the unit cell containing the defect.

The distinction between the two situations is simply that in one case there are well-defined internal states for the defect and its immediate environment and not in the other. Perhaps this can be made clearer by considering the specific heat. If there are internal energy levels, the specific heat will show Schottky anomaly when the temperature is such that kT is equal to the separation of the levels. This in general is not true of the defects we have been considering in the previous section.

We now present the general theory and appropriate comparison with experiment for the scattering of phonons from defects with internal states. In the hope of achieving as simple and physically understandable a derivation as possible, we base our treatment on that available in the standard texts on quantum mechanics.

Finally we should emphasize that the theory is applicable to all defects which display resonant scattering by transitions between internal states. These include spin systems in addition to the defects already mentioned.

4.2. Theoretical Background

We now consider what happens if there are well-defined internal energy levels associated with the defect. By internal levels we mean bound states associated with the defect and its immediate environment. We expect that there should be a strong scattering when the phonon energy is equal to the energy difference between two of these states. Since a large fraction of the information on such interactions is for low-energy (i.e., long-wavelength) phonons, we confine our attention to this case and treat the problem in the Debye approximation.

4.2.1. Partial Wave Analysis

Initially we briefly discuss the standard partial wave analysis of this problem,[41] which quickly yields the form of the scattering cross section and energy shift. Because of the short range of the defect potential and the long phonon wavelength, considerable simplification results because only the $l = 0$ partial wave, i.e., S-wave scattering, need be considered.[42] The scattering, then, is spherically symmetric.

We seek the solution of

$$[H_0 + V]\psi(r) = E\psi(r)$$

We assume that at a large distance r from the defect, $\psi(r)$ represents an incident plane wave (along the z direction) and an outgoing spherical scattered wave,

$$\psi(r) \to e^{ik \cdot z} + (e^{ik \cdot r}/r)f(\theta) \tag{33}$$

where θ is the angle between the direction of the scattered wave and that of the incident wave. The complex function $f(\theta)$ is the scattered amplitude.

Now the scattered amplitude and the incident plane wave are expanded in Legendre polynomials,

$$f(\theta) = \sum_l a_l(1/2k)i(2l + 1)P_l(\cos\theta)$$

$$e^{ikz} = \exp(ik \cdot r \cos\theta) = \sum_l (2l + 1)i_e j_l(kr)P_l(\cos\theta)$$

where a_l is a constant to be determined from a knowledge of the potential $V(r)$ and $j_l(kr)$ is the spherical Bessel function.

It is found that the a_l can be written in terms of phase shifts δ_l as

$$a_l = 1 - \exp(2i\delta_l)$$

whence

$$f(\theta) = (1/2ik) \sum_l (2l + 1)[\exp(2i\delta_l) - 1]P_l(\cos\theta)$$

If k is small enough for $1/k$ to be large compared with the range of the potential, $\delta_l \sim k^{2l+1}$ and only the $l = 0$ wave contributes.[8]

Following Landau and Lifshitz,[41] we can describe the internal states of the scattering system by an energy E_n and a width (an inverse lifetime) of the energy level Γ_n. In other words we assume that there are a number of levels of energy E_n above the ground state and that each of these levels is broadened by an amount Γ_n, where Γ_n is small compared with the spacing between the levels. Let us specialize to the case of a two-level system of separation E_0 and a particle of energy E.

The phase shift for scattering of a particle close to resonance is

$$\exp(2i\delta_l) = [\exp(2i\delta_l{}^0)]\left(1 - \frac{2i\Gamma}{E - E_0 + i\Gamma}\right)$$

where $\delta_l{}^0$ is the phase shift for scattering of a particle whose energy is much greater than E_0.

The scattering amplitude now becomes

$$f(\theta) = f^0(\theta) = \frac{2l+1}{k} \frac{\Gamma}{E - E_0 + i\Gamma} [\exp(2i\delta_l{}^0)]P_l(\cos\theta)$$

where $f^0(\theta)$ is the amplitude far from resonance (on the high-energy side). This corresponds to the so-called potential scattering. If we assume that this is small compared with the resonance amplitude, and if we consider only S-wave scattering ($l = 0$), then we obtain

$$f(\theta) \simeq \frac{1}{k} \frac{\Gamma}{E - E_0 + i\Gamma} \tag{34}$$

which is the well-known Breit–Wigner one-level formula.

In terms of this formula the differential scattering cross section becomes

$$d\sigma \simeq \frac{1}{k^2} \frac{\Gamma^2}{(E - E_0)^2 + \Gamma^2} d\Omega$$

and of course the total cross section is

$$\sigma = \frac{4\pi}{k^2} \frac{\Gamma^2}{(E - E_0)^2 + \Gamma^2} \tag{35}$$

or, if there is more than one level,

$$\sigma = \sum_n \frac{4\pi}{k^2} \frac{\Gamma_n{}^2}{(E - E_n)^2 + \Gamma_n{}^2} \tag{36}$$

To proceed further, it is necessary to calculate Γ. This can be accomplished more easily, and the meaning of Γ becomes somewhat clearer, if the problem is approached from a slightly different point of view.

4.2.2. *Formal Scattering Theory*

Our objective will be to calculate the phonon energy shift produced by the scattering. Of course we could also calculate the cross section. In either case the central problem is to obtain the elements of the T matrix, or more explicitly, we calculate the energy shift using Eq. (5),

$$\Delta E = \sum_i N_i \bar{T}_i \tag{37}$$

where $\bar{T}_i = \langle \phi \mid T_i \mid \phi \rangle$.

We now proceed to obtain an expression for $\bar{T}_i$ following the treatment by Messiah.[43]

The Hamiltonian for the system is $H = H_0 + V$. We decompose the unperturbed part further into a contribution from the lattice and one from the defect,

$$H_0 = H_l + H_d$$

Furthermore, we define internal state wave functions for the defect as

$$H_d \mid a \rangle = W_a \mid a \rangle$$

where W_a is the energy of a defect in state $\mid a \rangle$.

The phonon states are given by

$$H_l \mid k \rangle = (n_k + \tfrac{1}{2}) \hbar \omega_k \mid k \rangle$$

Thus we describe a state of the system by the occupation of phonon states and defect states as $\mid k, a \rangle$.

The T matrix is

$$T = V + VGV$$

Now if the interaction V is linear in the phonon operators, the first term in this equation cannot contribute to elastic scattering: A single-phonon operator either raises or lowers the number of excitations in the system, thus leaving it in a state which differs by a single quantum from the initial state. Thus the only process which is allowed by energy conservation consists in either the absorption or emission of radiation, which we will ignore since this can only occur at a single frequency and we need results far from resonance. We require the elements of T for the process whereby a quantum is absorbed and the same quantum reemitted. Thus, assuming that the scatterer's internal state is initially $|b\rangle$, and it scatters a phonon $|k\rangle$, making a virtual transition to another internal state $|a\rangle$, we have

$$\langle k_f, b \mid T \mid k, b\rangle = \sum_{a, a'} \{\langle k_f, b \mid V \mid a'\rangle\langle a' \mid G \mid a\rangle\langle a \mid V \mid k, b\rangle$$

$$+ \langle k_f, b \mid V \mid k, k_f, a'\rangle\langle k, k_f, a' \mid G \mid k, k_f, a\rangle$$

$$\times \langle k, k_f, a \mid V \mid k, b\rangle\}$$

where $|a'\rangle$ also refers to internal states of the scattering system, and $k_f = k$ for forward scattering. The second term is a result of the fact that the scattered phonon can be emitted before the incident phonon is absorbed.[13] For a two-level system $a' = a$, and we require

$$\langle a \mid G \mid a\rangle = G_a$$

It can be shown that[43]

$$G_a = 1/(E - W_a - \varepsilon + i\hbar\Gamma) \tag{38}$$

where ε is the energy shift of the defect level and Γ is the linewidth. In general ε is small, and henceforth W_a will be taken to include ε.

Equation (38) leads to the Breit–Wigner one-level formula, and Γ is now the lifetime of an "excited complex" formed by the incident particle and the scattering system.

The linewidth is equal to the transition probability per unit time, which is, using first-order perturbation theory and Fermi's golden rule,

$$\Gamma = (\Omega k^2/\pi\hbar^2 C_g) \, |\langle f \mid V \mid i\rangle|^2 \tag{39}$$

where Ω is the volume of the crystal, $|f\rangle$ refers to the metastable intermediate state, and C_g is the group velocity.

Including the energy shift $E = \hbar\omega_k - \varepsilon = \hbar\omega$, the matrix elements of T are now

$$\hbar T_{bb} = \hbar\langle k, b \mid T \mid k, b\rangle = \frac{\langle k, b \mid V \mid a\rangle\langle a \mid V \mid k, b\rangle}{\omega - \omega_0 + i\Gamma}$$

$$- \frac{\langle k, b \mid V \mid k, k, a\rangle\langle k, k, a \mid V \mid k, b\rangle}{\omega + \omega_0 - i\Gamma} \tag{40}$$

where $\hbar\omega_0 = W_a - W_b$. We observe that only the first term leads to a resonance, since ω_0 is positive. The second term leads to so-called potential scattering.

Substituting for Γ from (39), Eq. (40) can be written

$$T_{bb}^g = \frac{\pi\hbar C}{\Omega k^2} \left(\frac{\Gamma_{\text{res}}}{\omega - \omega_0 + i\Gamma_{\text{res}}} - \frac{\Gamma_{\text{pot}}}{\omega + \omega_0 - i\Gamma_{\text{pot}}} \right) \tag{41}$$

It should be noted that at finite temperature the excited level will also be populated. Therefore it will also be possible for the phonon to be scattered by a process which involves a virtual transition to the ground state. More specifically, we consider the scattering center in state $|a\rangle$ now, and we seek the T matrix elements for a transition to state $|b\rangle$, and back to $|a\rangle$ coupled with the absorption and reemission of a phonon. The matrix elements for this process can be immediately written as

$$\hbar T_{aa} = \hbar\langle k, a \mid T \mid k, a\rangle = \frac{\langle k, a \mid V \mid b\rangle\langle b \mid V \mid k, a\rangle}{\omega - [(W_a - W_b)/\hbar] + i\Gamma}$$

$$- \frac{\langle k, a \mid V \mid k, k, b\rangle\langle k, k, b \mid V \mid k, a\rangle}{\omega + [(W_a - W_b)/\hbar] - i\Gamma} \tag{42}$$

Now it is convenient to consider that the crystal contains two types of defect, those for which the scattering center is in the ground state, and those for which it is in the excited state. The concentrations would then be given by the appropriate statistical weights.

Making this assumption, we can write Eq. (37) as

$$\Delta E = N_d(\text{Re } T_{bb} + e^{-2\beta\omega_0} \text{ Re } T_{aa})(1 + e^{-2\beta\omega_0})^{-1} \tag{43}$$

where N_d is the number of defects and $\beta = \hbar/2k_{\text{B}}T$.

To proceed further, we now consider the interaction V.

4.2.3. *Form of the Interaction*

In general the defect's contribution to the energy of the crystal will depend on the positions of the neighboring ions. We can expand this in a Taylor series in the displacement of the neighbors about their equilibrium position

$$E_d = \sum_i E_d(x_i)$$

where the x_i are the coordinates of the neighbors,

$$E_d(x_i) = E_d(x_i^0) + (x_i - x_i^0)\frac{\partial E_d}{\partial x_i} + \frac{(x_i - x_i^0)^2}{2!}\frac{\partial^2 E_d}{\partial x_i^2} + \cdots$$

The first term yields the defect Hamiltonian H_d and the remaining terms constitute the interaction. Normally the effect of succeeding sets of neighbors dies off quickly enough that only the nearest need be considered. If we expand the displacements in phonon operators using Eq. (21), the term multiplying $\partial E_d/\partial x_i$ will lead to a term linear in the phonon operators, the next will be quadratic, then cubic, etc. Writing the displacement in terms of phonon operators, and assuming that $ka \ll 1$, where a is the nearest-neighbor separation, it can be shown that[44]

$$V = \sum_k k\left(\frac{\hbar}{2M\omega(k)}\right)^{1/2}(\alpha_k + \alpha_k^*)E_k$$
$$+ \sum_{k,k'} kk'\left(\frac{\hbar}{2M\omega(k)}\right)^{1/2}\left(\frac{\hbar}{2M\omega(k')}\right)^{1/2}(\alpha_k + \alpha_k^*)(\alpha_{k'}^* + \alpha_{k'})E_{kk'} + \cdots$$

(44)

where E_k and $E_{kk'}$ are, respectively, the Fourier coefficients of $\partial E_d/\partial x_i$ and $\partial^2 E_d/\partial x_i^2$.

The first term in V leads to transitions involving one phonon, the so-called direct process in the language of spin–lattice relaxation. The second and higher terms lead to higher order interactions, Raman processes, etc. We will ignore higher order effects here. It is probably clear at this point that their inclusion, while cumbersome, is no different in principle.

We will require matrix elements of V which, for the linear term are simply

$$\langle n_k + 1, b \mid V \mid n_k, a \rangle = (n_k + 1)^{1/2}(\hbar/2M\omega_k)^{1/2}k\langle b \mid V \mid a \rangle \qquad (45)$$

We note that all remaining uncertainties about the coupling mechanism are contained in $W = \langle b \mid V \mid a \rangle$. These matrix elements can also contain a frequency dependence. However, since they only connect states of the defect, this dependence can only be on the energy difference between the two states. The dependence of the matrix elements on the phonon's ω and k is always the same. For magnetic defects, for instance, W is independent of the energy difference (in this case the Zeeman splitting) for non-Kramers ions, whereas for Kramers ions it is proportional to the Zeeman splitting. Nevertheless, the dependence on the phonon's ω and k is the same. Since W is usually not known from first principles and it is a parameter to be determined from experiment, there does not seem to be any point in going into more detail than this.

We now return to Eqs. (40)–(43): Ignoring the difference between Γ_{res} and Γ_{pot}, we find

$$(\omega - \omega_k)(\omega^2 - \omega_0^2 - \Gamma^2) = \frac{W^2 N_d \tanh(\beta\omega_0)}{M\hbar C_0^2} \omega_0 \omega_k \qquad (46)$$

4.2.4. Dispersion Relations

Equation (46) can be rewritten as, letting $W^2/\varrho\hbar C_0^2 = A$,

$$\frac{\omega - \omega_k}{\omega_k} = \frac{C_p}{C_0} - 1 = [A\varrho_d \tanh(\beta\omega_0)] \frac{\omega_0}{\omega^2 - \omega_0^2 - \Gamma^2} \qquad (47)$$

where ϱ_D is the number of defects per unit volume, and C_p is a new phase velocity, $C_p \equiv \omega/k$. If we now let $\omega \to 0$, we observe that C_p tends to a constant value, which is less than C_0. In other words, the additional phonon dispersion introduced by the scattering extends all the way to zero frequency, and there is a change in the velocity of sound which extends all the way to $k = 0$. Since ultrasonic techniques for measuring changes in the velocity of sound are extremely sensitive, this would seem to be an excellent way to obtain W experimentally.

4.2.5. Scattering Cross Section

We now consider the scattering cross section, which we can obtain from the matrix elements in Eqs. (40) and (42). We observe that only one of the terms in each of these two expressions is singular. This term, then, leads to a resonance in the scattering, whereas the other, nonsingular in nature, leads to the so-called "potential scattering."

To compute the total scattering cross section, the simplest procedure at this point is to use the optical theorem, since we already have an expression for the T matrix:

$$\sigma = -(2\Omega/\hbar C_g) \, \text{Im} \, T \qquad (48)$$

where Ω is the crystal volume.

Since there are contributions from two sets of centers, we will require

$$l^{-1} = \sum_d \sigma_d \qquad (49)$$

It is important to perform the sum as outlined earlier and take adequate account of the phase relationships between waves scattered from defects in their ground and excited states,

$$l^{-1} = \frac{2\Omega}{\hbar C_g} \frac{\text{Im } T_{bb} + e^{-\beta\omega_0} \text{Im } T_{aa}}{1 + e^{-\beta\omega_0}} \varrho_D \tag{50}$$

From Eqs. (40) and (42) and using (45), we obtain

$$l^{-1} = \frac{2A^2\varrho_D \tanh(\beta\omega_0) C_0^2}{\pi C_g^2} \frac{\omega^2 + \omega_0^2 + \Gamma^2}{(\omega^2 - \omega_0^2 + \Gamma^2)^2 + 4\Gamma^2\omega_0^2} k^4 \tag{51}$$

This expression has three features we wish to emphasize at this point.

The first is that the cross section varies as k^4 at low frequencies, far from resonance.

The second is that at zero temperature and at resonance the cross section for a single defect is, since $\Gamma = (A/\pi)k^3$,

$$\sigma = 4\pi/k^2 \tag{52}$$

a standard result.[41]

Finally, we note that the scattering cross section is temperature dependent, becoming zero at the high-temperature limit. Another way of looking at this result is to recognize that the resonance cross section for centers in the ground state is precisely equal to the absorption cross section,[45] which for atoms in the excited state is equal to the emission cross section. Thus if the populations are equal, for every phonon that is absorbed, another is emitted.

4.3. Paramagnetic Ions

Since spin–lattice relaxation experiments are relatively easy to perform, there is quite a large body of information on the interaction between phonons and paramagnetic ions. However, this is usually not a clean experiment and the interaction constants obtained in this way are not normally reliable. All the information on the interaction comes from changes in the defect properties. Direct observation of the effect on the phonons is only available from ultrasonic work, which, of course, is confined to low frequencies.

There is an excellent review of the theory of the spin–lattice interaction and a description of the important techniques in an article by Tucker.[46] We will briefly summarize the main features of the interaction. For details the reader is referred to Tucker's article, the treatment of which we follow closely.

Qualitatively, the interaction can be understood as follows: The ionic displacements induce changes in the crystal field in which the defect ion finds itself. This in turn affects the electron orbit, which in turn is coupled to the electron spin through the spin–orbit interaction.

It is customary to represent the contribution of the spin system to the Hamiltonian by a spin Hamiltonian. This usually includes three important terms

$$\mu H \cdot g \cdot S + S \cdot D \cdot S + S \cdot T \cdot I + \cdots$$

where the tensors g, D, and T represent the interaction with the magnetic field H (μ is the Bohr magneton), crystal fields, and the nuclear moment I. Here we have neglected additional terms such as spin–spin, nuclear Zeeman, etc., which are uninteresting.

Now we allow for the fact that the tensors g, D, and T can depend on the positions of the neighbors of the paramagnetic defect. We can then define an interaction

$$V = \mu H \cdot \delta g \cdot S + S \cdot \delta D \cdot S + I \cdot \delta T \cdot S \tag{53}$$

where δg, δD, δT are second-rank tensors with a linear dependence on the strain. They can then be written

$$\delta g_i = F_{ij} e_j; \qquad \delta D_i = G_{ij} e_j; \qquad \delta T_i = Z_{ij} e_j \tag{54}$$

where the e_j represent strain components. These are written in Voigt notation such that the indices i and j can take on values from 1 to 6. The meanings of the values are as follows:

$$1 = xx, \quad 2 = yy, \quad 3 = zz, \quad 4 = yz, \quad 5 = xz, \quad 6 = xz$$

The strains are defined in standard fashion,[47] e.g., $e_{zz} = (\partial u_x/\partial y) + (\partial u_y/\partial x)$, where u_x and u_y are displacements in the x and y directions, respectively.

In addition to paramagnetic relaxation, the elements of these matrices can be determined by ultrasonic techniques and by static stress experiments.

Typically in the ultrasonic experiments, phonon pulses whose frequency is $\sim 10^9$ Hz are generated and their attenuation is measured. Each value of the attenuation yields a combination of the coupling constants. If sufficient data are available, these can be disentangled to yield the absolute value of each. The sign of the coupling constant, of course, is not available because the attenuation will depend on the interaction squared.

Static stress measurements, on the other hand, do yield the magnitude and sign of the coupling constants. In this experiment the shift in EPR frequency with stress is measured. It is clear from Eqs. (53) and (54) that the sign of the shift in resonance frequency depends on the sign of the coupling constants.

It is interesting that experimentally it has been found that the static coupling constants agree within experimental error with those obtained ultrasonically. We can conclude from this that in these cases the separation of electronic and ionic wave functions according to the Born–Oppenheimer approximation is valid. This may not be so, however, for those ions which display strong Jahn–Teller effects.

The measurements that have been made of the change in sound velocity introduced by the defect are also confined to the microwave frequency range. As we have already pointed out, the change in sound velocity is present at low frequencies and should provide a direct measure of the interaction.

We now discuss in detail the following question: What do we expect the effect of the interaction contained in Eq. (53) to be on the phonons, and to what extent has this been observed experimentally? First, then, we review the theory, and then deal with the experimental data.

4.3.1. Theory

We wish to consider a two-level system. For magnetic defects this corresponds to systems of effective spin $1/2$. Now this does not imply that the real spin is $1/2$. It means simply that, in a magnetic field, we have only two states whose separation is $\hbar\omega_0 = g\beta H$.

To conform to the existing literature, we write the spin–lattice interaction in the following phenomenological form.[48] The correspondence between these equations and Eqs. (44) and (45) is easily derived:

$$V = \sum_{i,j} \sum_{\alpha,\beta} S_{i\alpha} A_{\alpha\beta}(R_j - R_i) u_j^{\beta} \qquad (55)$$

where α and β denote Cartesian indices, x, y, z; $u_{j\beta}$ is the displacement of the atom at j; and $A(R_j - R_i)$ is a phenomenological coupling coefficient to the spin of the defect at i [see Eq. (21)]

$$u_j^{\beta} = \sum_{k} (\hbar/2M\omega_k)^{1/2}(\alpha_k + \alpha_{-k}^{+})e^{ik \cdot R_j} e_k^{\beta}$$

where e_k^{β} is the polarization vector, and k includes a branch index (a sum over k implies a sum over all branches).

The interaction can now be written

$$V = 2 \sum_k k^{1/2} A_k (\alpha_k + \alpha_k^+) S_x \qquad (56)$$

where now

$$A_k = \left(\frac{\hbar}{2MC_0}\right)^{1/2} \frac{1}{k} \sum_{j=1}^{N} [A_{x\beta}(R_j) e^{ik \cdot R_j}] e_k^{\beta} \qquad (57)$$

and we have ignored the coupling to the Z component of spin. Reference to Eq. (44) reveals that

$$2A_k S_x = k(\hbar/2M\omega_k)^{1/2} E_k \qquad (58)$$

While we have ignored the details of interaction, there is one feature we must preserve, and that is the difference in behavior between Kramers and non-Kramers ions.[49] Kramers' theorem forbids phonon-induced transitions between pairs of states which are time-reversal invariant. Those ions containing an odd number of electrons are only coupled to the extent that a magnetic field admixes other states. Thus A_k should be taken to be proportional to ω_0 for Kramers ions and is independent of ω_0 for non-Kramers ions. Since

$$\langle \tfrac{1}{2} | S_x | -\tfrac{1}{2} \rangle = \tfrac{1}{2} \qquad (59)$$

substitution in Eqs. (40), (42), and (43) yields for the frequency shift an expression similar to (47),

$$(\omega - \omega_k)/\omega_k = B\varrho_D(\tanh \beta\omega_0)[\omega_0/(\omega^2 - \omega_0^2 - \Gamma^2)] \qquad (60)$$

and

$$B = 2A_k^2 \Omega / C_0 \hbar^2 \qquad (61)$$

This result is similar to a result obtained by Stevens and Van Eekelen[50] and Elliott and Parkinson[51] using thermal Green's functions. Their calculation, however, was for a concentrated crystal, i.e., they assumed the presence of a spin on every site. Ours, on the other hand, is for a single spin; to take account of more than one spin in the crystal, we simply multiply by the number of defects. In doing so, we include no information about their distribution or, more importantly, about their separation. Therefore it makes no difference as far as the calculation is concerned if we replace the crystal by another with a spin on every site of magnitude $S_z = N_d/N$. Thus it is not surprising that our results are similar.

Our expression for Γ, when substituted in Eq. (60), leads to an effective shift in the resonant frequency.

We now turn our attention to the scattering cross section. Using Eq. (51), we obtain

$$l^{-1} = \frac{2}{\pi} \frac{C_0^2}{C_g^2} (B^4 k^4) \frac{\omega^2 + \omega_0^2 + \Gamma^2}{(\omega^2 - \omega_0^2 + \Gamma^2)^2 + 4\Gamma^2 \omega_0^2} \varrho_d \tanh(\beta\omega_0) \quad (62)$$

We observed that, because the coupling depends on the admixture of higher states for Kramers ions, the strength of the scattering is expected to increase with magnetic field. Our result indicates that it is proportional to the square of the magnetic field. For non-Kramers ions it is independent of the field.

4.3.2. Experiment

Our interest now lies in the extent to which experimental data confirm Eqs. (60) and (62). The data which bear on this question come from microwave ultrasonics and thermal conductivity. We first consider the dispersion relations.

A direct measure of the dispersion introduced by the spin–phonon scattering is the change in group velocity of the excitation. This has been observed for Ni^{2+} in MgO by Shiren.[52]

Guermeur et al.[53] have measured the phase velocity of the ultrasonic wave. This of course yields the dispersion relations directly. Their results are shown in Fig. 7 for the same system, Ni^{2+} in MgO. However, the interpretation is complicated by the presence of residual Fe^{2+} in the MgO and the fact that the acoustic paramagnetic resonance technique requires a constant frequency. Therefore the resonant frequency is varied in the experiment. Nevertheless, we observe that at high fields there is still quite a large change in phase velocity remaining.

Guermeur et al.[53] identify four transitions in all:

$$
\begin{array}{llll}
Fe^{2+} & \Delta m = 2 & \text{at} & 980 \text{ Oe} \\
Ni^{2+} & \Delta m = 2 & \text{at} & 1550 \text{ Oe} \\
Fe^{2+} & \Delta m = 1 & \text{at} & 1860 \text{ Oe} \\
Ni^{2+} & \Delta m = 1 & \text{at} & 3100 \text{ Oe}
\end{array} \quad (63)
$$

The last of these is the strongest by far, so we will ignore the first three.

At high fields the resonance frequency is above the acoustic paramagnetic resonance frequency. In the high-field limit, then, and if the strength of the interaction is independent of the resonance frequency (i.e., the magnetic field), there should be a decrease in sound velocity. The ion

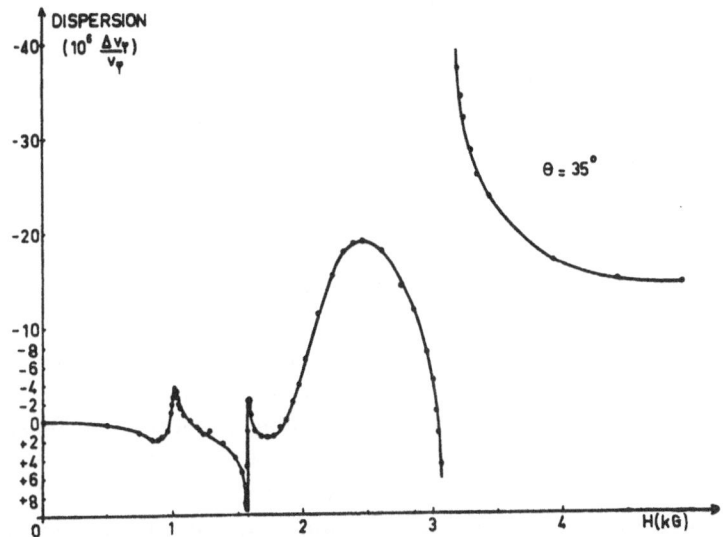

Fig. 7. Change in phase velocity for phonons propagating in MgO with 0.1% Ni as a function of magnetic field. Data from Ref. 53.

in question is Ni^{2+}, and is a non-Kramers ion; thus, according to Eq. (61), it should be frequency independent. Reference to Fig. 7 reveals that this is indeed the case, and, as expected, the sound velocity appears to become constant and less than the velocity at zero field.

The magnitude of the coupling constant is known for the Ni^{2+} $\Delta m = 1$ transition, and is about 50 cm^{-1} per unit strain. The concentration of Ni in their crystal was 10^{-3}. The experiment was performed at 4°K at a frequency of 9.4 GHz. Using these values and (60), the calculated decrease in the velocity of sound is two parts in 10^6, in reasonable agreement with the observed value of one part in 10^5.

Thermal Conductivity. Magnetic defects provide the investigator with an additional experimental variable, namely the magnetic field. In fact the temperature dependence of the thermal conductivity is a much less sensitive measure of the phonon damping than the magnetic field dependence at a constant temperature. The reasons are first experimental—a major source of uncertainty lies in the temperature calibration and measurement—and second that the intrinsic scattering can be temperature dependent but is not field dependent (at least not in a well-designed experiment).

Possibly, for this reason the success obtained by various investigators in computing a curve of thermal conductivity against magnetic field which will pass through the experimental points has been almost nil. What is done, of course, is to assume a form for the frequency dependence in the scattering cross section and attempt to reproduce the experimental data. The difficulties involved in such an exercise are twofold: First, most magnetic defects have more than two levels in a magnetic field; second, if they are sufficiently strongly coupled to the lattice, the levels are also affected by static strain. Taking a crystal with a high concentration of weakly coupled spins does not work, because the probability of defect pairs, triplets, etc. occurring becomes large enough for the effects of these multiplets to be important and even dominate.[54,55] Thus it is not surprising that, to date, there is only one complete analysis of magnetothermal conductivity data.[58] The temperature dependence, on the other hand, is usually easy to fit.[56,57]

Returning to the field dependence, probably the most successful analysis of experimental data was undertaken by McClintock et al.[58] for holmium ethyl sulfate. Unfortunately, however, in their analysis they were forced to assume that the presence of internal strains, hyperfine interaction, etc. led to the observed line shape for the resonant interaction. What we require, of course, is the answer to the question of whether or not Eq. (62) can account for the effect of a magnetic field on the thermal conductivity, and if the appropriate coupling constants agree with those determined by other means. At the time of writing this question still does not have a satisfactory answer.

4.4. "Molecular" Defects

Under this heading we wish to include all nonmagnetic defects which scatter by transitions between internal states. Some of these, such as the Li^+ ion, are not molecular at all, but behave as such with reference to their internal states.

There are a number of excellent reviews which discuss the properties of the defects themselves (e.g., Ref. 40). We will confine our attention to their effect on the phonons.

We have treated the coupling of the defect to the lattice in just the same general way we handled the spin problem: The interaction is expanded in a Taylor series in the displacements, and we assume that to lowest order it is linear in the defect and phonon coordinates.

4.4.1. Experimental

The resonance frequencies of most molecular defects are well out of reach of ultrasonic techniques. Neutron diffraction, on the other hand, has not had sufficient resolution to permit the observation of phonon widths or energy shifts. While a Brillouin scattering experiment can, in principle, detect the phase velocities and additional broadening due to some of the very lowest resonance frequencies, e.g., RbCl:CN$^-$, this experiment has yet to be performed. Therefore the only experimental information available comes from the thermal conductivity. See, however, Ref. 65.

While previous investigators had speculated that some of their results showed the effect of resonance scattering, Walker and Pohl[24] provided the first systematic study of these effects on the heat transport. Since then this subject has received a great deal of attention. We direct the reader to the excellent reviews[23,40] already in existence, however, with this caution: As we have already emphasized, the thermal conductivity behaves as a broadband spectrometer. Therefore it cannot yield the frequency dependence of a scattering law unless the significant change occurs over a range of frequencies wider than kT. In determining the frequency dependence of a scattering mechanism over a range of frequencies on the order of the spectrometer bandwidth or less, a trial and error approach must be employed: One attempts to fit the data with a specific scattering function. This can only be successful if the initial physical assumptions used with regard to the scattering mechanism are correct. In a sense, one must have the answer to begin with. The danger is that, since one is engaged in curve fitting, it is possible to fit the wrong scattering mechanism.

A phenomenological cross section of the form

$$\sigma = A\omega^2/(\omega^2 - \omega_0^2)^2 \tag{64}$$

has been used for a number of systems[59] (the damping term has been left out in the denominator because it has no effect on the curve fitting).

On the other hand, it has been concluded[59] that a law of the form

$$\sigma = A\omega^4/(\omega^2 - \omega_0^2)^2 \tag{65}$$

will not fit the data. This expression is deficient, though, in that it does not include the necessary temperature dependence expected of a two-level system [see Eq. (51)]. Furthermore, for a number of scatterers the interaction with the phonons is so strong that the energy shifts are significant. Since the phonon density of states has a strong influence on the conductivity, the perturbation of the density of states caused by the energy shift also can be significant.

If these two factors are included, we have found that we can fit the data for KCl:Li and KCl:CN with an equation like (65). We have also found that the resonance frequency ω_0 necessary to fit the data using an equation of the form of (51) may be considerably different from that obtained by using Eq. (64).

Which is the correct form to use? The ultimate decision must be made experimentally, and fortunately, techniques are becoming available that should resolve this question. However, we observe that a scattering law of the form of (64) leads to an unsatisfactory result if the real part of the index of refraction is calculated from dispersion relations and Lax's theory[9]: Using dispersion relations, we can relate the real part of the forward scattering amplitude to the imaginary part. The optical theorem connects the imaginary part and the total cross section. Thus from the total cross section we can, using Kramers–Kronig relations, obtain the real part of the forward scattering amplitude. If this is employed in Lax's expression for the real index of refraction, we find that because of the ω^2 in the numerator in Eq. (64), we obtain an index of refraction which diverges as the phonon frequency goes to zero.

To the extent that Eq. (51) is consistent with Lax's theory, then, we conclude that it is preferable.

We now outline in some detail the procedure we have used and the result of fitting Eqs. (47) and (51) to experimental data, obtained by Seward and Narayanamurti[60] for CN in KCl. We make this choice for two reasons: First, independent values are available for the strain-coupling coefficient from the work of Byer and Sack.[61] Second, we wish to compare the results of fitting thermal conductivity data with and without including dispersion.

Since phonons close to resonance are so strongly scattered that their contribution is negligible, we simplify our expressions for $\Delta\omega$ and $\sigma(\omega)$, but include the temperature dependence of a multilevel system.

For a multilevel system the frequency shift becomes

$$\frac{\Delta\omega}{\omega_k} = \varrho_D \sum_i \frac{A_i}{\omega_i} \left(\frac{\omega^2}{\omega_i{}^2} - 1 \right)^{-1} \frac{1 - g_i \exp(-\beta\omega_i)}{1 + \sum_i g_i \exp(-\beta\omega_i)} \qquad (66)$$

where g_i is the degeneracy of level i. A similar generalization can be performed on Eq. (51).

4.4.2. Comparison with Experiment

There are two features of Eq. (66) which are confirmed directly by experiment: the temperature dependence of the fractional phonon frequency shift, and the independence of this shift with respect to frequency in the limit of low frequencies.

Byer and Sack[61] have measured sound velocities and the ultrasonic attenuation at frequencies between 30 and 200 MHz for modes of E_g and T_{2g} symmetry for a variety of alkali halide crystals doped with impurities which are resonant scatterers. Their results for KCl:CN⁻ directly confirm Eq. (66) in the low-frequency limit: The change in velocity of sound is independent of frequency and the observed T^{-1} temperature dependence at high temperatures is also in agreement with (66).

The level structure of the KCl:CN⁻ system is discussed fully in an article by Narayanamurti and Pohl.[40] Of interest here are the three lowest tunnel-split levels and the librational levels at 13.5 cm⁻¹. The tunnel splittings have been measured by Hetzler and Walton,[62] and lie at 1.1 and 2.0 cm⁻¹ above the ground state.

Byer and Sack observed that the E_g mode was essentially only coupled to the libration level, and the T_{2g} mode only to the tunneling levels. Unfortunately, their data do not go to sufficiently low temperatures to allow a complete separation of the contribution from each tunneling state: The coupling to the level at 2.0 cm⁻¹ would be expected to saturate below about 1.5°K, becoming temperature independent, and the remaining temperature dependence would yield the magnitude of the coupling constant to the 1.1 cm⁻¹ level. A rough estimate of the fraction to be assigned to each level is obtainable from the data of Hetzler and Walton, and the contribution from the higher transition appears to be about twice that from the lower. When allowance is made for the ω^4 frequency dependence of the scattering cross section, the constants A_i/ω_i in Eq. (66) are approximately in the ratio 6:1. The magnitude of the coupling constants were then obtained from Byer and Sack's data, in the high-temperature limit: $A/\omega_{1.1} = 1.2 \times 10^{-20}$ cm³ and $A/\omega_{2.0} = 2 \times 10^{-21}$ cm³.

The resulting dispersion relations are plotted in Fig. 8 for a CN⁻ concentration of 4.9×10^{19} cm⁻³. It is clear that there is considerable perturbation of the phonons in the vicinity of both the resonant frequencies at 1.1 and 2.0 cm⁻¹. What effects do the resulting large changes in sound velocity have on the thermal conductivity?

In an attempt to answer this question, the thermal conductivity for KCl doped with CN⁻ at concentrations of 9×10^{17}, 8.4×10^{18}, and 4.9×10^{19} cm⁻³ will be calculated and compared with the data of Seward and Narayanamurti.[60] The constants required for this calculation were obtained in the following way: In conventional fashion a fit was obtained to the experimental conductivity of the pure KCl crystal. Then additional scattering terms due to the CN⁻ ion were included in the denominator of Eq. (1).

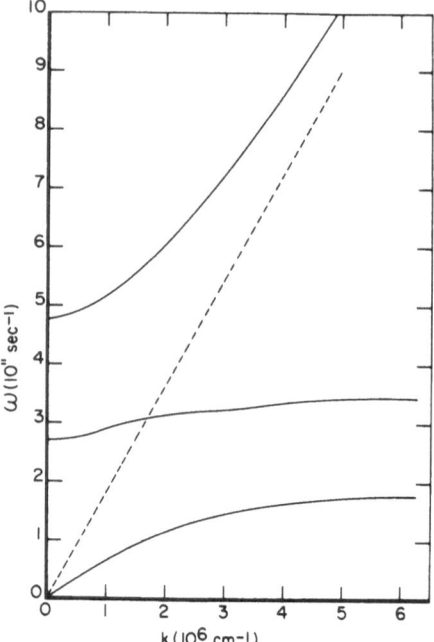

Fig. 8. Calculated dispersion relation for
$KCl + 4.9 \times 10^{19}$ CN$^-$ per cm^3.

It is characteristic of the effect of resonance scattering that quite a large band of phonons about the resonance frequency is ineffective in heat transport. Thus, long before a frequency is reached where the damping term $4\Gamma'^2\omega_0^2$ in Eq. (51) becomes important, the phonons in question have ceased to contribute. For this reason Eq. (51) was rewritten as

$$l^{-1} \simeq \varrho_d \sum_i \frac{4A_i^2 C_0^2}{\pi C_p^2 C_g^2} \frac{\omega^4}{(\omega^2 - \omega_i^2)^2} \frac{1 - \exp(-\beta\omega_i)}{1 + \sum_i g_i \exp(-\beta\omega_i)}$$

where ϱ_d is the number of CN$^-$ ions per unit volume.

The phase velocity C_p was obtained from Eq. (24) and

$$\Delta\omega/\omega_k = (C_p/C_0) - 1$$

The group velocity C_g was obtained from

$$C_g = \partial\omega/\partial k = C_p + k(\partial C_p/\partial k)$$

The known values of the resonant frequencies listed above were used. The coupling constants A_i to the two tunneling levels were computed as outlined above, and that to the librational levels at $13.5 \, \text{cm}^{-1}$ was treated as an adjustable parameter. It should be noted that there are three librational levels; however, they were lumped together and treated as a single level for the purposes of this calculation. This is justified since it is the effect of the tunneling states that we seek. With these values the solid lines shown in Fig. 9 were computed. The concentrations employed were the spectroscopically determined values, listed in the figure.

The computed conductivities which would have been obtained if the dispersion had been neglected, i.e., $C_p = C_g = C_0$, are shown as the dashed lines in Fig. 9. It is clear that for the highest doping the changes in velocity of sound resulting from the scattering make a substantial difference to the calculated conductivity at the lowest temperature.

In calculating thermal conductivities, the imaginary part of the T

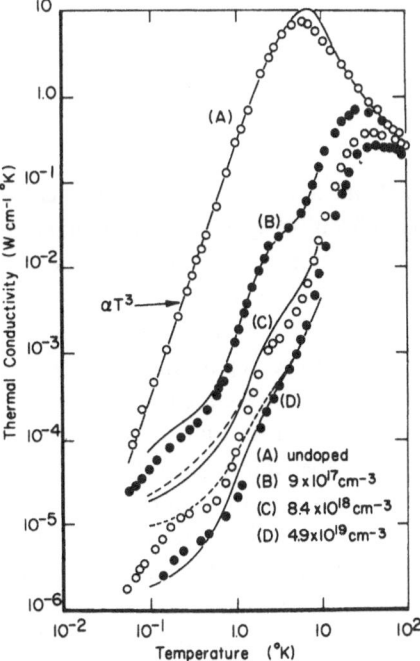

Fig. 9. Thermal conductivity as a function of temperature for KCl:CN⁻. The solid curves are calculated including dispersion, the dashed curves neglecting dispersion. The points are experimental data taken from Ref. 60.

matrix, which leads to damping of the phonons, is most important. Some confidence in the expressions derived here for resonant systems is inspired by the fact that they fit the measured thermal conductivities with only one adjustable parameter, the coupling to the librational level. *All other parameters were determined independently by other techniques.* It is true that the agreement is not as good for the KCl + 8.4×10^{18} cm^{-3} CN$^-$ crystal as that obtained by Seward and Narayanamurti.[60] However, it should be noted that in addition to varying the resonance frequencies and coupling constants, they allowed the concentration of CN$^-$ to vary in order to obtain their curves. This was not done here since it was felt that the temperature dependence of the thermal conductivity is too insensitive to the details of a scattering mechanism to be able to yield these if such a procedure is followed. Needless to say, when such a procedure was followed, agreement with all the measured curves was readily obtained which was well within the experimental error. However, it was also found that a variety of combinations of the parameters yielded essentially identical results.

Excepting the intermediate-concentration, low-temperature data, the expressions used here in calculating the conductivity yield adequate agreement between theory and experiment with only one adjustable parameter. It is felt therefore that they are an adequate approximation. Using these expressions, it was found that the inclusion of dispersive effects made a difference of as much as a factor of six in the calculated conductivity. Thus it appears that effects of dispersion can make an important difference, particularly at low temperature, and must be included.

A straightforward curve-fitting procedure was followed in order to fit data obtained by Baumann *et al.*[63] for Li in KCl and is shown in Fig. 10: The four tunneling levels were replaced by a two-level system, and the coupling constant and resonance frequency were varied until a best fit was obtained. This fit was as good as that obtained by the original authors. Again it was found that dispersion made a large difference to the calculated conductivities at the low temperatures, in this case a factor of three.

Finally we again observe that the temperature dependence of the thermal conductivity is clearly severely limited in its ability to provide detailed information and is incapable of yielding the frequency and temperature dependence of the scattering cross section: It is, after all, a broadband spectrometer whose bandwidth is many times greater than the width of the resonance being studied. Furthermore, in using the temperature dependence of the thermal conductivity as a test for independently obtained expressions, it is important that the dispersion introduced by the scattering be included.

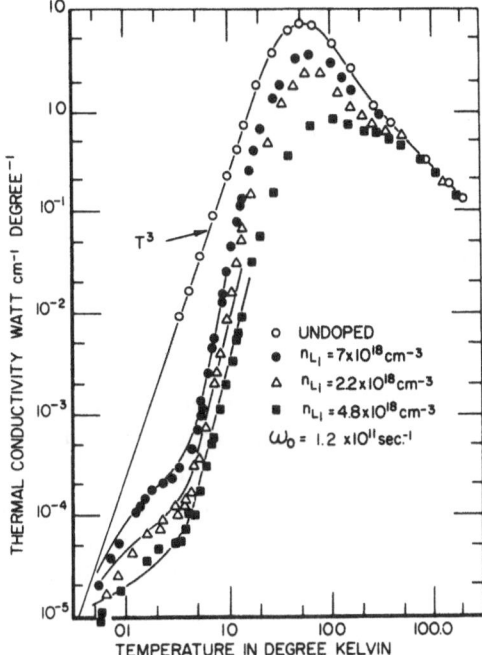

Fig. 10. Thermal conductivity as a function of temperature for KCl with different concentrations of LiCl. The solid lines are calculated curves, as described in the text, the points are data taken from Ref. 63.

4.4.3. *Some Other Probable Consequences*

The dispersion introduced by the resonant scattering is frequency dependent, and leads to dispersion relations for phonons which are curved. Thus three-phonon processes which are forbidden when there is a linear relationship between ω and k now become allowed. The result is that normal processes should become important at longer wavelengths, and become effective at lower temperatures.

A second consequence arises from the decreased velocity of sound. In the long-wavelength limit the phase velocity becomes constant and less than the velocity in the pure crystal. Thus the thermal conductivity at sufficiently low temperature should become greater than that for the pure crystal. For a crystal with a high concentration of strongly interacting defects the measured K vs. T should eventually cross the pure crystal curve at low temperatures. There are some indications that this effect has already been observed for spin systems.[54]

5. SUMMARY AND CONCLUSION

In the foregoing we have presented a simple model of the phonon–defect interaction which is capable of adequately accounting for presently available experimental results.

We have presented an analysis of the interaction of phonons with defects which depends on an interaction linear in the phonon operators. The formalism is identical for all defects, whether the transition involves spin states or tunneling states. To the extent that experimental data are available, the expression we have obtained, which is essentially the Breit–Wigner one-level formula, is in excellent agreement with experiment. However, the experimental evidence rests largely on thermal conductivity data. This type of experiment is insensitive to the details of the scattering mechanism (for instance, it is unable to resolve the difference between our expression and those used by Pohl and co-workers). Furthermore, it cannot yield more detailed information: the shape of the dispersion curves, magnitude of the scattering-induced dispersion, etc. This information is necessary to properly test any theory. For this reason our treatment of the interaction between phonons and defects has been quite elementary. While more refined theories are available, presently available experimental results are simply not sensitive to them.

Specifically, the following questions, among others, are clearly unresolved at the time of writing:

Is the bilinear interaction the most important for all defects? When must higher order interactions be taken into account?

What is the effect of the strong curvature in the phonon dispersion curve near a resonance on three-phonon interactions?

When do multiple scattering effects become important?

The answers to these and many other questions await the development of better "phonon spectrometers" for the thermal frequency range.

ACKNOWLEDGMENTS

The author wishes to express his appreciation to Dr. D. W. Taylor for many helpful discussions during the preparation of this review. He is also grateful to Dr. R. J. Elliott for helpful comments on the manuscript in its initial stages.

REFERENCES

1. R. F. Wallis, ed., *Localized Excitation in Solids* (Plenum, New York, 1968).
2. W. Eisenmenger and A. H. Dayem, *Phys. Rev. Lett.* **18**, 125 (1967).
3. N. S. Shiren, *Phys. Rev. Lett.* **11**, 3 (1963).
4. P. A. Carruthers, *Rev. Mod. Phys.* **33**, 92 (1961).
5. P. L. Taylor, *A Quantum Approach to the Solid State* (Prentice Hall, Englewood Cliffs, New Jersey, 1970).
6. P. M. Morse and H. Feshbach, *Methods of Theoretical Physics* (McGraw-Hill, New York, 1953), p. 791.
7. N. F. Mott and H. S. W. Massey, *The Theory of Atomic Collisions* (Clarendon Press, Oxford, 1965), p. 198.
8. T. Y. Wu and T. Ohmura, *Quantum Theory of Scattering* (Prentice Hall, Englewood Cliffs, New Jersey, 1962).
9. M. Lax, *Rev. Mod. Phys.* **23**, 287 (1951).
9a. A. Messiah, *Quantum Mechanics*, II (Wiley, New York), p. 850.
10. D. W. Taylor, *Phys. Rev.* **156**, 1017 (1967).
11. P. Soven, *Phys. Rev.* **156**, 809 (1967).
12. F. Yonezawa, *Progr. Theor. Phys. (Kyoto)* **40**, 734 (1968).
13. Y. Onodera and Y. Toyozawa, *J. Phys. Soc. Japan* **24**, 341 (1968).
14. B. Velicky, *Phys. Rev.* **184**, 614 (1969).
15. R. N. Aiyer, R. J. Elliott, J. A. Krumhansl, and P. L. Leath, *Phys. Rev.* **181**, 1006 (1969).
16. R. J. Elliott and D. W. Taylor, *Proc. Roy. Soc. A* **296**, 161 (1967).
17. P. G. Klemens, *Proc. Phys. Soc. A* **68**, 1113 (1955).
18. P. A. Carruthers, *Rev. Mod. Phys.* **33**, 105 (1961).
19. T. H. Geballe and G. W. Hull, *Phys. Rev.* **110**, 773 (1958).
20. J. Callaway, *Phys. Rev.* **113**, 1046 (1959).
21. R. Berman and J. C. F. Brock, *Proc. Roy. Soc. (Lond.) A* **289**, 46 (1965).
22. P. Thacher, *Phys. Rev.* **156**, 975 (1967).
23. C. T. Walker and G. A. Slack, to appear in *Progress in Solid State Physics*, Ed. by D. Turnbull.
24. C. T. Walker and R. O. Pohl, *Phys. Rev.* **131**, 1433 (1963).
25. R. F. Caldwell and M. V. Klein, *Phys. Rev.* **158**, 851 (1967).
26. L. G. Radosevich and C. T. Walker, *Phys. Rev.* **171**, 1004 (1968).
27. R. J. Elliott and D. W. Taylor, *Proc. Phys. Soc.* **83**, 189 (1969).
28. R. J. Elliott and A. A. Maradudin, *Inelastic Scattering of Neutrons* (International Atomic Energy Agency, Vienna, 1965), Vol. 1, p. 231.
29. E. C. Svensson, B. N. Brockhouse, and J. M. Rowe, *Phys. Rev.* **155**, 619 (1967).
30. E. C. Svensson and B. N. Brockhouse, *Phys. Rev. Lett.* **18**, 858 (1967).
31. E. C. Svensson and W. A. Kamitakahara, *Can. J, Phys.* **49**, 2291 (1971).
32. R. Bruno and D. W. Taylor, *Can. J. Phys.* **49**, 2496 (1971).
33. F. C. Baumann and R. O. Pohl, *Phys. Rev.* **163**, 843 (1967).
34. R. Berman, C. L. Bounds, and S. J. Rogers, *Proc. Roy. Soc. (Lond.) A* **289**, 66 (1965).
35. B. Bertman, H. A. Fairbank, R. A. Guyer, and C. W. White, *Phys. Rev.* **142**, 79 (1966).

36. D. Walton, to be published.
37. R. Smoluchowski, *Colloque Ampere XV* (North-Holland, Amsterdam, 1969), p. 120.
38. M. Gomez, S. P. Bowen, and J. A. Krumhansl, *Phys. Rev.* **153**, 1009 (1969).
39. D. Walton, *Phys. Rev. Lett.* **19**, 305 (1967).
40. V. Narayanamurti and R. O. Pohl, *Rev. Mod. Phys.* **42**, 201 (1970).
41. L. D. Landau and E. M. Lifshitz, *Quantum Mechanics, Non-Relativistic Theory* (Pergamon, London, 1958).
42. J. Callaway, *J. Math. Phys.* **5**, 786 (1964).
43. A. Messiah, *Quantum Mechanics*, II (Wiley, New York, 1965), pp. 994–1007.
44. K. W. H. Stevens, *Rep. Prog. Phys.* **30**, 189 (1967).
45. M. L. Goldberger and K. M. Watson, *Collision Theory* (Wiley, New York, 1964), p. 478.
46. E. B. Tucker, *Proc. IEEE* **53**, 1547 (1965).
47. L. D. Landau and E. M. Lifshitz, *Theory of Elasticity* (Pergamon, London, 1959).
48. V. F. Sears, *Proc. Phys. Soc.* **84**, 953 (1964).
49. R. Orbach, *Proc. Roy. Soc. A* **264**, 458 (1961).
50. K. W. H. Stevens and H. A. M. Van Eekelen, *Proc. Phys. Soc.* **92**, 680 (1967).
51. R. J. Elliott and J. B. Parkinson, *Proc. Phys. Soc.* **92**, 1024 (1967).
52. N. S. Shiren, *Phys. Rev.* **128**, 2103 (1962).
53. R. Guermeur, J. Joffrin, A. Levelut, and J. Penné, *C.R. Acad. Sci. (Paris)* **260**, 108 (1965).
54. D. Walton, *Phys. Rev. B* **1**, 1234 (1970).
55. R. T. Harley and H. M. Rosenberg, *Proc. Roy. Soc. A* **315**, 551 (1970).
56. G. T. Fox, M. W. Wolfmeyer, J. R. Dillinger, and D. L. Huber, *Phys. Rev.* **165**, 898 (1968).
57. L. J. Challis, M. A. McConachie, and D. J. Williams, *Proc. Roy. Soc. A* **308**, 355 (1968).
58. P. V. E. McClintock, J. P. Morton, R. Orbach, and H. M. Rosenberg, *Proc. Roy. Soc. (Lond.) A* **298**, 359 (1967).
59. R. O. Pohl, in *Localized Excitations in Solids*, Ed. by R. F. Wallis (Plenum, New York, 1968), p. 443.
60. W. D. Seward and V. Narayanamurti, *Phys. Rev.* **148**, 476 (1968).
61. N. E. Byer and H. S. Sack, *Phys. Stat. Sol.* **30**, 569 (1968).
62. M. C. Hetzler and D. Walton, *Phys. Rev. B* **8**, 1801 (1973).
63. F. C. Baumann, J. P. Harrison, W. D. Seward, and R. O. Pohl, *Phys. Rev.* **159**, 691 (1967).
64. D. W. Alderman and R. M. Cotts, *Phys. Rev. B* **1**, 2870 (1970).

Chapter 6

POINT DEFECTS IN MOLECULAR SOLIDS

A. V. Chadwick

University Chemical Laboratory
University of Kent, Canterbury
Kent, England

and

J. N. Sherwood

Department of Pure and Applied Chemistry
University of Strathclyde
Glasgow, Scotland

1. INTRODUCTION

The economic significance of metallic, ionic, and valence crystals, coupled with the availability of specimens of high quality, has led to the extensive examination of the structure and nature of point defects in these materials. Consequently, as will be seen from other chapters in this Treatise, the basic properties of these defects in these solids are well defined. In comparison, the detailed study of the defect structure of molecular solids is in its adolescence. The initial report of a point-defect-controlled property, self-diffusion in solid hydrogen, was made by Cremer[1] in 1938. No further studies were attempted until the early 1950's, when publications appeared on self-diffusion in sulfur[2,3] and α-white phosphorus.[4] Following this protracted infancy, the first report of self-diffusion in an organic solid—anthracene[5]—has been followed by a gradually increasing number of studies of self-diffusion and other defect-controlled properties of organic and inorganic molecular solids. The initial lack of interest in this facet of the molecular solid state can be ascribed to the fact that until recently the effects of lattice imperfections on other physical properties was not generally appreciated.

In molecular solids the lattice molecules are bound in the crystal by van der Waals and dipole interactions. The relative weakness of the intermolecular forces compared with those which bind the atoms in the molecular unit leads, in some instances, to properties more representative of the free molecule. Intermolecular lattice effects are secondary and the effects of imperfections seemed minor. Recent technical improvements have indicated that this is not always so and that often defect effects can dominate. Also, the molecular units which comprise the solids vary considerably in complexity, from the simple atoms to the highly complicated polyatomic units of the infinitely large class of organic solids. Thus, in spite of their relative weakness, there is a considerable variation in the complexity of the intermolecular forces. Overall, this results in the characteristic highly varied structural behavior which makes these materials so much more interesting than some other solid systems (see, e.g., Ref. 6). By analogy point defects should play a major role in both dynamic and static structural properties. Thus detailed information on their nature must be forthcoming if a proper appreciation of their role in these phenomena and of the phenomena themselves is to be obtained.

At a more fundamental level, it is not always possible to make a direct and detailed association between the more complex physical properties and the nature of intermolecular forces in the solid. Point defect properties are relatively simple and hence are more amenable to theoretical calculation. This is most apparent for the monoatomic solids, where molecular and geometric simplicity permit the theoretical calculation of these and other physical properties. The availability of experimental formation and migration parameters to compare with theoretically derived values offers an excellent test of the reliability of the intermolecular potentials used and hence their suitability for more complex calculations.

2. SELF-DIFFUSION

Currently, the most used experimental technique for assessing point defect properties is self-diffusion. Since considerable emphasis is placed on the results of these measurements, it is of interest to consider the performance and reliability of the various experiments.

2.1. Radiotracer Measurements

The most widely applicable method for self-diffusion studies is the tracer method.* Several variants of the basic technique have been developed for application to these materials. As for other solid systems, the serial-sectioning technique is the most reliable, providing as it does indications of simultaneous interfering processes. Integrating methods such as the surface decrease technique and gas/solid, solution/solid exchange reactions have also been successfully used. As might be expected, for all types of experiments highly perfect specimens and pure tracers must be used if reliable results are to be obtained.

2.1.1. Tracers

For the elemental solids and other inorganic crystals, tracers of high specific activity or isotopic abundance can usually be purchased directly. Similarly, simple organic compounds of moderately high specific activity are commercially available. With these tracers, diffusion concentration profiles can be followed over a reasonable range of specific activity and penetration to yield the precision attainable for metals and ionic solids. Often, however, labeled forms of the more interesting molecules are unavailable and these have to be synthesized, sometimes from rather elementary starting points. The products are often of relatively low specific activity, which places some limitations on the ease of performance of accurate experiments. With care in experiment design, results of similar accuracy to those obtainable with more active tracer can be achieved. The synthetic techniques used are basically those of semimicro preparative organic chemistry.[8] Some guidance to particular methods of preparation and purification can be found in the source books in isotopic synthesis.[9-11]

2.1.2. Specimen Preparation

Single crystals of high purity provide the best substrates. The presence of even small concentrations of dislocations and grain boundaries, which act as short-circuiting paths, can lead to the evaluation of diffusion coefficients which are considerably higher than the true lattice values. Crystals of both inorganic and organic molecular solids can be grown by the usual basic techniques which are adequately described in other, more specific

* Reference 7 provides the most recent summary of methods for the performance and analysis of diffusion experiments.

reviews.[12] Most methods yield crystals which are adequate for diffusion experiments. Melt-grown crystals are usually the most defective. Small amounts of residual impurity readily segregate to yield an extensive grain-boundary substructure and excessive concentrations of immobile disloca-tions.[13] Also, due to the thermal strains developed during growth and cooling, induced dislocation concentrations can be high ($\sim 10^{10}$ cm^{-2}).[13] The use of ultrapure material and care during growth can reduce the former to zero and the latter to $\sim 10^5$ cm^{-2}.[14] Large vapor-grown crystals are usually much more perfect. Impurities do not seem to be included in the same manner and dislocation contents are usually less than 10^4 cm^{-2}. The most perfect specimens are those obtained from solution. If grown slowly, these contain little or no solvent and are virtually dislocation free.[15] In all cases the concentration of mobile dislocations can be reduced by annealing[13,14,16] and since handling and preparation inevitably result in the multiplication or introduction of defects, the crystals should always be annealed before use. For those solids in which the self-diffusion coefficients are low, these periods can be lengthy.[16]

For the rare gas solids, which have to be prepared *in situ* in the experi-mental apparatus, modifications of the normal techniques have to be employed. The methods are essentially similar, however, to those used for melt and vapor growth at the higher temperatures.[17–19]

The gross perfection of all crystals can be assessed by optical examina-tion, usually using polarized light microscopy. Dislocation and grain bound-ary contents can be assessed by etching techniques or X-ray topography.[15,20] The former is the most convenient method and the correspondence between dislocations and etch pits has been well defined.[14,20]

2.1.3. Sectioning Experiments

Following the preparation of the crystal, the radioactive tracer can be applied by vacuum deposition or from a suitable saturated solution. Either process yields the same results but we favor the latter since it avoids waste of what is sometimes precious synthetic tracer. Either method permits the application of a sufficiently thin deposit to approximate to the usual thin source conditions.[7] After a suitable isothermal annealing period during which organic solids should be maintained *in vacuo* or an inert atmosphere to avoid oxidation, the crystal is cooled, the sides removed to avoid con-tamination due to surface diffusion or vaporization, and the sample sec-tioned parallel to the initial face. The last is best carried out on a calibrated microtome which permits the removal of the thin sections ($\sim 2~\mu$m) necessary in some experiments. In some cases where the material is low melting or highly volatile we have found it convenient to carry out all operations in a deep freeze unit converted to a glove-box to minimize the loss of the specimen. Alignment of the crystal with the microtome blade can be effected by use of an autocollimator or optical lever. The assessment of slice thickness can be made by weighing or, where this is inadvisable due to volatility, by spectroscopic methods.

Finally the radioactivity can be measured by any of the standard techniques. For organic solids liquid scintillation counting provides the most convenient method, giving high efficiencies for the low-energy particles emitted from ^{14}C and ^{3}H.*

2.1.4. Penetration Profiles

The integration of Fick's law for the usual geometry of diffusion couple, i.e., a thin source of total activity Q diffusing into a semi-infinite solid, yields an expression[7] for the concentration of radioactive tracer (specific activity A) as a function of distance x into the crystal after an annealing period t of the form

$$A = [Q/(\pi Dt)^{1/2}] \exp(-x^2/4Dt) \tag{1}$$

Ideally, a log A vs. x^2 plot should be linear and the diffusion coefficient D obtained from the slope of the line. Early studies of the aromatic hydrocarbon crystals yielded markedly curved plots[5,21,22] as depicted in Figs. 1 and 2. With increasing sample purity the excessive curvature decreased (Fig. 1). It was possible to associate this with a decrease in the density of subgrain boundaries in the solid.[23] That is to say, diffusion along these boundaries should be considerably more rapid than through the lattice and hence a high proportion of the diffusing activity could follow these paths. The reliability of this conclusion was tested by evaluating the self-diffusion coefficients in the boundaries from the deep penetration portion of the profiles, using polycrystalline samples.[21,22,25,26] The results are shown in Table I. The influence of impurities on the subgrain boundary structure was tested by etching techniques and Fig. 2 shows the consequence of small changes in purity on the deep penetration portion of the diffusion profile.

Where the curvature persists even in the purest material, it is common practice to evaluate the self-diffusion coefficient from the initial linear portion of the profile. Provided that this involves a penetration $x \leq 3(Dt)^{1/2}$,

---

* References 10 and 11 provide extensive bibliographies for counting techniques.

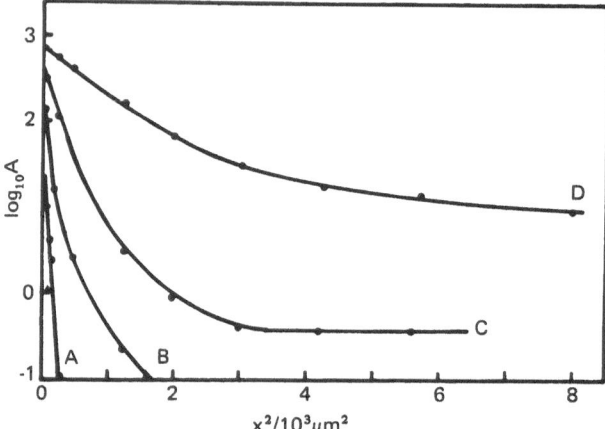

Fig. 1. Diffusion profiles for a series of naphthalene crystals of increasing purity and perfection. A is the purest and most perfect sample. For all specimens $t \simeq 100$ hr, $T \simeq 50°C$.

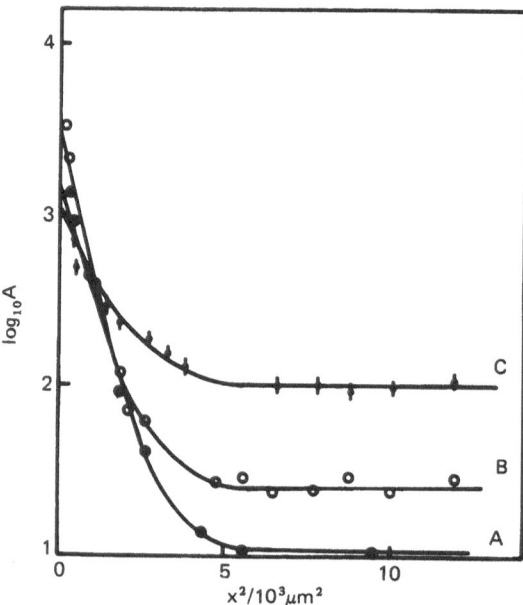

Fig. 2. Diffusion profiles for anthracene crystals containing (A) 1–10 ppm general impurity (anthraquinone 1 ppm), (B) 1 ppm anthraquinone, (C) 2 ppm anthraquinone ($t = 60$ hr, $T = 195°C$).

Table I. Subgrain Boundary Self-Diffusion in Aromatic Hydrocarbon Solids

	$(D_g/D_l)_m$ [a]	D_0, m²/sec	E, kJ/mole	Ref.
Anthracene	10^5	1×10^2	113	26
Naphthalene	10^6	9×10^3	87	24
Benzoic acid	10^6	1×10^8	133	25

[a] At the melting point. D_g refers to the grain boundaries and D_l to the lattice.

then lattice self-diffusion is dominant[27] and acceptable lattice self-diffusion coefficients result. However, in some cases it is difficult to ensure that the true lattice diffusion coefficient and not an enhanced value is being used in the assessment. Consequently, the earlier measurements led to the acceptance of diffusion coefficients $\sim 10^3$–10^4 times higher than the true values.[5,21,22] The best test of the data is to repeat the experiment for successively increasing times. Then the lattice diffusion will become progressively more pronounced and a more accurate coefficient is obtained.[26] A measured diffusion coefficient which decreases with increasing annealing period is indicative of grain boundary enhancement.[28]

Unfortunately a value thus obtained is not necessarily the true diffusion coefficient and even with crystals which yield linear profiles it is not always correct to assume that lattice diffusion dominates. This is particularly important for those solids where the self-diffusion coefficients are extremely low. Under these circumstances at $x < 3(Dt)^{1/2}$ diffusion along dislocations can still strongly influence the overall diffusion process. Hart[29] has shown that under conditions where the true lattice diffusion coefficient is considerably lower than that in the dislocation, or when the dislocation concentration is rather high, the diffusing tracer will spend a relatively high proportion of its diffusing lifetime in the dislocation. Consequently an overall diffusion coefficient D_{obs} will be observed which is compounded of those in the lattice D_l and dislocation D_d,

$$D_{obs} = (1 - g)D_l + gD_d \tag{2}$$

where g is the fraction of the diffusing tracer in the dislocation. Thus a linear $\log A$ vs. x^2 plot results. In order to test the absolute reliability of the data, D_d and g should be measured.

No attempts have been made to determine the self-diffusion coefficients in the dislocations. These should, however, be similar or at least no larger than those for the grain boundaries D_g. Table I shows the relative diffusion rates in the lattice and the boundary. These will diverge with decreasing temperature. The quantity g can be related to the dislocation content. For a typical aromatic hydrocarbon solid containing 10^5 dislocations cm^{-2}, then, $g \sim 10^5$ dislocations cm$^{-2} \times 10^{-15}$ cm^2 molecule$^{-1} \times 10$ molecules dislocation^{-1}, i.e., $g \sim 10^{-9}$. Since g is small, we can rewrite Eq. (2) as

$$D_{\text{obs}}/D_l = 1 + (D_g/D_l)g \qquad (3)$$

and the factor $(D_g/D_l)g$ is a measure of the degree of dislocation enhancement. If we assume that this will become visible when $(D_g/D_l)g > 0.01$, then this limits us to temperatures at which $D_g/D_l < 10^7$ or temperatures greater than 0.95 of the melting point. Thus, at worst, reliable measurements can only be made in a limited temperature range near the melting point. Obviously this range will increase for more perfect specimens and for materials where there is a smaller differential between D_g and D_l. In an ideal diffusion experiment, some assessment should be made of this temperature range and diffusion measured over an adequate range of annealing times. In this way both dislocation and grain boundary enhancement can be obviated[26,30] and true lattice diffusion coefficients obtained.

2.1.5. Integrating Techniques

Among these we class all three methods which follow the total increase or decrease of tracer at the surface of the solid during the course of a diffusion experiment. These methods do not directly distinguish between simultaneous diffusion processes and there is always a danger that a true lattice diffusion coefficient will not be observed.

Surface decrease experiments, in which the progress of the activity diffusing from a surface deposit is monitored during the course of the experiment, have been tried for cyclohexane[32,33] and naphthalene.[34] Theoretically, the sensitivity of such methods should be high since they rely on the absorption of the emitted radiation by the solid as the tracer diffuses beyond the maximum β range in the sample. Both ^{14}C and ^{3}H have extremely low-energy beta radiation. Also, the relatively low experimental temperatures permit the samples to be mounted in sealed Geiger–Müller or scintillation counting assemblies so that the progress of diffusion can be monitored continuously. These advantages are outweighed, however, by the volatility of the samples and by counting difficulties. The result is that at best the method is useful as a check that surface activity decreases are compatible with those measured by sectioning and hence that vaporization losses are negligible.[32]

The rate of exchange between an isotopically enriched solid and its molecules in solution[2] or the gas phase[33,35-37] has been extensively used. This latter method is ideal for the rare gas solids, where the low critical point temperatures preclude sectioning experiments. For these the crystal can be grown *in situ* in the exchange apparatus by adaptations of the normal vapor and melt growth techniques.[17-19,37] A considerable expertise has been built up in the growth and examination of these crystals for this and other purposes and there is no doubt that they can be of adequate perfection for diffusion studies.

Following the preparation of the crystal it is necessary to equilibrate it with its vapor (isotopically enriched vapor if the crystal is inactive and vice versa). This is a major problem but as can be seen from the original papers, equilibrium can be satisfactorily achieved and maintained without evaporation–condensation dominating the process. Once established, the experiment can be continued as a function of temperature.

The most ideal conditions involve the equilibration of a radioactive solid with an equivalent amount of inactive molecules in the gas phase. The gradual increase of radioactivity in the gas phase is more sensitively detected than the reverse process and the exchange kinetics is described by[37]

$$d(n^2)/dt = 4A^2C_0^2D/\pi \tag{4}$$

where n is the number of tracer atoms in the gas phase at time t, A is the surface area of the solid, and C_0 is the initial concentration of tracer in the solid. Other geometric arrangements of solid, gas, and tracer yield more complicated equations[35] but all are similar in that $D \propto d(n^2)/dt$. Thus the measurement involves the assay of the number of tracer atoms in the gas/solution as a function of time. Either radioactive or stable tracers can be used and n determined by gas flow counting or mass spectroscopy, the former having obvious advantages. The evaluation of D relies upon a knowledge of the surface area of the sample, which is difficult to assess with accuracy. In spite of this drawback the method has the particular advantage that the exchange rate can be measured as a function of temperature and hence an activation energy obtained for one particular sample.

As noted above, the possibility of simultaneous processes must always be considered. The influence of these can be assessed in a similar manner to that used for sectioning experiments, i.e., by examining the time dependence of the evaluated D. For the rare gases, however, lattice self-diffusion rates are high ($\sim 10^{-12}$ m²/sec at the triple point) and presumably the temperature range for reliable measurement will be proportionately longer than for the aromatic hydrocarbons. In the only attempt to compare the exchange method with a sectioning study, the activation energy for the exchange of cyclohexane vapor with the solid[33] was found to be 71.1 ± 4.2 kJ/mole compared with 68.4 ± 1.3 kJ/mole. Absolute values of the diffusion coefficient could not be evaluated due to the lack of an accurate value for the surface area of the sample. It was possible to assess, however, that these would be similar to those predicted by extrapolation of the sectioning data.

2.1.6. Results

The repetition of diffusion measurements as a function of temperature yields an equation for the temperature dependence of the diffusion coefficient of the usual form[7,27]

$$D = D_0 \exp(-E/RT) \tag{5}$$

Table II. Tracer Self-Diffusion in Molecular Solids

	D_m,[a] m²/sec	D_0, m²/sec	E, kJ/mole	E/L_s	Ref.
Monatomic solids					
Argon	10^{-14}	2×10^{-5}	15.1	1.9	36
Krypton	10^{-12}	5×10^{-4}	20.1	1.9	37
Rotator phase solids					
Cyclohexane	10^{-11}	4×10^2	68.4	1.9	32
Pivalic acid	10^{-12}	5×10^2	91.2	1.6	39
Camphene	10^{-11}	2×10^4	96.1	1.9	40
Hexamethylethane	10^{-13}	2.2	85.7	2.0	42
Limited rotator solids					
Benzene	10^{-13}	1×10^5	96.8	2.1	30
Nonrotator solids					
Naphthalene	10^{-15}	2×10^{11}	179	2.4	23
Anthracene	10^{-16}	1×10^6	202	2.3	26
Phenanthrene	10^{-15}	3×10^{13}	202	2.4	31
Biphenyl	10^{-15}	4×10^{10}	169	2.3	38
Benzoic acid	10^{-16}	2×10^8	184	2.0	25
Imidazole	10^{-14}	—	250	—	62

[a] Self-diffusion coefficient at the melting point.

The results of all recent tracer experiments expressed in terms of the parameters D_0 and E are noted in Table II. The compounds examined have been collected into groups which might be expected, by virtue of their other physical properties, to show similar behavior. There is a gradation in self-diffusion rate across the series, the larger molecules which form rigid lattices yielding the slowest rates. With increasing general molecular freedom the self-diffusion rate increases until for the globular molecules and the rare gas solids values are similar to those found for other solid systems at the melting point. This variation in self-diffusion coefficient is reflected more in the preexponential factor than in the activation energy. For most single-crystalline solids the latter is approximately double the latent heat of sublimation of the solid L_s. Those experiments that have yielded different results are the early studies of anthracene,[5,21,22,28] which are now known to include pronounced boundary effects, orthorhombic sulfur,[3] and phosphorus.[4] Both of the latter yield similar peculiar characteristics in the activation plots [Eq. (5)] which could be associated with dislocation or grain boundary effects. More recent studies yield results which promise to follow the generally observed pattern and consequently there seems little point in discussing these anomalies and the values have been omitted from the table.

2.2. Nuclear Magnetic Resonance Measurements

The relaxation times observed in nuclear magnetic resonance experiments are the inverse rate constants for the return of the nuclear spin system to equilibrium. These are associated with the internal equilibrium of the spin system—the spin–spin relaxation time T_2—and the equilibrium of the spin system with the external heat bath—the spin–lattice relaxation time T_1. The latter may be dominated by the constant external field (T_1), the radio-frequency field ($T_{1\varrho}$) or the local nuclear dipole field (T_1^*). At higher temperatures the basic lattice motions which control the rates of relaxation are the rotational and translational motions of the lattice atoms and molecules. Thus the observed relaxation times can be regarded as being compounded of contributions from rotational (r) and diffusional (d) processes, e.g.,

$$T_1^{-1} = T_{1r}^{-1} + T_{1d}^{-1}$$

The separate contributions can be related to the jump times for the two processes. Thus, if a distinction can be made between the temperature regions where the rotational and translational motions dominate the relaxation, the jump times τ can be obtained. Detailed mathematical analysis of the distinction between and behavior of the relaxation processes are given in the background texts[43] and in those papers quoted in Table III. The evaluation of a translational jump time τ_d and the temperature dependence should permit the calculation of the corresponding behavior of the diffusion coefficient (since $D = \frac{1}{6}a^2/\tau_d$, where a is the nearest neighbor distance in the lattice). Since the measurements are made on a short time scale, the influence of physical imperfections could be negligible and the technique could yield valuable confirmation for the tracer experiment. The method has been used with success for ionic and metallic systems,[44-46] good agreement being found with radiotracer experiments.

For molecular solids, the situation is not yet well defined. NMR evidence for translational motions has only been found for xenon,[47] neon,[57] and for those rotator phase organic solids with equally high diffusion coefficients.[33,48] Presumably this indicates that the translational motions in the other solids are too slow to dominate the spin relaxation, and confirms the distinction shown by the tracer measurements (Table II). Further than this there is little agreement. Most NMR studies have yielded diffusion parameters which differ from those evaluated from tracer studies, e.g., activation energies approximately one-half the tracer values. Consequently there has been some controversy as to which method truly measured self-diffusion. Recent studies of a range of organic solids[42,49-55,56] promise the resolution

Table III. Comparison of Tracer and NMR Correlation Times and Activation Energies

	NMR[a]		Tracer[b]		
	τ_0, sec	E, kJ/mole	τ_0, sec	E, kJ/mole	Ref.
Neon	3.4×10^{-15}	3.9	$4.8 \times 10^{-17\ c}$	4.6[c]	57
Xenon	4.5×10^{-17}	31.1	$6.5 \times 10^{-17\ c}$	28.7[c]	47
Adamantane	1.6×10^{-21}	153	—	151	49
Hexamethylethane	2.4×10^{-20}	85.7	3.3×10^{-20}	85.7	42
Pivalic acid	7.1×10^{-16}	55.6	1.1×10^{-22}	91.2	39, 55, 56
Cyclohexane	1.2×10^{-15}	45.1	1.2×10^{-22}	68.4	32, 50, 58

[a] Values obtained by fitting the correlation time τ to $\tau = \tau_0 \exp(E/RT)$.
[b] Values obtained from the tracer diffusion coefficients from the relation $D = a^2/6\tau$, where a is the nearest neighbor distance.
[c] Values obtained from a corresponding states calculation using tracer data for krypton.[87]

of this controversy but also indicate that the range of applicability of the NMR method may be limited. Some of the recent data shown in Table III show that there is excellent agreement between the NMR data and those measured or predicted by tracer experiments for the solids xenon,[47] neon,[57] adamantane,[49] and hexamethylethane.[51–53] In comparison, the data for pivalic acid[55,56] and cyclohexane[50,58] show no agreement and exhibit the characteristic behavior shown by many other organic solids.[48] Most of the latter solids have extremely low entropies of fusion compared with the more normal organic solids and there is abundant evidence from other sources[59] to suggest that they are rotationally and translationally disordered. Presumably the observed differences are a consequence of this disorder. Motional effects other than simple vacancy hopping influence the diffusion process, whereas in the more ordered solids, vacancy molecule exchange dominates. Further evidence for this distinction is presented in Section 5. The possibility that the gradual increase in lattice disorder leads to a gradual divergence between the NMR and tracer data,[60] coupled with the observation that NMR has yet to detect translational motion in nonrotator phases, means that the range of application of the NMR techniques to polyatomic molecular solids is currently very limited. These anomalies are presently under detailed study by a number of groups. Their resolution will lead no doubt to a much wider use of the NMR technique for diffusion measurements in molecular systems and also shed further light on the nature of point defects in the solids of very low entropy of fusion.

2.3. Radical Recombination Studies

High-energy irradiation of organic solids yields radical species of two basic kinds, small fragments which rapidly recombine at low temperatures with adjacent molecules or radicals, and larger fragments. Early studies[61] showed that in nonrotator-phase organic solids the latter persisted to the melting point, whereupon they recombined. Reports also indicate that benzene radicals recombined at high temperatures, but below the melting point,[61] and cyclohexane recombined immediately on entering the rotator phase.[61] This behavior parallels the relative self-diffusion rates quoted in Table II. Recent, more detailed studies have confirmed that this process is self-diffusion controlled. Following the irradiation of imidazole[62] and anthracene,[41] McGhie et al. identified the residual radical as that of the parent molecule. By measuring the kinetics of recombination as a function of temperature, they were able to obtain activation energies for the recombination process in excellent agreement with those for tracer self-diffusion.

Table IV. Activation Energies and Self-Diffusion Coefficients Derived from Radical
Recombination Experiments

	E, kJ/mole	$D_m,^a$ m²/sec	Ref.
Cyclohexane	84	—	61
Anthracene	210 ± 8	~10^{-16}	41
Imidazole	230 ± 21	~10^{-16}	62

^{a} Extrapolated to the melting point.

Similar agreement was found between the self-diffusion coefficients calculated from the rates of recombination. These results provide an excellent
test of the tracer data and indicate that this method should be applied more
widely. The data are summarized in Table IV.

2.4. Summary

Rates of self-diffusion in molecular solids vary widely, being low in
the more rigid lattices and extremely rapid in the more disordered systems.
This reflects closely the variation in intermolecular forces in the systems.
The reliable measurements are typified by an activation energy which is
approximately double the lattice energy of the solid. The small variation
of this (1.9–$2.4L_s$) is inadequate to account for the larger variation in the
diffusion coefficient. This is rather better reflected in the preexponential
factor. We can express the diffusion coefficient as

$$D = \tfrac{1}{6}\gamma a_0^2 \nu \exp(S_d/R) \exp(-E/RT) \qquad (6)$$

γ is a geometric factor related to the lattice geometry. The lattice parameter
a_0 and lattice vibrational frequency ν will be similar for most organic solids.
Thus the variations in D_0 reflect variations in the entropy term S_d, which
can be regarded as the sum of the entropy terms associated with the formation and migration of the diffusing defect. Some typical values are quoted
in Table V. Those for the organic solids are extremely large compared with
values for the simpler solids. There seems no reason, however, why the
motion of polyatomic molecules and the formation of even simple point
defects in such lattices should not be accompanied by relatively large
entropy changes. The considerable variation within the series presumably

Table V. Entropies of Self-Diffusion for Molecular Solids

	S_d, J mole^{-1} $^\circ$K^{-1}	S_d/S_f
Anthracene	263	4
Phenanthrene	367	7
Naphthalene	334	6
Biphenyl	318	6
Benzoic acid	280	7
Benzene	209	6
Adamantane	146	7
Hexamethylethane	121	6
Pivalic acid	167	25
Cyclohexane	167	19
Argon	59	4
Krypton	71	4

reflects the nature of the intermolecular forces and comparisons are meaningless unless the data can be normalized. This is not possible in absolute terms but it is interesting to note that the ratio of S_d to the entropy of fusion S_f yields a value of 4–7 for most solids. Since the entropy of fusion approximately represents the residual lattice order, it must provide some relative measure of the intermolecular forces in the system. The constancy of the ratio could imply an overall similarity in defect structure for most solids. The ratio is much higher for the two organic solids of low entropy of fusion, which again suggests a different behavior.

3. EXPERIMENTAL DETERMINATION OF THE FORMATION AND MIGRATION PARAMETERS FOR POINT DEFECTS

The statistical thermodynamics of crystals containing point defects has been treated in a number of books and reviews.[27,63-67] It will be useful at this point to outline the treatment for the case of a pure, monatomic crystal containing vacancies.[67-69] The results of this treatment should be valid for rare gas solids, since interstitials are not favored energetically,[70,71] and they will be applied, with some reservations, to other molecular solids.

The Gibbs free energy of a real crystal $G(T, P)$ containing N atoms and n_v mobile, noninteracting vacancies at temperature T and uniform pressure

P can be written as

$$G(T, P) = G^0(T, P) + n_v g_v - T S_c \qquad (7)$$

where $G^0(T, P)$ is the Gibbs free energy of a perfect crystal containing N atoms and no vacancies, g_v is the Gibbs free energy of formation of a vacancy, and S_c is the configurational entropy, given by

$$S_c = k \ln[(N + n_v)!/n_v!] \qquad (8)$$

At thermal equilibrium the value of n_v will be such that G is a minimum. Hence, by applying the condition

$$(\partial G / \partial n_v)_{T,P,N} = 0 \qquad (9)$$

and using Stirling's approximation, Eq. (7) yields

$$n_v/(N + n_v) = c_v = \exp[(s_v T - h_v)/kT] \qquad (10)$$

c_v is the site fraction of vacancies, and s_v and h_v are the entropy and enthalpy, respectively, associated with g_v.

Equations (7)–(10) can now be used to determine other thermodynamic functions of the crystal. Full details of the treatment can be found in the review by Howard and Lidiard.[67] If we continue to use the nomenclature X, X^0, and x_v to represent the functions of the real crystal, perfect crystal, and the vacancy, then the results are summarized by the equation

$$X = X^0 + n_v x_v \qquad (11)$$

Equation (11) holds for the volume V, enthalpy H, and internal energy U. Similar equations to those above can be derived for crystals containing interstitials.[67]

Experimental determinations of vacancy formation parameters are based on Eq. (11).

At the present time no direct measurements have been made of the vacancy migration parameters. Quenching experiments of the kind used in other types of solid would be hampered by the low thermal conductivity of molecular solids. Usually the entropy s_m and enthalpy h_m of migration are evaluated from diffusion data and experimental values of s_v and h_v, i.e., via the equations

$$s_m = S_d - s_v \qquad (12)$$

and

$$h_m = E - h_v \qquad (13)$$

3.1. Excess Specific Heat Studies

The isobaric specific heat of a crystal is given by

$$C_P = (\partial H/\partial T)_P \qquad (14)$$

Hence from Eqs. (10) and (11) we can write

$$C_P - C_P{}^0 = \Delta C_P = n_v h_v{}^2/kT^2 \qquad (15)$$

Experimentally, the excess specific heat ΔC_P due to vacancy formation is seen as an "anomalous" upward curvature in plots of C_P versus T at temperatures close to the melting point. ΔC_P is determined by extrapolating the C_P curve from low temperatures ($n_v \simeq 0$, $C_P \simeq C_P{}^0$) and subtracting the extrapolated values from the measured C_P.[72,73] A plot of $\ln(\Delta C_P \times T^2)$ versus $1/T$ should yield a straight line of slope $-h_v/k$ and intercept $(s_v/k + \ln h_v{}^2)$.

A major point of uncertainty in this analysis is the extrapolation to obtain $C_P{}^0$.[74] C_P can be rising near the melting point for other reasons as well as vacancy formation, i.e., due to impurities and anharmonicity. It has even been suggested[75] that one does not need to invoke vacancy formation to interpret the high-temperature specific heat curve. An alternative approach, due to Foreman and Lidiard,[74] is to work from the isochoric specific heat C_V, which can be calculated from the measured C_P by using the relationship

$$C_V = C_P - \beta^2 VT/\chi \qquad (16)$$

where β is the volume coefficient of thermal expansion $[1/V(\partial V/\partial T)_P]$ and χ is the isothermal compressibility $[-1/V(\partial V/\partial P)_T]$. The excess specific heat ΔC_V is given by

$$\begin{aligned}
\Delta C_V &= C_V - C_V{}^0 \\
&= (n_v/kT^2)(h_v - \beta^0 v_v T/\chi^0)(h_v - \beta v_v T/\chi)
\end{aligned} \qquad (17)$$

The uncertainty can now be removed to some extent by calculating $C_V{}^0$ theoretically from a model which includes an anharmonic contribution.[74,76] The difference between $\beta^0 v_v T/\chi^0$ and $\beta v_v T/\chi$ can be neglected and the term evaluated from reasonable estimates of β, v_v, and χ.

Most of the studies of excess specific heat have been confined to rare gas solids. Recently, Baughman and Turnbull[77] analyzed selected C_P data for a number of organic solids. For these, one has to consider the added complication of the internal vibrational and rotational contributions. This was overcome by calculating these terms from spectroscopic data and subtracting them from C_P before analysing for ΔC_P.

Table VI summarizes the vacancy formation parameters that have been obtained from specific heat data.

3.2. Thermal Expansivity Measurements

The volume of a crystal can be determined by measurement of the bulk dimensions or from measurements of the X-ray lattice parameter. These two values will not be identical for a crystal containing vacancies and the difference between "bulk" and "X-ray" values can be used to give a direct measure of the vacancy concentration.

The volume of formation of a vacancy v_v may not be the same as the atomic volume v_a due to relaxation of the atoms surrounding the vacancy. In this case we can represent v_v by

$$v_v = v_a(1 + x) \tag{18}$$

where x is a constant. From Eq. (11) the volume of the real crystal is given by

$$V = V^0 + n_v v_a(1 + x) \tag{19}$$

The value of the X-ray volume V_X will not be equal to V^0 due to this relaxation but is given by

$$V_X = V^0 + n_v v_a x \tag{20}$$

Thus the difference between V and V_X yields an expression for the number of vacancies in a crystal:

$$n_v = (V - V_X)/v_a \tag{21}$$

and provided $N \gg n_v$, c_v is given by

$$c_v = (V - V_X)/V^0 \tag{22}$$

In the case of a crystal which contains both vacancies and interstitial defects the relationship equivalent to Eq. (21) is

$$n_v - n_i = V - V_X/v_a \tag{23}$$

Table VI. Vacancy Formation Parameters from Specific Heat Data

	$T_{t.p.},$[a] °K	c_v at t.p.,[b] %	$s_v,$ J mole^{-1} °K^{-1}	$h_v,$ kJ/mole	$L_s,$[c] kJ/mole	Method[d]	Ref.
Monatomic solids							
Argon	83.810	—	—	5.06	7.779	C_P	72
		1.38	$28.4^{+4.2}_{-9.2}$	5.35 ± 0.54	—	C_P	73
		0.25	45.1	7.94	—	C_P	74
		2.78	57.7	7.32	—	C_V	74
		1.10	42.6	6.69	—	C_V	76
		0.45	46.0	7.61	—	C_P	(e)
Krypton	115.776	1.36	$28.4^{+4.2}_{-9.2}$	7.40 ± 0.84	10.78	C_P	73
Rotator phase solids							
cis-1,2-Dimethylcyclopentane	219.45	0.46 ± 0.08	52.7 ± 14.2	21.4 ± 3.4	41.4	C_P	77
Cyclooctane	287.98	0.36 ± 0.11	54.8 ± 11.7	29.2 ± 4.2	46.2	C_P	77
1,1-Dimethylcyclohexane	239.81	0.22 ± 0.05	73.6 ± 12.1	29.8 ± 3.3	43.1	C_P	77
2,2,3-Trimethylbutane	248.57	0.76 ± 0.13	37.6 ± 5.0	19.4 ± 1.6	36.9	C_P	77
Perfluoropiperidine	274.12	0.33 ± 0.07	61.4 ± 14.6	29.8 ± 4.5	33.7	C_P	77
Pentaerythrityl fluoride	367.43	0.60 ± 0.25	74.4 ± 18.4	43.1 ± 8.4	39.3	C_P	77
Hexamethylethane	373.97	0.53 ± 0.20	72.7 ± 20.9	43.5 ± 9.2	39.0	C_P	77
Limited rotator solids							
Benzene	278.691	0.09 ± 0.02	129.6 ± 17.1	52.3 ± 4.6	44.8	C_P	77

[a] Triple point temperature.
[b] c_v at the triple point; where values are not given in the original reference an estimate has been made from the reported h_v and s_v.
[c] Latent heat of sublimation at the triple point.
[d] C_P, analysis of isobaric specific heat. C_V, analysis of isochoric specific heat.
[e] Reference 74 includes a reestimate by Morrison allowing for an anharmonic contribution to C_P.

Thus if $V - V_X$ is positive, it is proof that vacancies are the dominant point defects.

Equation (22) can be rearranged in a number of ways to give relationships between experimental and vacancy parameters. Since $c_v \ll 1$, it can be shown that

$$c_v = (\varrho_X - \varrho)/\varrho_X \qquad (24)$$

and

$$c_v h_v / kT^2 = \beta - \beta_X \qquad (25)$$

where ϱ and β are the bulk density and volume coefficients of expansion, respectively, and ϱ_X and β_X are the corresponding X-ray values. If the crystal is isotropic, Eq. (22) can be written as

$$c_v = (L/L^0)^3 - (a/a^0)^3 \qquad (26)$$

where L and a are the length and lattice parameter of the crystal, respectively. Equation (26) can be rearranged to the form

$$c_v = 3[(\Delta L/L^0) - (\Delta a/a^0)] \qquad (27)$$

where $\Delta L = L - L^0$ and $\Delta a = a - a^0$.

Equations (24), (25), and (27) have all been used as the basis of experimental determinations of c_v, h_v, and s_v in molecular crystals.

It should be emphasized that the effect being observed is extremely small and the accuracy of the measurements needs to be extremely high. For example, if c_v is 0.1%, the bulk and X-ray quantities must each be measured to 0.01% to give an error in c_v of around 10%. Thus great care must be taken in the preparation of the specimen used in the bulk determinations since gross bulk imperfections will give anomalous results.

The most reliable method is to determine c_v from simultaneous measurements of L and a on one crystal. This removes possible errors due to sample variation and thermometer calibration. L^0 and a^0 in Eq. (27) can be replaced by low-temperature values, where c_v is negligible. This technique, originally developed by Simmons and Balluffi,[78] has been used in the study of a variety of solids.

The results of differential thermal expansion studies are listed in Table VII. For some of the values quoted the bulk and X-ray parameters were not determined by the same set of workers and the reader should consult the references given for the details. Only the measurements of Losee and

Table VII. Vacancy Formation Parameters from Differential Thermal Expansion Measurements

	$T_{t.p.}$,[a] °K	c_v at t.p.,[b] %	s_v, J mole^{-1} °K^{-1}	h_v, kJ/mole	L_s,[c] kJ/mole	Method[d]	Ref.
Monatomic solids							
Argon	83.810	≤0.13	—	—	7.779	ρ, ρx	79
	—	<0.10	—	(e)	—	ΔL, Δa	80
	—	0.37	—	—	—	ρ, ρx	81
	—	<0.2	—	(f)	—	ρ, ρx	82
Krypton	115.776	0.32	$16.7^{+4.2}_{-2.1}$	7.44 ± 0.84	10.78	ΔL, Δa	19
	—	0.28	$23.4^{+6.7}_{-8.8}$	8.30 ± 0.84	—	ΔL, Δa	69
Rotator phase solids							
Carbon tetrachloride	250.17	≤2.0	—	—	34.7	ρ, ρx	83
t-Butyl chloride	247.8	≤2.0	—	—	28.8	ρ, ρx	83
t-Butyl bromide	257.0	≤2.0	—	—	—	ρ, ρx	83
Cyclohexane	279.71	≤2.0	—	—	35.5	ρ, ρx	83
	—	≤0.5	—	—	—	ρ, ρx	84
Cyclohexanol	298.31	≤2.0	—	—	—	ρ, ρx	83
	—	≤0.3	—	—	—	ρ, ρx	84
dl-Camphene	323	7.7[g]	146 ± 8[g]	52.3 ± 2.1[g]	43.5	ρ, ρx	84
Succinonitrile	328	≤0.1	—	—	50.6	ΔL, Δa	77
Cyclooctane	287.98	0.4 ± 0.1	—	—	46.2	ΔL, Δa	77
Limited rotator solids							
Benzene	278.69	≤0.2	—	—	44.8	ρ, ρx	83

[a] See Table VI, footnote a.
[b] See Table VI, footnote b.
[c] See Table VI, footnote c.
[d] ρ, ρx—comparison of bulk and X-ray density; ΔL, Δa—comparison of bulk length and X-ray lattice parameter.
[e] From this experiment $g_v > 5.10$ kJ/mole.
[f] From this experiment $g_v \geq 4.35$ kJ/mole.
[g] These values must be regarded as doubtful. The temperature dependences of ρ and ρx were inferred from data outside the temperature region where c_v is significant. Also, no impurity levels were quoted.

Simmons[19] on krypton were of the simultaneous type described in the last paragraph.

3.3. Compressibility Measurements

Differentiating Eq. (22) with respect to P and substituting with the isothermal compressibilities yields the relation[69]

$$v_v c_v / kT = \chi - \chi_X \qquad (28)$$

where χ_X is the compressibility computed from changes in lattice parameter. This equation has been used as the basis of an experimental determination of vacancy parameters for krypton.[69] The results obtained were consistent with those obtained from differential thermal expansion.

Combining Eqs. (25) and (28) gives

$$v_v = \frac{h_v}{T} \frac{\chi - \chi_X}{\beta - \beta_X} \qquad (29)$$

and an experimental determination of v_v is possible. The value obtained for krypton[69] was $v_v = 1.5^{+0.3}_{-0.1} v_a$.

3.4. Summary

The information currently available on the magnitudes of the formation parameters of defects in molecular solids is sparse. The differential thermal expansivity data do indicate that the predominant point defects are lattice vacancies. The concentration of these seems to be similar to that found in other solid types. Confirmation of the magnitude of the formation parameters is given by the results of the excess specific heat measurements. It must be noted, however, that this method is fundamentally less reliable. In particular the experimental data used for the evaluation of the formation parameters of the organic solids are probably inadequate for this purpose. Consequently the results can only be taken as a general guide to behavior. Overall the results indicate that the data presented in Section 2 should represent vacancy-associated self-diffusion. It is particularly noteworthy that the values of s_v obtained from the present experiments are much higher for the organic solids than for the rare gas solids. This difference parallels that noted for S_d in Section 2.4 and confirms the speculation that such high values are reasonable for vacancy self-diffusion in organic solids.

4. THEORETICAL CALCULATIONS OF POINT DEFECT
PARAMETERS

4.1. Results

The apparent simplicity of the rare gas solids has led to their frequent use as a test of theoretical models. Consequently a number of calculations have been made of the point defect parameters of these crystals. Table VIII summarizes these calculations. As can be seen, a variety of models have been used. Since a law of corresponding states can be applied to the properties of rare gas solids,[85] the calculations on one specific solid can be used to predict the parameters for others. Some of the authors listed in Table VIII have extended their calculations in this way. The result of the basic calculation alone is listed. The results of the calculations show a reasonable agreement for h_r and, as might be expected from simple arguments, this parameter is approximately equal to L_s. The values are much higher and the evaluations of c_v much lower than the experimental values presented in the previous section. As indicated there, the most reliable comparison is with the krypton experiment[19,69] and the considerable disagreement between theory ($h_v \simeq 1L_s$) and experiment ($h \simeq 0.7L_s$) here has been a cause of considerable concern. It has been proposed that for argon the relative difference could result from the neglect of many-body terms in the calculation.[74] Losee and Simmons[19] stress the need for the inclusion of many-body forces as a result of their experiments on krypton and show that the value of h_v is very sensitive to such effects. The origin and magnitude of the many-body forces and the mode of their inclusion in the calculation are still matters of debate.[97] A significant reduction of h_v can be obtained if a three-atom electron exchange term is included[98] but the details of this approach have been questioned.[99,100] The calculations, which include non-pairwise contributions to the van der Waals interaction energy,[93,95] yield values of h_r for argon which are lower than those from the two-body calculations. However, they are still significantly higher than those predicted by the krypton expansivity experiment. Obviously further experimentation to confirm the sole experimental value is essential to the resolution of this controversy. In view of the difficulties faced in the calculations for the rare gas solids, it is perhaps not surprising that similar calculations for organic solids have yet to be attempted.

Theoretical calculations of h_m have been made for solid argon. These have been combined with the corresponding values of h_r to yield the activa-

Table VIII. Calculated Vacancy Parameters for Rare Gas Solids

	$T_{p.t.}$, °K [a]	c_v at t.p., % [b]	s_v, J mole⁻¹ °K⁻¹	h_v, kJ/mole	L_s, kJ/mole [c]	Method of calculation [a]	Ref.
Argon	83.810	—	—	8.50	7.779	Lattice statics, central forces	86
		0.010	50.2	10.62		Static model, 12–6 potential	87
		0.030	67.7	11.29		Dynamic model, central forces	88
		—	—	8.36		Static model, 12–6 potential	89
		—	—	8.42		Static model, 12–6 potential	90
		0.00755	26.8	8.78		Quasiharmonic, 12–6 potential	91
		—	—	8.54		Quasithermodynamic approximation	92
		—	—	7.48		Static model, zero-point and three-body terms included	93
		0.04	41.8	8.57		Monte Carlo simulation, 12–6 potential	94
		—	—	7.28		Lorentz oscillator model	95
Krypton	115.776	—	—	10.96	10.78	Static model, Morse potential	70
		0.003	27.6	13.29		Mayer method	96
Xenon	161.364	—	—	15.63	14.42	Static model, Morse potential	71

[a] See Table VI, footnote a.
[b] See Table VI, footnote b.
[c] See Table VI, footnote c.
[a] In some of the calculations a number of variants of the basic model were also reported. Details can be found in the original papers.

Table IX. Calculated Self-Diffusion Parameters for Solid Argon

h_v, kJ/mole	h_m, kJ/mole	E_v,^{a} kJ/mole	E_v/L_s	D_0,^{a} m²/sec	Method of calculation	Ref.
—	—	13.35	1.72	4.2×10^{-8}	Dynamic model, 12–6 potential; absolute rate theory	101
8.42	7.51	15.93	2.05	—	Static model, 12–6 potential; absolute rate theory	90
8.78	7.10	15.88	2.04	2.3×10^{-5}	Quasiharmonic model, 12–6 potential; absolute rate theory	91
7.48	6.35	13.83	1.78	—	Static model, zero-point and three-body terms included; absolute rate theory	93

^{a} Experimental values $D_0 \simeq 4 \times 10^{-4}$ m²/sec, $E_v = 16.16 \pm 0.82$ kJ/mole.[85] $D_0 = 2^{+3.5}_{-1.4} \times 10^{-5}$ m²/sec, $E_v = 15.05 \pm 0.63$ kJ/mole.[86]

tion energy for vacancy self-diffusion E_v. The results are listed in Table IX. Calculations of the preexponential factor D_0 are few.

The agreement between the theoretical values of E_v and the experimental activation energies for self-diffusion (Table II) is reasonable and suggests a vacancy diffusion mechanism.

4.2. Summary

The situation is rather complicated. Theoretical calculations of the formation parameters yield $h_v \simeq h_m$ and $E_v \simeq 2L_s$. Experimental diffusion studies yield values of $E_v \sim 1.9$–$2.5L_s$. These results provide a measure of consistency. Against these we must set the fact that h_v (experimental) for krypton is $0.7L_s$, which would yield $E_v \simeq 1.4L_s$. On the basis that the theoretical calculations may be in error, Burton[93,102] has attempted to rationalize the experimental data by proposing that the experimental diffusion activation energy represents divacancy self-diffusion. Divacancies, although low in concentration ($\sim 0.006\%$ at the triple point), may be more mobile than monovacancies. Theoretical estimates do indicate that the activation energies for monovacancy and divacancy diffusion may be similar. Thus linear Arrhenius plots might be found for a mixed diffusion mechanism with the observed result.

It would appear that self-diffusion must involve a vacancy migration of some type. The present controversy concerning the true mechanism is based on a comparison of the results of one technically reliable expansivity experiment with those of the tracer diffusion experiments and the theoretical calculations. It can only be resolved by further experimentation.

5. JUMP CORRELATION EXPERIMENTS

Phenomenological comparisons of the type discussed above have been invaluable in assessing the nature of the point defect structure of numerous solids. The most definitive experiments, however, are those which rely on the measurement of the degree of correlation between successive diffusional jumps.[103] The simple theories of diffusion assume that the process is random. While this is true for the point defects that have a random choice of exchanging with any of the neighboring molecules, it is not so for the counter-diffusing molecules which we usually follow. For these, immediately following the molecule–defect exchange, the next most likely event for the molecule is a reexchange since there is only a small probability that there will

Table X. Correlation Factors for Self-Diffusion

	f
Vacancy diffusion	
fcc	0.7815
bcc	0.7272
Divacancy diffusion	
fcc	0.475
Exchange or interstitial diffusion	1.0

be a second defect in a neighboring position. Thus consecutive molecule jumps are correlated with each other and the observed diffusion coefficient D_X is related to the true random diffusion coefficient D by

$$D_X = f_X D \tag{30}$$

The value of the correlation factor f_X is dependent on the crystal lattice, the nature of the defect, and, to some extent, the technique involved. The factor can be calculated for the various possible diffusion mechanisms and several examples are quoted in Table X.[104] The values are quite distinct. Comparison of these with experimentally derived values could lead to the definition of a diffusion mechanism and hence the predominant point defect. For molecular solids there are two possible techniques for the estimation of these factors: the comparison of tracer and NMR diffusion coefficients and isotope-mass effect measurements.

5.1. Comparison of NMR and Tracer Diffusion Measurements

At the present time few attempts have been made to use this method because, as noted in Section 2.2, the relationship between the two techniques is still not well defined. Also, there is a lack of accurate calculations of the correlation factors f_n for NMR diffusional relaxation. The estimates which have been made indicate that for vacancy diffusion in fcc lattices[105,106]

$$f_n = 0.55\text{–}0.67 \tag{31}$$

Since for a radiotracer study $f_t = 0.78$ (Table X), we have

$$D_n \simeq 0.70\text{–}0.86 D_t \tag{32}$$

We have no indication at present as to how this ratio would vary as we move to other lattices or mechanism. The only reliable test of the relationship between the coefficients has been made for two bcc solids, succinonitrile[54] and hexamethylethane,[42] for which separate measurements were made using the same material. As will be seen below, vacancy diffusion occurs in the latter. For both solids, D_n was approximately 30% lower than D_t. Within the error of the measurement, we cannot foresee any marked difference in correlation factors between bcc and fcc systems. Thus we believe that this confirms the proposed difference, and indicates the potentiality of the method. The availability of a greater range of values of f_n could at least permit a distinction to be made between mechanisms. Improvements in techniques, namely the continuous monitoring of the NMR signal during the course of a tracer measurement in the cavity, could rule out sample and temperature errors and allow an even more detailed analysis of the results.

5.2. Isotope-Mass Effect Measurements

This is currently the most reliable and widely applicable method for the assessment of correlation factors for molecular solids. Adequate, detailed accounts of the background theory are presented elsewhere.[103] Briefly, one examines the relative diffusion rates of two isotopic species of mass m_a and m_b in a host of mass m. Both the diffusion coefficient and the correlation factor are dependent on the jump frequency and hence on the mass of the species. Thus we have diffusion coefficients D_a, D_b, and D_t related by correlation factors f_a, f_b, and f_t to D, the true random lattice diffusion coefficient. If it is considered that the migrating molecule moves without affecting the surrounding lattice, these parameters can be related by the equation

$$D_a/D_b - 1 = f_t[(m_b/m_a)^{1/2} - 1] \qquad (33)$$

and f_t could be obtained by comparing D_a with D_b. The assumption behind the derivation of this equation is probably unrealistic since it is highly unlikely that the surrounding lattice will remain unaffected during the jump step. There will be some cooperative motion which must result in the reduction of the mass effect. To allow for this, a fractional factor ΔK is introduced to the equation so that

$$\frac{D_a/D_b - 1}{(m_a/m_b)^{1/2} - 1} = f \Delta K = E_t^{ab} \qquad (34)$$

$\Delta K < 1$ and tends toward zero as more surrounding lattice molecules are disturbed by the diffusive motion processes.

Equations (33) and (34) refer to single molecule jump processes. To complete the picture, account has also been taken of the possibility of more than one molecule (n) being involved in the diffusion step, to yield the equation

$$\frac{D_a}{D_b} - 1 = E_t^{ab}\left[\left(\frac{m_b + (n-1)m}{m_a + (n-1)m}\right)^{1/2} - 1\right] \qquad (35)$$

This allows for all possible variations in defect structure and self-diffusion mechanism, and for $n = 1$ reduces to Eq. (34).

5.2.1. Experimental Measurement of the Mass Factor

The basic requirements for the experiments are two isotopic species of sufficiently different mass ($\geq 5\%$ difference). These can be stable or radioactive, but in the latter case the emitted radiations should be sufficiently different to allow the simultaneous assay of both. In the former case this can be achieved mass spectrometrically. In either situation both isotopes can be diffused simultaneously and the experiment cannot be invalidated by the variations in sample, time, and temperature which could result from two separate experiments.

In an exchange-type experiment Parker et al.[107] measured the relative rates of diffusion of $^{78}Kr/^{86}Kr$ into doped argon crystals by studying the variation of 78/86 ratio in the surrounding gas phase as a function of time. From this they were able to calculate the relative diffusion coefficients and hence the mass factor. The results were subject to a wide error but the success of the experiment indicates that it should be extended to other solids and to self-diffusion.

The sectioning technique has been applied successfully to some organic solids[30,42,108] and is capable of wider application. Use was made of deuteration to provide the mass difference. Labeling of the separate species with ^{14}C and ^3H permitted discrimination; for benzene[30] the tracers were $^{14}C_1^{12}C_5^1H_6$ and $^{12}C_6^2H_5^3H_1$. The mass differences achieved in this way can be quite large. The difficulties of distinguishing between the rather weak radiations of the two β-emitters is a limitation to accuracy but the accuracies achieved were adequate to permit mechanistic assessment. For the simultaneous diffusion of the two species each will obey Eq. (1). Thus

$$\ln A_a = \text{const} - x^2/4D_a t \qquad (36)$$

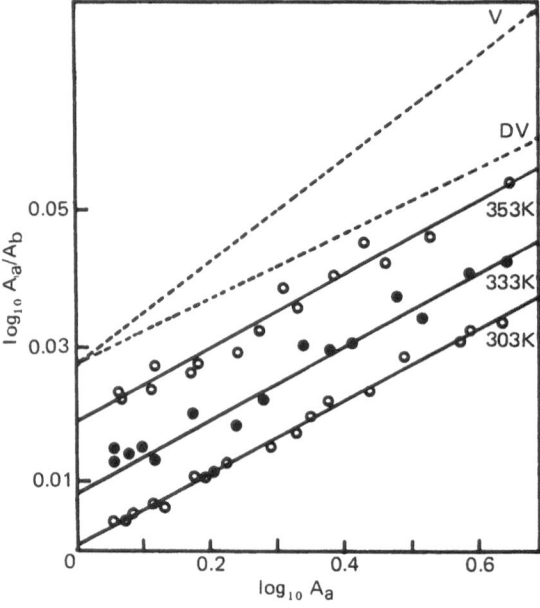

Fig. 3. Plots of $\log(A_a/A_b)$ versus $\log A_a$ for hexamethylethane at various temperatures in the rotator phase.[42] The dashed lines refer to the expected variation for vacancy (V) and divacancy (DV) diffusion.

and

$$\ln A_b = \text{const} - x^2/4D_b t \tag{37}$$

Subtraction of the two equations yields

$$\ln A_a/A_b = \text{const} - (D_a/D_b - 1) \ln A_a \tag{38}$$

and $D_a/D_b - 1$ can be evaluated from a plot of $\ln(A_a/A_b)$ versus $\ln A_a$. Figure 3 depicts a typical series of results and shows the precision attainable.

5.2.2. Results

The results for all experiments to date are shown in Table XI. Due to the rather large error on the experimental value, the Kr/Ar experiment contributes little to the present controversy concerning the diffusion mechanism in the rare gas solids. It has been argued that this result points to a divacancy mechanism, but it is not distinct from the value for a monovacancy process with an acceptable ΔK contribution. Furthermore, theoretical calculations[109] indicate the possibility that the inclusion of krypton in the argon lattice may have some effect on the point defect structure and we cannot be sure that the experiment reflects intrinsic self-diffusion. It does, however, prove the feasibility of the experiment.

The values obtained for the organic solids are quite distinctive. Although benzene is orthorhombic, the lattice is closely related to the fcc structure and within the present error (0.2%) the correlation factors for the latter system should provide an adequate description. Self-diffusion occurs by vacancy migration with little cooperation with the surrounding lattice. The experimental value is quite distinct from that for interstitial or divacancy processes. The more disordered hexamethylethane yields a value intermediate between those for vacancy and divacancy diffusion. Mixed processes are unlikely since the isotope effect is temperature independent[42] and the most reasonable conclusion is that this result represents vacancy motion with some cooperation between the moving molecule and the surrounding lattice ($\Delta K = 0.71$). Values of ΔK of this order are reasonable and have been justified on theoretical grounds.[110]

Cyclohexane, in spite of the very favorable mass ratio, yields an almost zero isotope effect. This indicates a cooperative process involving many molecules. It is of interest to assess the number. From Eq. (35), using $f = 1$ and $\Delta K = 1$, we find that $n = 12$–20 would yield a value of E_t^{ab} approximately equivalent to that found experimentally. These are only the nearest and next nearest neighbors. This and other interpretations, e.g., multivacancy defects ($f_t \to 0$) or highly relaxed vacancies ($\Delta K \to 0$), can be collectively taken to indicate a highly disordered solid. This view is com-

Table XI. Isotope Mass Effects for Diffusion in Molecular Solids

Isotopes	$f \Delta K$		Ref.
	Calculated[a]	Experimental	
Argon	^{78}Kr, ^{86}Kr	0.78 (0.48[b])	0.48 ± 0.25	107
Benzene	$^{12}C_5{}^{14}C_1H_6$, $^{12}C_6D_5T$	0.78	0.78 ± 0.03	30
Hexamethylethane	$^{12}C_7{}^{14}C_1H_{18}$, $^{12}C_8H_5D_{12}T$	0.73	0.52 ± 0.02	42
Cyclohexane	$^{12}C_5{}^{14}C_1H_{12}$, $^{12}C_6D_{11}T$	0.78	0 ± 0.03	108

[a] Monovacancy with $\Delta K = 1$.
[b] Divacancy with $\Delta K = 1$.

patible with the known physical properties of this solid and other rotator phase solids of low entropy of fusion.

5.3. Summary

Jump correlation experiments confirm that self-diffusion in molecular solids proceeds via vacancy-associated processes. Divacancy motion may dominate in argon. The isotope mass effect measurements on the two organic solids benzene and hexamethylethane yield convincing evidence for the predominance of monovacancy diffusion in these solids. This is confirmed by the less accurate NMR/tracer comparisons.

The lack of an isotope effect for cyclohexane indicates its highly disordered nature and confirms the apparent differences between solids of this type and other organic solids noted from the results of the self-diffusion experiments (Section 2.1.5).

6. CONCLUSIONS

On the basis of the presently available data, it seems most reasonable to conclude that the basic point defects in molecular solids are molecule vacancies. The structure, complexity, and aggregation of these entities seem to vary considerably from one type of molecular solid to another. Self-diffusion apparently reflects the dominant motion of vacancies, divacancies, and disordered or multivacancies in particular solids. These variations must reflect the nature of the intermolecular forces in the solids. This confirms the proposition made at the outset that defect parameters may yield as good a property as any to test intermolecular potentials and theoretical calculations in the molecular solid state. There is obviously considerable scope for further theoretical and experimental studies in this area. We hope that this chapter, which emphasises the current lack of information on defect properties as much as reviewing that already in existence, may stimulate others to work in this field.

Note added in proof: In the text (p. 451) it was noted that the early diffusion work on sulfur[3] and phosphorus[4] appeared to be anomalous. Recent tracer measurements on sulfur [E. M. Hampton and J. N. Sherwood, *Phil. Mag.* **29**, 763–769 (1974)] and phosphorus [E. M. Hampton, P. McKay, and J. N. Sherwood, *Phil. Mag.* **30**, 853–868 (1974)] have shown that self-diffusion in these materials is consistent with the general pattern found in other molecular solids.

REFERENCES

1. E. Cremer, *Z. Phys. Chem. B* **39**, 445–464 (1938).
2. M. M. Haissinsky and D. Peschanski, *J. Chem. Phys.* **47**, 191–197 (1950).
3. R. D. Cuddeback and H. G. Drickamer, *J. Chem. Phys.* **19**, 790–791 (1951).
4. N. H. Nachtrieb and G. S. Handler, *J. Chem. Phys.* **23**, 1187–1193 (1955).
5. J. N. Sherwood and S. J. Thomson, *Trans. Faraday Soc.* **56**, 1442–1451 (1960).
6. D. Fox, M. M. Labes, and A. Weissberger, eds., *Physics and Chemistry of the Organic Solid State*, Vols. 1–3 (Wiley, New York, 1963).
7. Y. Adda and J. Philibert, *La Diffusion dans les Solides*, Vols. 1 and 2 (Presses Universitaire, Paris, 1966).
8. N. D. Cheronis, *Techniques of Organic Chemistry*, Vol. 6, *Micro and Semimicro Methods* (Interscience, New York, 1954).
9. A. Murray and D. L. Williams, *Organic Synthesis with Isotopes*, Vols. 1 and 2 (Interscience, New York, 1958).
10. J. F. Raaen, G. A. Ropp, and H. P. Raaen, *Carbon 14* (McGraw-Hill, New York, 1968).
11. E. A. Evans, *Tritium* (Butterworth, London, 1966).
12. G. F. Reynolds, in *Physics and Chemistry of the Organic Solid State*, Ed. by D. Fox, M. M. Labes, and A. Weissberger (Wiley, New York, 1963), Vol. 1, pp. 224–286.
13. J. N. Sherwood, in *Purification of Organic and Inorganic Materials*, Ed. by M. Zief (Dekker, New York, 1969), pp. 157–168; N. T. Corke, J. N. Sherwood, and R. C. Jarnigan, *J. Cryst. Growth* **34**, 766–770 (1968).
14. J. N. Sherwood, *Mol. Cryst. and Liquid Cryst.* **9**, 37–57 (1969).
15. E. M. Hampton, R. M. Hooper, B. S. Shah, and J. N. Sherwood, J. Di-Persio, and B. Escaig, *Phil. Mag.* **29**, 743–761 (1974).
16. J. N. Sherwood, in "Crystal Growth," Suppl. to *J. Phys. Chem. Solids* **1967**, 839–842.
17. G. Pollack and E. N. Farabaugh, *J. Appl. Phys.* **36**, 513–518 (1965).
18. D. N. Batchelder, D. L. Losee, and R. Simmons, in "Crystal Growth," Suppl. to *J. Phys. Chem. Solids* **1967**, 843–847.
19. D. L. Losee and R. O. Simmons, *Phys. Rev.* **172**, 934–943 (1968).
20. J. M. Thomas and J. O. Williams, *Prog. Solid State Chem.* **6**, 119–154 (1971).
21. C. H. Lee, H. K. Kevorkian, P. J. Reucroft, and M. M. Labes, *J. Chem. Phys.* **42**, 1406–1410 (1965).
22. P. J. Reucroft, H. K. Kevorkian and M. M. Labes, *J. Chem. Phys.* **44**, 4416–4420 (1966).
23. J. N. Sherwood and D. J. White, *Phil. Mag.* **15**, 745–753 (1967).
24. J. N. Sherwood and D. J. White, *Phil. Mag.* **16**, 975–980 (1967).
25. A. R. McGhie and J. N. Sherwood, *J. Chem. Soc. Faraday I* **68**, 533–538 (1972).
26. G. Burns and J. N. Sherwood, *J. Chem. Soc. Faraday I* **68**, 1036–1040 (1972).
27. P. G. Shewmon, *Diffusion in Solids* (McGraw-Hill, New York, 1966).
28. A. R. McGhie, A. V. Vaschenkov, P. J. Reucroft, and M. M. Labes, *J. Chem. Phys.* **48**, 186–190 (1968).
29. E. W. Hart, *Acta Met.* **5**, 597 (1957).
30. R. Fox and J. N. Sherwood, *Trans. Faraday Soc.* **67**, 3364–3371 (1971).
31. G. Burns and J. N. Sherwood, *Mol. Cryst. Liquid Cryst.* **18**, 91–94 (1972).
32. G. M. Hood and J. N. Sherwood, *Mol. Cryst.* **1**, 97–112 (1966).
33. G. M. Hood and J. N. Sherwood, *J. Chim. Phys.* **63**, 121–126 (1966).

34. D. J. White, Ph.D. Thesis, Univ. of Glasgow.
35. A. Berne, G. Boato, and M. M. De Paz, *Nuovo Cimento* **46B**, 182–208 (1966).
36. E. H. C. Parker, H. R. Glyde, and B. L. Smith, *Phys. Rev.* **176**, 1107–1110 (1968).
37. A. V. Chadwick and J. A. Morrison, *Phys. Rev. B* **1**, 2748–2753 (1970).
38. N. T. Corke and J. N. Sherwood, *J. Materials Sci.* **6**, 68–73 (1971).
39. H. M. Hawthorne and J. N. Sherwood, *Trans. Faraday Soc.* **66**, 1783–1791 (1970).
40. N. T. Corke, N. C. Lockhart, and J. N. Sherwood, unpublished work.
41. A. R. McGhie, H. Blum, and M. M. Labes, *Mol. Cryst.* **5**, 245–255 (1969).
42. N. C. Lockhart and J. N. Sherwood, *Faraday Symposium, Chem. Soc.*, no. 6, 57–65 (1972).
43. A. Abragam, *The Principles of Nuclear Magnetism* (Oxford Univ. Press, 1961); C. P. Slichter, *Principles of Magnetic Resonance* (Harper and Row, New York, 1963); T. C. Farrar and E. D. Becker, *Pulse and Fourier Transform N.M.R.* (Academic Press, New York, 1971).
44. T. G. Stoebe and R. A. Huggins, *J. Materials Sci.* **1**, 117–126 (1966); I. M. Hoodless, J. H. Strange, and L. E. Wylde, *J. Phys. C.* **4**, 2742–2748 (1971).
45. D. F. Holcomb and R. E. Norberg, *Phys. Rev.* **98**, 1074–1091 (1955).
46. J. G. Spokas and C. P. Slichter, *Phys. Rev.* **113**, 1462–1472 (1959).
47. W. M. Yen and R. E. Norberg, *Phys. Rev.* **131**, 269–275 (1963).
48. A. V. Chadwick and J. N. Sherwood, in *Diffusion Processes*, Ed. by J. N. Sherwood, A. V. Chadwick, W. M. Muir, and F. L. Swinton (Gordon and Breach, New York, 1971), pp. 475–491.
49. H. A. Resing, *Mol. Cryst. Liquid Cryst.* **9**, 101–132 (1969); H. A. Resing, N. T. Corke, and J. N. Sherwood, *Phys. Rev. Lett.* **20**, 1227–1228 (1968).
50. S. B. W. Roeder and D. C. Douglass, *J. Chem. Phys.* **52**, 5525–5530 (1970).
51. J. M. Chezeau, J. Dufourcq, and J. H. Strange, *Mol. Phys.* **20**, 305–316 (1971).
52. S. Albert, H. S. Gutowsky, and J. A. Ripmeester, *J. Chem. Phys.* **56**, 1332–1336 (1972).
53. R. H. Baughman and D. Turnbull, *J. Phys. Chem. Solids* **33**, 121–128 (1972).
54. P. Bladon, N. C. Lockhart, and J. N. Sherwood, *Mol. Phys.* **22**, 365–368 (1971).
55. R. L. Jackson and J. H. Strange, *Mol. Phys.* **22**, 313–323 (1971).
56. G. M. Hood, N. C. Lockhart, and J. N. Sherwood, *J. Chem. Soc. Faraday I* **68**, 736–743 (1972).
57. R. Henry, *Bull. Am. Phys. Soc.* **14**, 188 (1969).
58. P. Bladon, N. C. Lockhart, and J. N. Sherwood, *Mol. Phys.* **20**, 577–584 (1971).
59. J. G. Aston, in *Physics and Chemistry of the Organic Solid State*, Ed. by D. Fox, M. M. Labes, and A. Weissberger (Wiley, New York, 1963), Vol. 1, pp. 543–583.
60. N. C. Lockhart and J. N. Sherwood, to be published.
61. N. N. Semenov, *Pure Appl. Chem.* **5**, 353–376 (1952); H. Swarc and R. Marx, *J. Chim. Phys.* **57**, 680–681 (1960).
62. A. R. McGhie, H. Blum, and M. M. Labes, *J. Chem. Phys.* **52**, 6141–6144 (1970).
63. H. G. van Bueren, *Imperfections in Crystals* (North-Holland, Amsterdam, 1961).
64. F. A. Kroger, *Chemistry of Imperfect Crystals* (North-Holland, Amsterdam, 1965).
65. W. Jost, *Diffusion in Solids, Liquids and Gases* (Academic, New York, 1952).
66. R. A. Swalin, *Thermodynamics of Solids* (Wiley, New York, 1961).
67. R. E. Howard and A. B. Lidiard, *Rep. Progr. Phys.* **27**, 161–240 (1964).
68. J. Holder and A. V. Granato, *Phys. Rev.* **182**, 729–741 (1969).
69. P. Korpiun and H. J. Coufal, *Phys. Stat. Sol. A* **6**, 187–199 (1971).

70. R. M. J. Cotterill and M. Doyama, *Phys. Lett.* **25A**, 35–36 (1967).
71. M. Doyama and R. M. J. Cotterill, *Phys. Rev. B* **1**, 832–833 (1970).
72. D. L. Martin, in *Melting, Diffusion and Related Topics*, Report of 2nd Symposium NRC (Ottawa, Ontario, Canada, 1957), p. 31.
73. R. H. Beaumont, H. Chihara, and J. A. Morrison, *Proc. Phys. Soc. (Lond.)* **78**, 1462–1481 (1961).
74. A. J. E. Foreman and A. B. Lidiard, *Phil. Mag.* **8**, 97–103 (1963).
75. M. L. McGlashan, *Disc. Faraday Soc.* **40**, 59–68 (1965).
76. J. Kuebler and M. P. Tosi, *Phys. Rev.* **137**, 1617–1620 (1965).
77. R. H. Baughman and D. Turnbull, *J. Phys. Chem. Solids* **32**, 1375–1394 (1971).
78. R. O. Simmons and R. W. Balluffi, *Phys. Rev.* **117**, 52–61 (1960).
79. B. L. Smith and J. A. Chapman, *Phil. Mag.* **15**, 739–743 (1967).
80. R. G. Pritchard and D. Gugan, *Phys. Lett.* **32A**, 124–125 (1970).
81. W. van Witzenburg, *Phys. Lett.* **25A**, 293–294 (1967).
82. O. G. Peterson, D. N. Batchelder, and R. O. Simmons, *Phil. Mag.* **12**, 1193–1201 (1965).
83. P. F. Higgins, R. A. B. Ivor, L. A. K. Staveley, and J. J. des C. Virden, *J. Chem. Soc.* **1965**, 5762–5768.
84. J. R. Green and D. R. Wheeler, *Mol. Cryst. Liq. Cryst.* **6**, 13–21 (1969).
85. G. L. Pollack, *Rev. Mod. Phys.* **36**, 748–791 (1964), and references therein.
86. H. Kanzaki, *J. Phys. Chem. Solids* **2**, 24–36 (1957).
87. G. F. Nardelli and A. Repanai Chiarotti, *Nuovo Cimento* **18**, 1053–1071 (1960).
88. G. F. Nardelli and N. Terzi, *J. Phys. Chem. Solids* **25**, 815–826 (1964).
89. G. L. Hall, *J. Phys. Chem. Solids* **3**, 210–222 (1957).
90. J. J. Burton and G. Jura, *J. Phys. Chem. Solids* **28**, 705–710 (1967).
91. H. R. Glyde and J. A. Venables, *J. Phys. Chem. Solids* **29**, 1093–1098 (1968).
92. K. Mukherjee, *Phys. Lett.* **25A**, 439–440 (1967).
93. J. J. Burton, *Phys. Rev.* **182**, 885–891 (1969).
94. D. R. Squire and W. G. Hoover, *J. Chem. Phys.* **50**, 701–706 (1969).
95. S. R. Druger, *Phys. Rev. B* **3**, 1391–1396 (1971).
96. A. R. Allnatt and L. A. Rowley, *J. Phys. Chem. Solids* **30**, 2187–2199 (1969).
97. *Disc. Faraday Soc.* No. 40 (1965).
98. L. Jansen, *Phil. Mag.* **8**, 1305–1311 (1963).
99. C. E. Swenburg, *Phys. Lett.* **24A**, 163–164 (1967).
100. E. Lombardi and L. Jansen, *Phys. Rev.* **167**, 822–828 (1968).
101. R. Fieschi, G. F. Nardelli, and A. Repanai Chiarotti, *Phys. Rev.* **123**, 141–147 (1961).
102. J. J. Burton, *Comments Solid State Phys.* **3**, 82–87 (1971).
103. A. D. LeClaire, in *Physical Chemistry—An Advanced Treatise*, Ed. by W. Henderson, H. Eyring, and W. Jost (Academic, New York, 1970), Vol. X, pp. 261–330.
104. K. Compaan and Y. Haven, *Trans. Faraday Soc.* **52**, 786–801 (1956); K. Compaan and Y. Haven, *Trans. Faraday Soc.* **54**, 1498–1508 (1958).
105. M. Eisenstadt and A. G. Redfield, *Phys. Rev.* **132**, 635–643 (1963).
106. T. G. Stoebe, T. O. Ogurtani, and R. A. Huggins, *Phys. Stat. Sol.* **12**, 649–657 (1965).
107. E. H. C. Parker, B. L. Smith, and H. R. Glyde, *Phys. Rev.* **188**, 1371–1375 (1969).
108. A. V. Chadwick and J. N. Sherwood, *J. Chem. Soc. Faraday I* **68**, 47–50 (1972).
109. J. J. Burton and G. Jura, *J. Phys. Chem. Solids* **27**, 961–974 (1966).
110. B. N. N. Achar, *Phys. Rev. B* **2**, 3848–3855 (1970).

INDEX

A center, 270, 273, 291, 302, 303, 304, 312, 314, 317, 326, 328, 360, 362, 367, 376, 378, 379, 382, 389
A tensor, 344
Absorption coefficient, 61
Activation energy, 170, 451, 452, 454, 465
Adamantane, 452
Agglomerates; *see* Aggregates
Aggregates, 8, 14, 15, 27, 195
Aluminum III–V compounds 203, 223
Anisotropy; *see* Orientation effects
Annealing stages; *see* Recovery
Anthracene, 441, 447, 450, 453
Attenuation, phonon, 395, 425

Benzene (solid), 450, 453, 459, 461, 469, 471
Benzoic acid, 447, 450
Beryllium oxide, 120, 127
Binding energy, 5, 40
Biphenyl, 450
Bohr magneton, 335
Boltzmann–Matano analysis, 177, 191, 193, 207, 215, 218, 220
Boron III–V compounds, 203
Bourgoin mechanism, 6, 15, 29, 49
Breit–Wigner formula, 418, 420
Brillouin scattering, 431
Brower diagram, 243
Built-in field, 209
Butyl bromide, 461
Butyl chloride, 461

Cadmium chalcogenides, 35, 123, 227, 235, 237, 245, 274, 316
Camphene, 450, 461
Capture probability, 282, 283, 288, 294

Carbon tetrachloride, solid, 461
Carrier concentration, 259
Channeling, 45, 71
Chemical potential, 209
Close-pairs; *see* Pairs
Coherent potential approximation, 401
Complexes (of defects), 195, 200, 222, 248, 270, 277, 278, 283, 301
Compressibility (effect of vacancies), 462
Compound semiconductor, 21, 32, 35
Compton scattering, 66
Conductivity
  electrical; *see* Resistivity
  thermal, 128, 394, 395, 408, 410, 413, 429, 432, 437
Congruent evaporation, 232
Conwell–Weisskopf relation, 275
Correlation effects (in diffusion) 466
Crowdion, 79
Crystalline electric field, 342, 370
Cyclohexane, solid, 448, 452, 453, 471
Cyclohexanol, 461
Cyclooctane, 459, 461

D tensor, 356, 425
D_i defect, 315
Damping, phonon, 394, 408
Degeneracy factor, 260
Diamagnetic solid, 336
Diamond, 5, 13, 111, 179
Dichroism, 304, 308, 310, 314, 315, 378
Diffusion, 27, 29, 34, 36, 49, 57, 124, 133, 163, 384, 441; *see also* Self-diffusion
Dimethyl cyclopentane, solid, 459
Dimethylcyclo-hexane, 459
Dipole interaction (in bonding), 442

Direct process, phonon, 422
Dislocations, 54, 55, 191, 222, 443, 444,
    447, 451
Dispersion, phonon, 393, 398, 412, 423, 428,
    432
Displacement damage, 2, 38, 80, 84, 92, 95,
    126, 134
Displacement field, 407
Displacement spike; see Spikes
Dissociative diffusion mechanism, 199, 207,
    208, 219, 221, 236, 248
Divacancy, 5, 7, 9, 25, 36, 55, 59, 107,
    109, 138, 270, 308, 314, 319, 326,
    367, 376, 382, 389
Divacancy diffusion mechanism, 165, 182,
    466
Double-stream diffusion mechanism, 200

E center, 325, 326
ENDOR, 350, 360
EPR (electron paramagnetic resonance), 26,
    31, 34, 36, 54, 55, 107, 110, 120, 127,
    133, 187, 270, 300, 302, 306, 333
Effective charge, 68
Effective diffusion coefficient, 211, 216, 236
Effective mass, 244
Einstein relation, 209
Electric field, effect on energy levels, 381
Electroluminescence, 202
Energy losses, 81
Energy shift, phonon, 398, 408, 412, 419
Enthalpy
    of formation, 168, 184, 459, 464, 465
    of motion, 169, 185, 465
Entropy
    of diffusion, 455
    of formation, 168, 459, 464
    of fusion, 453
    of motion, 169
Equilibrium diagram; see Brouwer diagram
Exchange reaction (for determination of
    diffusion), 443, 449
Exciton, 63, 121, 322

Fermi contact term, 348
Fick's first law, 168
Fick's second law, 171, 445
Fine structure terms, 334, 355
Fluorescence, 121, 123
Focusing collisions, 78, 80
Force constant change, effect on phonon,
    405, 410, 412

Formation energy, 5
Frenkel pair, 48, 49, 145
Frequency shift, 394, 395, 397, 432

GR1 optical absorption, 111
G tensor, 334, 351, 425
Gallium arsenide, antimonide, phosphide,
    33, 118, 119, 203, 204, 207, 223,
    224, 225, 233, 273, 279, 311, 320
Gaussian profile, 172
Germanium, 5, 28, 58, 61, 97, 138, 175,
    179, 181, 261, 273, 276, 280, 282,
    283, 284, 301, 312, 323
Grain boundary, 443, 444
Grain boundary diffusion, 448, 451
Graphite, 5, 20, 37, 43, 112
Green's function, 398, 401, 406
Group velocity, phonon, 395, 399, 434

Hall effect, 103, 117, 237, 259, 269, 274,
    275, 278, 313
Hall–Shockley–Read equation, 280, 281,
    283
Helium, solid, 409, 413
Hexamethylethane, 450, 452, 459, 468, 471
Hexavacancy, 11
Hund's rule, 371
Hydrogen, solid, 441
Hyperfine interaction, 334, 342

ITD (irradiation temperature dependent)
    defects, 109, 271, 305
ITI (irradiation temperature independent)
    defects, 109, 271, 279
Illumination, effects on EPR, 382
Imidazole, 450, 453
Indium antimonide, 33, 115, 128, 203, 227,
    273, 311
Indium arsenide, phosphide, etc., 117, 203,
    226, 273
Infrared absorption, 28, 30, 100, 109, 133,
    300, 378
Internal states, of defects, 415
Interstitital, 2, 5, 10, 22, 25, 27, 36, 37,
    53, 124, 143, 145, 240, 263, 306,
    316, 327, 346, 357, 360, 372
Interstitial diffusion mechanism, 165, 197,
    207, 235, 237, 242
Interstitial–substitutional diffusion; see
    Dissociative diffusion mechanism
Interstitialcy diffusion mechanism, 166,
    235

Ionization, 47, 49, 60, 64, 67, 80, 107, 120, 128
Ion explosion spike; see Spikes
Ionization spike; see Spikes
Isoconcentration diffusion, 217
Isotope effect (in diffusion), 468
Isotopes, effect on phonons, 408

Jahn–Teller coupling coefficient, 375, 386
Jahn–Teller effect, 359, 366, 367, 373, 380

Knock-on, 40, 133
Kramers' theorem, 427
Kramers–Kronig relations, 394, 397, 432

Lambda tensor, 353
Lifetime, minority carrier, 280, 283, 291
Lifetime, phonon, 393, 395, 397, 408
Limited rotator solid, 450
Linewidth
  EPR, 385
  phonon transition, 418, 420
Lithium fluoride, 351, 409
Longitudinal relaxation time, T_1, 387
Luminescence, 125, 316

Magnesium oxide, 126, 428
Magnetic dipole transition, 337
Mass defect, effect on phonons of, 404, 408, 410
Mean free path, phonon, 397
Migration; see Diffusion
Migration energy, 5, 378
Minimum pressure, 232, 237
Mobility, 209, 233, 260, 274, 278, 279
Molecular defects, 430
Molecular solids, 441
Mössbauer-type emission, 317, 319
Motional narrowing, 385
Multiple scattering, 396, 401, 412

Naphthalene, 446, 447, 448, 450
Neutron diffraction, 394, 395, 408, 412
Nonrotator solid, 450
Nuclear magnetic resonance, in diffusion, 451, 467
Nuclear magneton, 342

Off-center ions, 415
Optical absorption, 111, 123, 127, 300; see also Infrared
Optical aligment, 383

Optical theorem, 423, 432
Orbital angular momentum, 335
Orientation effects, 99, 103, 113, 116, 133, 135, 136
Oxygen–vacancy pair; see A center

Pairs (of defects), 16, 18, 30, 38, 107, 109, 116, 120, 126, 142, 263, 271, 305, 362, 364, 373, 385, 403
Pair production (electron-hole), 66
Paramagnetic ions, effect on phonons, 424
Paramagnetism, 335
Partial wave analysis, 417
Penetration profile, 445
Pentaerythrityl fluoride, 459
Pentavacancy, 11
Perfluoropiperidine, 459
Phase rule, 206
Phase velocity, phonon, 434
Phenanthrene, 450
Phonon, 393
Phosphorus, white, 441, 451
Phosphorus–vacancy pair, 364, 373, 385
Photoconductivity, 300, 311
Photon interactions, 61
Photovoltaic effect, 102
Pivalic acid, 450, 452
Potassium chloride, 346, 351, 383, 411, 415, 431, 432, 433, 436
Potential scattering, 418, 421, 423
Precipitation, 222
Pre-exponential factor, 451, 465

Q (quality factor), 338
Quadrivacancy, 10, 369
Quenching,
  of crystals, 59
  of orbital angular momentum, 340

Radiation-induced annealing, 100, 264, 305, 326, 327
Radical recombination technique, for diffusion, 453
Radiotracer technique, 443
Raman process, 422
Rare gas solids, 444, 449, 450, 452, 455, 457, 459, 461, 462, 464, 465, 469, 471
Recoil, 2, 39, 54, 73, 80, 82, 90, 92
Recombination, 45, 280, 283, 284, 291, 305, 316, 322
Recovery, 27, 28, 32, 33, 37, 45, 113, 116, 117, 118, 128, 262, 277, 296,

Recovery *(continued)*
 301, 309, 360, 378, 384
Reorientation time, 376, 385
Replacement collision, 41, 79, 113, 139
Resistivity, conductivity (electrical), 31, 97, 103, 109, 112, 117, 128, 133, 260, 275
Resonance, 336
Resonance mode, 406, 411, 412, 414
Resonant scattering, 414, 416, 431
Rotator phase solid, 450
Rutherford scattering, 73, 83

Saturation, phonon effects, 395
Scattering, phonon, 397, 413, 423, 428
Schottky anomaly, 416
Schottky reaction, constant, 238, 243
Sectioning technique, 443, 444
Selection rules, 359, 360
Self-diffusion, 441, 442, 447, 450
Semivacancy pair, 326
Short-term annealing, 299
Silicon, 5, 13, 26, 54, 55, 56, 102, 139, 179, 181, 265, 269, 273, 278, 280, 291, 303, 313, 317, 324, 325, 345, 360, 369, 376, 382, 389
Silicon carbide, 36, 179, 274
Sodium fluoride, 351
"Soft spots," 45
Specific heat, excess, 457
Spectrometer, EPR, 339
Spikes, 17, 43, 45, 47, 53
Spin, 335
Spin–lattice relaxation, interaction, 424, 426, 451
Spin–orbit interaction, 353, 425
Spin–phonon scattering, 428, 437
Spin–spin relaxation, 451
Split-interstitial, 13, 15, 24, 26, 143
Sponge defects, 8, 9
Strain-coupling coefficient, 432
Strain field scattering, 407, 410, 413
Stress, effects of, 304, 308, 311, 319, 321, 363, 373, 386, 394, 425,
Subgrain boundaries; *see* Substructure
Substitutional diffusion; *see* Vacancy mechanism
Substructure, 444, 445
Subthreshold damage, 105, 115, 128

Succinonitrile, 461, 468
Sulfur, 441, 451
Superhyperfine interaction, 360
Surface decrease technique (for diffusion), 443, 448
Surface effects, 133, 190
Susceptibility, magnetic, 337

T operator, matrix, 398, 406, 419, 425, 435
Tellurium, 37
Tetravacancy; *see* Quadrivacancy
Thermal conductivity; *see* Conductivity, thermal
Thermal expansivity, 458
Thermal spike; *see* Spikes
Thermal vibration, 43
Thermoluminescence, 123
Threshold energy, 39, 84, 95, 96, 102, 111, 112, 115, 117, 118, 120, 121, 127, 133, 136
Transverse relaxation time, T_2, 387
Traps, trapping, 265, 280, 284, 312
Trimethylbutane, 459
Trivacancy, 10
Tunnel splitting, 433, 436

Ultrasonic techniques, 394, 395, 423, 425, 433

V_k center, 346, 351, 354, 383
Vacancy, 2, 5, 22, 25, 27, 36, 53, 124, 145, 221, 230, 263, 357, 376, 455, 464
Vacancy diffusion mechanism, 165, 169, 182, 235, 237, 248, 453
Van der Waals interactions, 442, 463
Varley mechanism, 47, 64
Vibrational bands, 307

Wurtzite lattice, 21, 25, 229

X-ray absorption, damage, 63, 127, 128

Zero-phonon band, 318
Zinc-blende lattice, 21, 25, 202
Zinc chalcogenides, 35, 120, 121, 227, 230, 236, 245, 367
Zinc oxide, 127